THE CHEMISTRY OF WINE

The Chemistry of Wine

FROM BLOSSOM TO BEVERAGE... AND BEYOND

David R. Dalton

OXFORD
UNIVERSITY PRESS

Oxford University Press is a department of the University of Oxford. It furthers
the University's objective of excellence in research, scholarship, and education
by publishing worldwide. Oxford is a registered trade mark of Oxford University
Press in the UK and certain other countries.

Published in the United States of America by Oxford University Press
198 Madison Avenue, New York, NY 10016, United States of America.

© Oxford University Press 2017

All rights reserved. No part of this publication may be reproduced, stored in
a retrieval system, or transmitted, in any form or by any means, without the
prior permission in writing of Oxford University Press, or as expressly permitted
by law, by license, or under terms agreed with the appropriate reproduction
rights organization. Inquiries concerning reproduction outside the scope of the
above should be sent to the Rights Department, Oxford University Press, at the
address above.

You must not circulate this work in any other form
and you must impose this same condition on any acquirer.

Library of Congress Cataloging-in-Publication Data
Names: Dalton, David R., 1936– author.
Title: The chemistry of wine : from blossom to beverage . . . and beyond /
David R. Dalton.
Description: New York, NY : Oxford University Press, [2017] |
Includes bibliographical references and index.
Identifiers: LCCN 2017030132 | ISBN 9780190687199
Subjects: LCSH: Wine and wine making—Chemistry. | Viticulture.
Classification: LCC TP548 .D254 2017 | DDC 663/.2—dc23
LC record available at https://lccn.loc.gov/2017030132

9 8 7 6 5 4 3 2 1

Printed by Sheridan Books Inc., United States of America

To Cecile

"A Book of Verses underneath the Bough
A Jug of Wine, a Loaf of Bread—and Thou
　Beside me singing in the Wilderness–
　Oh, Wilderness were Paradise enow"

> XII Quatrain, Fourth (1879) and Fifth (1889) editions
> Rubáiyát of Omar Khayyám
> (translated by Edward Fitzgerald)
> Fine Editions Press, New York, 1957, p 116

"So, we'll go no more aroving
So late into the night
Though the heart be still as loving
And the moon be still as bright."

> George Gordon 1788–1824

Edmund J. Sullivan, Windsor Press, 1938. This graphic is from the Wikimedia Commons and is in the public domain.

"And much as Wine has play'd the Infidel,
And robb'd me of my Robe of Honour—Well,
    I wonder often what the Vintners buy
    One half so precious as the stuff they sell."

XCV Quatrain, Fourth (1879) and Fifth (1889) editions
Rubáiyát of Omar Khayyám
(translated by Edward Fitzgerald)
Fine Editions Press, New York, 1957, p 137

"Darwin was right to hedge his bets, but today we are pretty certain that all living creatures on this planet are descended from a single ancestor. The evidence . . . is that the genetic code is universal, all but identical across animals, plants, fungi, bacteria, archaea and viruses. The 64-word dictionary, by which the letter DNA words are translated into twenty amino acids and one punctuation mark, which means 'start reading here' or 'stop reading here' is the same 64-word dictionary wherever you look in the living kingdoms . . ."

Dawkins, Richard "The Greatest Show on Earth"
Free Press, New York, 2009, pp 408–9

Cabernet Harvest, 2012 courtesy G. Hamel, Hamel Family Ranch, Sonoma Valley, CA (photo by S. Duncan). Used with permission of George Hamel, Jr.

וַתּוֹצֵא הָאָרֶץ דֶּשֶׁא עֵשֶׂב מַזְרִיעַ זֶרַע, לְמִינֵהוּ, וְעֵץ עֹשֶׂה-פְּרִי אֲשֶׁר זַרְעוֹ-בוֹ, לְמִינֵהוּ; וַיַּרְא אֱלֹהִים, כִּי-טוֹב.

וַיְהִי-עֶרֶב וַיְהִי-בֹקֶר, יוֹם שְׁלִישִׁי.

And the earth brought forth grass, herb yielding seed after its kind, and tree bearing fruit, wherein is the seed thereof, after its kind; and God saw that it was good.

And there was evening and there was morning, a third day.

Contents

Acknowledgments xiii
Author's Note xv
Prologue xix

SECTION I | GROWING THE GRAPES
1. *General Comments* 3

2. *Grapevine from Seed* 6

3. *Grapevine from Grafting* 10

4. *Grapevine from Hardwood Cuttings* 14

5. *The Soil* 16

6. *The Roots of the Grape Vine* 18

SECTION II | CELLS
7. *Roots, Shoots, Leaves, and Grapes* 23

8. *The Leaf* 35

9. *The Light on the Leaves* 39

10. *Harvesting the Light* 44

11. *Working in the Dark* 58

12. *Flowers* 67

SECTION III | BERRIES
13. *The Grape Berry* 91

SECTION IV | A SAMPLE OF GRAPE VARIETIES
14. *A Selection of Grapes* 123

SECTION V | FROM THE GRAPE TO THE WINE
15. *General Comments* 177

16. *More Than Skin Deep* 179

17. *Adding Sulfur Dioxide (SO_2)* 185

18. *Yeasts* 192

19. *Finishing the Wine* 231

20. *Sealing the Bottles* 252

SECTION VI | SPECIAL WINES
21. *Specialized Wines* 259

SECTION VII | DRINKING THE WINE
22. *Drinking the Wine* 297

EPILOGUE 305
APPENDICES
   APPENDIX 1: A CHEMISTRY PRIMER 307
   APPENDIX 2: BIOSYNTHETIC PATHWAYS OF ODOR, COLOR, AND FLAVOR COMPOUNDS 356
   APPENDIX 3: LIST OF ESTER ODORANTS 375
   APPENDIX 4: COMPOUNDS AND COLORS 382
   APPENDIX 5: IMPACT ODORANTS 395
GLOSSARY 399
INDEX 437

Acknowledgments

THAT THIS WORK came into being is largely due to the influence of Professor Robert J. Levis, Department of Chemistry, Temple University. Having invented and taught the initial offering of a General Education course in the Chemistry of Wine, Robert invited me to participate in the creation of the laboratory for the course and then, from time-to-time, replace him as lecturer while he was away.

I quickly discovered how difficult it was to talk about chemistry at the appropriate General Education level. Indeed, I was very uncomfortable and floundered when it was clear that the class members either ignored or believed everything I said. I was ignored when it was "too complicated" and believed when I made it "too simple." And so, having tried to teach the whole course for a semester (inviting Professor Levis in to help) I began to listen when others of my colleagues, Eric Borguet and Robert Rarig, had a go at teaching the course. Despite their sterling efforts, I found myself more or less unhappy whatever was done. This work is an attempt to salve my conscience; it could be used in a chemistry course.

It is clear that my late wife, to whom this is dedicated, and my extended family suffered neglect while this work was in progress. I apologize to them.

My friends and colleagues at Temple University and other institutions where I have had the pleasure of spending time have been supportive beyond expectation, and I need to particularly thank my longtime laboratory coworker and friend Professor Linda M. Mascavage of Arcadia University for her suggestions and criticism.

The editorial staff at Oxford University Press has done more than could be expected. Indeed, in addition to much hand holding early on by Jeremy Lewis and his able coworker Anna Langley, I am grateful for the efforts of the group headed by Sasirekka Nijanthan and Janani Thiruvalluvar, the superb editorial work of Linda Mamassian who converted obscure to clear and the efforts of Beth Bauler to bring the work to you.

I take comfort from the poetic prose of Alexander Pope who, in his Essay on Criticism pointed out...

"Whoever thinks a faultless Piece to see,
Thinks what ne'er was, nor is, nor e'er shall be.
In ev'ry Work regard the Writer's End,
Since none can compass more than they Intend..."

Despite efforts to correct errors and eventually repair the repairs, flaws for which I alone am responsible will be found. I apologize to you, reader, and ask that you let me know what needs fixing as your time and interest permits.

David R. Dalton

Author's Note

THIS IS NEITHER the book I intended to write nor the book I began to write.

Poets extol the burst of aroma when the bottle is opened, the wine poured, the flavor on the palate as it combines with the olfactory expression detected and the resulting glow realized. But what is the chemistry behind it? What are the compounds involved and how do they work their wonder?

What do we know?

Events in what can best be described as Analytical Chemistry[1] have moved faster than most (myself included) had anticipated, and even now, the science and art of analysis is developing at an accelerating pace. We find that is it is fairly certain that distinct and measurable differences in *terroir*,[2] coupled to the plasticity of the grape berry genome and the metabolic products,[3] as well as the work of the vintner, are critical to the production of the symphony of flavors found in the final bottled product.

Analytical chemistry can inform us about the chemical differences and similarities in the grape berry constituents with which we start and what is happening as the grape matures. Indeed, during maturation, we can begin to see how the constituents change in the grape as well as what transpires as it is exposed to fungi (such as one or more of the strains of wild yeast found in most vineyards) and bacteria that are always present. Compounds can be observed following the grape's crushing, both before and after the addition of a strain of *Saccharomyces cerevisiae*,[4] a yeast commonly used for fermentation. Maturation of the wine can also followed whether or not oak casks are used. Once bottled, little is known until the bottle is opened. Then differences in headspace and liquid constituents can be quickly measured, before—and after—oxidation has affected the contents.

The details of the grape and its treatment produce substantive detectable differences in each wine. But there are clear generalities. The finished product, wine, is mostly water. Some ethanol (CH_3CH_2OH, ethyl alcohol), usually more than 10% but less than 20% of the volume, and (again usually) about 4 % of nonvolatile and volatile organic compounds (VOCs)[5] make up the remainder of what is found in the bottle. The VOCs include low-molecular-weight (a few hundred atomic weight units [awu] at best) alcohols, aldehydes, esters, ethers, ketones, terpenes, thiols, acids, etc. The nonvolatile compounds are mostly sugars such as glucose and its isomer fructose (the quantities of which are generally low but may be substantial in some specially treated wines), and some nonvolatile acids such as tartaric acid, malic acid, citric acid, gallic acid, and oxalic acid. There are some, more complex, secondary metabolites such as anthocyanins, phenols, tannins, sesqui- and di-terpenes, etc., which may or may not have sugars attached, and they make up additional minor components.

But it is the details, shown to us by Analytical Chemistry and structural analysis accompanying it, that clearly allow one wine to be distinguished from another. Then, as will be discussed, compounds belonging to the general classes noted above are shown to vary in their structural details as well as their respective concentrations. And this makes all the difference.

Although students or lovers of chemistry may derive the most pleasure from this work, it is my hope that all readers—even those who lack any specialized knowledge of chemistry—will be able to enjoy the stories that follow as much as they enjoy a glass of fine wine.[6]

Finally, the amount of information published appears to grow faster than a grape vine in summer. Thus it is inevitable that the knowledge in this book cannot be comprehensive. I have attempted to choose what I believe is substantive, reliable, and of lasting value. The pages that follow represent the results of that endeavor.[7]

NOTES AND REFERENCES

1. Analytical Chemistry is that subdiscipline of chemistry recognized as being mainly concerned with identifying and measuring the quantity of a substance or components of a substance.

2. If the concept of *terroir* is only slightly expanded beyond the environment (normally including soil, atmosphere, etc.) to include the effects of the vintner postharvest up to and through bottling, it becomes clearer how ostensibly identical vintages can be different. See, in particular, Anesi, A.; Stocchero, M.; Dal Santo, S.; Commisso, M.; Zenoni. S.; Ceoldo, S.; Tornielli, G. B.; Siebert, T. F.; Herderich,; M. Pezzotti, M.; Guzzo, F. *BMC Plant Bio*. **2015**, *15*, 191.

3. The "metabolome" is generally recognized as the **complete** set of metabolites—products of metabolism—of the organism or some part of the organism as dictated by its genome. The "genome" (an organism's complete deoxyribonucleic acid [DNA] expression) is the set of instructions passed, often with some changes, from generation to generation that allows maintenance of life and reproduction of kind.

4. *S. cerevisiae* was the first (1996) eukaryotic genome to be completely sequenced. All organisms whose cells contain a nucleus and other bodies called organelles enclosed within membranes are called *eukaryotes*.

5. The types of compounds listed are briefly discussed in Appendix 1, and many examples are to be found in the text.

6. Because I want to communicate to nonchemists as well as to chemists, and since this is a work dealing with chemistry, I have struggled with the naming of compounds (otherwise called

"nomenclature"). Chemists have systematic ways (often appearing arcane) of naming compounds. Systematic ways of naming compounds have been developed by The International Union of Pure and Applied Chemistry (IUPAC), the Chemical Abstracts Service (CAS) of the American Chemical Society (ACS), and others. Many databases utilize IUPAC and CAS (frequently overlapping but not completely identical) systems. Nonchemists have nonsystematic ways (often appearing arcane) of naming compounds. The latter are called "trivial" names by chemists. I have used trivial names but inserted a glossary at the end to satisfy myself that the requirements of appropriate nomenclature are met and to meet the personal requirement of subsequent database acquisition.

7. "How can the events in space and time which take place within the spatial boundary of a living organism be accounted for by physics and chemistry? . . . The obvious inability of present-day physics and chemistry to account for such events is no reason at all for doubting that they can be accounted for by those sciences" (E. Schrodinger, 1944).

Prologue

WHERE TO BEGIN?

Of course the wine comes from the grape and the grape from the vine. But how did the grape get there, and which grape will provide the best wine under the conditions found where the vine is growing. Is the soil good? Is there enough (not too much!) water? Is there enough (not too much!) sun in the right season? In short, is the climate conducive to the grape the vintner wants?

It appears you need to know the "territory."

Terroir (from the French *terre* for "land") refers to the fitness of the place for the ends to which it is to be used. Territorial fitness has resulted in specific vines preferentially being grown in specific places. While it is necessary that the general climate is temperate enough to allow a grape's maturation, it is also clear that where the vine is planted will affect the grape and thus the taste of the wine made from that grape. So it is reasonable to ask what it is about the territory that makes the difference. Are there, for example, minerals in the soil that add to producing robust plants and flavorful grapes? Can soil emendation be effective? What other plants are growing there?

Subsequent to issues surrounding terroir, it will be important to know what yeast can best digest the sugars (carbohydrates) present in the mature grapes that have been grown and what else will happen to the constituents of those grapes in the fermentation process. What is the best way to treat the reaction mixture in which the alcohol is being produced? In what vessels should the product (first as it matures and subsequently for transport and then use) be held? And, as an aside, what is this maturation process all about and why bother?

With each passing year more information about wine grapes has become available. Genetic markers that relate varieties of grapes and the determination of the genome* of a *Vitis vinifera,* closely related to the Pinot Noir grape, have been finished.

Connections between the ancient arts and the modern practices of the vintner along the path from vine to fruit to beverage are being made. I believe that travel down this path is in the spirit of enriching our understanding rather than an attempt to deny the value of centuries of careful work devoted to production of excellent wines. A superior wine can be compared to any other work of art. Both a glass of wine and fine piece of art can be enjoyed knowing nothing about the details of their production, but some art lovers may find their appreciation enhanced knowing more about the canvas, the pigment, and the lighting that shows the work in all its glory.

NOTE AND REFERENCES

* The genome of any life form defines its heredity and contains the instructions for reproduction, growth, maturation, and senescence. The grapevine gene bank (with more than 2,300 varieties stored) is under threat (Butler, D. *Nature* **2014**, *506*, 18). Further, since the grape is suggested to be one of the earliest, widely cultivated domesticated fruit crops, it is not surprising that characterization of genome-wide patterns of more than a thousand (1,000) samples of the domestic grape (*Vitis vinifera* subsp. *Vinifera*) shows it to be genetically diverse. However, the diversity may be less than it could have been since it appears to have arisen by "crosses among elite cultivars." (Myles, S.; Boyko, A. R.; Owens, C. L.; Brown, P. J.; Grassi, F.; Aradhya, M. K.; Prins, B.; Reynolds, A.; Chia, J.-M.; Ware, D.; Bustamante, C. D.; Bukler, E. S. *Proc. Nat. Acad. Sci. (US)* **2011**, *108*, 3530). At this writing, the Agricultural Research Service of the United State Department of Agriculture in cooperation with the University of California, Davis has a large (and growing) data set.

THE CHEMISTRY OF WINE

SECTION I
Growing the Grapes

1

General Comments

VITICULTURE IS THE art and science of vine-growing and grape-harvesting.

In general, the portions of our planet lying between 20° and 50° latitude on either side of the equator (Figure 1.1) are considered suitable for the vines. In these temperate climates, the tilt of the earth's axis in relation to the sun results in temperatures in the Northern hemisphere in March (and the Southern hemisphere in September) that rarely fall below about 50° Fahrenheit (F) (10° Celsius [C]).[1]

As analytical chemistry[2] and possibilities for genomic[3] and epigenomic[4] analysis have evolved, it has become possible to begin to monitor and understand, in detail, how climate affects wine production. For example, a recent Italian study[5] investigated the changes, over a three-year period, in phenotype[6] of a *Vitis vinifera* cultivar[7] by looking at the transcriptome.[8] They found that most responses they could follow could be attributed to local early spring weather patterns. While a complete analysis was not reported, it was concluded that weather patterns in the previous year were correlated with current year growth.

Much of the biology for raising grapes has been discussed, and while 80% of the world's grape crop is used for production of wine there have been and remain formidable technical difficulties in breeding grapevines.[9] Part of the breeding problem derives from the fact that grapevines are highly heterozygous outcrossers,[10] and so they do not breed true from seed. But they are good cultivars[7] and are polygenic (*poly* = many; *genic* = genes). Their inheritance appears to be controlled by large numbers of genes of minor effect. As a consequence, traditional cultivars, grown in their accustomed places, appear to possess subtle combinations of genes whose totality can be preserved by grafting techniques but not in other ways. That is why it appears that the wines produced by such longstanding and traditional cultivars in the hands of vintners accustomed to their nature have unique characteristics of style and quality.[10] Indeed, it seems that it is largely to encourage and maintain such subtle differences that grapevines are seldom propagated from seedlings (but see below) unless a "new" variety is sought by breeders. Once established, new varieties are continued with cuttings.

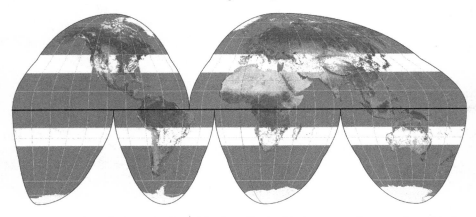

FIGURE 1.1 A homolosine projection of the planet Earth with the equator shown as a heavy black line and the regions from 20 to 50 degrees latitude as white bands on either side of the equator. In this projection, the land masses are shown approximately to their relative scale. Figure adapted from image in the public domain provided by NASA.

The art of identification and description of the grapevine cultivars is called ampelography. Widespread movement of cultivars in the last several centuries has resulted in confusion in naming cultivars as the environment (terroir) seriously affects plant properties.

NOTES AND REFERENCES

1. On the Fahrenheit scale (named after its developer, Daniel Gabriel Fahrenheit) pure liquid water and pure solid water (ice) are in equilibrium—with solid water (ice) melting to liquid and liquid water freezing solid (to ice) at the same time—at exactly 32.0 °F. On this scale, liquid water and steam (water at the temperature at which it boils) also are in equilibrium—with steam condensing to liquid water and liquid water converting to steam at exactly 212.0 °F. On the Celsius scale (developed by Anders Celsius), these values are, respectively, exactly zero (0.0 °C) and 100.0 (100.0 °C) degrees. Both scales are widely used, although the Celsius scale (occasionally referred to as the Centigrade scale—*centi* being a prefix in the metric system denoting a value of one hundred [100]), is more common. The state or condition called "equilibrium" is present when opposing actions, conditions, or events occur with equal facility. Here, at 32.0 °F or 0.0 °C, the amount of water freezing is exactly the same as the amount melting at the same time. At the boiling temperature, 212.0 °F or 100.0 °C, the amount of water becoming steam is exactly the same as the amount of steam becoming liquid water at the same time.

2. *Analytical chemistry* is that branch of chemistry involved in separation, identification, and quantitative measurement of constituents of natural and synthetic mixtures and materials.

3. A *genome* is an expression in chemical and biochemical terms of the entirety of the makeup of a life form's heredity. It appears that the genome of a "generic" wine grape, *Vitis vinifera*—said to be closely related to the Pinot Noir grape—was among the first determined (Jaillon, O. et al. *Nature* **1997**, *449*, 463). Since that time, significant progress in additional genomes has been effected (Mallikarjuna, K. A. A.; Dangl, G. S.; Prins, J.-M. B.; Walker, M. A.; Meredith, C.;

Simon, C. J. *Genet. Res. Camb.* **2003,** *81,* 179) and beyond that even more recent studies completed. Some of the more recent work will be addressed in discussion of specific grape varieties (*vide infra*).

4. The epigenome is the record of modification to the bases that make up the genome before its expression. The epigenome thus reflects the expression of the genome as actually occurring rather than the simple straightforward organized list of bases.

5. Dal Santo, S.; Tornielli, G. B.; Zenoni, S.; Fasoli, M.; Farina, L; Anesi, A.; Guzzo, F.; Delledonne, M.; Pezzotti, M. *Genome Biol.* **2013,** *14,* r54 (DOI:10.1186/gb-2013-14-6-r54). See also Hunter, P. *EMBO Reports* **2013,** *14,* 769.

6. A phenotype is the composite of the observable characteristics of an organism that result from the expression of an organism's genes as well as the influence of environmental factors and the interactions between the two.

7. Cultivars are a group or set of groups of plants chosen to be cultivated and maintained by propagation.

8. The transcriptome is the set of all RNA (ribonucleic acid) molecules, including messenger RNA (mRNA), ribosomal RNA (rRNA), transfer RNA (tRNA), and other non-coding RNA produced in cells.

9. Mullins, M. G.; Bouquet, A.; Williams, L. E. *Biology of the Grapevine*; Cambridge University Press: Cambridge, UK, 1992; p 2.

10. They contain two different alleles (different forms of the same gene) for the same trait.

2

Grapevine from Seed

IT IS WIDELY claimed that growing the vines that will produce good wine grapes starting from seed is difficult.[1] In part, as noted above, this is apparently due to the presence of different alleles expressed differently as a function of environmental factors. As a consequence, most wine is produced from grapes arising from a graft of a vine that already produces desirable product. However, it is possible to plant seeds to generate vines—although the product is not always what is expected!

The fact that parent varieties (the flower of one parent and pollen of another) will generally produce a variety different from either parent is generally sought to be avoided in commercial enterprise. However, since grape flowers (as will be discussed in Chapter 12) are often found as tight clusters, hermaphroditic reproduction either naturally or by intervention can be effective. Adventures in crossing, such as with the *Vitis vinifera* varieties Cabernet franc and Sauvignon blanc can be profitable. They are reported to have led to the formation of Cabernet Sauvignon.

The grape seed needs to germinate.[2] Germination is evidenced by the forming of the plant within the seed and the opening of the seed coat to produce a seedling (Figure 2.1). The plant embryo responds, as dictated by the genome, to the warmth of the soil and the availability of water, and continues to grow from the first cell division until the plant sprouts.

It is not uncommon for seeds of many species to have set a genetically dictated timer. The setting of the timer may, for example, require that the ground be frozen and subsequently thawed (a process called vernalization). Once moistened,[3] by a thaw or rain, the dry seed takes up water that passes through channels in cell walls and membranes[4] (the inside of the cell being drier than the outside) that apparently open in response to the "timer" and in response to soil constituents. Ions[5] found in the soil are washed in with the water. The water and nutrients in the soil are now available to put the enzymes[6] and their cofactors, previously lying fallow in the seed, to work. That is, chemical reactions can be carried out utilizing the food reserves (sugars, fats, and proteins) that were stored in the seed. These initial processes build intermediates needed for production of roots that are destined to go down, stems that are destined to go up, and eventually leaves that are destined to be above the soil, so that

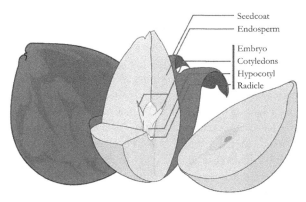

FIGURE 2.1 A representation of a seed. The embryo is the multicellular, membrane-enclosed structure containing the genetic information (the genome) of the plant that is germinating (i.e., sprouting). The genome and the food required for the energy needed to grow has been stored in the seed when it was propagated from the plant. The seed coat is tough enough to protect its contents from invaders who might wish to use those contents as food. The coat is mostly cellulose, a *polymer* (i.e., "*poly* = many, *mer* = body") compound largely composed of linked carbohydrate units and about which much more will be said later. Figure adapted from the Wiki Commons image for the seed of an avocado which was released into the public domain by its author, M. R. Villarreal.

energy can be drawn from the sun for subsequent processes. Oxygen (O_2) is needed in the beginning, and so the soil must be fairly loose to allow oxygen transport to the (yet underground) plant. Further, although moisture is necessary, too much water will prevent oxygen from getting to the growing plant. In addition, excess water will encourage bacteria to multiply and attack the embryo as well as the forming roots. Rot will set in. Indeed, hillsides are often sought because the drainage is generally better there than on flat plains.

As the plant embryo begins to grow, a shoot called the radicle (which will become the root meristem)[7] grows downward into the surrounding soil. At about the same time, the first leaves or cotyledons (which, as will be discussed later, are not considered true leaves) push out of the soil, and the hypocotyl emerges just behind them. The hypocotyl, due to become the apical meristem or plumule, lifts the growing tip, often carrying the seed coat out of the ground. The apical meristem is thus the primary above-ground organ of extension of the young plant, and it develops into the stem, which will bear the first true leaves and, eventually, flowers.

Among the more interesting questions posed by deep thinkers over the years has been that of how the seed, underground and away from light, knows how the roots should go down and the shoots up. In principle, of course, once up *or* down is determined, the opposite direction is also determined. Charles and Francis Darwin[8] were among those who studied this question of gravitropism (the sensing of gravity) and who found by experimentally removing them that the tips of the roots of plants were critical for the plant to know which way was down. Since then, hard, dense bodies called stratoliths have been found at the tips (root caps) of simple plants presumably drawn there by gravity. Parenchyma cells, which are the cells that store the starches, water, and oils, are known to contain dense starch bodies called amyloplasts. The latter are largely composed of carbohydrates and referred to as tylose as it forms from the parenchyma cells. The amyloplasts are transported to the root tips through the phloem cells, the transport system for moving sugars and other molecules, but not water and minerals, which move through the xylem. It can be argued that this transport and root tip storage system of the heavy bodies is the basis of gravitropism.

Indeed, the long-distant transport of sugars, photoassimilates, and signaling molecules which are transported through the phloem clearly require special transport sieve elements as living cells. This material is currently being actively studied.[9,10]

By showing that the sensing system for the presence of gravity fails when plants are permitted to germinate in the absence of normal gravity (i.e., on the International Space Station), affirmation of the general idea of gravitropism has been validated.[11]

Most recently, it was noted that adaptation to gravity, thus allowing plants to move to land, must have represented a significant evolutionary step and genomic modification. A genomic search for evidence that a plant's shoots which would need to gain "a negative gravitropic response and roots to gain a positive gravitropic response" has begun.[12]

Growing the vine from seed is also complicated by the intrusion of root feeders (insects that attack the roots of plants) such as Phylloxera (*Daktulosphaira vitifoliae*).[13]

NOTES AND REFERENCES

1. However, it has been noted numerous times that with proper treatment, particularly with holding the seeds at low temperature for some period of time, favorable results can be obtained. See for examples (a) Flemion, F. *Contrib. Boyce Thomp. Inst.* **1937**, *9,* 7; (b) Kachru, R. B.; Singh, R. N.; Yadav, I. S. *Vitis* **1972**, *11,* 289; (c) Conner, P. J. *Hort. Sci.* **2008**, *43,* 853; (d) Reisch, B. I. [Online] **2017**. https://hort.cals.cornell.edu/people/bruce-reisch (2017).

2. Germination is the process leading from seed to plant and can be interrupted by ground-dwelling insects that feed on the forming plant as well as by molds, yeasts, and attack by various viruses (or vira).

3. Interestingly, it has recently been pointed out that "the acquisition of water-conducting tissue enabled the transition of plants . . . to land . . . (and) provide(d) mechanical strength to the stem, while allowing efficient water conduction." Xu, B.; Ohtani, M.; Yamaguchi, M.; Toyooka, K. P.; Wakazaki, M.; Sato, M.; Kubo, M.; Nakano, Y.; Sano, R.; Hiwatashi, Y.; Murata, T.; Kurata, T.; Yoneda, A.; Kato, K.; Hasebe, M.; Demura, T. *Science* **2014**, *343,* 1505.

4. *Membranes* are barriers designed to allow passage of some materials but not others.

5. As discussed more fully in Appendix 1, *ions* consist of positively or negatively charged atoms or groups of atoms in which the number of protons and electrons (equal to each other in neutral species) are not the same. An excess of electrons yields an *anion*. A *cation* results with too few electrons. Since processes are electrically neutral, cations do not occur without anions and *vice versa*. The accompanying ion is not always specified and is, generically, known as a *gegenion*.

6. *Enzymes* are large naturally occurring molecules that, in turn, produce the molecules needed for metabolic processes (i.e., processes that sustain life) to occur in reasonable time.

7. Meristem is the name we give to cells that are presumed to be undifferentiated and thus capable, in principle, of becoming any part of the plant. Meristem cells as they become different (as dictated by the DNA of the plant) give rise to different plant tissues.

8. Darwin, C.; Darwin, F. *The Power of Movement in Plants;* John Murray: London, 1880.

9. Furuta, K. M.; Yadav, S. R.; Lehesranta, S.; Belevich, I.; Miyashima, S.; Heo, J.; Vatén, A.; Lindgren, L.; De Rybel; B.; Van Isterdael, G.; Somervuo, P.; Lichtenberger, R.; Rocha, R.; Thitamadee, S.; Tähtiharju, S.; Auvinen, P.; Beeckman, T.; Jokitalo, E.; Helariutta, Y. *Science* **2014,** *345,* 933.

10. Groover, A. *New Phytologist* **2016.** DOI: 10.1111/nph.13968.

11. Paul, A.-L.; Amalfitano, C. E.; Ferl, R. J. *BMC Plant Bio.* **2012,** *12,* 232.

12. Ge, L.; Chen, R. *Nature Plants* **2016,** *2,* 16255. DOI: 10.1038/nplants.2016.155.

13. As discussed more fully in the section on grafting, it has been pointed out (Powell, K. S. In *Root Feeders, an Ecosystem Perspective;* Johnson, S. N., Murray, P. J., Eds.; Center for Agricultural Biosciences International: Engham, UK, 2008) that Phylloxera introduced into Europe by British botanists in the period 1850–1860 caused major damage and in some cases extirpation throughout the European vineyards as the insect spread. Subsequently, it became clear that the rootstock in North America was more resistant to attack either because upon attack the roots produced compounds that deterred further attack (i.e., anti-feedants) or that such compounds were already present in the roots. While the exact details remain unknown, successful grafting (*vide infra*) of remaining vines onto the North American rootstock allowed wine production to prosper.

3

Grapevine from Grafting

THE HISTORY OF cutting deeply into the vascular tissues of one growing, strong, host plant and then inserting a part of another plant in such a way that they join together is called grafting (originally from the Greek, "*graphion*" referring to the sharpened end of piece to be inserted, the scion). The original cut into the host plant, the rootstock, is made into the vascular cambium of the plant (i.e., that part of the plant stem that contains the meristem, which is the plant tissue made up of undifferentiated cells where growth can take place). The piece to be grafted, the scion, is also cut to its vascular tissue. The vascular joining is called inosculation, and the process can be traced back to the early cultivation of fruit trees. Healthy, fruit-bearing crops from the stock of the scion rather than that of the rootstock are known to result. That is, the meristem adapts.[1] The vascular cambium itself consists of cells that are already partially specialized (e.g., the "xylem" for the woody tissue that carries water and some water soluble mineral nutrients and the "phloem" for carrying carbohydrates and other similar nutrients). The plan is that the undifferentiated cells, as well as those partially differentiated, will accommodate the scion to the rootstock, and the phloem from the rootstock will learn to feed the growing scion graft. Should the graft "take," the matured scion will, with the advent of photosynthesis (*vide infra*), return the favor to the rootstock. Both will profit.

One story of the grafting process and the interaction between plants and the insects that feed on the plants as applied to the wine industry has been told often.[2] A family of plant parasitic insects which are native to North America, the Phylloxeridae (Genus: *Daktulosphaira*; Species: *vitifolia*, Fitch, 1855, commonly called "phylloxera") were involved.

Grapevines in North America had built resistance to some members of the Phylloxeridae family and had, apparently, been able to match genetic changes in the insect with their own changes over the years. However, late in the nineteenth century (about 1880) commerce between England and North America increased, and botanists in England collected specimens of American grape vines for return to England. Along with the vines, specimens of phylloxera were also taken. The biology of phylloxera is complex, with a number of different forms of the insect found on different parts of plants at different times.[3] Thus, roots, vines, and leaves can all be infected. The plague of phylloxera introduced in this way spread from England to the vineyards of Europe, and most of the French vineyards, where varieties of *Vitis vinifera* were abundant, were ravaged. As much as nine-tenths of all European vineyards were destroyed.[4]

The eventual solution to the problem was to graft a *Vitis vinifera* scion from a local vineyard onto the rootstock of a resistant American native species (e.g., *Vitis aestivalis, V. arizonica, V. californica*) which was also already acclimated to terroir where the "new" grafted rootstock was to be grown.[5,6]

Careful alignment of the meristem portions of the scion and rootstock, binding of the joint, and treatment of the bound area with appropriate growth accelerators all promote attachment of the two living parts. Growth accelerators include naturally occurring organic messenger compounds called phytohormones, plant hormones[7] that control or regulate germination, growth, metabolism, or other physiological activities. Naturally occurring phytohormones include the gibberellins, the cytokinins, abscisic acid, and some vitamins. Some specific messengers include naturally occurring hormones called auxins, such as heteroauxin (indole-3-acetic acid) as well as synthetic analogues like the corresponding butanoic acid derivative of indole and α-naphthylacetic acid.

The auxins appear to promote growth in plants and are composed of a group of aromatic compounds that also bear carboxylic acid side chains (aromatic compounds and carboxylic acids are considered functional groups and are discussed in Appendix 1). It has been reported[8] that auxins promote root growth. The structural representations of Figure 3.1 correspond to some of these auxins.

The hormones called gibberellins, such as gibberellic acid (Figure 3.2), are a special subgroup of a very large class of compounds that either contain twenty carbon atoms or are derived from other compounds that contained twenty carbon atoms. The large group of compounds is called diterpenes, and thousands are known. Some of the chemistry of

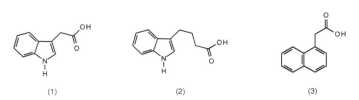

FIGURE 3.1 Representations of some auxins: (1) indole-3-acetic acid; (2) indole-3-butanoic acid; (3) α-naphthylacetic acid.

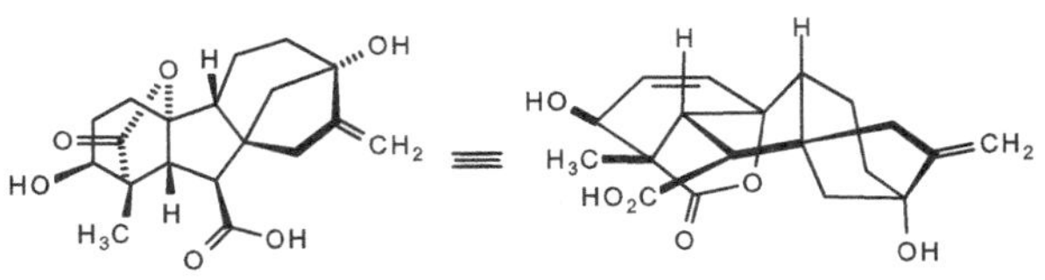

FIGURE 3.2 Representations of the structure of (+)-gibberellic acid.

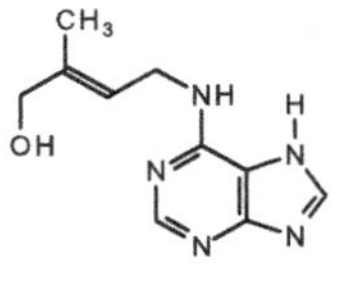

FIGURE 3.3 A representation of zeatin.

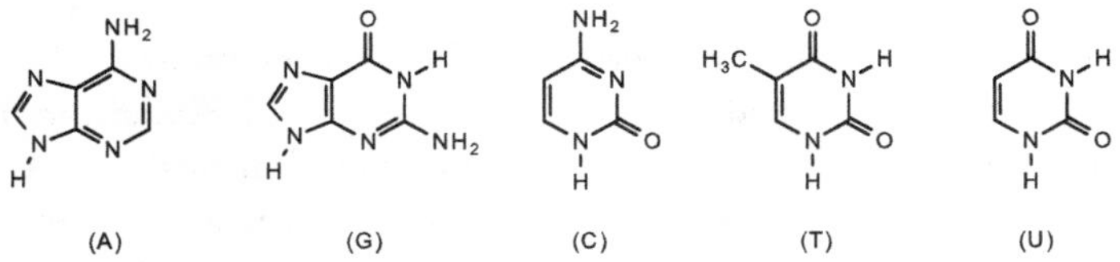

FIGURE 3.4 Representations of adenine (A); guanine (G); cytosine (C); thymine (T); uracil (U).

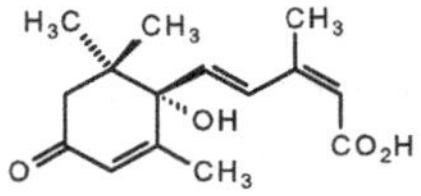

FIGURE 3.5 A representation of abscisic acid.

terpenes (10 carbons), sesquiterpenes (15 carbons), diterpenes (20 carbons), and related materials will be discussed subsequently (and may be found in the Appendices) since, among other features, they are associated with the odors of the flowers of many varieties of grape as well as the grapes and wines derived from them.

The phytohormones that produce signals that promote cell division or cytokinesis are called cytokinins. It appears that, in contrast to auxins (that promote root growth) cytokinins promote bud growth although they are reported to be biosynthesized in the roots! These nitrogen-rich compounds, such as zeatin (Figure 3.3) from corn, belong to a general class of compounds called purines and are related to the same subset of bases making up ribonucleic acids (RNAs) and deoxyribonucleic acids (DNAs), representations of which are shown in Figure 3.4.[9]

Finally, the sesquiterpene phytohormone, abscisic acid (Figure 3.5), appears to be involved in a wide variety of signaling duties, including abscission (the normal process by which deciduous plants defoliate) and other situations that put stress on plants.

NOTES AND REFERENCES

1. Meristem cells, while not actually immortal, appear to have the ability to grow into different plant cells within the host as well as to remain unchanged over many cycles of plant growth and senescence. As already noted, there are shoot apical meristems as well as root apical meristems, differentiated by their respective ability to produce shoots or roots (Jones, R; Ougham, H.; Thomas, H.; Waaland, S. *The Molecular Life of Plants*; Wiley-Blackwell: Oxford, UK, 2013).

2. Campbell, C. *The Botanist and the Vintner: How Wine Was Saved For the World*; Algonquin Books: Chapel Hill, NC, 2005.

3. Powell, K. In *Root Feeders An Ecosystem Perspective*; Johnson, S. N.; Murray, P. J., Eds.; Center for Agricultural Biosciences International: Engham, UK, 2008.

4. Boubals, D. *Prog. Agric. Vitic.,* Montpellier, **1993**, *110,* 416.

5. See www.vivc.de for the International Variety Catalogue.

6. Although the genetics of the rootstock have long been thought to be divorced from those of the scion, it has recently been found that some exchange does, apparently, occur. Stegemann, S.; Bock, R. *Science* **2009**, *324,* 649 reported that there is exchange of genetic material between cells in the plant tissue grafts and that this may lead to gene transfer. Subsequently, Yin, H.; Yan, B.; Sun. J.; Jia, P.; Zhang, Z.; Yan, X.; Chai. J.; Ren, Z.; Zheng, G.; Liu, H. *J. Exp. Botany* **2012**, *63,* 4219 reported signal exchange processes between scion and rootstock as soon as 2–3 days after grafting. Most recently, entire nuclear genome across graft junctions has been determined, see Fuentes, I.; Stegemann, S.; Golczyk, H.; Karcher, D.; Bock, R. *Nature* **2014**, *511,* 232.

7. *Hormones* are vaguely defined as "messenger" molecules, generated by biological processes in one place in a living system for transmission to another place where an effect is produced. Any list of such materials is incomplete because new members are being added as more is learned.

8. Vannests, S.; Friml, J. *Nature Chem. Bio*. **2012**, *8,* 415 and Villalobos, L. I. A. C.; Lee, S.; de Olivera, C.; Ivetac, A.: Brandt, W.; Armitage, L.; Sheard, L. B.; Tan, X.; Parry, G.; Mao, H.; Zheng, N.; Naier, R.; Kepinski, S.; Estelle, M. *Nature Chem. Bio*. **2012**, *8,* 477.

9. More information about the structural features and the various groups (e.g., $-OH$, $-CO_2H$, $-NH_2$ hanging on to the rings as well as the cyclic compounds themselves can be found in the Appendices. The bases of RNA (A, G, C, U) and DNA (A, G, C, T) are adenine, A; guanine, G; cytosine, C; thymine, T; uracil, U.

4

Grapevine from Hardwood Cuttings

ROOTED PLANTS CAN often be obtained and transferred from one environment into another either in order to increase the number of vines producing a specific grape in a vineyard or to introduce a new variety or propagate a new cultivar.[1] It has been found that some vines can be grown from hardwood cuttings.

The technique of hardwood cutting involves removing a cane (Figure 4.1, a and b) from a successful vine once the vine has gone dormant for the winter, trimming it appropriately, and then planting it in well-fertilized soil either with or without growth stimulants (i.e., phytohormones, *vide supra*). It is clear that the conditions of planting, reported by various sources, are a function of variety and terroir. Interestingly, it appears that the cutting, which may have been grown on a rootstock different from the variety of grape produced, will produce roots that are true to the variety of grape.

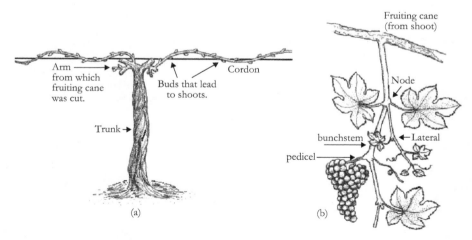

FIGURE 4.1 Cartoon representations of (a) a dormant vine after trimming and (b) a fruit-bearing vine before trimming.

Once the vine, from seed, grafting, or cane begins to grow, it must be "trained" so that its growth can be monitored and successful grape crops harvested. The training includes proper spacing of vines and the establishment of a trellis system or posts for each vine.

Trellis systems are set up during the first or second year of the growth of the vine since harvesting of grape crops before the third year is rare. The trellis, which will need to bear the weight of the vine and grapes, is built much like a fence. Thus, the row of grape vines is held up by end posts at the end of the row and line posts about 20 feet apart between the ends. Usually, there is a line post for every two or three vines with some species needing more space than others. Generally the end posts are thick treated wood, concrete, or steel and are strongly anchored. The line posts are thinner, and the trellis itself is made of twelve (12) gauge or heavier wire with the number of wires a function of the weight to be supported and the height to which the grapes are to be grown.

In place of a trellis system, each vine itself may be held to a post with appropriate cross pieces to allow the cordons to stretch and bear the weight of the fruiting vine.

In the meantime the analysis of the soil and the work to improve its fertility can begin.

NOTE AND REFERENCE

1. A cultivar is generally defined as a plant or group of plants that has (or have) been selected for cultivation because of a specific trait or group of traits. For deciduous plants (i.e., plants that lose their leaves), once leaves have been lost and the plants are dormant, straight, young cane shoots (preferably at least one year old) can be cut with well-cleaned, sharp shears. These canes are referred to here as "hardwood cuttings."

5

The Soil

THE WIDESPREAD PRACTICES of viniculture (the study of production of grapes for wine) and oenology (the study of winemaking) affirm the generalization that grapevines have fewer problems with mineral deficiency than many other crops. Only occasionally is the addition of iron (Fe), phosphorus (P), magnesium (Mg), and manganese (Mn) supplements to the soil needed. Addition of potassium (K), zinc (Zn), and boron (B) to the soil is more common. And, of course, nitrogen (N) is critical for the production of proteins. Over the years, various transition metals (metals in groups three through twelve [3–12] of the periodic table, Appendix 1) have been shown to be generally important. These groups include iron (Fe), magnesium (Mg), manganese (Mn), zinc (Zn), and copper (Cu). Many metals are bound to organic molecules that are important for life.[1]

Some of the metals, such as copper (Cu) and iron (Fe), are important in electron transport while others, including manganese (Mn) and iron (Fe), inhibit reactive oxygen (O) species (ROSs) that can destroy cells.

Metals serve both to cause some reactions to speed up, called positive catalysis while causing others (e.g., unwanted oxidation) to slow down (negative catalysis).

It is not uncommon to add nitrogen (N), in the form of ammonium salts such as ammonium nitrate (NH_4NO_3), as fertilizer to the soil in which the vines are growing. It is also common to increase the nitrogen (N) content in the soil by planting legumes (legumes have roots that are frequently colonized by nitrogen-fixing bacteria). Nitrogen-fixing bacteria convert atmospheric nitrogen (N_2), which plants cannot use, to forms, such as ammonia (NH_3) or its equivalent, capable of absorption by plants. Nitrogen, used in plant proteins, tends to remain in the soil after harvest or decomposition. With sufficient nitrogen present in the soil the growth cycle can begin again in the following season without adding too much fertilizer.[2]

In a more general sense, however, it is clear (as mentioned earlier) that the soil must be capable of good drainage so the sub-soil parts of the plant do not rot and it must be loose enough to permit oxygen to be available to the growing roots.

Interestingly, it has recently been suggested that edaphic factors within vineyards outweigh biogeographic trends.[3] Since soil bacterial assemblages have extensive local heterogeneity even across small distances, the immediate surroundings of the vines in the vineyard may be more important than the overall locale. So, as our tools and understanding become infused with more knowledge, it becomes clearer that the local soil microbiome has a major influence on grapevine-associated microbiota.[4]

NOTES AND REFERENCES

1. In eukaryotic cells (complex cells, cells with many enclosed features within walls), metals are found in the mitochondria, which are the membrane-enclosed regions where most of the adenosine triphosphate (ATP), needed to power chemical reactions in the plant, is synthesized. Many other necessary compounds are also synthesized there. A more thorough discussion of cells and their components is treated later.

2. The greenhouse gas nitrous oxide (N_2O) and the run-off pollutant nitrate anion (NO_3^{2-}) are often found when too much fertilizer is applied.

3. Zarraonaindia, I.; Owens, S. M.; Weisenhorn, P.; West, K.; Hampton-Marcell, J.; Lax, S.; Bokulich, N. A.; Mills, D. A.; Martin, G.; Taghavi, S.; van der Lelie, D.; Gilberta, J. A. [Online] **2015.** http://mbio.asm.org/content/6/2/e02527-14 (accessed Apr 5, 2017) (edaphic, "the nature of the soil" (http://www.definitions.net/definition/edaphic).

4. The microbiome corresponds to the body of collective genomes of the microorganisms that reside in the environment under investigation. The microbiota are the microorganisms themselves.

6

The Roots of the Grape Vine

ASIDE FROM GRAFTING onto already established rootstock or the development of roots[1] from a planted cane (*vide supra*), root systems develop from the radicle in the plant's seed. Both as roots begin to form from the cane, and as the sprouting seed coat opens in response to soil temperature, moisture, and genetic programming left in place when the seed formed, the roots begin to grow and interact with the rhizosphere.[2]

Similarly, signals received by rootstock where grafting has been effected also occur. The roots begin to bring moisture and food to produce and support the stem and, eventually, the leaves, flowers, and fruit. Heavily fruited plants such as grapes require additional support for the stems. In the roots, epidermal (surface) cells elongate and develop into root hairs. Beneath the epidermal cells it appears that the phloem cells which bring the starch bodies (amyloplasts) to the root tips and help direct which way "down" is, develop first. Then xylem elements develop in order to move the minerals into the system. Most of the minerals are absorbed through channels developing in the walls of the growing undifferentiated cells (the meristems). Because of concentration gradients (i.e., there is less on one side of a cell membrane than on the other), some minerals appear to be actively transported into the cells of the xylem (presumably through similar channels) in response to signals emanating from the plant. From the xylem cells, the minerals and water move upward into the apical meristem and get distributed to other regions. Interestingly, although most of the cells are derived from the same group of meristems which thus might be considered true stem cells, it is genetic programming which permits that differentiation. Thus, the derivatives of the meristems undergo transformation and develop into various cell types that perform the different functions (Figure 6.1).

Relatively recently there has been an increased interest in what has been the largely unexplored biology of roots. The major difficulty, of course, is that the roots are underground (although some effort to grow them on artificial above-ground soils has begun), and to work with them and the surrounding soil the roots must be removed from the ground. Thus, although their functions regarding uptake of nutrients and water, as well as exchange with the soil, and serving to hold the plant in place are clear, it has been rare that sufficient

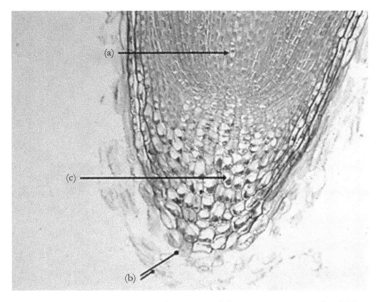

FIGURE 6.1 A representation of a *root* apical meristem: (a) quiescent center cells which are a type of stem cell that are thought to prevent surrounding cells from changing their nature and controlling the growth of new root cap cells as the root grows; (b) epidermal cells just ahead of the (c) root cap cells. The oldest cells are sloughed off and are dead. Adapted from Emmanuel Boutet File-Root Tip Creative Commons.

experimental evidence involving other aspects of their function (e.g., in the synthesis of specific compounds transported from the root to the leaf and *vice versa*) is known.

As we now know and as pointed out in a recent review,[3] there is an entire root-specific metabolism,[4] as well as a variety of cell types which, at least in the system examined, may be organized around a radial axis.[5] These root-specific processes involve not only signal exchange with the soil and other local plants but also some protective functions against pathogens and pests (e.g., nematodes). Indeed, it is now clear that there are specific compounds that are synthesized in the roots and perhaps elsewhere in the plant that are transported to the roots and then secreted from the roots to the soil. Some of these compounds protect the roots, some communicate with the environment, and a few actually appear to aid in converting nutrients into a form that can then be absorbed into the roots themselves. In this vein, and again with regard to a few well-studied systems and part of the above noted report, it appears that some of the genes in the variety of cell types had expression patterns that fluctuated. That is, they were turned on early in development, shut off, and then were turned back on later. This would allow for more or less protein production as required by growth and environmental change. Adjustment of the rhizosphere to accommodate the transport appears to be needed and might be accommodated in this way. Some of the compounds actively involved in defense processes (shown in Figure 6.2) include phytohormones such as abscisic acid (already mentioned); salicylic acid (the O-acetyl derivative at the phenol is "aspirin"), and the signaling pathway agents jasmonic acid and ethylene that are used to alert the plant that pathogens are present.

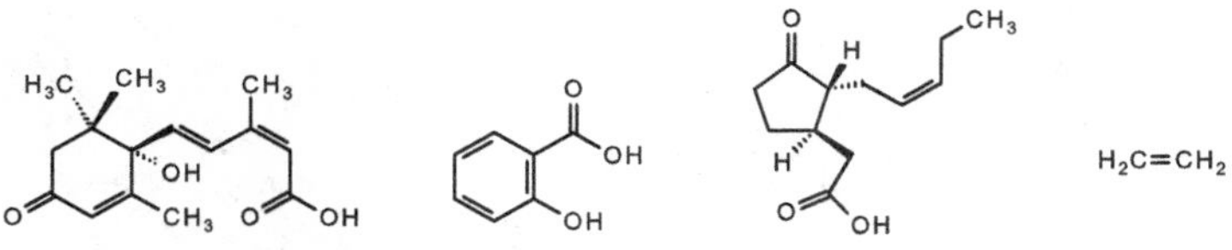

FIGURE 6.2 Representations of some compounds thought to be actively involved in defense of roots: (1) abscisic acid; (2) salicylic acid; (3) jasmonic acid; (4) ethylene.

Interestingly, there has been a recent efflorescence in the microbiology of the rhizome[6] and the entire rhizosphere,[7] and these have carried over into efforts with specific reference to grape.[8,9]

NOTES AND REFERENCES

1. "A plant may be rooted in place, but it is never lonely. There are bacteria in, on and near it, munching away on their host, on each other, on compounds in the soil. Amoebae dine on bacteria, nematodes feast on roots, insects devour fruit—with consequences for the chemistry of the soil, the taste of a leaf or the productivity of a crop." Ledford, H. *Nature* **2015**, *523*, 137.

2. The rhizosphere is the immediate area that surrounds the roots of plants, where the interactions between the soil, the organisms and minerals in the soil, and the plant take place. It is generally argued that the interactions are complex and, as a result, are currently poorly understood (*vide supra* regarding the soil-related microbiome).

3. Bais, H. P.; Loyola-Vargas, V. M.; Flores, H. E.; Vivanco, J. M. *In Vitro Cell. Dev. Biol.-Plant* **2001**, *37*, 730.

4. Vernalization (the process by which exposure for enough time to cold temperature promotes flowering) is among the more interesting observations in this regard. It appears that "if a vernalized shoot tip is grafted to nonvernalized stock, it will flower, but a nonvernalized shoot grafted to a vernalized stock will not flower." Amasino, R. *Plant Cell* **2004**, *16*, 2553. DOI: 10.1105/tpc.104.161070.

5. Brady, S. M.; Orlando, D. A.; Lee, J.-Y.; Wang, J. Y.; Koch J.; Dinney, J. R.; Mace, D.; Ohler, U.; Benfey, P. N. *Science* **2007**, *318*, 801.

6. Philippot, L.; Raaijmakers, J. M.; Lemanceau, P.; van der Utten, W. H. *Nat. Rev.* **2013**, *11*, 789.

7. Mendes, R.; Garbeva, P.; Raaijmakers, J. M. *FEMS Microbiol. Rev.* **2013**, *37*, 634.

8. Zarraonaindia, I.; Owens, S. M.; Weisenhorn, P.; West, K.; Hampton-Marcell, J.; Lax, S.; Bokulich, N. A.; Mills, D. A.; Martin, G.; Taghavi, S.; van der Lelie, D.; Gilbert, J. A. *mBio* **2015**, *6*, e02527–14, (doi: 10.1128/mBio.02527-14).

9. Barber, N. A.; Gorden, N. L. S. *J. Plant Ecol.* **2015**, *8*, 1.

SECTION II
Cells

7

Roots, Shoots, Leaves, and Grapes

AS NOTED EARLIER and as anticipated by Charles and Francis Darwin[1] it has been argued that plants sense the direction of gravity (gravitropism) by movement of starch granules found in cells called statocytes that contain compartments (organelles)[2] called statoliths. The synthesis of statoliths appears to occur in the plastid (plant organelle) compartments called *amyloplasts* (Figure 7.1, 1). It has been suggested that this gravitropic signal then leads to movement of plant hormones such as indole-3-acetic acid (auxin) (Figure 7.2), through the phloem opposite to the pull of gravity to promote stem growth.

Chloroplasts (Figure 7.1, 2) are cell compartments (plastids or organelles) in which photosynthesis is carried out. The process of photosynthesis, discussed more fully later, is accompanied by the production of adenosine triphosphate (ATP) from adenosine diphosphate (ADP) and inorganic phosphate (P_i) (Figure 7.3). ATP is consumed and converted to ADP and P_i in living systems. The cycle of production and consumption allows ATP to serve as an "energy currency" to pay for the reactions in living systems. Beyond this generally recognized critical function of chloroplasts, it has recently been pointed out that light/dark conditions affect alternative splicing of genes which may be necessary for proper plant responses to varying light conditions.[3]

The organelles or plastids which contain the pigments for photosynthesis and the amyloplasts that store starch are only two of many kinds of plastids. Other plastids, leucoplasts for example, hold the enzymes for the synthesis of terpenes, and elaioplasts store fatty acids. Apparently, all plastids are derived from proplastids which are present in the pluripotent apical and root meristem cells.

The *cell wall* (Figure 7.1, 3) is the tough, rigid layer that surrounds cells. It is located on the outside of the flexible cell membrane, thus adding fixed structure. A representation of a portion of the cell wall (as made up of cellulose and peptide cross-linking) is shown below in Figure 7.7. The cells will have different sizes as a function of where they are found (e.g., leaf, stalk, root), but in every case, the cell wall limits the size of the membrane that lies within.

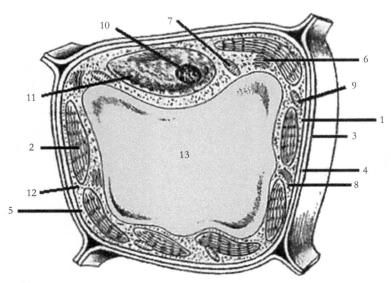

FIGURE 7.1 A representation of a plant cell. The numbered items, *viz.* (1) an amyloplast; (2) a chloroplast; (3) the cell wall; (4) the cell membrane; (5) cytoplasm; (6) endoplasmic reticula, both rough and smooth; (7) Golgi bodies; (8) a microtubule; (9) a mitochondrion; (10) the nucleolus; (11) the nucleus; (12) ribosomes; and (13) a vacuole are discussed in the following paragraphs. The representation is from Harold, F.M. *The Way of the Cell*, Oxford University Press, 2001, p 29 and is reproduced courtesy of Oxford University Press, USA. Used by permission of Oxford University Press, USA.

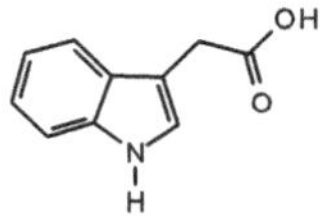

FIGURE 7.2 A representation of indole-3-acetic acid (auxin).

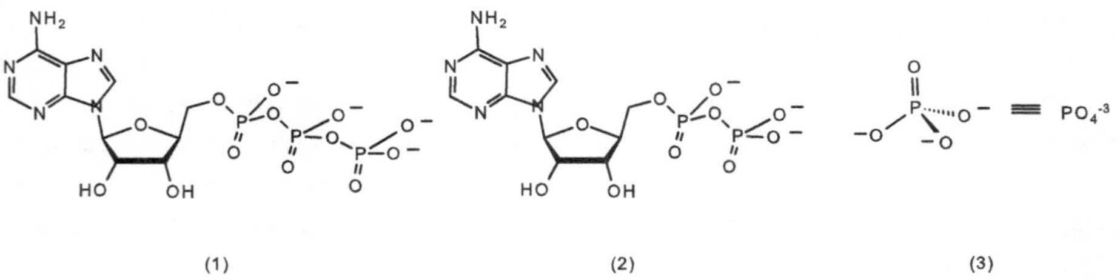

FIGURE 7.3 Representations of adenosine triphosphate, ATP (1), adenosine diphosphate, ADP (2), and the phosphate anion, PO_4^{3-}, often abbreviated as P_i (3).

The ***cell membrane*** (Figure 7.1, 4) separates the inside of the cell (cytoplasm) from the outside of the cell. The membrane controls the movement of substances in and out of cells and, in that way, protects the cell from its surroundings. As described in greater detail in Figures 7.4 and 7.9 through 7.12, the membrane is made up of nonpolar fatty acids attached

Roots, Shoots, Leaves, and Grapes

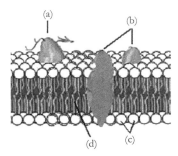

FIGURE 7.4 A cartoon representation of a plasma membrane structure: (a) a membrane-spanning glycoprotein with carbohydrate side chains (used for recognition of approaching entities); (b) membrane-spanning proteins; (c) hydrophilic polar head groups; (d) hydrophobic fatty acids. Modified with permission from Mariana Ruiz Villarreal.
https://commons.wikimedia.org/wiki/File:Phospholipids_aqueous_solution_structures.svg

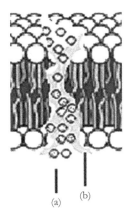

FIGURE 7.5 A cartoon representation of an "aquaporin," which is a proteinaceous cross-membrane pore through which water can be transported. The outside of the membrane is considered at the top and the inside at the bottom. Water (a) moves through the walls of the proteinaceous membrane represented by the grey area (b). Charged species that may be present promote the movement by helping to hold the aquaporin open. Modified with permission from Mariana Ruiz Villarreal.
https://commons.wikimedia.org/wiki/File:Phospholipids_aqueous_solution_structures.svg

to polar head groups and is interrupted by proteins that pass through the membrane and which, under appropriate conditions, are signaled to allow channels to open and close, allowing or forbidding transport of materials.

A representation of a membrane-spanning aquaporin protein is shown in Figure 7.5 to indicate how water moves across a membrane in response to osmotic pressure differences that arise from differences in concentrations of ions. Interestingly, it appears that some small molecules such as carbon dioxide (CO_2) and oxygen (O_2) simply diffuse through the membrane.

The *cytoplasm* (Figure 7.1, 5) is comprised of the cytosol, the gel-like substance enclosed within the cell membrane which, even now, is best described as a complex aqueous mixture of simple ions and macromolecules. The latter are composed of whole or pieces of proteins, carbohydrates, polyphenols, etc. It is also clear that some enzyme-catalyzed synthetic processes also occur there. In addition to the cytosol, the cytoplasm also contains organelles devoted to various functions. Much of the chemistry that goes on in the cell occurs within the organelles of the cytoplasm.

The *endoplasmic reticulum* (Figure 7.1, 6) (the ER) is one of the organelles in the cytoplasm. The ER is comprised of an interconnected network of membrane vesicles (i.e., a supramolecular assembly of lipids). The ER is broadly divided into two types: rough endoplasmic reticulum (RER) and smooth endoplasmic reticulum (SER). The RER is studded with ribonucleic acid (RNA) derived ribosomes (the sites of protein synthesis) on the cytosolic face. The SER is a smooth network of vesicles, lacking ribosomes, where the reactions defining the metabolism (chemical transformations) of lipids and carbohydrates are found to occur.

The *Golgi bodies* (Figure 7.1, 7) are the entities that package proteins inside the cell before they are sent to their destination, whether within the cell or secreted from the cell.

Microtubules (Figure 7.1, 8), made up of protein filaments, are part of the structural network within the cell's cytoplasm. The primary role of the microtubule cytoskeleton may be structural support helping to organize the layout of the cell.

Mitochondria (Figure 7.1, 9) (singular mitochondrion) are organelles which have been described as "cellular power plants" in the biology literature. They are called "power plants" because they generate most of the cell's *de novo* supply of ATP, the currency of chemical energy. Many of the functions of a cell, including life and death, are controlled by actions within mitochondria.

The *nucleolus* (Figure 7.1, 10) captures and immobilizes proteins. The captured proteins are then unable to diffuse and to interact with their binding partners. This appears to be one of the pathways for post-translational regulation (i.e., modification of a protein after it has been formed). It is now held that non-coding RNA originating from stretches of deoxyribonucleic acid (DNA) between recognized genes is responsible for the nucleolus.

The *nucleus* (Figure 7.1, 11) (the kernel, plural nuclei) is an organelle containing most of the cell's genetic material as DNA. The DNA, complexed with appropriate proteins, forms chromosomes, the thread-like structures of DNA and accompanying proteins that carry genetic information. The information carried in specific stretches of DNA, called genes, codes for specific proteins. The genes within these chromosomes are the cell's nuclear *genome*. The information in the nucleus is in control of the cell and thus the plant and all that subsequently results to distinguish one plant from another.

The *ribosome* (Figure 7.1, 12), part of which is derived from the sugar ribose (*vide infra*), rather than deoxyribose as in DNA, is a large and complex molecular forge used for the production of proteins. That production involves choosing the sequence of amino acids and then linking them together into peptides. The choice is made by messenger RNA (mRNA), and the amino acid is delivered by transfer RNA (tRNA). Ribosomes in most plants consist of two major subunits: one large (about 50S) and one small (about 20S) bound together with proteins into one 70S particle. The "S" refers to a Svedberg unit, a measure of the rate of sedimentation in a centrifuge. It appears that the small ribosomal subunit reads the mRNA, while the large subunit joins amino acids to form the polypeptide chain (protein).

A *vacuole* (Figure 7.1, 13) is a compartment within the cell, with its own membrane, filled with water containing inorganic ions and some water-soluble organic species. The materials present help control the hydrogen ion concentration (pH, see Appendix 1), the

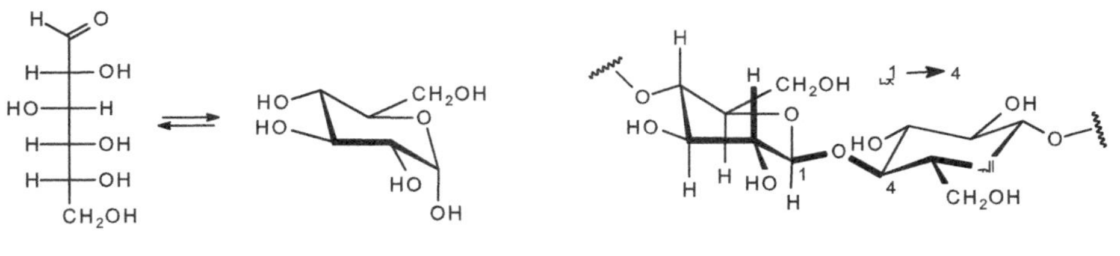

FIGURE 7.6 (1) A representation of the acyclic form of D-(+)-glucose; (2) A representation of the cyclic pyranose form of the α-anomer of D-(+)-glucose; (3) A representation of a fragment of the polymer cellulose showing the connection between C-1 (β-anomer) of one unit to C-4 of the next. Thousands of such connections define the observed structure.

internal overall cell pressure, proteins needed for future growth, small carbohydrates needed for rapid deployment, some ions such as magnesium (Mg^{2+}), calcium (Ca^{2+}), sodium (Na^+), and potassium (K^+). Vacuoles are formed by fusion of membrane-bound bubbles or vesicles.

The walls of the root and the walls of the seed and cuttings are cross-linked cellulose.[4] The long glucose[4] polymer with all β-1,4-linked D-glucose units that constitutes cellulose (Figure 7.6) cross-links polymer strands with hydrogen bonds.[4] Other bonds of saccharides[4] (e.g., new carbon–oxygen bonds to form hemiacetals[4] and acetals[4]) form glycans.[4] Glycans can also be composed of β-1,4-linked D-glucose units, as well as homo- or heteropolymers[4] of monosaccharide[4] residues, and can be linear or branched.

The loosening and, in some cases, the breaking of the bonds between the tightly held cellulose polymeric strands allows nutrients (e.g., nitrogenous materials, minerals and water) to enter the plant.

The entry of the nutrients process appears to be governed by enzymatic[5] processes that are responding to differences in pressure and temperature.

The repeating glucose units in cellulose are shown in Figure 7.6, (3). Despite the fact that cellulose is the most common and widespread organic compound on the planet, it is really only a special glucose-based polysaccharide. The cellulose chains are found to be linked to more chains (and thus cross-linked), as shown in Figures 7.7 and 7.8, using additional glucose units or units derived from or related to glucose.

The polysaccharide chains are also held by protein and lignin units. Proteins are composed of amino acids, whereas lignins are polyphenolic acid–based materials. Both of these structural units are capable of cross-linking with carbohydrates.

As noted earlier, there is significant and growing interest in learning more about roots. With advances in the genomic details of *Arabidopsis thaliana* and the ability to subsequently modify the genome, much has been learned about exactly how roots grow laterally through the established and hardened walls of the main root.

Interestingly, it has recently been pointed out that while chemical signals from growth agents (e.g., auxins) are required for lateral root growth, those primordia grow through overlying cell layers that accommodate the incursion.[6]

Lateral root formation in plants can be studied as the process of interaction between chemical signals and physical forces during development. Lateral root primordia grow through overlying cell layers that must accommodate this incursion. Those overlying cell layers that

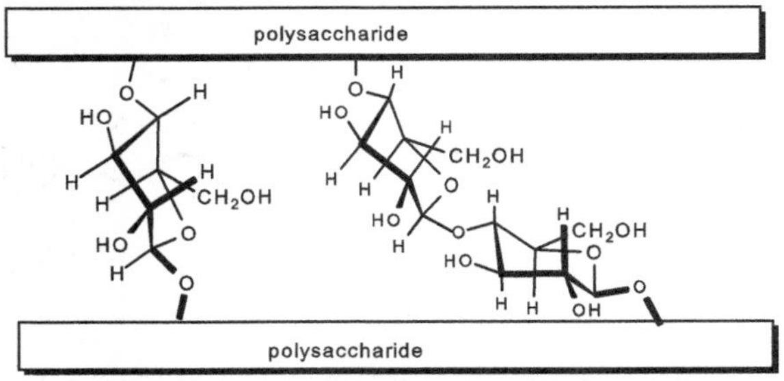

FIGURE 7.7 A cartoon representation of the cross-linking of polysaccharide chains by glucose units.

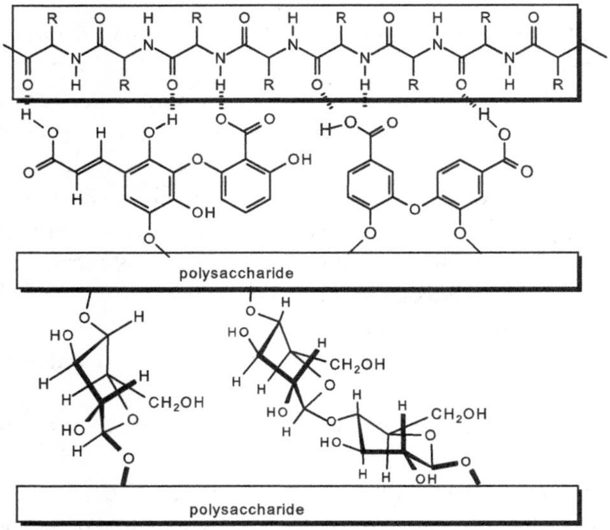

FIGURE 7.8 A cartoon representation of cross-linking of polysaccharide units with lignin and protein units. As represented here, the topmost linkage represents a protein where the groups "R" are either the same or different and are, in principle, representative of the amino acids (e.g., R = H, glycine [Gly, G]; R = CH$_3$, alanine [Ala, A]) present in the plant. The aryl phenolic acids, either with a three-carbon side chain (e.g., cinnamic acid derived from phenylalanine [Phe, F]) and with or without a side chain and co-linked one to another, are drawn as representatives of lignin.

lose some of their volume grow thinner and, in that process, also allow the tight junction barrier, which both keeps the shape of the root meristem and protects the contents, to open and make way for the emerging lateral root.

It also appears to be true that below ground organisms interact with the roots and influence plant–pollinator interactions,[7] but as discussed later, the issue with grapes is somewhat more complicated as they tend to be self-pollinating (i.e., hermaphroditic with both male and female organs—stamens and ovaries—present in the flowers).[8]

Inside the cell wall, composed of the polysaccharide and polyamide units, there is the plasma membrane (seen earlier) of the cell. The plasma membrane is generally composed of

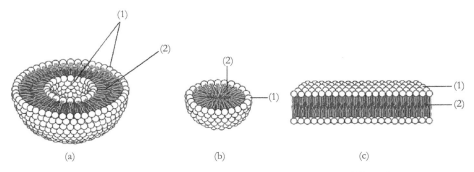

FIGURE 7.9 Idealized representations of: (a) a liposome, with hydrophilic (1) head groups on the outside and inside and hydrophobic (2) units (e.g., fatty acid chains) separating the two sets of hydrophilic groups; (b) a micelle, with only one such layer with hydrophilic (1) without and hydrophobic (2) within and; (c) a bilayer sheet with the hydrophilic layers (1) without and the hydrophobic layers (2) within. Modified with permission from Mariana Ruiz Villarreal. https://commons.wikimedia.org/wiki/File:Phospholipids_aqueous_solution_structures.svg

lipids (i.e., waxes, with or without polar head groups that contain phosphorus), fats, oils, and steroids. The wall also contains various peptide and protein amino acid polymers.

When the cells are not structural members of roots or stems (branches), plasma membranes can serve as walls that separate cells from each other as well as from the surrounding environment. Such cells can migrate through channels in the plant and can be carried by the phloem transport system.

The lipids are arranged so that polar hydrophilic ends face outward and hydrophobic ends face into the layer. By having two such layers (a lipid bilayer) both the inside and outside hydrophilic regions are kept apart (Figures 7.9 and 7.10). The lipid bilayer, a liposome, is studded with various proteins and signaling components as well as many receptors whose function is to allow cells to communicate, to absorb nutrients, and to eliminate products of reactions that have occurred in the cell.

The compounds involved on the outside of the membrane—sphingomylein (Figure 7.11, b), glycolipids (Figure 7.11, d), phosphatidylcholine and cholesterol (Figure 7.11, c)—are not, generally, same as those on the inside of the cell except for cholesterol. However, these compounds are similar and include cholesterol, phosphatidylserine, phosphatidylinositol and phosphatidylethanolamine (choline) as shown in Figure 7.12.

The striking similarity from one cell to the next in overall function and geometry, at least at the beginning of life, is frequently observed upon.[9] But, of course, there are differences since not all cells serve all functions.

The unanswered remaining questions of how a cell knows where it is and what its specific role overall is supposed to be are only now being approached. Thus, as stem cells must meet root cells, the former sensitive to light and the latter not so much, at ground level, the cells must know when to be a root or when to be a shoot. And, of course, when the leaves come forth from the stems, the stem cells must stop being stems and begin to become leaves.

Since the early work with the genome of *Vitis vinifera*, which was completed in 1997,[10] additional studies have been undertaken. Thus, it has recently been found[11] that the grapevine genome contains a set of genes that produce a protein family containing transcription

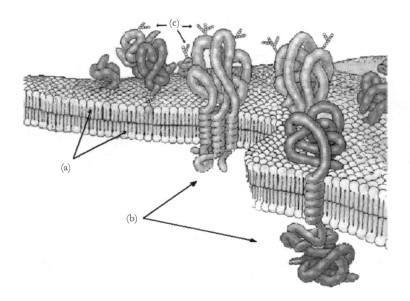

FIGURE 7.10 A cartoon representation of a lipid bilayer (a) as a plasma membrane with imagined additional detail of inserted peptide transmembrane members (b), several of which have carbohydrate appendages (c) attached. Modified with permission from Mariana Ruiz Villarreal. https://commons.wikimedia.org/wiki/File:Phospholipids_aqueous_solution_structures.svg

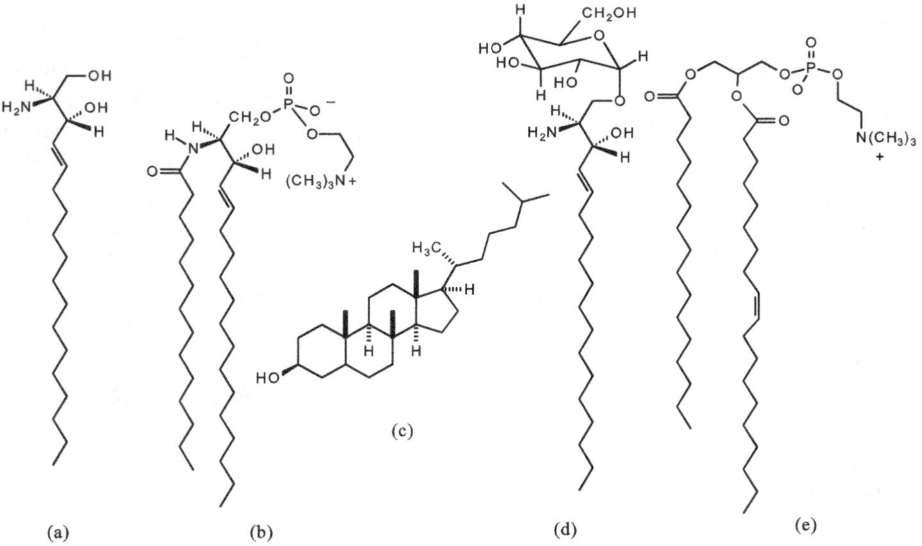

FIGURE 7.11 A representation of some of the structures which constitute the **outside cell** membrane (a) a sphingosine derived from serine with a palmitic acid (hexadecanoic acid)–derived side chain [(2S,3R)-2-aminooctadec-(E)-4-ene-1,3-diol]; (b) a sphingomylein with a myristic acid (tetradecanoic acid) amide attachment; (c) a representation of cholesterol; (d) a representation of a glycolipid with a glucose head attached to a serine precursor and a palmitic acid–derived side chain; (e) a representation of phosphatidylcholine where a glycerol head is attached (left-to-right) to palmitic acid (hexadecanoic acid) as an ester, to an oleic acid [(Z)-octadec-9-enoic acid] as an ester and a phosphate ester of N,N,N-trimethylethanolamine (choline).

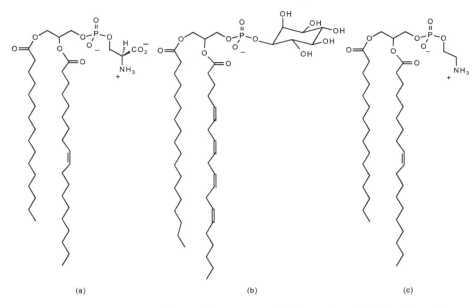

FIGURE 7.12 A representation of some of the structures which constitute the **inside cell** membrane (a) a representation of phosphatidylcholine where a glycerol head is attached (left-to-right) to palmitic acid (hexadecanoic acid) as an ester, to an oleic acid [(Z)-octadec-9-enoic acid] as an ester and a phosphate ester of serine; (b) a phosphatidylinositol where a glycerol head is attached (left-to-right) to stearic acid (octadecanoic acid) as an ester, to arachidonic acid [(5Z,8Z,11Z,14Z)-icosa-5,8,11,14-tetraenoic acid], and finally, at the third hydroxyl, to a phosphate ester of inositol [(1R,2R,3S,4S,5R,6S)-cyclohexane-1,2,3,4,5,6-hexol]; and (c) a representation of phosphatidylethanolamine where, again, a glycerol is attached (left-to-right) to palmitic acid (hexadecanoic acid) as an ester, to an oleic acid [(Z)-octadec-9-enoic acid] as an ester and a phosphate ester of ethanolamine.

factors that play a crucial role in plant growth and development. These genes are comparable to that of other dicotyledons, a grouping used for the flowering plants whose seed typically has two embryonic leaves or cotyledons (Figure 7.13). The particular transcription factors found are among those that regulate a number of biological processes including development, reproduction, responses to hormones, and adaptation to biotic and abiotic stresses. It is anticipated that new work with the same set of genes in grapevines is likely to provide the basis for studying the molecular regulation of berry development and the ripening process.

At about the same time in 2010, but using *Arabidopsis thaliana*, where more genomic information was available, overall embryogenesis involving primary shoot meristem formation and cotyledon initiation was undertaken.[12] It was found that shoot meristem and cotyledon specification occur concomitantly with the primary shoot meristem forming between the two out-growing cotyledon. Further, although the processes underlying the two different specification events are linked, they are not dependent on each other. Interestingly, the processes that establish the primary meristems of shoot and root share related features. The two form at opposite ends of the provasculature, the root meristem at its basal end and the shoot meristem at its apical end.

Although cotyledons and true leaves may both have chloroplasts in the mesophyll cells, it is held that the light gathering process(es) are substantially different[13] in these different

FIGURE 7.13 A photographic representation of embryonic leaves (cotyledons). Victor M. Vicente Selvas, Wiki Commons.

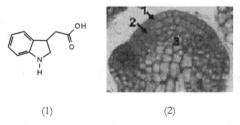

FIGURE 7.14 Representations of (a) folic acid and (b) flavin adenine dinucleotide (FAD).

FIGURE 7.15 (1) A representation of the plant hormone auxin (indole-3-acetic acid). (2) Modified representation from a photograph of a shoot apical meristem showing (easily seen as cells of different sizes) different cells. The fastest growing and relatively small are at the surface (site 1), while those behind continue to swell and push outward (site 2). The more mature cells (site 3) are larger in volume and are, it is argued, growing with a different set of metabolic products. Modified from Wiki Commons presentation by Emmanuel Boutet based on JPG by Clematis.

growths. Cotyledons and shoot apical meristems, as first observed by Charles and Francis Darwin,[1] appear particularly sensitive to blue light (light with wavelengths from about 400 to 450 nm). Thus, the Darwins[1] found that if a growing plant was irradiated with light the plant would bend toward the light because, as we now know,[14] auxins are transported to the opposite side and cause cell enlargement there (allowing for the bending). Irradiation of the cotyledons appears to result in increase of chloroplast size and accumulation of starch. Thus, cotyledons can be considered functionally similar to leaves although it appears that the photoreception is even more poorly understood than it is for leaf photosynthesis. For

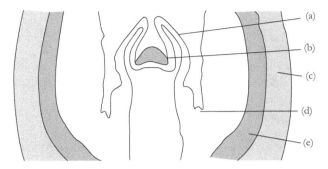

FIGURE 7.16 A cartoon representation that shows older as well as younger primordium parts of the growing plant and, along with the apical meristem, the beginning of leaves. The parts shown are (a) the young leaf primordium (i.e., the earliest recognizable differentiated state); (b) the apical meristem; (c) the older leaf primordium; (d) lateral bud primordium; and (e) vascular tissue. Somewhat more is known about leaf cells than is known about cells of the cotyledons.

cotyledons, it is known that proteins involved, called *cryptochromes* because of their mysterious way (i.e., crypto-) of light absorption, have at least two chromophores.[15] One such chromophore is a pterion (5,10-methenyltetrahydrofolic acid) and the other a flavin (flavin adenine dinucleotide, FAD) (Figure 7.14).[16] Both can absorb light in the blue region of the spectrum.

In the meantime, at the **shoot** apical meristem, cells are apparently undifferentiated. These so called primordial cells (i.e., cells capable of triggering growth of leaf or other organs) seem to start off as a crease or indention and, only later, push out.[17] The primordial leaf, now a true leaf, and primordial flower (to be discussed later) are regulated by sets of genes whose investigation remains underway, but again, the plant hormone auxin (indole-3-acetic acid) has been implicated, since growth appears most rapid where auxin concentration is highest (Figures 7.15 and 7.16).

NOTES AND REFERENCES

1. Darwin, C.; Darwin, F. *The Power of Movement in Plants;* John Murray: London, 1880.

2. Organelle is the name given to any of the specialized, isolated (with their own membrane) subunits within a cell.

3. Petrillo, E.; Herz, M. A. G.; Fuchs, A.; Reifer, D.; Fuler, J.; Yanovsky, M. J.; Simpson, C.; Brown, J. W. S.; Barta, A.; Kalyna, M.; Kornblihtt, A. R. *Science* **2014**, *344,* 427.

4. The meanings of terms in this paragraph (some of which are general to the subject of organic chemistry and can be found in Appendix 1) will be discussed in more detail subsequently.

5. Enzymatic processes are functions carried out by the large proteinaceous molecules called enzymes that are responsible for the thousands of chemical reactions that occur in living systems and that define life.

6. Vermeer, J. E. M.; von Wangenheim, D.; Barberon, M.; Lee, Y.; Stelzer, E. H. K.; Maizel, A.; Geldner, N. *Science* **2014**, *343,* 178. DOI: 10.1126/science.1245871.

7. Barber, N. A.; Soper-Gorden, N. L. *J. Plant Ecology*, **2015**, *8,* 1.

8. A more complete discussion of self-pollination is provided in Chapter 12. See also Considine, J. A.; Knox, R. B. *Ann. Bot.* **1979,** *43,* 11 and Lebon, G.; Duchêne, E.; Brun, O.; Clêment, C. *Ann. Bot.* **2005,** *95,* 943.

9. The initial observations are attributed to Schwann, T.; Schleyden, M. J. *Microscopical Researches into the Accordance in the Structure and Growth of Animals and Plants;* Printed for the Sydenham Society: London, 1847.

10. Jaillon, O. et al. *Nature* **1997,** *449,* 463. And see Simon, C. J., et al. *Genet. Res. Camb.* **2003,** *81,* 179.

11. Licausi, F.; Giorgi, F. M.; Zenoni, S.; Osti, F.; Pezzotti, M.; Perata, P. *BioMedCental (BMC) Genomics* **2010,** *11,* 719.

12. De Smet, I.; Lau, S.; Maer, U.; Jurgens, G. *The Plant Journal* **2010,** *61,* 969.

13. Bertamini, M.; Nedunchezhian, N. *Vitis* **2003,** *42,* 13.

14. Heisler, M. G.; Ohno, C.; Das, Pl; Sieber, P.; Reddy, G. V.; Long, J. A.; Meyerowitz, E. M. *Curr. Bio.* **2005,** *15,* 1899. DOI: 10.1016/j.cub.2005.09.052.

15. A *chromophore* is that group of atoms in molecule, or the whole molecule itself, that gives rise to the absorption of light.

16. Information about the functional groups and some of the structural details for these and other molecules can be found in Appendix 1. For light absorption and another pathway that can be followed, please see Dodson, C. A.; Hore, P. J.; Wallace, M. I. *Trends Biochem. Sci.* **2013,** *38,* 435.

17. Gross-Hardt, R.; Laux, T. *J. Cell Sci.*, **2003,** *116,* 1659. DOI: 10.1242/jcs.00406.

8

The Leaf

GRAPE LEAVES ARE thin and flat. As is common among leaves in general, they are composed of different sets of specialized cells. Today, on average, sunlight reaching their surface is about 4% ultraviolet (UV) (<400 nm), 52% infrared (IR) (>750 nm) and 44% visible (VIS) radiation. Little of the UV and IR are used by plants.[1]

As with other leaves that are green, only the red and blue ends of the visible part of the electromagnetic spectrum are absorbed, thus leaving green available by reflection and transmission.

On the surface of the leaf (Figure 8.1), the cells of the outermost layer (the epidermis) are designed to protect the inner cells where the workings needed for gathering the sunlight used for photosynthesis and other chemistry necessary to the life of the plant are found. That is, the more delicate cells, beneath the epidermis, are involved in production of carbohydrates as well as the movement of nutrients in and products out of the leaf. The epidermis, exposed to the atmosphere, has cells that are usually thicker and are covered by a waxy layer made up of long-chain carboxylic acids that have hydroxyl groups (–OH) at or near their termini. These so-called omega hydroxy acids can then form esters using the hydroxyl group of one and the carboxylic acid of the next. This yields long-chain polyester polymers called "cutin."

As indicated in the earlier discussion of cells and, in particular, regarding the fatty acids of cell walls, the fatty acids found in the epidermis generally consist of an even number of carbon atoms, and for cutin, the sixteen carbon (palmitic acid) family (Figure 8.2) and the eighteen carbon family (oleic acid bearing a double bond or the saturated analogue stearic acid) are common. While one terminal hydroxyl group is usual (e.g., 16-hydroxypalmitic acid, 18-hydroxyoleic acid, or its saturated analogue 18-hydroxystearic acid) more than one (allowing for cross-linking) is not uncommon (e.g., 10,16-dihydroxypalmitic and 9,10,18-trihydroxystearic acid). This waxy cuticle inhibits the loss of water from across the leaf surface, and since there are few double bonds—none of them conjugated—near ultraviolet, visible, and longer wavelength light easily passes through the cutin to reach the chloroplast-containing mesophyll cells.

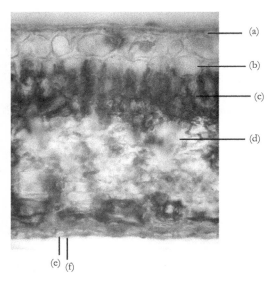

FIGURE 8.1 A. A photomicrograph of leaf cells.(a) The upper leaf surface (epidermis); (b) a leaf vein; (c) palisade mesophyll cells; (d) spongy mesophyll cells; (e) stoma entrance/exit where respiration (carbon dioxide, CO_2, enters and oxygen, O_2, exits) takes place and where transpiration also occurs; (f) guard cell. Modified from a photo by Mnolf provided on the Creative Commons.

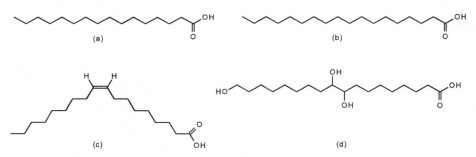

FIGURE 8.2 Representations of (a) palmitic acid, (b) stearic acid, (c) oleic acid; and (d) 9,10,18-trihydroxystearic acid.

In addition to the outer covering of epidermal cells, grape leaves have pores on their undersurface where, as also shown, the waxy epidermis is either absent or missing. These pores are called, (*plural*) stomata. A (*singular*) stoma is an opening (pore) which is bounded by two bean-shaped cells called "guard cells." The guard cells differ from normal epidermal cells in that they have chloroplasts (*vide infra*) and the cell walls are of uneven thickness (i.e., the outer wall is thin and the inner wall, nearest the opening, is thick). These stomata allow the escape of oxygen (O_2) and water (H_2O) (for transpiration) and entry of carbon dioxide (CO_2).

The enclosed area of the leaf (the lamina) itself is a sandwich of photosynthetic mesophyll ("middle leaf") cells, and these parenchyma ("visceral flesh," the so-called functional part of the cell) are distinguished from the stoma, or structural part. The parenchyma is differentiated into palisade cells and, as shown, spongy mesophyll. Air space to allow for CO_2 to reach the individual cells is present among the mesophyll cells. The veins or vascular bundles containing the xylem and the phloem also run through it.

Interestingly, as is the case with roots that form root hairs, similar growths may also be found in the epidermis of leaves and stems.

Even cursory examination of leaf material shows large and small veins throughout the structure of the leaf blade. The leaf connects to the stem at a node (the space between nodes is called the internode) through the leaf stem or petiol and serves to hold the leaf in place as well as to transport fluids between the stem and the cells in the leaf. The net-like pattern of the veins provides some support for the blade, while the veins themselves serve to transport nutrients.

The transport occurs through this vascular bundle of veins which, as already noted, exists in two forms: (1) the xylem for the tissue that carries water and some water soluble mineral nutrients; and (2) the phloem for carrying carbohydrates and other similar nutrients. Both these tissues are present in the vascular bundle, which, in addition, will include supporting and protective cells. Ultimately, these allow nutrients to be brought from the roots to the leaves and the carbohydrate products produced in the leaves to be returned for building more roots, shoots, stems and leaves. Eventually, the same systems yield flowers and fruit wherein are the seeds for their progeny. The leaves also serve to shade those same grapes.

At the node, where the leaf comes forth from the stem, there is found a small growth called a stipule. The tendrils (stem-like structures used by the vines to attach to a support) grow from the stipule, and they are commonly considered to be modified shoots from which a leaf blade fails to form. Tendrils wrap around supports, and when a tendril comes into contact with a support, there is faster growth for cells on the opposite side to the support so that the tendril rapidly forms a coil around the support. The signaling process using auxins for bending of shoots toward light appears to operate here too. The tendrils function to pull the vine close to the main axis of growth and to hold it in place when the plant is buffeted in the wind.

Leaves, vital for the life of the plant, also need to be protected by the plant bearing them from insects and herbivores. As a consequence some plant products (Figure 8.3), such as tannins (e.g., tannic acid), bitter principles (e.g., aloin), and even simpler phenols (see Appendix 1), to name a few, are also present and appear to serve as antifeedants. Leaves also defend

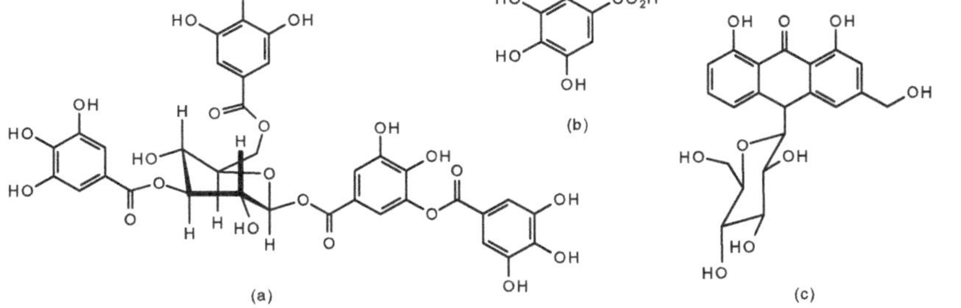

FIGURE 8.3 Representations of: (a) a tannic acid. Glucose esterified with gallic acids makes up a family of compounds called tannic acids; (b) gallic acid; and (c) aloin.

themselves in other ways.² Interestingly, the vascular bundle of veins generally has the xylem in the center of the bundle or closer to the interior and the phloem on (or closer to) the exterior of the bundle. One consequence of this is that with the phloem closer to the surface and transporting sucrose, aphids and other sucking insects find it advantageous to seek the underside of the leaf where, in addition, the waxy epidermis is absent or thin. Finally, since grape vines are among deciduous plants and are grown in temperate climates, they will shed their leaves in autumn. The shed leaves may be expected to contribute their retained nutrients to the soil where they fall.

The increasing application of analytical technology has permitted the question of trimming the leaves and the subsequent effect of that treatment on the wine! A subset of the vast number of constituents found in Chardonnay, Gewurztraminer, Merlot, and Tempranillo wines (to be discussed subsequently) have been measured as a function of two different operations in the vineyard. One involved removing leaves at the base of the vine (basal leaf plucking) to encourage leaf growth where sunshine more readily struck, while the other involved "head trimming" of new leaf growth, again to encourage more fulsome light on mature leaves. Glycosidic (i.e., sugar bearing) precursors were analyzed in grapes, and volatile compounds were studied in the wines. Correlations were found between precursors of volatile organic compounds that included terpenes, norisoprenoids, phenols, and vanillin derivatives (described more fully within) where more of each, generally, were found in vineyards that had been trimmed!³

NOTES AND REFERENCES

1. See Newport Corporation [Online] **2016**. http://www.newport.com/Introduction-to-Solar-Radiation/411919/1033/content.aspx (accessed Apr 5, 2017).

2. Farmer, E. E. *Leaf Defense;* Oxford University Press: New York, 2014.

3. Hernandez-Orte, P.; Boncerjero, B.; Astrain, J.; Lacau, B.; Cacho, J.; Ferreira, V. *J. Sci. Food Agric.* **2015,** *95,* 688.

9

The Light on the Leaves

AS NOTED EARLIER in the general description of the plant cell, there is a site at which photosynthesis, the process which allows plants to capture sunlight and convert it into energy, occurs. It is this process which has produced oxygen on the planet, food for herbivores, and the cool green hills of Earth we enjoy today. The capture of sunlight allows the grape vine to grow and produce fruit. Of course, while the discussion of the "light reactions" (capture of sunlight) and the subsequent so-called "dark reactions" (producing carbohydrates) is necessarily brief here, it is, nonetheless, an exciting story. We are only now beginning to understand a little of it.

The earlier picture (Figure 7.1) of the plant cell is repeated here (Figure 9.1) so that the position of the chloroplast is seen. Refer to page 24 for a discussion of the numbered items.

As the leaves begin to develop alongside the apical meristem, proplastids, which are present in the meristematic regions of the plant, are formed. Proplastids grow into plastids (such as amyloplasts and chloroplasts) as they mature in different ways dictated by the plant's DNA. Some plastids (e.g., chloroplasts) carry pigments, discussed more fully below, that allow them to carry out photosynthesis. Others are used for storage of fat, starch (amyloplasts) or specialized proteins. Still other plastids are used to synthesize specialized compounds needed to form different tissues or to produce compounds for protection (e.g., tannins).

Each plastid builds multiple copies of its DNA as it grows. If it is growing rapidly, it makes more genome copies than if it is growing slowly. The genes, ignoring epigenetic (literally "above the gene") and postgenetic (literally "after the gene") modifications, about which we still have much to learn, encode plastid proteins, the regulation of whose expression controls differentiation and thus which plastid is eventually formed.

However, despite the differentiation of plastids, it appears that many plastids remain connected to each other by tubes called stromules through which proteins can be exchanged. This is particularly important because membrane-associated plastid DNA, where protein–DNA complexes are clumped in groups called nucleoids, can then remodel proplastids for future use.

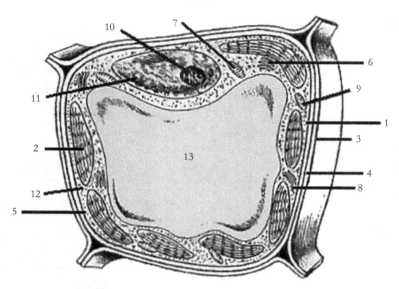

FIGURE 9.1 (Figure 7.1 repeated): A "Generic" Plant Cell. The representation is from Harold, F. M. *The Way of the Cell*, Oxford University Press, 2001, p 29 and is reproduced with permission and courtesy of Oxford University Press, USA. The numbered items are fully discussed beginning on page 24

Finally, many plastids, particularly those responsible for photosynthesis (chloroplasts), possess numerous internal membrane layers.

A representation of the chloroplast, where the reactions associated with the absorption of light (photosynthesis) take place, is shown in Figure 9.2. It is here that photons are absorbed. A photon is the name we give to that species of electromagnetic radiation which, depending upon the experiment performed, behaves as a particle or a wave. It has neither rest mass nor charge. Photons carry the force of the electromagnetic field. Their energy is a function of the wavelength of light. Shorter wavelengths have higher energies (Appendix 1).

The absorption of photons (light) ultimately results in the conversion of water (H_2O) to oxygen (O_2) and protons (H^+) and sets in motion the processes that reconvert adenosine diphosphate (ADP) to adenosine triphosphate (ATP) (Figure 9.3). It is here that the energy necessary to power the growth of the plant, the formation of flowers, and the setting of the fruit to make the wine is found.

COMPONENTS OF THE CHLOROPLAST

The chloroplast envelope consists of an outer and an inner membrane. The membranes themselves consist of, for the most part, the same lipid bilayers but with different concentrations of three components, *viz.* phospholipids, such as one containing phosphatidylcholine, galactolipids, and sulfolipids (Figure 9.4). However, the two membranes, because of the different concentrations of these materials, behave very differently.

The outer membrane transports ions and metabolites and serves to protect the inner membrane. Eventually it must transport products produced in the chloroplast into the cell. The inner membrane contains transport proteins that export products formed in the chloroplast into the intermembrane space. The products are carbohydrates produced in photosynthesis and, in particular, D-(+)-glyceraldehyde 3-phosphate (triose phosphate). Other transport proteins transport inorganic phosphate back through the inner membrane. The phosphate is used to make more triose phosphate as well as to convert adenosine diphosphate (ADP) into adenosine triphosphate (ATP), the latter necessary to form the triose phosphate as well as to drive the dark reactions needed for carbohydrate production.

In the colorless fluid (the stroma) that fills the chloroplast and is involved in transport of materials within the chloroplast, there are found thylakoids where the light-dependent photosynthetic processes actually occur. The thylakoids are disk-like bodies found in stacks (called grana) held together by stroma thylakoids. A thylakoid membrane, containing those

FIGURE 9.2 An electron micrograph of a section of leaf mesophyll cells showing the chloroplast; (1) starch granule; (2) granum (grana); (3) stroma thylakoid; (4) stroma; (5) outer membrane; (6) inner membrane. The electron micrograph is reproduced from Halliwell, B.; *Chloroplast Metabolism*, Oxford University Press, 1981, p 2 and is used with permission of Professor Halliwell.

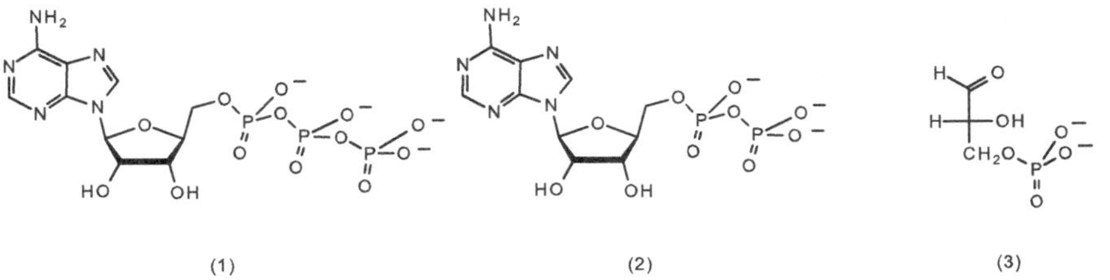

FIGURE 9.3 Representations of: (1) adenosine triphosphate (ATP); (2) adenosine diphosphate (ADP) and; (3) D-(+)-glyceraldehyde 3-phosphate (triose phosphate).

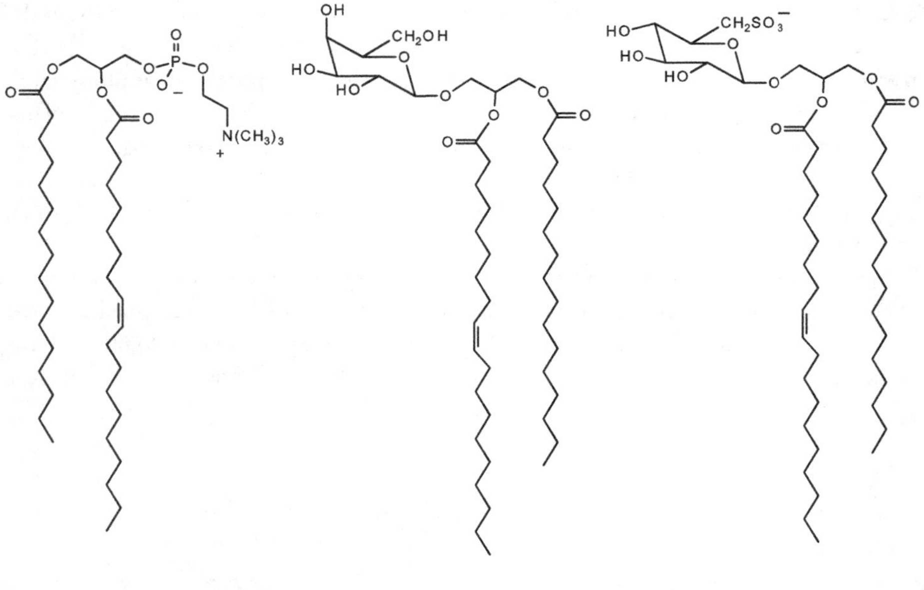

FIGURE 9.4 A representation of some of the structures which constitute membranes of the chloroplast envelope: (a) a representation of phosphatidylcholine where a glycerol head is attached (left-to-right) to palmitic acid (hexadecanoic acid) as an ester, to an oleic acid [(Z)-octadec-9-enoic acid] as an ester and a phosphate ester of choline; (b) a galactolipid where a glycerol head is attached (left-to-right) to (1) the β-anomer of galactose at the anomeric carbon, (2) to an oleic acid [(Z)-octadec-9-enoic acid] as an ester and finally, (3) at the third hydroxyl, to palmitic acid (hexadecanoic acid) as an ester; (c) a representation of a sulfolipid where, again, a glycerol is attached (left-to-right) to the β-anomer at the anomeric carbon of a sulfoglucopyranose (6-deoxy-6-sulfo-D-glucopyranose) and, as above to an oleic acid [(Z)-octadec-9-enoic acid] as an ester and finally, at the third hydroxyl, to palmitic acid (hexadecanoic acid) as an ester.

proteins that are involved in the light-dependent processes so necessary for life, surrounds the thylakoid lumen where oxygen evolution occurs. Two different thylakoids (grana and stroma) are shown in Figure 9.2. The different thylakoids differ in protein structure (other distinguishing characteristics are still open to debate), and it is clear that the smaller number of stroma thylakoids generally have higher protein/chlorophyll ratios and greater photosystem and NADH dehydrogenase activities than the grana thylakoids.[1]

The nucleoid is a separate body that has the DNA of the plant and its own wall. There may be more than one copy present.

The ribosomes produce the proteins needed by the system as dictated by the DNA and the RNA derived from it.

The plastoglobuli appear to be attached either to a thylakoid or to another plastoglobulus attached to a thylakoid. These "floating" bodies contain both structural proteins and enzymes involved in lipid synthesis and lipid metabolism, and it is reported[1] that they can exchange their contents with the thylakoids to which they are attached. There is some evidence that they contain some chlorophylls as well as plastoquinones (*vide infra*).

NOTE AND REFERENCE

1. Cuello, J.; Quiles, M. J. *Methods Mol. Biol.* **2004**, *274*, 1.

10

Harvesting the Light

PRODUCTS OF REACTIONS are separated from reactants by a barrier or barriers. If this were not so we could not have any reactants—everything would already be products!

In order for the grapevine to grow beyond the materials provided in the seed, the rootstock, or the cutting, it is necessary for the reactants obtained from the environment (i.e., nutrients in the soil and air) to be converted to plant material. The energy for this conversion comes from the sun, and it is the chloroplasts that take the light and, using the aforementioned materials, convert it to useful energy in the plant.[1] So, overall, for processes to occur within the plant, a high energy species must be formed and then used. Subsequent regeneration of the high energy species can use more sunlight. The currency of energy is adenosine triphosphate (ATP). When it is used, it is converted to adenosine diphosphate (ADP) and inorganic phosphate (P_i), and in that conversion (or those conversions as more than one can be used to accomplish the same end) the barrier between reactant and product can be overcome (Figure 10.1). Additionally, for moving electrons and protons around where simple solvation (the use of—and interactions with—solvents) will not work, a cofactor (a "factor" that needs to be present in addition to an enzyme to enable the catalyzed reaction to occur) is often needed. These movements of electrons and protons are simply oxidations and reductions (see Appendix 1), and it is common to find oxidation and reduction being effected by using, as cofactors, either the oxidized or reduced forms of the phosphate ester of nicotinamide adenine dinucleotide ($NADP^+$) to/from (NADPH) and/or the related conversion of the oxidized/reduced forms of flavin adenine dinucleotide (FAD)/($FADH_2$) (Figure 10.2).

A cartoon representation of the chloroplast wall, with the stroma (the colorless fluid filling the chloroplast through which materials move) shown on the top and the lumen of the thylakoid body (where the light-dependent photochemistry occurs) on the bottom is provided in Figure 10.3. The working agents in the membrane are shown. These agents take the energy from the sunlight, absorb it, and power the enzymes that transform it into energy the plant can use. The transformation effects, among other things, the catalytic oxidation of water (H_2O) to oxygen (O_2) and protons (H^+) and the reduction of nicotinamide adenine dinucleotide 2′-phosphate, $NAD(P)^+$, to reduced nicotinamide adenine dinucleotide

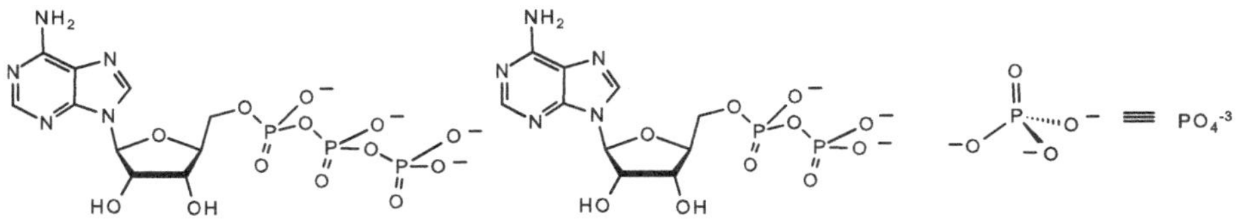

FIGURE 10.1 Representations of adenosine triphosphate, ATP (1), adenosine diphosphate, ADP (2), and the phosphate anion, PO_4^{3-}, often abbreviated as Pi (3).

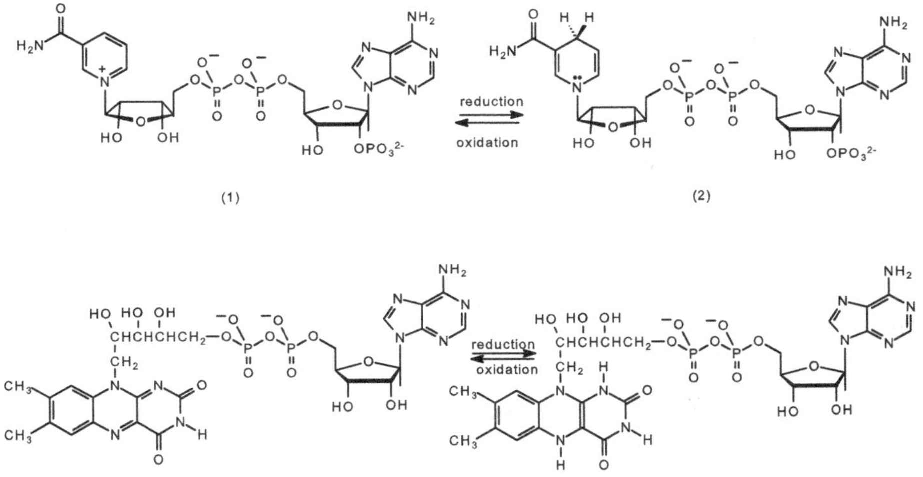

FIGURE 10.2 Representation of the oxidation-reduction pairs of: (1) nicotinamide adenine dinucleotide 2′-phosphate ($NADP^+$); (2) reduced nicotinamide adenine dinucleotide 2′-phosphate (NADPH); (3) flavin adenine dinucleotide (FAD); and (4) reduced flavin adenine dinucleotide ($FADH_2$).

2′-phosphate, NAD(P)H, as well as proton transfer from within the thylakoid to the stroma and thence through the system. The enzymatic system that catalyzes the conversion of adenosine diphosphate (ADP) and inorganic phosphate (P_i) into adenosine triphosphate (ATP), also gets its energy here.

The story begins on the left with photosystem II (PSII), so called because it was discovered after photosystem I (PSI).

It is generally agreed that, in plants, the protein complex of PSII is where photosynthesis and the reactions utilizing sunlight in the conversion of water to oxygen and protons begin. Indeed, although it not yet understood in detail how water (H_2O) is converted into oxygen (O_2) and protons (H^+), some of the broad picture is beginning to emerge. A representation of PSII from the cyanobacterium *Thermosynechococcus elongatus* is shown in Figure 10.4.

Photosystem II (PSII) is held in the thylakoid membrane and is a large multi-subunit dimeric complex, as shown in Figure 10.4. Commonly, in a PSII, there are about twenty (20)

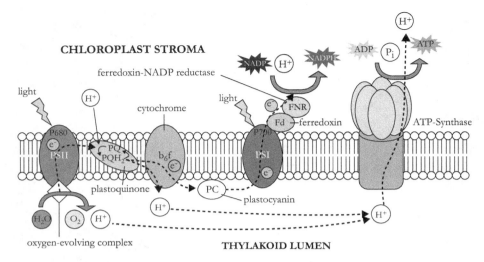

FIGURE 10.3 A cartoon representation of the chloroplast wall with embedded agents. As shown, the stroma (the colorless fluid filling the chloroplast through which materials move) is on the top, and the lumen of the thylakoid cell (where the light-dependent photochemistry occurs) is on the bottom. The working agents of (left-to-right) photosystem II (PSII), where light enters to provide energy for the oxygen (O_2)-evolving complex, plastoquinone, the iron (Fe)-containing b_6f cytochrome needed for oxidation, photosystem I (PSI) where ferredoxin reductase (EC 1.18.1.2) promotes the conversion of NAD(P)$^+$ to NAD(P)H, and finally, the beautifully complex machine-like ATP synthase (EC 3.6.3.14) where the conversion of adenosine diphosphate (ADP) and inorganic phosphate (P_i) to adenosine triphosphate (ATP) takes place along with proton (H$^+$) transfer across the membrane. This representation is modified from Somepix, licensed under the Creative Commons Attribution-Share Alike 4.0 International.

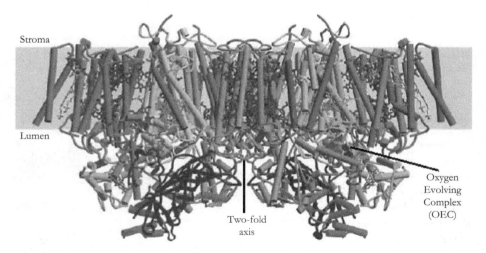

FIGURE 10.4 A representation of the molecular structure of the photosystem II dimer proposed by the X-ray diffraction model from *Thermosynechococcus elongates* (PDB 1S5L as reported by Ferreira, K. N.; Iverson, T. M.; Maghlaoui, K.; Barber, J.; Iwata, S. *Science* **2004**, *303*, 1831). For information about the amino acids that comprise the "barrels" and "sheets" in protein structure, please see Appendix 1. The PDB archive data is used in accord with the PDB Charter. A full color representation can be found in the Protein Data Bank (PDB) as entry 1S5L (http://www.rcsb.org/pdb/home/home.do).

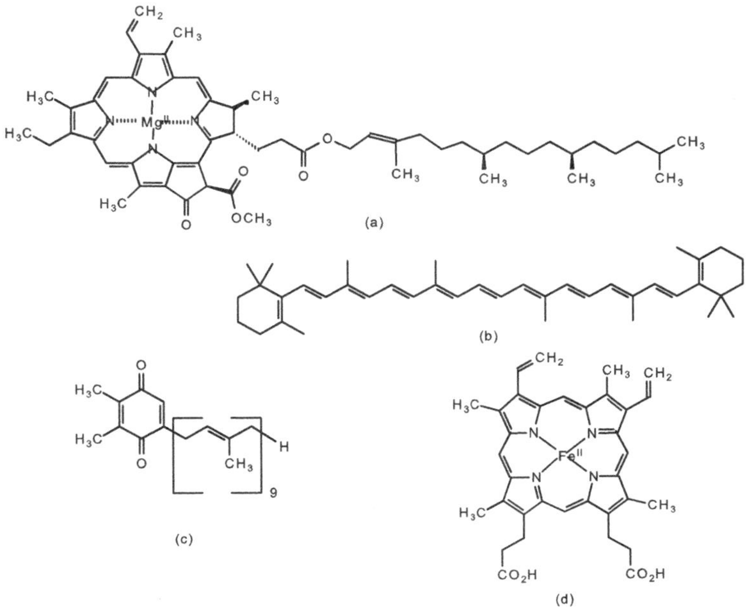

FIGURE 10.5 Representations of some cofactors held in place by the proteins of Photosystem II (PSII): (a) chlorophyll-a and pheophytin, which is (a) without the metal Mg^{2+}; (b) β-carotene; (c) plastoquinone-b (PQB); and (d) heme b.

subunits and about nine (9) membrane-spanning polypeptides (for a total mass of about 400,000 atomic mass units (amu) or 400 kiloDaltons (400 kDa). There is also a group of cofactors (Figure 10.5) held in place by the proteins that include many copies of chlorophyll-a, many copies each of β-carotene, phaeophytin, plastoquinone, heme b, lipids, at least one bicarbonate (HCO_3^-), at least one Mn_4CaO_5 "oxygen-evolving center" (or OEC), at least one Fe(II) ferrous iron, a chloride anion, and two calcium (Ca^{2+}) ions in each monomeric unit.

The Chlorophyll-bearing-Proteins of masses corresponding to about 43 and 47 kDa (called CP 43 and CP 47, respectively), of which each monomer in the dimer is made, are common to many PSII systems. There are, however, variations in the specific details of a given PSII as a function of the plant or cyanobacterium from which it is obtained. There are a number of different crystal structures for (different) PSII complexes.[2] Further, not only are there some differences depending upon the source, but different research groups succeeded in getting crystals differently (e.g., from different solvents and with different ions present), and the analyses thus also differ.

The reaction with light begins with the excitation of so-called antenna chlorophylls—which, in many systems, are two protein-held chlorophyll-a molecules whose absorption maxima are near 680 nm (in the red region of the visible spectrum). These are frequently noted as P680's for the absorption band in the protein. The absorption is attributed to the conjugated system of alternating single and double bonds around the macrocycle (i.e., large ring) which is made up of co-joined smaller five-membered rings containing nitrogen (pyrrole rings) (Figure 10.5).

EQUATION 10.1 A representation of the reduction of (a) plastoquinone PQB undergoing reduction to (b) plastoquinol while water is oxidized to oxygen and protons in the stroma are sacrificed to produce protons in the lumen.

Then, continuing with the current view, absorption of the energy of the photon moves a low-lying electron in the chlorophyll-a into an excited state. Relaxation from the excited state (i.e., emission) results in recovery of the initial state of the chlorophyll-a and the rapid transfer of the absorbed energy, associated with the excited state (as an electron), to a nearby phaeophytin molecule. The phaeophytins are the same as chlorophyll-a molecules, but are missing the magnesium in the center and are purple in color rather than green. Subsequently, the electron travels through two (2) different plastoquinone molecules (PQA) and (PQB), where the first is more tightly bound than the second. The second plastoquinone (PQB) is then reduced to plastoquinol as shown in Equation 10.1 (A more detailed picture may be found in the Plant Metabolic Network at http://pmn.plantcyc.org/ARA/NEW-IMAGE?object=PWY-5864).

Since a reduction has occurred, there must have been an oxidation. Indeed, it is water that is oxidized. The oxidation of water occurs in the oxygen-evolving complex (OEC) (Figures 10.6 and 10.7) to produce oxygen and protons. Details of this process remain debatable, but it is clear that the OEC lies at a proteinaceous interface on the lumen side of the membrane. It is argued that four (4) protons (H^+) are moved from the stromal side to the lumen side to accommodate the protonation of the plastoquinol, and in the process, the oxidation state of manganese (Mn) in the (OEC) is also changing. It is unclear what is happening at the calcium (Ca) in the Mn_4CaO_5 of the OEC and if the chloride anion (Cl^-) is present in every OEC. The interested reader is directed to a site specific to the subject. (http://www.chm.bris.ac.uk/motm/oec/motm.htm).

In all three representations of the oxygen-evolving complex (OEC) from the thermophilic cyanobacteria *Thermosynechococcus elongatus* the same amino acids surround the calcium cation (Ca^{2+}), and the four (4) manganese (Mn^{2+}) atoms of the OEC and water molecules (H_2O) are explicitly shown. In the second view of the OEC (at 190 pm) (Figure 10.7), recognizing that the water is the source of one of the two oxygen atoms in the oxygen molecule (O_2) formed, they are emphasized.

Again, as shown in the first representation (at 350 pm) (Figure 10.6) of the amino acids helping to hold the OEC and beginning clockwise on the left at about nine o'clock, the protein backbone is composed of histidine (His, H190) and then alanine (Ala, A344). A tyrosine (Tyr, Y161) and an aspartic acid (Asp, D170) with the nearby CP43 protein and a glutamic acid (Glu, E354) follow. The protein holding the OEC site continues with glutamic acid (Glu, E333), histidine (His, H332), and another glutamic acid (Glu, E189) at around 6 o'clock before the aspartic acid (Asp, D 342) to take us back to where we began. The structures of the amino acids are given in Figure 10.8.

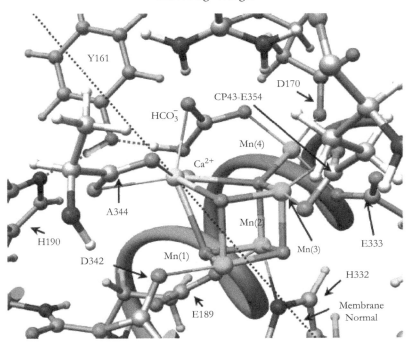

FIGURE 10.6 A view of the oxygen-evolving complex (OEC) from the thermophilic cyanobacterium *Thermosynechococcus elongatus* at 350-pm resolution after Sproviero, E. M.; Gascon, J. A.; McEvoy, J. P.; Brudvig, G. W.; Batista, V. S. *Coord. Chem. Rev.* **2008,** *252,* 395. The amino acids surrounding the complex are defined by their one-letter names (Y = tyrosine; D = aspartic acid; E = glutamic acid; H = histidine; A = alanine) and their positions by the number following the one-letter name. Ca = calcium, Mn = manganese. HCO_3^- is the bicarbonate anion. The original literature cited should be examined for additional detail. Used with permission of Elsevier B.V.

There is currently some debate[3] concerning which oxygen of the OEC combines with an oxygen of water to form the oxygen–oxygen bond.

Oxygen (O_2) and protons having been generated, it is important that the plastoquinol generated be oxidized to plastoquinone, so that the cycle can be continued.

The cytochrome b6f complex (plastoquinol–plastocyanin reductase; EC 1.10.9.1) (Figure 10.9) is an enzyme that is employed to fulfill the role of the oxidation of the plastoquinol back to plastoquinone so that the cycle can be continued. In the process, two protons (H^+) are moved into the lumen.

As was the case for the PSII, the cytochrome b6f complex is also a dimer. Here, again depending upon the source, each monomer supports a number of membrane-bound subunits. The subunits include: different cytochromes which differ in the substituents around the heme ring; an [2Fe–2S] cluster; and a series of small subunits. A total mass of about 220 kDa for these units is not uncommon. Indeed, a crystal structure of cytochrome b6f complex from the thermophilic cyanobacterium *Mastigocladus laminosus* (PDB 2zt9) has been determined, and it has eight (8) protein chains with various cytochrome complexes and an [2Fe–2S] cluster. Cytochrome b_6f catalyzes the transfer of electrons from plastoquinol (QH_2) to the copper (Cu)-containing plastocyanin, as well as moving protons from the stroma to the lumen. A representation of these reactions is shown below in Equation 10.2.

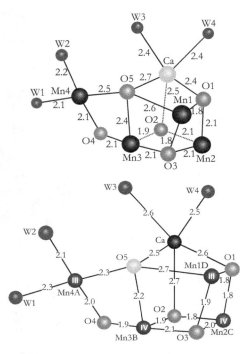

FIGURE 10.7 A view of the oxygen-evolving complex (OEC) from the thermophilic cyanobacterium *Thermosynechococcus vulcanus* at 190-pm resolution at the top. Reprinted by permission from Macmillan Publishers Ltd after Umena, Y.; Kawakami, K.; Shen, J.-R.; Kamiya, N. *Nature*, **2011**, *473*, 55. (DOI: 10.1038/nature09913); and, at the bottom, after Suga, M.; Akita, F.; Hirata, K.; Ueno, G.; Murakami, H.; Nakajima, Y.; Shimizu, T.; Yamashita, K.; Yamamoto, M.; Ago, H.; Shen, J.-R. *Nature*, **2015**, *517*, 99. (DOI: 10.1038/nature13991) using the newer femtosecond X-ray pulses The numbers along the bonds are distances, and those nuclei marked "W" refer to waters surrounding the OEC within the protein matrix. It is here in the OEC that the step $2\,H_2O \rightarrow 4\,H^+ + 4\,e^- + O_2$ occurs.

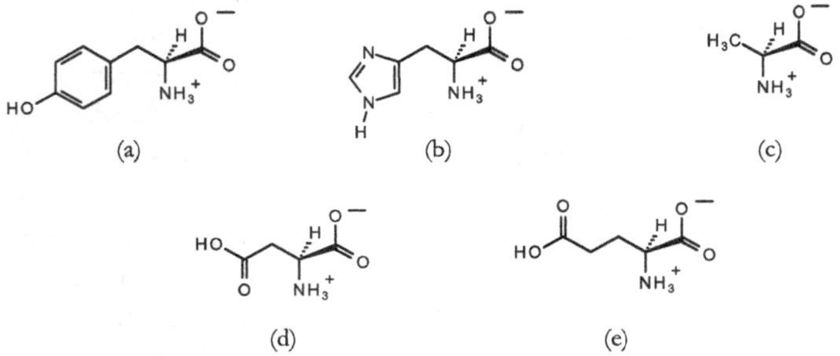

FIGURE 10.8 The amino acids surrounding the active oxygen-evolving complex (OEC) from the thermophilic cyanobacterium *Thermosynechococcus elongatus*: (a) tyrosine, (Tyr, Y); (b) histidine, (His, H); (c) alanine, (Ala, A); (d) aspartic acid, (Asp, D); (e) glutamic acid, (Glu, E).

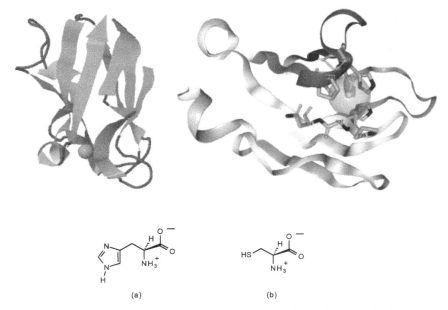

FIGURE 10.9 A representation of a plastocyanin, from *spinacia oleracea* (PDB 1ag6), where the copper (Cu^{2+}) is shown as a small sphere on the lower left, embedded in the protein, and on the right as a larger sphere, it is shown as held by histidines (His, H)(a) and a cysteine (Cys, C) (b) from the protein surrounding it. See Xue, Y.; Okvist, M.; Young, S. *Protein Sci.* **1998**, *7*, 2099. The PDB archive data is used in accord with the PDB Charter.

EQUATION 10.2 A representation of the oxidation of plastoquinol (a) to plastoquinone (b) in the presence of the enzyme (plastoquinol–plastocyanin reductase; EC 1.10.9.1).

Now, at this point, there are four (4) protons (H^+) in the lumen from the oxidation of water, and there are another four (4) from the oxidation of plastoquinol. This proton build-up provides the proton gradient that drives the synthesis of ATP in chloroplasts.

Further, the plastocyanin needs to be reoxidized.

Light, shining on photosystem I (PSI), a system composed of many polypeptides and only slightly smaller than PSII (about 356 kDa rather than 400 kDa), also uses chlorophyll-a and a quinone system (vitamin K) as well as iron-sulfur-containing (4Fe–4S) ferredoxin proteins to effect chemistry.

A ferredoxin, such as the oxidoreductase (PDB 1a8p; Figure 10.10), possesses an N-terminal six-stranded antiparallel beta-barrel domain. It is in this domain that the oxi-dized (and reduced) forms of the cofactor flavin adenine dinucleotide ($FAD/FADH_2$) are

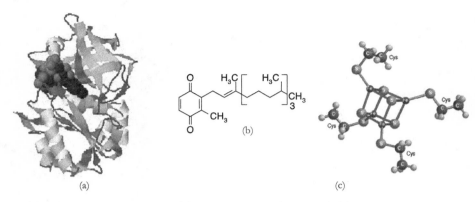

FIGURE 10.10 Representations of: (a) an oxidoreductase (PDB 1a8p); (b) vitamin K_1 (phylloquinone); and (c) an iron-sulfur (4Fe–4S) domain, held in place by cysteine residues which are part of the cofactor. See Prasad, G. S.; Kresge, N.; Muhlberg, A. B.; Shaw, A.; Jung, Y. S.; Burgess, B. K.; Stout, C. D. *Protein Sci.* **1998**, *7*, 2541. The PDB graphic is used in terms of the PDB Charter.

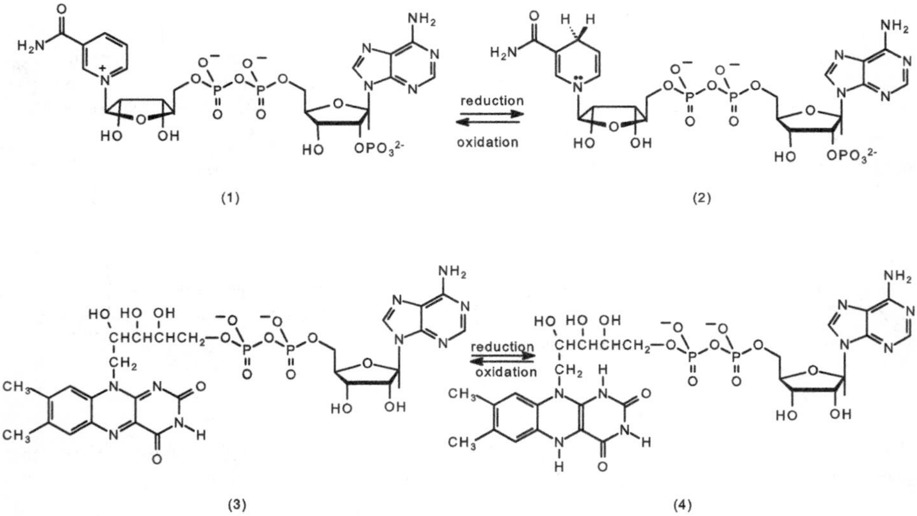

FIGURE 10.11 Representation of the oxidation-reduction pairs of: (1) nicotinamide adenine dinucleotide 2′-phosphate ($NADP^+$) and (2) reduced nicotinamide adenine dinucleotide 2′-phosphate (NADPH); and (3) flavin adenine dinucleotide (FAD) and (4) reduced flavin adenine dinucleotide ($FADH_2$). As noted above, these two redox systems are held in proximity in photosystem I so that both are available for subsequent use on the stroma side of the thylakoid membrane.

found. There is also a C-terminal five-stranded parallel beta-sheet domain, which binds the reduced and oxidized forms of phosphate-bearing nicotinamide adenine dinucleotide ($NADPH/NADP^+$) as shown in Figure 10.11. These two redox systems are held in proximity so that both redox systems are available for subsequent use on the stroma side of the thylakoid membrane.

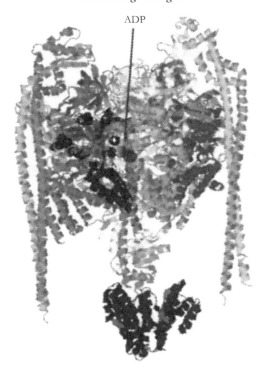

FIGURE 10.12 A representation of the structure of the intact *Thermus thermophilus* proton (H^+)-driven ATP synthase after Lau, W. C. Y.; Rubinstein. J. L. *Nature*, **2012**, *481*, 214 (PDB 3j0j). A full color representation is available in the Protein Data Bank as PDB 3j0j. The PDB archive data is used in accord with the PDB Charter.

A significant amount of information dealing with the structural basis for energy transfer in the photosystem I (PSI) light-harvesting complex (LHC) has recently become available and, while exciting, is beyond the discussion here.[4]

With $NADP^+$ and NADPH finally in hand, we are now prepared to deal with the "currency of energy." As noted earlier, when adenosine triphosphate (ATP) is used, it is converted to adenosine diphosphate (ADP) and inorganic phosphate (P_i), and in that conversion (or those conversions as more than one can be used to accomplish the same end) the barrier between reactant and product can be overcome. The ATP that has been used up now must be regenerated. The machinery for this feat is truly spectacular.

In the regeneration of ATP from ADP and inorganic phosphate (P_i), water must be lost as an anhydride is forming. The details of that process on the enzyme surface (ATP synthase or ATPase, e.g., EC 3.6.3.14) remain unclear at this writing, but it is clear that ATPase is highly conserved[5] across all living systems. It is found not only in the thylakoid membranes of chloroplasts but also in membranes of bacteria and mitochondria of animals (e.g., PDB 3j0j).

In 1960, Racker and coworkers[6] reported that the oxidative phosphorylation system (found in beef heart mitochondria and, by analogy, everywhere else) could be separated into two fragments or factors (F_0 and F_1). One fragment, membrane associated and insoluble (F_0), catalyzed electron transport without phosphorylation. The other fraction

(F_1, the "soluble component"), when added to the particulate fraction, "restored oxidative phosphorylation."

A year later, Mitchell[7] proposed (in Nobel Prize–winning work) that phosphorylation of ADP to ATP (and the reverse) was tightly coupled to proton transfer across a membrane. The Mitchell proposal was affirmed in 1971 by Kagawa and Racker[8] by reconstituting the ATPase, and then in 1977 Kagawa and coworkers[9] showed that phosphorylation occurred when a proton gradient was established.

ATP synthases are highly conserved very large proteins (e.g., molecular masses ~500 kDa), and a typical one from *Thermus thermophilus* (PDB 3j0j) is shown above (Figure 10.12).

The operation of the synthase is suggested to be as described in Figure 10.13, and as shown, the F_1 fraction section above the membrane is linked to the membrane-bound F_0 section.

To begin, a proton enters a channel in the a-subunit F_0, and it moves from that channel on the outside to the center where it binds to one c-subunit of the membrane unit c-ring. Then

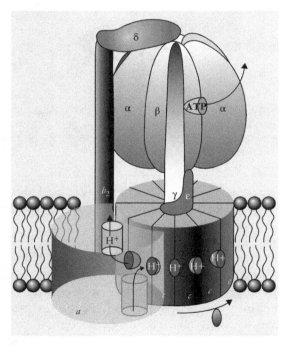

FIGURE 10.13 A representation of an F_0F_1 ATPase viewed from the side. It is currently held that the a-subunit in F_0 contains two partial channels. For a proton to traverse the membrane, it moves from one channel to the center, where it binds to one c-subunit of the c-ring. Then, the c-ring subunit rotates, and the proton moves out. The c-subunits are anchored to the γ-subunit (part of a rotor), whereas the a-subunit is anchored through the $b_2δ$ to the $α_3β_3$ hexamer in F_1. So, rotation of the c-ring relative to the a-subunit in F_0 will drive the rotation of the γ-subunit relative to the $α_3β_3$ hexamer. This rotation induces conformational changes in the proteins making up the α and β subunits where ADP, ATP, and phosphate are bound. Adapted with permission from Macmillan Publishers Ltd: Cross, R. L. *Nature*, **2004**, *427,* 407. This has since been elaborated upon in greater detail. See Junge, W.; Sielaff, H.; Engelbrecht, S. *Nature* **2009**, *459,* 354.

the c-ring subunit rotates, and the proton moves out. The c-subunits are anchored to the γ-subunit (part of a rotor), whereas the a-subunit is anchored through the $b_2\delta$-fragment to the $\alpha_3\beta_3$ hexamer in F_1. So rotation of the c-ring relative to the a-subunit in F_0 will drive the rotation of the γ-subunit relative to the $\alpha_3\beta_3$ hexamer. This rotation induces conformational changes in the proteins making up the α- and β-subunits where ADP, ATP, and phosphate are bound.

The data are consistent with three states of the three αβ pairs making up the $\alpha_3\beta_3$ hexamer (Figure 10.14) where, largely in the β-strand, the adenine and ribosyl units are held. As shown, these three states are referred to as "open" or (O), "loose" (L), and "tight" (T). As the γ-subunit turns, the three states in the three αβ pairs alternate, and ATP is formed from ADP and inorganic phosphate (P_i) (i.e., PO_4^{3-}) as it turns in one direction, or ADP and inorganic phosphate (P_i) (i.e., PO_4^{3-}) is formed from ATP as it turns in the opposite direction. In forming ATP then, ADP and inorganic phosphate (P_i) enter into the open state, are brought together to react in the loose state, and the reaction to phosphorylate the ADP to ATP (Equation 10.3), as shown, is consummated in the tight state. The next rotation into the open state allows the ATP to leave and the vacancy to be recharged with additional ADP and inorganic phosphate (P_i) for the next round. In the hydrolysis of ATP to ADP and inorganic phosphate (P_i) the process is reversed.

With ATP/ADP, $NADP^+$/NADPH in the stroma, we can leave the light and move over to the dark reactions. The so-called "dark" reactions are those that occur in the stroma and do not, apparently, require the direct energy input by photons.

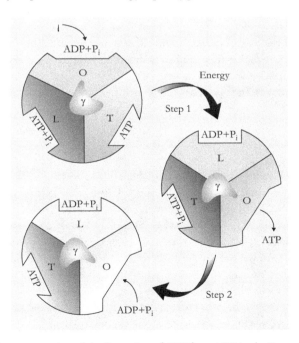

FIGURE 10.14 A representation of the formation of ATP from ADP in the F_1 portion of ATPase. As the γ-subunit turns, the three states (O, T, and L) in the three αβ pairs alternate, and ATP is formed from ADP + PO_4^{3-}. Adapted with permission from Macmillan Publishers Ltd: Cross, R. L. *Nature* **2004**, *427,* 407.

EQUATION 10.3 A representation of the conversion of adenosine diphosphate (ADP) and inorganic phosphate (P_i) on the left to adenosine triphosphate (ATP) on the right. The process is catalyzed by the enzyme ATP synthase (ATPase, EC 3.6.3.14). The inorganic phosphate (P_i) and an accompanying proton are shown as being delivered without specification as to their source. The proton is subsequently returned to its source in this representation.

In the meantime, oxygen (O_2) that has been produced in the lumen needs to exit the leaf, and carbon dioxide (CO_2) needs to enter the leaf. At the cellular level, the leaving of oxygen and entry of carbon dioxide appear to occur mainly by diffusion through the walls. To leave the plant, however, both gases move through pores (i.e., the stomata) found on the underside of the leaves.

The absorption of CO_2 occurs by diffusion through the walls and in the stroma of the chloroplasts where carbohydrates are produced. The adenosine triphosphate (ATP) generated earlier is used to provide the energy to do that work, and adenosine diphosphate (ADP) results. As seen in the next chapter, it is here that glucose and its isomer fructose are produced, and it is also here that they can be combined to yield their dimer sucrose. The latter is transported to make structural elements as well as food for the yeast in the production of ethanol following its return to glucose and fructose.

NOTES AND REFERENCES

1. The interested reader is encouraged to examine the subject in greater detail as found in (1) Bertini, I.; Gray, H. B.; Stiefel, E. I.; Valentine, J. S. *Biological Inorganic Chemistry, Structure and Reactivity;* University Science Books: Sausalito, CA, 2007; (2) Nield, J.; Barber, J. *Biochem. Biophys. Acta* **2006,** *1757,* 353; (3) Scholes, G. D.; Fleming G. R.; Olaya-Castro, A.; van Grondelle, R. *Nature Chem.* **2011,** *3,* 763; and (4) Orr, L; Govindjee. http://www.life.illinois.edu/govindjee/photoweb/ (2016) a website which, at this writing (2017) is devoted to the photosynthesis process. During the preparation of this work, a marvelous review article on light absorption and energy transfer has appeared. Mirkovic, T.; Ostroumov, E. E.; Anna, J. M.; van Grondelle, R.; Govindjee; Scholes, G. D. *Chem. Rev.* **2016.** DOI: 10.1021/acs.chemrev.6b00002.

2. At this writing, some of the protein data bank (PDB) accession codes for the protein from some different sources include 1S5L, 2AXT, 3ARC, 3BZ1, and 3BZ2. A fuller and very recent discussion of photosynthesis can be found at Quin, X.; Suga, M.; Kuang, T.; Shen J.-R. *Science* **2015,** *348,* 989. The membrane protein of the photosystem II (PSII)-light harvesting complex

II (LHCII) from spinach (*Spinacia oleracea*) at 320 pm has been reported in Wei, X.; Su, X.; Cao, P.; Liu, X.; Chang, W.; Li, M.; Zhang, X.; Liu, Z. *Nature* **2016**, *534*, 69. DOI: 10.1038/nature18020.

3. Siegbahn, P. E. M. *Phil. Tran. R. Soc. B* **2008**, *363*, 1221. DOI: 10.1098/rstb.2007.2218; and Suga, M.; Akita, F.; Hirata, K.; Ueno, G.; Murakami, H.; Nakajima, Y.; Shimizu, T.; Yamashita, K.; Yamamoto, M.; Ago, H.; Shen, J.-R. *Nature*, **2015**, *517*, 99. DOI: 10.1038/nature13991.

4. Qin, X.; Suga, M.; Kuan, T.; Shen, J.-R. *Science* **2015**, *348*, 989.

5. Noji, H.; Yoshida, M. *J. Biol. Chem.* **2001**, *276*, 1665.

6. Pullman, M. E.; Penefsky, H. S.; Datta, A.; Racker, E. *J. Biol. Chem.* **1960**, *235*, 3322.

7. Mitchell, P. *Nature* **1961**, *191*, 144. Nobel Prize in Chemistry, 1978.

8. Kagawa, Y.; Racker, E. *J. Biol. Chem.* **1971**, *246*, 5477.

9. Sone, N.; Yoshida, M.; Hirata, H.; Kagawa, Y. *J. Biol. Chem.* **1977**, *252*, 2956.

11

Working in the Dark

PART A. THE CALVIN-BENSON-BASSHAM (CBB) CYCLE

Three turns of the Calvin cycle[1] (Figure 11.1), allow the conversion of three (3) equivalents of carbon dioxide (CO_2) (i.e., 3 C_1 units) along with three (3) equivalents of the five-carbon carbohydrate derivative, ribulose-1,5-bisphosphate (i.e., 3 C_5 units) to yield three (3) not yet isolated six-carbon adducts, 2-carboxy-3-ketoribitol-1,5-bisphosphate (3 C_1 + 3 C_5 = 3 C_6) to form. The three (3) C_6 species then undergo fragmentation to yield six (6) equivalents of the three (3) carbon dihydroxy monocarboxylate, 3-phosphoglycerate (i.e., 3 C_6 = 6 C_3). A cartoon representation of this process is shown in Scheme 11.1 for one of the three CO_2 units.

Of the six (6) three-carbon unit equivalents, five (5) are used to regenerate three (3) equivalents of ribulose-1,5-bisphosphate (i.e., 5 C_3 = 3 C_5), while the sixth three-carbon fragment is now available to combine with another to make a six (6) carbon sugar (2 C_3 = 1 C_6) such as glucose ($C_6H_{12}O_6$) (Figure 11.2).

Additionally, as shown in Figure 11.3, 3-phosphoglycerate can be used to make other small compound building blocks such as glyceric acid, lactic acid, pyruvic acid and even acetic acid (after decarboxylation).

Ribulose-1,5-bisphosphate (often abbreviated as RuBP), using the enzyme ribulose-1,5-bisphosphate carboxylase (EC 4.1.1.39, carboxydismutase, rubisco), catalyzes the Mg^{2+}-dependent conversion of the 1,5-bisphosphate ester of the carbohydrate ribulose with carbon dioxide (CO_2) to produce two (2) equivalents of 3-phosphoglycerate (PGA). As shown in the Schemes 11.1 and 11.2. A hypothetical the six carbon intermediate, 2-carboxy-3-ketoribitol-1,5-bisphosphate, is often written.

It is important to keep in mind that we want the 3-phosphoglycerate for purposes of construction of other important compounds. But, as noted above, three turns of the cycle are necessary to produce six (6) equivalents of 3-phosphoglycerate, and five (5) of them are reused in making the three (3) ribulose-1,5-bisphosphates necessary to turn the cycle three (3) times.

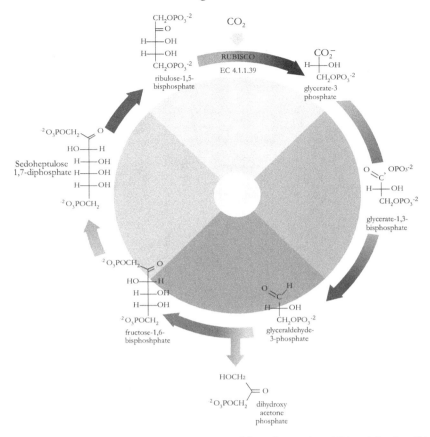

FIGURE 11.1 An abbreviated cartoon representation of the Calvin cycle. Additional details will be presented subsequently and a more complete representation is found in Appendix 2.

As shown, after the absorption of CO_2 and the formation of two (2) equivalents of glycerate-3 phosphate a further phosphorylation occurs (Scheme 11.3). The 3-phospho-D-glycerate is now phosphorylated on the carboxyl oxygen to make a mixed carboxylate-phosphate anhydride, 1,3-bisphospho-D-glycerate (3-phospho-D-glyceroyl phosphate). The phosphate exchange involves the removal of a phosphate group from adenosine triphosphate (ATP) to form the bisphosphoglycerate and adenosine diphosphate (ADP). The process is catalyzed by the enzyme phosphoglycerate kinase (EC 2.7.2.3).

Next, reduction can be effected to produce 3-phosphoglyceraldehyde (glyceraldehyde 3-phosphate). The reduction is accomplished enzymatically with glyceraldehyde-3-phosphate dehydrogenase (EC 1.2.1.13) by using the reduced form of the cofactor nicotinamide adenine dinucleotide phosphate [NAD(P)H], which is oxidized in the process to the oxidized form nicotinamide adenine dinucleotide phosphate [NAD(P)$^+$].

With the formation of glyceraldehyde-3-phosphate (3-phospho-D-glyceraldehyde), the rebuilding of ribulose-1,5-diphosphate to continue turning the Calvin cycle by combining fragments continues.

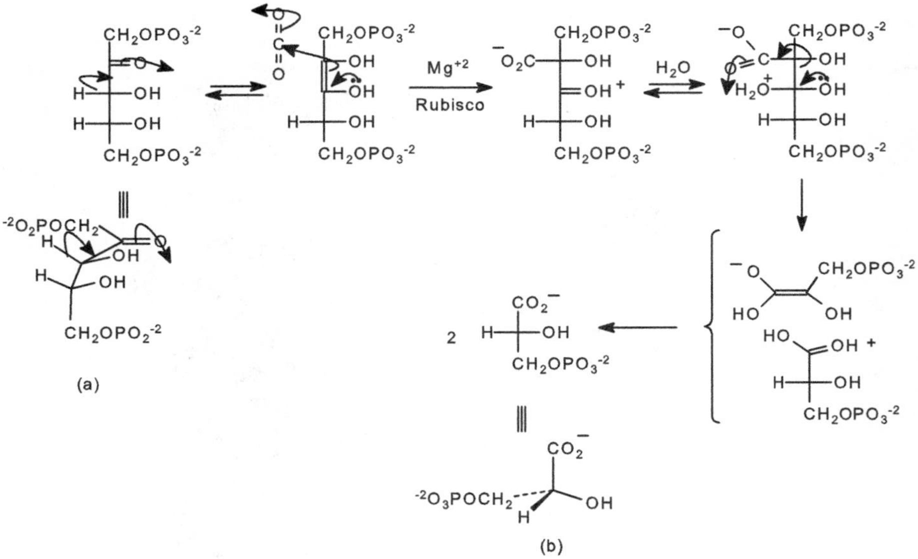

SCHEME 11.1 A cartoon representation of the addition of carbon dioxide (CO_2) to the five-carbon keto-sugar diphosphate, ribulose-1,5-bisphosphate (a), in an imagined process in the presence of Mg^{2+} and the enzyme ribulose-1,5-bisphosphate carboxylase (EC4.1.1.39). Since 2-carboxy-3-ketoribitol-1,5-bisphosphate is not isolable, the sense of addition, i.e., to which face of the double bond the addition occurs, (*re* or *si*) cannot be established. Fragmentation is presumed to occur to yield, after suitable proton adjustments, two equivalents of (b) 3-phosphoglycerate ($C_1 + C_5 = 2\,C_3$). For a definition of the terms used here, consult Appendix 1. The curved arrows are a representation of the presumed flow of electrons in an accounting of the bonds that have been made and broken.

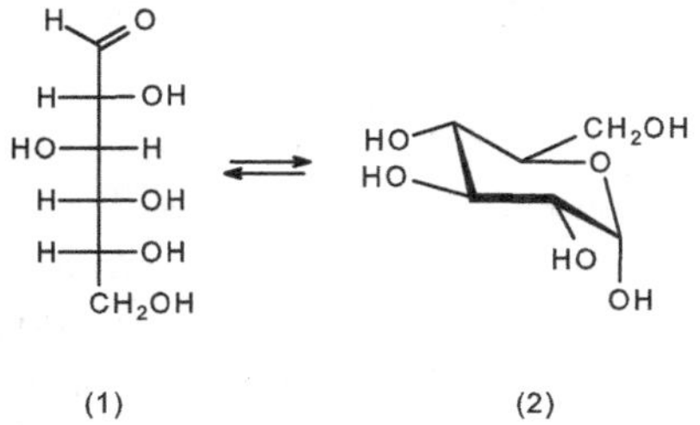

FIGURE 11.2 Representations of: (1) glucose drawn in an acyclic form; and (2) glucose drawn as the α-anomer of the cyclic pyranose form.

The value of the C_3 fragment, glyceraldehyde-3-phosphate, to the chemistry of wine cannot be underestimated. It will be used to produce the six-carbon compounds called hexoses ($2 \times C_3 = C_6$), such as glucose, which will then be used by yeast to make the wine. Glucose is also used to make cellulose, which is needed for stems, cell walls, etc. Furthermore, phospho-3-glycerate, the carboxylic acid precursor to glyceraldehyde 3-phosphate, is also

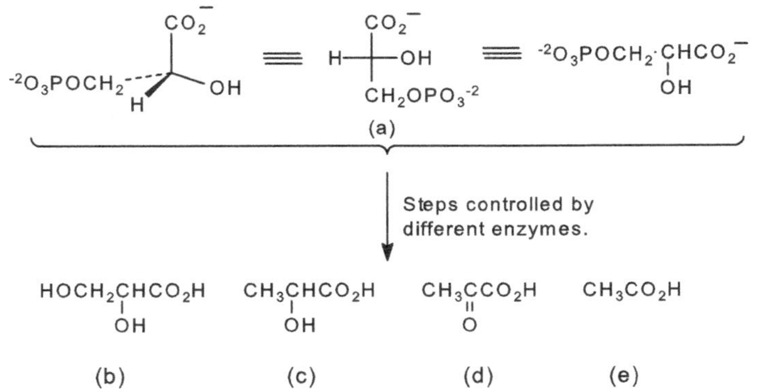

FIGURE 11.3 A cartoon showing that (a) 3-phosphoglyceric acid can be converted by suitable enzymes into a host of similarly sized fragments: (b) glyceric acid, (c) lactic acid, (d) pyruvic acid, and (e) acetic acid. The fragments can also be used as "building blocks" for other compounds.

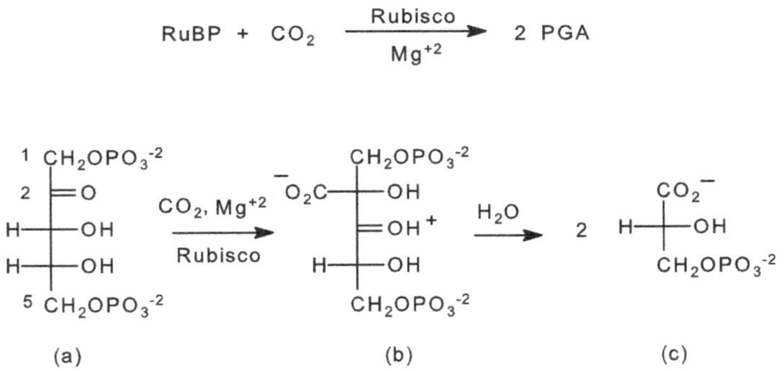

SCHEME 11.2 A representation of the overall process of converting (a) ribulose-1,5-diphosphate (RuBP) to the presumed intermediate (b) 2-carboxy-3-ketoribitol-1,5-bisphosphate by addition of carbon dioxide (CO_2) in the presence of magnesium (Mg^{2+}) and the enzyme ribulose-1,5-bisphosphate carboxylase (rubisco, EC 4.1.1.39). The addition is followed by cleavage of the shown six-carbon presumed intermediate into two (2) equivalents of (c) 3-phosphoglycerate (PGA).

converted to phospho-2-glycerate and then, by dehydration, to phosphoenolpyruvate (PEP) and subsequently to pyruvate and finally acetate—for acetyl coenzyme A (acetyl-CoA) and products such as fatty acids, sterols, and phenols built therefrom.

Glyceraldehyde 3-phosphate (a C_3 fragment) is isomerized to dihydroxyacetone monophosphate in a process catalyzed by triosphosphate isomerase (EC 5.3.1.1). The isomerization is followed by a condensation reaction with glyceraldehyde 3-phosphate (a C_3 fragment) to produce the six-carbon ketosugar fructose 1,6-bisphosphate (catalyzed by fructose bisphosphate aldolase, EC 4.1.2.13) (Scheme 11.4).

Fructose and glucose are isomeric (they have the same composition, but the former is a ketone and the latter an aldehyde). They are interconvertible. When they combine with each other sucrose results. The discussion in the next several paragraphs presents a fuller picture of the Calvin cycle so that the cycle can be made complete. We will return to sucrose, glucose, and fructose.

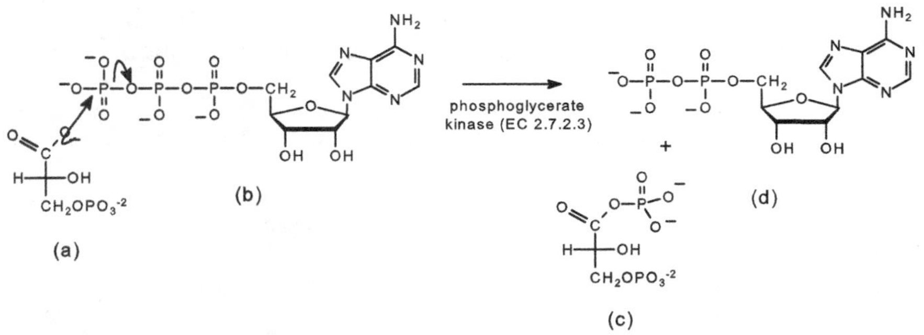

SCHEME 11.3 A representation of the phosphorylation of the carboxylate group of (a) 3-phosphoglycerate to produce a mixed carboxylate-phosphate anhydride, (c) 1,3-bisphosphoglycerate. The phosphoglycerate kinase (EC 2.7.2.3) catalyzed formation of the anhydride involves the removal of a phosphate group from (b) adenosine triphosphate (ATP), yielding (d) adenosine diphosphate (ADP) and the 1,3-bisphosphoglycerate (c). The stereochemistry of the attack at phosphorus is backside to the leaving group.

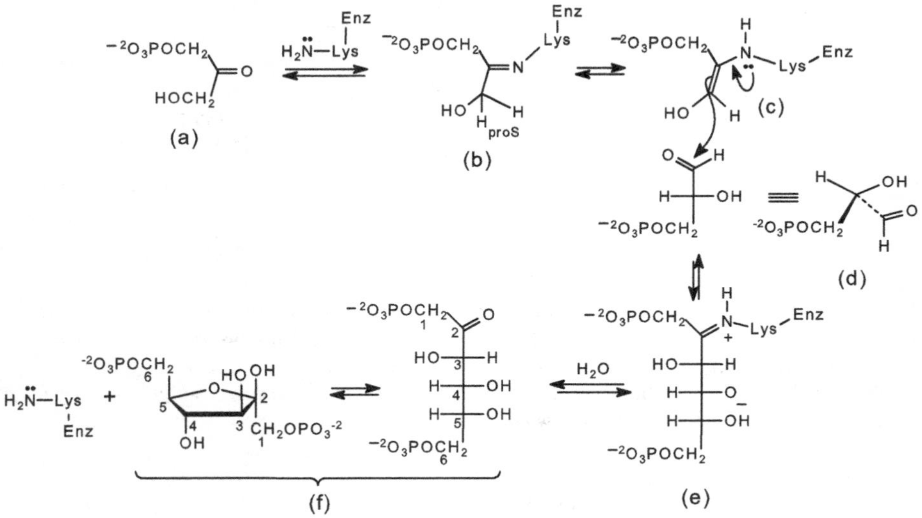

SCHEME 11.4 A cartoon representation of the catalyzed (fructose bisphosphate aldolase, EC 4.1.2.13) condensation between dihydroxyacetone monophosphate (a) and glyceraldehyde 3-phosphate (d) to produce the six-carbon ketosugar fructose 1,6-bisphosphate (f). The path shown involves reaction between the dihydroxyacetone monophosphate keto group with an active site lysine (Lys, K) on the enzyme (Enz) fructose bisphosphate aldolase (EC 4.1.2.13). The imine (b) (formed by loss of the pro S proton) in this way is presumed to isomerize to the corresponding enamine (c), which then undergoes the condensation reaction with glyceraldehyde 3-phosphate (d). The imine that results (e) is hydrolyzed to regenerate the active site lysine (Lys, K) and fructose 1,6-bisphosphate (f), which is drawn both as an open chain and its cyclic furanose form.

This six-carbon sugar (fructose 1,6-bisphosphate) is used to return to ribulose 6-phosphate as well as to isomerize to glucose. The glucose is then a building block for starch or cellulose. (As will be seen in Chapter 11, Part B, [Scheme 11.6] glucose combines with fructose to yield sucrose[2] which is transported[3] in the vine for structural use and for storage in the grape.)

With the intervention of the enzyme transketolase (EC 2.2.1.1), an enzyme that utilizes thiamine diphosphate (Vitamin B_1) cofactor, two carbons are removed from fructose 6-phosphate ($C_6 - C_2 = C_4$) leaving behind the four-carbon sugar erythrose 4-phosphate. The two-carbon unit that is removed is then added to one of the 3-phosphoglycerates made available initially from reaction of carbon dioxide (CO_2) with ribulose-1,5-bisphosphate ($C_2 + C_3 = C_5$). The product, xylulose 5-phosphate, is an isomer of ribulose 5-phosphate which, in the presence of the enzyme ribulose-phosphate 3-epimerase (EC 5.1.3.1), can be isomerized to the latter.

The erythrose 4-phosphate (C_4) generated in the same step as the derivative of thiamine diphosphate is then available for enzyme-catalyzed condensation with dihydroxyacetone monophosphate (C_3) to produce a seven-carbon sugar, sedoheptulose 1,7-bisphosphate ($C_3 + C_4 = C_7$). On hydrolysis, catalyzed by the enzyme sedoheptulose-bisphosphatase (E.C. 3.1.3.37), sedoheptulose 7-phosphate is formed.

Now, finally, sedoheptulose 7-phosphate undergoes a transketolase (E.C.2.2.1.2) catalyzed process to remove two carbon atoms using the enzyme cofactor thiamine diphosphate to yield ribose 5-phosphate and a two-carbon fragment. When the two-carbon fragment is added to glyceraldehyde 3-phosphate, xylulose 5-phosphate results. The xylulose 5-phosphate isomerizes to ribulose 5-phosphate.

So, as the circle turned, two ribulose 5-phosphates have been regenerated from two xylulose 5-phosphates, and one ribulose 5-phosphate has come from ribose 5-phosphate. Phosphorylation returns all three to ribulose-1,5-bisphosphate, and there is one C_3 unit left over.

Unlike carbon dioxide (CO_2) and oxygen (O_2), the products of the Calvin cycle needed for plant growth, etc. are too large to diffuse through the cell walls, and they must be actively removed by wall-penetrating pumps or through specific channels or pores. In addition, as the apical meristem grows, moving products (such as sucrose) up the plant requires acting against gravity, and the walls of the xylem and phloem systems play a critical role in such transport.[4]

PART B. GLUCONEOGENESIS AND THE FORMATION
OF SUCROSE AND CELLULOSE

D-Glyceraldehyde[5], the "simplest sugar" (as the 3-phosphate), has been produced in the turning of the Calvin cycle. Hydrolysis of the phosphate generates the free alcohol.

The synthesis of glucose ($C_6H_{12}O_6$) from smaller fragments (gluconeogenesis, Scheme 11.5) which is then stored as starch (a mixture of α-glucose polymers, Figure 11.4a) and/or cellulose (a mixture of β-glucose polymers, Figure 11.4b) occurs by condensation of 3-phosphoglyceraldehyde (glyceraldehyde 3-phosphate, a C_3 fragment) with another C_3 fragment, dihydroxyacetone monophosphate, as shown in Scheme 11.5.

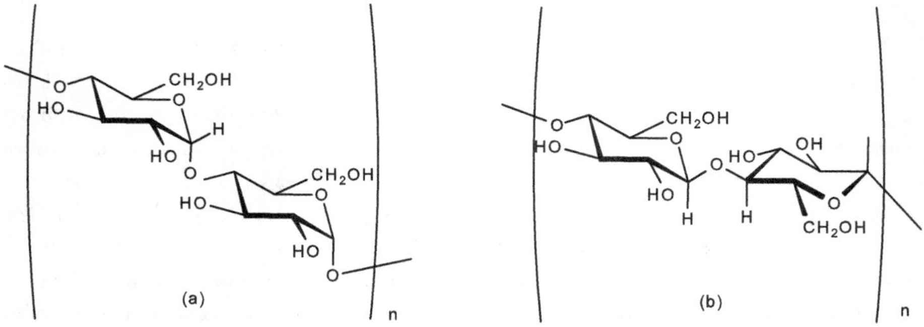

FIGURE 11.4 First approximate representations of fragments where n > 10^4 of (a) starch, the α-D-1→4-glucopyranose polymer and (b) cellulose, the β-D-1→4 glucopyranose polymer. The actual, more complicated, polymers have additional ligands and varying amounts of cross-linking.

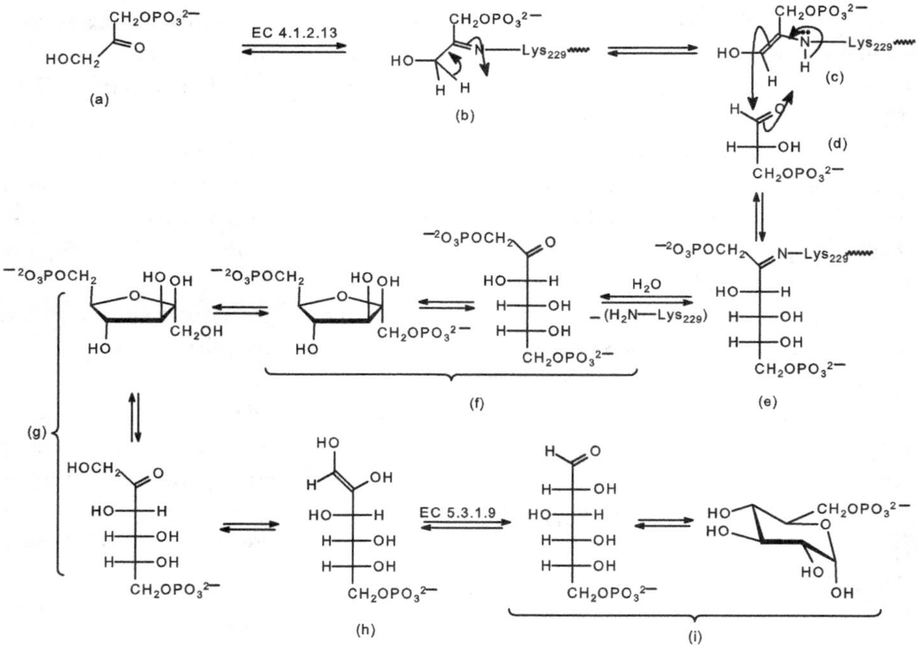

SCHEME 11.5 A representation of a portion of the pathway of gluconeogenesis. The three (3) carbon fragment (a) dihydroxyacetone monophosphate is shown as forming an imine (b) with the amino group of lysine (Lys, K) 229 of the enzyme fructose bisphosphate aldolase (EC 4.1.2.13). Water is lost in the process. The imine (b) is then shown in equilibrium with its corresponding enzyme-bound enamine (c). On aldol-type condensation of the enamine (c) with glyceraldehyde 3-phosphate (d) a new six (6) carbon imine (e) is generated. Hydrolysis of the imine regenerates the free lysine (Lys, K_{229}) and fructose 1,6-bisphosphate (f). The fructose bisphosphate (f) is shown in both open and closed (furanose) forms. Hydrolysis of one phosphate generates the 6-phosphate of fructose (g) (again shown both in cyclic and acyclic forms). Enolization of fructose 6-phosphate (g) produces the enol common to both fructose and glucose (h), and then isomerization (glucose-6-phosphate isomerase, EC 5.3.1.9) yields glucose 6-phosphate (i), shown as the open chain and cyclic α-D-pyranose forms. The 6-phosphate of glucose is considered capable of hydrolysis to glucose with the appropriate hydrolase (glucose-6-phosphatase, EC 3.1.3.9) catalyst. The anomer shown is defined as α (because the hydroxyl on the anomeric carbon is *cis* or *syn* to the hydroxyl at C5).

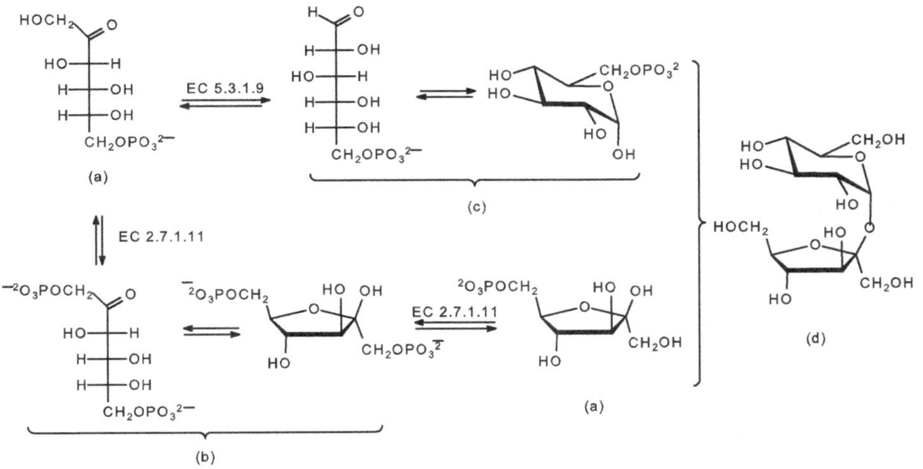

SCHEME 11.6 A representation of the formation of sucrose from glucose and fructose. As shown, fructose 1,6-bisphosphate (b) is formed from fructose 6-phosphate (a) by phosphorylation through the agency of the enzyme phosphofructokinase, EC 2.7.1.11. Tautomeriztion of fructose 6-phosphate (a) via the common enol (*vide supra*, Scheme 11.5) produces glucose 6-phosphate (c) drawn first in the open form and then as the α-D-anomer in the pyranose form. Then, fructose 6-phosphate (a) drawn as the β-anomer in the furanose form combines with glucose 6-phosphate (c), losing water to produce sucrose (d) As shown in Scheme 11.7 the question as to which oxygen remains in the bridging position has largely been resolved.

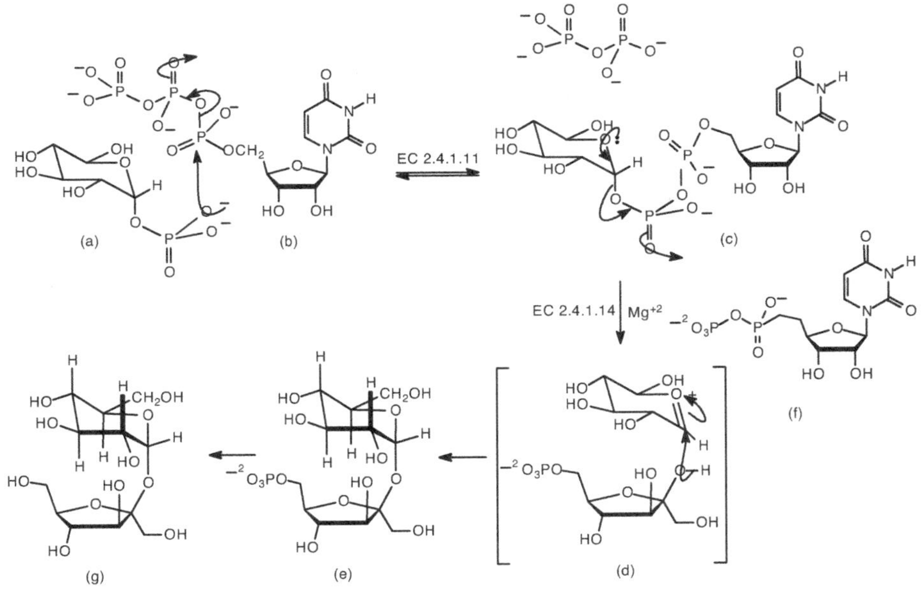

SCHEME 11.7 As commented in the enzyme database (http://www.enzyme-database.org). "Sucrose synthase (EC 2.4.1.14) has a dual role in producing both UDP-glucose (necessary for cell wall and glycoprotein biosynthesis) and ADP-glucose (necessary for starch biosynthesis)." So, as shown in this cartoon representation, α-D-glucose phosphate (a) undergoes condensation with uridine triphosphate (UTP) (b) to produce the activated adduct (c) where two (2) equivalents of inorganic phosphate (P_i) are lost and the phosphate group attached to glucose is activated for elimination. It is presumed an enzyme such as glycogen synthase (EC 2.4.1.11) is involved. Then, in the presence of sucrose-phosphate synthase (EC 2.4.1.14), the activated phosphate (c) condenses with β-D-fructofuranose 6-phosphate (d) to generate sucrose 6F-phosphate (e) and uridine diphosphate (f). Finally, sucrose (g) itself is formed by hydrolysis of the phosphate ester from what was C-6 of the fructose.

As noted earlier and as seen below in Schemes 11.6 and 11.7, sucrose, the sugar formed from isomeric glucose and fructose hexoses, is the species transported through the growing plant from where it is synthesized to where it is needed for growth.

As noted earlier (page 27) cellulose is a linear (1→ 4) glucose polymer. It is synthesized and secreted by a membrane-integrated cellulose synthase. The details of its synthesis and transport are under active investigation.[6,7,8]

NOTES AND REFERENCES

1. The Calvin cycle (after Bassham, J.; Benson, A.; Calvin, M. *J. Biol. Chem.* **1950,** *185,* 781) is a cycle. Cycles have neither beginnings nor ends, for then it would not be a cycle. But materials are added and subtracted as the cycle turns. The abbreviation EC found herein refers to the Enzyme Commission numbering system. Details can be found at http://www.enzyme-database.org/ (accessed Apr 5, 2017).

2. Delmer, D. P.; Albershein, P. *Plant Physiol.* **1970,** *45,* 782.

3. Liu, D. D.; Chao, W. M.; Turgeon, R. *J. Exper. Bot.* **2012,** *63,* 4315. DOI: 10.1093/jxb/ers127.

4. Chen, L.-Q.; Qu, X.-Q.; Hou, B.-H.; Sosso, D.; Osorio, S.; Fernie, A. R.; Frommer, W. B. *Science* **2012,** *335,* 207. DOI: 10.1126/science.1213351.

5. D, because when the Fischer projection structure is drawn vertically, with the most highly oxidized carbon at the "top," the –OH on the penultimate carbon is on the right. In that vein, it will be recalled that the Fischer projection is a representation in which the groups that lie on the horizontal lines are to be thought of as lying above the plane of the paper and the groups on the vertical lines as those lying below the plane of the paper, thus defining a tetrahedron view "edge on" (Appendix 1).

6. Olek, A.; Rayon, C.; Makowski, L.; Kin, H. R.; Ciesielski, P.; Badger, J.; Paul, L. N.; Ghosh, S.; Kihara, D.; Crowley, M.; Himmel, M. E.; Bolin, J. T.; Carpita, N. C. *Plant Cell* **2014,** *26,* 2996. DOI 10.1105/tpc.114.126862.

7. Morgan, J. L. S.; McNamara, J. T.; Fischer, M.; Rich, J.; Chen, H.-M.; Withers, S. G.; Zimmer, J. *Nature* **2016,** *531,* 329.

8. Kumar, M.; Wightman, R.; Atanassova, I.; Gupta, A.; Hurst, C. H.; Hemsley, P. A.; Turner, S. *Science* **2016,** *353,* 166.

12

Flowers

PART A. BUDS TO FLOWERS

Generally, grape vines produce extraneous shoots ("suckers")[1] on the plant in addition to those growing beyond the few desired on the cordon wanted for proper vine growth. Generally, again, suckers are less fertile than the primary shoots, they crowd the canopy of the vine, and their growth utilizes resources required for proper growth of the primary shoots. Further, the chaotic growth makes it difficult to manage the harvest. A crowded canopy (as will be discussed subsequently) is not a healthy one for grape growth.

As shown in Figures 12.1 and 12.2 and noted earlier, buds (the small part of the vine that lies between the vine's stem and the leaf stem or petiole) can start alongside the beginning of leaves at the base of the apical meristem. The buds swell and eventually produce shoots. As the shoot grows the flowers appear on a stem from the node, from where leaves have also sprung. That is, grape nodes hold buds that grow into leaves and inflorescences or "clusters

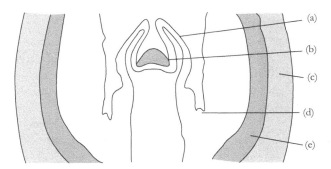

FIGURE 12.1 (Figure 7.16 repeated). A cartoon representation that shows older as well as younger primordium parts of the growing plant and, along with the apical meristem, the beginning of leaves. The parts shown are (a) the young leaf primordium; (b) the apical meristem; (c) the older leaf primordium; (d) lateral bud primordium; and (e) vascular tissue. Somewhat more is known about leaf cells than is known about cells of the cotyledons.

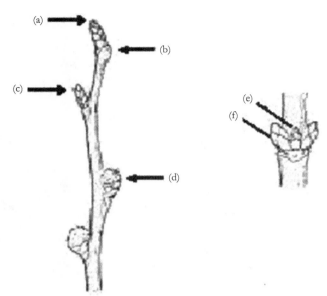

FIGURE 12.2 A cartoon representation showing leaf and flower buds along the same branch Shown are (a) a pseudoterminal growth or pseudoterminus: (b) a terminal bud scar; (c) a leaf bud; and (d) a flower bud. Also shown are (e) collateral buds, which are accessory buds arranged on either side of a lateral bud and (f) an axillary bud, one that is borne at the axil of a leaf and is capable of developing into a branch shoot or flower cluster.

of flowers" (i.e., the reproductive portion of a plant) arranged on a smaller stem growing from the node. It is not yet clear, despite recognizing the flow of nutrients and auxins as well as changes in proteins, how, after vernalization (i.e., the ability to flower so that fruit can be set—but only after exposure cold), the plant decides which, leaf or stem bearing flowers, should sprout from the node.

The fundamentals of the coming forth of the buds are often outlined[2,3] as a three-step process. First there is the formation of uncommitted primordia (primordia refer to tissues in their earliest recognizable stages of development) called "anlagen" (from German, in English, "*assets*" or "*facilities*") at the apices of lateral buds. Second, differentiation of anlagen to form inflorescence primordia or tendril primordia occurs. Finally, flowers form from the inflorescence primordia when activated by phytohormones, nutrients, and growth regulators and when the external conditions of light and temperature are correct.

The energy to facilitate this growth initially comes from reserves of carbohydrates stored in roots and wood of the vine from the last growth cycle or from the seed or root stock of a graft.

Eventually, of course, and as noted above, the shoots sprout tiny leaves that can begin the process of photosynthesis, producing the energy to accelerate growth. With warm, sunny days growth of the shoots rapidly accelerate and might grow as much as 3 cm (1 in) a day!

The shoots bearing the small green unopened flowers continue to grow until the individual flowers open (Figure 12.3). The timing is a function of the particular variety of grape (as dictated by its genome), the *terroir*, and weather conditions. Once the flower has opened, the cap, or calyptra, which fused the petals together, is shed, and pollen is released to fertilize the ovary, forming seeds and ultimately grape berries. Most wine grapes have hermaphroditic flowers—that is, with well-developed and functional male and female parts so that

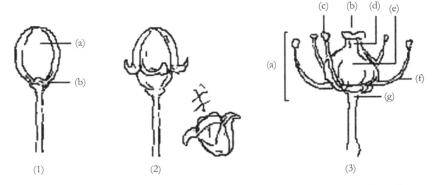

FIGURE 12.3 A representation of the bloom sequence of the grape flower. On the left, the flower (1) before blooming, the calyptra (cap) not yet fully formed over the closed petals of the (a) corolla [attached to (b) the pedicel]. In the center (2), the calyptra having formed is then separated while the flower (3) forms. The (a) stamen of the open flower is shown with (b) the stigma; (c) an anther; (d) the style above (e) the ovary. The anther is held in place by the (f) filament, all of which is held above and thus passed through on the way to the (g) nectary. Illustration by Jason Stafne. Used with permission.

self-fertilization is possible. However, because the flowers are so close together, it is possible for pollen from one flower to fertilize an adjacent flower.

Initially, as shown in Figure 12.3, there is a corolla (the closed petals) which contains the unopened flower and which is attached to the branch by the pedicel.[4]

The corolla lies within the calyptra (a cap-like covering). The calyptra is formed by closing or condensation of the petals of the unopened flower where the stamen (consisting of a filament, topped by an anther where pollen is produced) and the stigma are protected from the environment until signaled from the auxins and other transmitters to open and blossom.

The pollen is produced by meiosis—cell division where the number of sets of chromosomes is reduced by half.

The stigma is the receptive tip of the carpel in the gynoecium of the flower.

The gynoecium is the innermost whorl of the flower and consists of a carpel which forms the hollow ovary wherein megaspores, also produced by meiosis, develop into female gametophytes (i.e., egg cells).

Thus, there is an ovary surrounded by style and stigma, together called a pistil. The supportive stalk, the style, becomes the pathway for pollen tubes to grow from pollen grains adhering to the stigma. After fertilization, the plant is said to have set, and veraison (the onset of ripening) is observed.

The grapes (fruit of the vine) are the mature ovaries of the flowers. The edible flesh from which the wine is derived and that part of the grape which lies around the seeds and beneath the skin of the grape is called the pericarp. The pericarp develops from the ovary wall of the flower.

As described in the poetic literature, the odors of the flowers of the grape vines have been shown to be attractive to humans as well as other animals.[5] However, since the flowers are so close together, the role of pollinators is generally less important than with flowers of other plants. Indeed, it is widely acknowledged that wind and insects generally play only a small role in aiding pollination of grape flowers, although human intervention is not uncommon.

As shown in Table 12.1, the vapor above the grape flowers has been examined.[6,7] The sesquiterpenes (a discussion of terpenes, sesquiterpenes, etc. is provided in Appendix 2) and

TABLE 12.1

A list of compounds found in the headspace above the flowers of *V. vinifera* (Vitaceae) species "Grüner Veltliner." The data in this table is taken from Buchbauer, G.; Jirovetz, L.; Wasicky, M.; Nikiforov, A. *J. Essent. Oil Res.* **1994**, *6*, 311. For compounds marked with (*) the stereochemistry was not specified and more than one isomer is known to occur naturally.

Alphabetical list of compounds	Structure	Headspace as relative percent
acetone	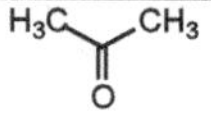	2.3
benzyl alcohol	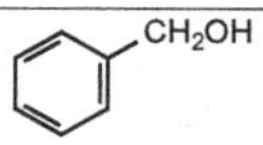	3.8
benzyl benzoate	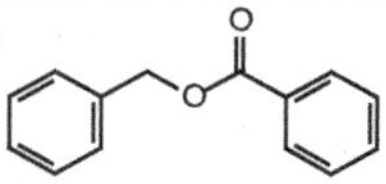	2.1
β-bisabolene	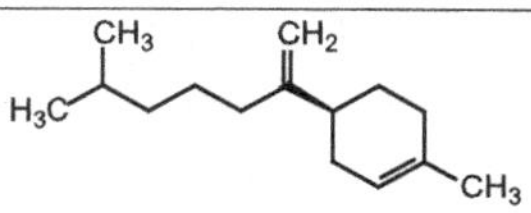	1.9
1-butanol		0.7
γ-cadinene	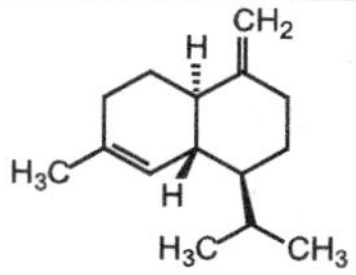	trace
β-caryophyllene	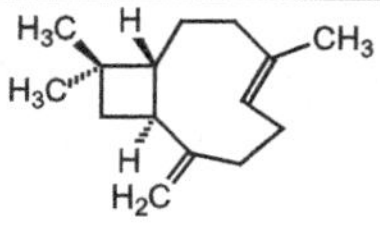	1.7
β-caryophyllene oxide	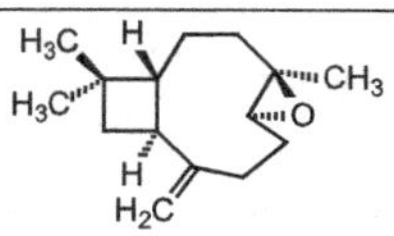	0.3
ethanol	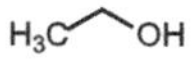	0.7
ethyl acetate	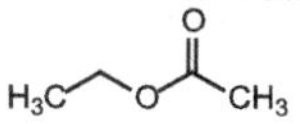	2.4
α-farnesene	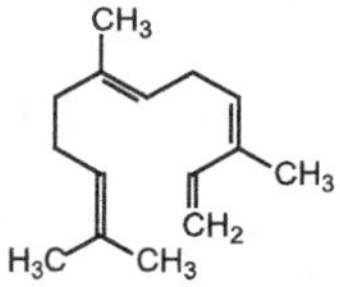	2.2

(Continued)

TABLE 12.1 CONTINUED

β-farnesene	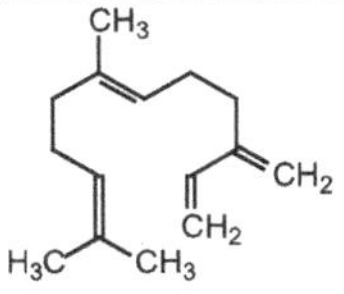	2.0
geraniol	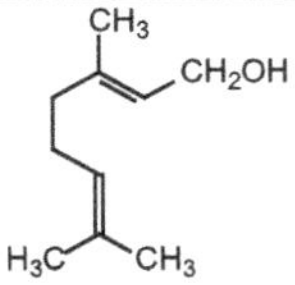	0.3
geranylacetone	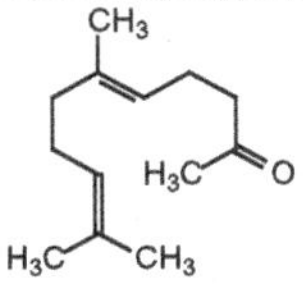	1.4
α-guaiene	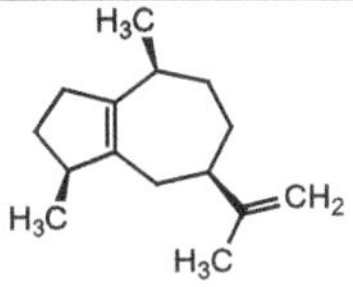	1.1
δ-guaiene	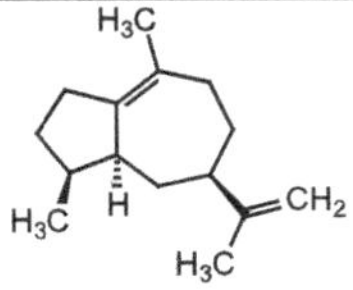	0.8
hexadecanoic acid	$CH_3(CH_2)_{14}CO_2H$	1
hexanol		2.3
1-hexene	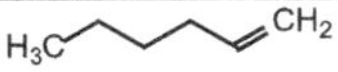	1.1
1-hexene-3-ol*	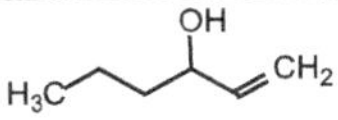	2.7
α-humulene	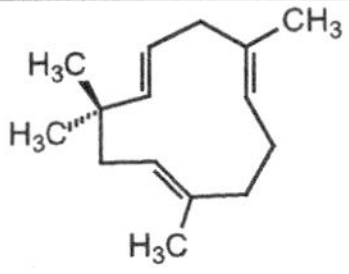	3.3

(Continued)

TABLE 12.1 CONTINUED

Name	Structure	Value
limonene*	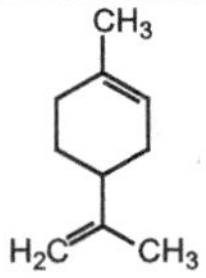 (structure)	0.2
linalool*	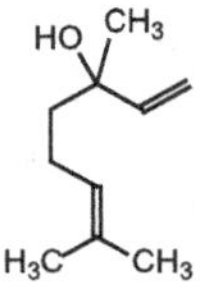 (structure)	0.3
6-methyl-5-hepten-2-one	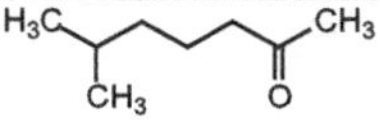 (structure)	2.3
methyl hexadecanoate	$CH_3(CH_2)_{14}CO_2CH_3$	1
nerolidol*	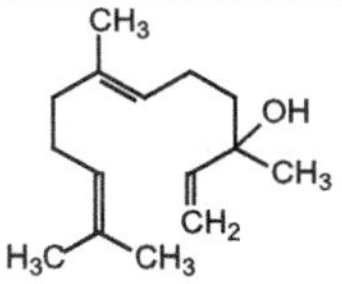 (structure)	4.3
nonadecane	$CH_3(CH_2)_{17}CH_3$	0.4
1-nonadecene	$CH_3(CH_2)_{16}CH=CH_2$	0.2
1-octanol	$CH_3(CH_2)_6CH_2OH$	1.7
pelargonaldehyde (nonanal)	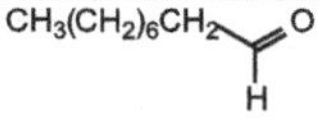 $CH_3(CH_2)_6CH_2CHO$	1.2
1-pentadecene	$CH_3(CH_2)_{12}CH=CH_2$	3.0
2-phenylethanol	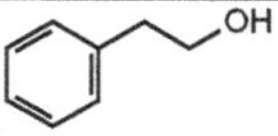 (structure)	2.2
β-selinene*	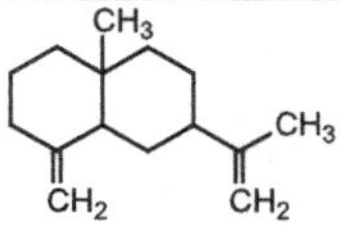 (structure)	0.5
δ-selinene*	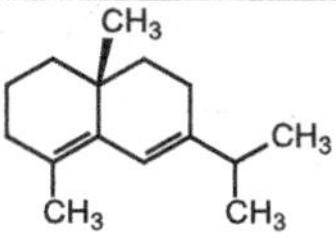 (structure)	0.2

(Continued)

TABLE 12.1 CONTINUED

2-tridecanone	$CH_3(CH_2)_9CH_2\text{–CO–}CH_3$	4.6
1-tridecene	$CH_3(CH_2)_9CH_2CH=CH_2$	0.7
valencene	(structure with CH_3, CH_3, CH_3, CH_2 groups)	35.7
hydrocarbons (higher than C_{16})		2.4

sweet-smelling esters of carboxylic acids (Appendices 1 and 3) are considered to account for the odors enjoyed.

Given that grape flower headspace[8] often has the same components from cultivar-to- cultivar (albeit in different concentrations)[7] and given the observation that humans can discriminate between many (perhaps more than one trillion!)[9] olfactory signals[10] it is likely that a vast neural network of signals contributes to the symphony of recognition of the odors of flowers. The same human neural network is used to detect the compounds in the headspace above the wine—although the compounds found in there (and in the wine) are often very different than those in the headspace above the flowers.

It is reasonable of course to ask how the nose does its job (Figure 12.4).

The nose airway surface, like other surfaces of organs and vessels, is lined with a layer of cells called the epithelium. These epithelial tissues receive information from below (nourishment) and information from above (odorants). The detection of an odor by the sensory cells in the epithelium, where molecular reception occurs, is often site specific. Subsequent transduction of the message to a neuron embedded in the epithelium (for signaling to the brain) is effected by G-protein coupled receptors (GPCRs). These receptors are both embedded in the neuron as well as in numerous hair-like projections from the epithelium called cilia. When a receptor (or set or series of receptors) receives a signal (e.g., an organic molecule that fits or nearly fits the receptor[s]), a series of events, outlined in more detail below,[11] is triggered. The neuron or neurons transmit the signal to the brain, and it is perceived as a smell.

The structure of a G (for guanosine) protein coupled receptor (PDB 3t33) from *Arabidopsis thaliana* found in the plasma membrane of the plant and (apparently) used for the regulation of an abscisic acid signaling pathway is shown in Figure 12.5.

It has recently been claimed and (in the nature of science advances) subsequently refuted[9] that the number of possible olfactory stimuli that can be discriminated by interaction of the odorant with the thousands of G-protein coupled receptors available is of the order of one trillion (1×10^{12}). How, regardless of the magnitude of the actual large number, is such a thing possible? The value for the discrimination possibilities resides in the observation that when the odorant (signal) interacts with a receptor or receptors, channels open and

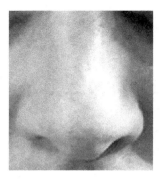

FIGURE 12.4 A representation of a nose.

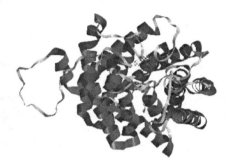

FIGURE 12.5 A G-protein coupled receptor from *Arabidopsis thaliana* both acetate ($CH_3CO_2^-$) and zinc (Zn^{2+}) are present in this representation from the Protein Data Bank (PDB 3t33) (http://www.rcsb.org/pdb/home/home.do) where a full color representation can be seen. Chen, J.-H.; Guo, J.; Chen, J.-G.; Nair, S. K. (to be published). The PDB graphic is used in terms of the PDB Charter.

additional signals result. The variable strength and thus the number of possibilities is a function of the number of receptors activated and extent of receptor involvement.

G-protein coupled receptors (GPCRs) are generally seven-segment trans-membrane proteins with hydrophobic (water avoiding) peptide segments in the membrane and the potentially ionized groups amino on the outside of the membrane and carboxylic acid on the inside (Figure 12.6) of the membrane. That the GPCRs are very important is suggested by the observation that a large number of genes in the genome of every animal examined (about 4% in humans) are devoted to their production.

In addition to their role in olfaction considered here, the GPCRs are also involved in vision[12] and while different GPCRs are different, aside from their physical transmembrane similarities (they are a superfamily), they all appear to act in a similar fashion. All of the processes involved begin after the signal, starting outside the cell, passes from the receptor (GPCR) in the wall to the inside, where guanosine triphosphate (GTP) is hydrolyzed to guanosine diphosphate and inorganic phosphate (Figure 12.7).

Thus, on the outside, the receptor is activated (fully or partially by binding or other means), and the bundle of helices responds. It may be that the bundle opens—or opens part way—and a signal (protons [H^+], sodium ions [Na^+], calcium ions [Ca^{2+}], etc.) is

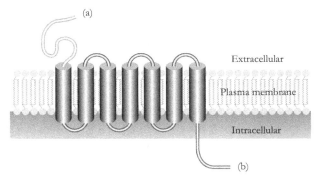

FIGURE 12.6 A cartoon representation of a seven (7) segmented G-protein, each of the seven connected to the others by protein strands in loops. The "outside" of the membrane (a) has the ammonium terminus, and the inside (b) the carboxylate. Also see Kroeze, W. K.; Sheffler, D. J.; Roth, B. L. *J. Cell Sci.* **2003**, *116,* 4867 (DOI: 10.1242/jcs.00902).

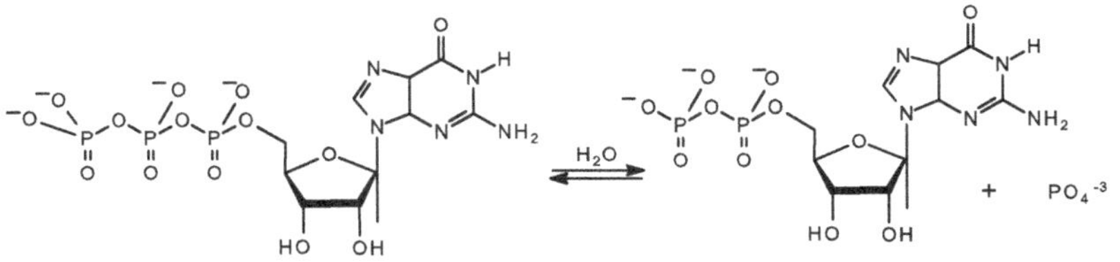

FIGURE 12.7 A representation of an equilibrium between guanosine triphosphate, GTP (a), and guanosine diphosphate (b) and inorganic phosphate.

permitted to pass through. While there may be other means of activation too, the result is that G-proteins located within the cell are activated.

The G-proteins then, in turn, activate entire cascades of further signaling events that can result in a change in cell function. Many hormones, neurotransmitters, and other signaling entities are regulated along with metabolic enzymes, transporter helpers, and other parts of the cell machinery by G-proteins. It appears that G-proteins also participate in transcription of DNA to RNA as well as cell motility and secretion of factors that ultimately govern homoeostasis and learning and memory.

In summary, with regard to olfaction detection, thousands of different olfactory receptors, each encoded by a different gene and each recognizing different odorants, are involved. Each of the receptors is encoded by a specific gene in the DNA of the animal detecting the odor. If a gene is missing or damaged the odor cannot be detected. When one smells many fruits or flowers, what is being smelled corresponds to members of the terpene, sesquiterpene, and other hydrocarbon species (Appendices 2 and 5), as well as esters (Appendix 1 for synthesis from carboxylic acids and alcohols and Appendix 3 for a list of esters), ketones, aldehydes, alcohols, and even thiols and other sulfur-containing compounds evaporating from the fruit or flower.[13]

Interestingly, after the above summary was prepared, yet another new set of receptors, the membrane-spanning 4-domain family, subfamily A (MS4A) was found active in odor detection. These detectors are apparently sensitive to unsaturated fatty acids as well as specialized odors. Details as to how they function have yet to be described.[14]

PART B. THE COLORS OF FLOWERS

As demonstrated in about 1670 by Sir Isaac Newton,[15] white light is composed of all of the colors of the visible spectrum (Figure 12.8) which, using a prism, can be deconstructed into the colors we see and, not shown in the Figure, using another prism, reconstituted as white light.(Appendix 1).

From the earlier discussion of absorption of radiation by leaves, it will be recalled that it is largely in the red and blue regions of the visible region of the spectrum that the leaves absorb visible light. When the red and blue regions are removed by absorption from the visible, green is left for reflection and transmission. This allows humans, who can detect green, to enjoy the green hills of Earth in our present verdant environment.

However, unlike leaves, flowers can apparently take advantage of the wealth of colors available in the remainder of the visible spectrum (as well as regions beyond our vision). Thus, with regard to the visible spectrum, flowers either absorb none of the light (remaining white) or absorb part or parts of the visible spectrum that allow light which was not absorbed to be available for the pleasure of the viewer (and, presumably, to other ends and for reasons we do not yet fully understand or know about).

So, there are two (2) aspects that must be considered. First, as a consequence of absorption in the visible region of the spectrum, only certain frequencies are available to be seen by an observer. Second, the observer must be able to detect those frequencies not absorbed.

In general, there are three major (and other minor) different classes of compounds that consist of uncounted members that give rise to flower color, *viz.*, betalains, anthocyanidins, and carotenoids.

For the latter, the carotenoids, compounds such as lutein, a major pigment in marigolds (*Tagetes patula*), are present. Such compounds are clearly related to β-carotene (Figure 12.9). The biosynthetic pathways shown in Appendix 2 provide a way to show how this material is derived from simpler five-carbon precursors.

Although both betalains and anthocyanidins (Figure 12.10) are derived, in part, from the amino acid tyrosine (Tyr, Y) they are formed in ways that differ from each other and from carotenoids. Their biosyntheses are provided in Appendix 4. Betalains have not yet been reported in *Vitis vinifera*.

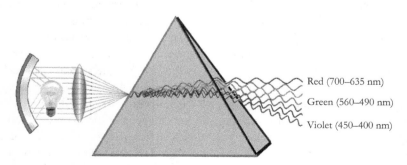

FIGURE 12.8 A representation of the colors of the visible spectrum observed by passing white light through a prism. The wavelength regions corresponding to red, green, and violet colors are shown.

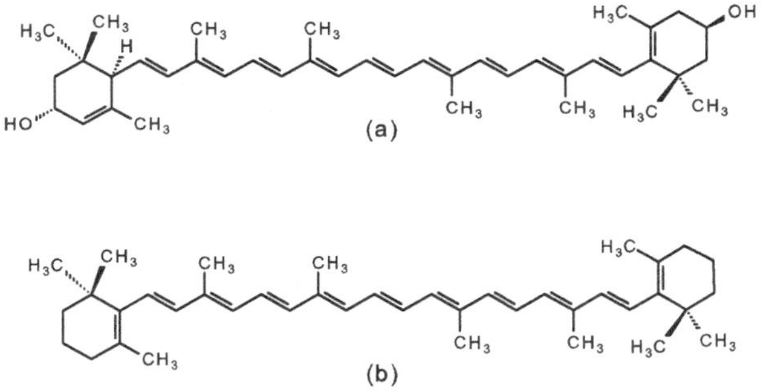

Flowers

FIGURE 12.9 Representations of the carotenoids (a) lutein and (b) β-carotene. Their close relationship, presumably involving oxidation of the latter to produce the former, is evident.

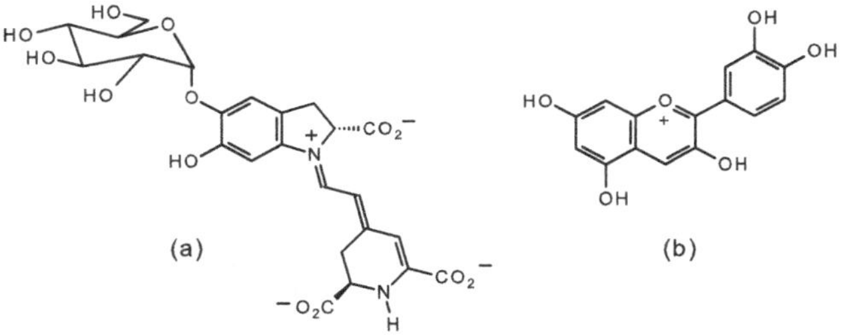

FIGURE 12.10 Cationic water-soluble plant pigments (a) betanin (a betalain pigment) found in dark red plants and (b) cyanidin (an anthocyanidin pigment) found in many flowers. Colors of anthocyanidins are determined, in part, by the substitution pattern of hydroxyl groups (–OH) on the aromatic rings and also by whether or not the phenolic hydroxyls are themselves free of further substitution (Appendix 4).

Indeed, only the aglycone anthocyanidins (and their corresponding anthocyanin glycones) have been found. Therefore the information on the betalains biosynthesis is provided with the understanding that they may yet be found, albeit in low concentrations.

Some anthocyanidin pigments are shown in Table 12.2, and one is elaborated in greater detail in Figure 12.11. Like the betalains and carotenoids, these compounds are highly conjugated. Indeed, the extensive array of alternating double and single bonds in these pigments, along with further modifications to their structures by addition of various substituents, largely accounts for the colors of the flowers that give rise to the grapes. Such conjugated systems absorb portions of the visible light spectrum as a function of the extent of conjugation, and our optics detect what is not absorbed.

With the chloride anion as the counter ion, chrysanthemin has been reported to be blue-to-purple at neutral pH. However, as is typical of such phenolic compounds and already noted, the color is often a function of the pH as well as what other substituents are present (Figure 12.12).

A less complicated (because fewer substituents are present) description of the function of pH with regard to color is exemplified by phenolphthalein, whose structure(s) are shown in Figure 12.13. Under acidic conditions or at nearly neutral pH, a solution containing phenolphthalein is colorless, the lactone is closed, and the phenolic groups are protonated. At basic pHs the color of the solution changes to pink (the anion being present at low concentration),

TABLE 12.2

A representation of some of the known and more common anthocyanidins

Fundamental Structure	Anthocyanidin	R_1	R_2	R_3	R_4	R_5	R_6
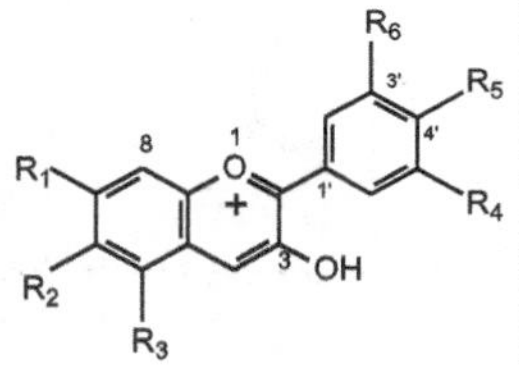	aurantinidin	OH	OH	OH	H	OH	H
	cyanidin	OH	H	OH	H	OH	OH
	delphinidin	OH	H	OH	OH	OH	OH
	europinidin	OH	H	OCH_3	OH	OH	OCH_3
	malvidin	OH	H	OH	OCH_3	OH	OCH_3
	pelargonidin	OH	H	OH	H	OH	H
	peonidin	OH	H	OH	H	OH	OCH_3
	petunidin	OH	H	OH	OCH_3	OH	OCH_3
	rosinidin	OCH_3	H	OH	H	OH	OCH_3

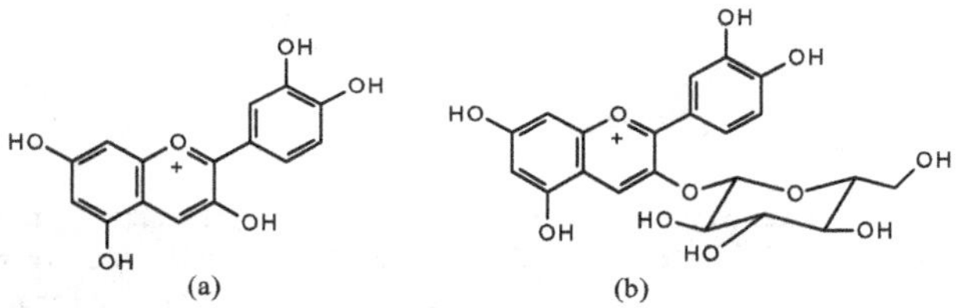

FIGURE 12.11 Representations of the 3-glucoside of the anthocyanidin, cyanidin (a), is shown as the anthocyanin, chrysanthemin (b) where, as is common, the anomeric carbon of the sugar glucose is attached to the 3 position of the anthocyanidin.

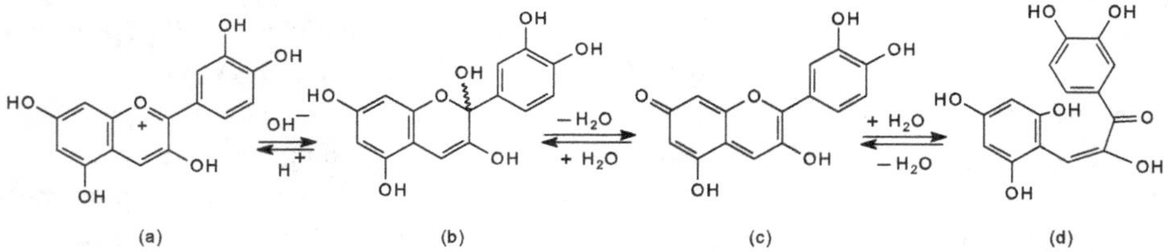

FIGURE 12.12 A representation of the anthocyanidin, cyanidin, with the chloride anion as the gegenion, as a function of changes in acid concentration (pH); (a) the flavylium cation, pH between 1 and 2, color = red; (b) the carbinol pseudo-base, pH between 4 and 5, color = colorless; (c) the quinoidal base, pH between 6.0 and 6.5; color = blue; (d) the chalcone, pH > 7; color = pale yellow.

and as the solution becomes more basic, the color becomes deeper, approaching an intense fuchsia at very high pH.

The acidic form of phenolphthalein has a strong absorption peak <250 nm so that visible light (400–800 nm) passes through unabsorbed and the solution is colorless. The anion, on the other hand, absorbs strongly at 550 nm (i.e., in the green region of the spectrum). Thus the observed color is a mixture of red and blue (i.e., fuchsia).

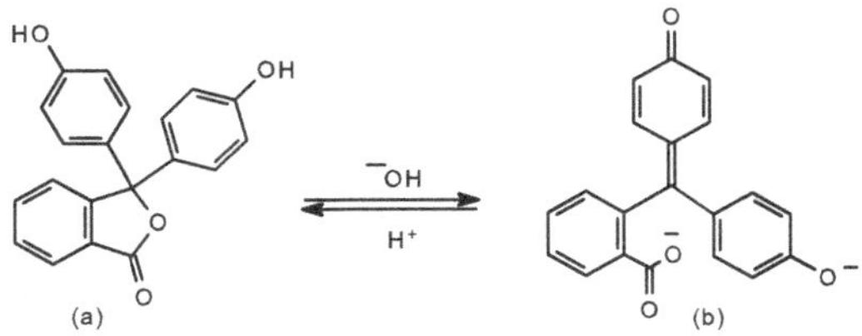

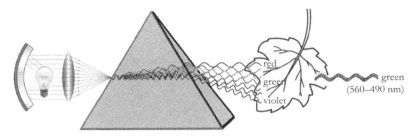

FIGURE 12.13 Representations of phenolphthalein as a function of hydrogen ion concentration (a) pH < 8.2; (b) pH > 8.2.

FIGURE 12.14 A representation of the colors of the visible spectrum observed by passing white light through a prism. The wavelength regions corresponding to red, green, and violet colors are shown striking a representation of a green leaf. Colors in the visible region of the spectrum that are not green are absorbed so that only green is seen.

The conjugated system of single and double bonds in the molecule is responsible for the color. Such portions of molecules against which such responsibility can be laid are called chromophores. The unique feature of the chromophore in this case is that the energy difference between the two different molecular orbitals (Appendix 1) involved lies in the visible spectrum. Thus, when visible light is absorbed by the chromophore the system loses that light and only that which was not absorbed (i.e., that which is transmitted) can be seen.[16]

As shown in Figure 12.14, the colors falling on an object can be absorbed and only the colors not absorbed are seen. So, for example, the absorption of red and blue light by plant leaves is perceived as the color we call green.

Humans have five (5) different visual receptors. The receptors are called opsins, and all of them are trans-membrane G-protein coupled receptors. They differ from each other in some amino acids, but nonetheless, they all have similar shapes. All five are sensitive to photons, and while their respective absorption bands are different and quite broad, together they cover the spectrum over which our eyes work. What we call the visible spectrum.

Rhodopsin absorbs green-blue light ($\lambda_{max} \approx 500$ nm), photopsin I absorbs in the region 500–570 nm ($\lambda_{max} \approx 464$–480), photopsin II absorbs in the region 450–630 nm ($\lambda_{max} \approx 534$–545), and photopsin III absorbs between 400 and 500 nm ($\lambda_{max} \approx 420$–440). The fifth opsin, melanopsin, absorbs blue and appears tied to human circadian rhythm. It has been estimated that humans can distinguish roughly 10 million different colors; a small fraction (*vide supra*) of the skill that our olfactory sense appears to possess.

The opsins are present in the structures in the eye called rods and cones that are attached to the retina. Rods are both more numerous and more sensitive than cones (it has been suggested that a single photon of the appropriate energy can serve to trigger them), but the rods are monochromatic. A view of bovine night blindness causing rhodopsin (PDB 4bez) with carbohydrates and fatty acid attached is provided (Figure 12.15).[17]

Rhodopsin consists of the protein opsin and the covalently bound cofactor, retinal. This G-protein coupled receptor (opsin) has the usual set of connected trans-membrane helices and binds the vitamin A derivative, 11-*cis*-retinal (a polyunsaturated aldehyde) via imine formation using the terminal nitrogen of a lysine (Lys_{296}, Lys, K). The light-catalyzed isomerization of 11-*cis*-retinal into 11-*trans*-retinal induces a conformational change (called bleaching) in opsin. That change then activates the associated G-protein and triggers an intracellular guanosine (G) cascade similar to that occurring with the sense of smell.

A representation showing the isomerization of the imine of *cis*-retinal to the imine of *trans*-retinal (Figure 12.16) is shown below (after Jäger, S.; Palczewski, K.; Hofmann, K. P. *Biochem.* **1996**, *35*, 2901)

So, while the color of an object is the result of its interaction light, the color of an object perceived by the brain involves the optics of the perceiver as well as those properties possessed by the object. The different cells in the retina are excited by light of different wavelengths striking the different opsins in the different cones.

PART C. FROM FLOWER TO GRAPE

Flowers on a grape vine are found in clusters (an inflorescence), and when the individual flowers on the inflorescence open, the small clusters make them look different than the blooms of most flowers. In the period prior to flower opening (the anthesis) flower clusters consist of a rachis, made up of all the cap stems (pedicles) and flowers which are held to the branch or shoot by a peduncle (Figure 12.17).

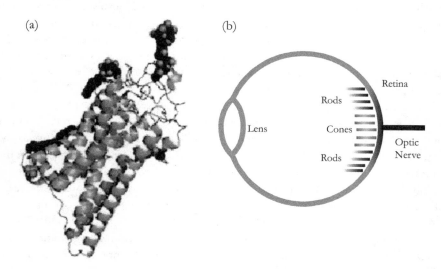

FIGURE 12.15 Representations of (a) night blindness causing g90d rhodopsin (PDB 4bez), and (b) a minimalist cartoon of a human eye. See Singhal, A.; Ostermaier, M. K.; Vishnivetskiy, S. A.; Panneels, V.; Ho, K.T.; Tesmer, J. J. G.; Veprintsev, D.; Deupi, X.; Gurevich, V. V.; Schert, G. F. X.; Standfuss, J. *EMBO Rep.* **2013**, *14*, 520. The PDB graphic is used in terms of the PDB Charter.

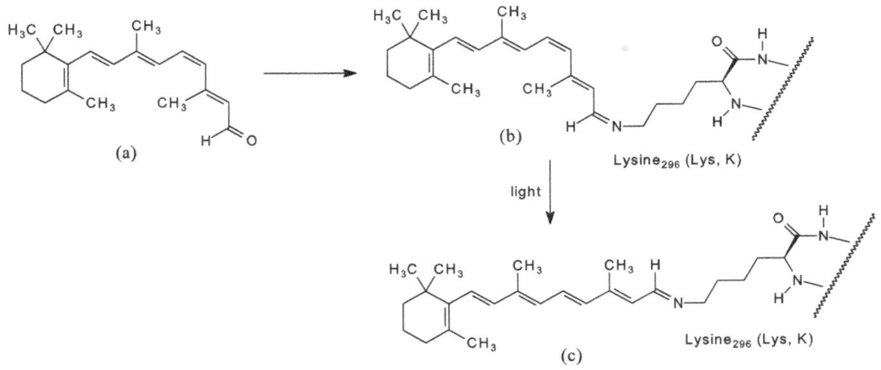

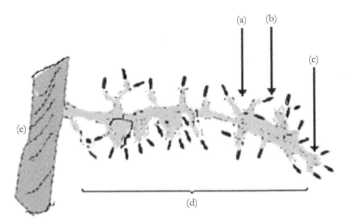

FIGURE 12.16 A cartoon representation of the conversion of 11-*cis*-retinal (a) to the corresponding imine (b) at lysine (Lys, K) 296 of rhodopsin, followed by the photochemical (hν) conversion to the imine of 11-*trans*-retinal (c). The lysine is found in the center of the helices of the enzyme. After Jäger, S.; Palczewski, K.; Hofmann, K. P. *Biochem.* **1996**, *35*, 2901.

FIGURE 12.17 A representation of a grape cluster prior to anthesis. The parts shown are: (a) an inflorescence; (b) a flower; (c) a flower attached to a pedicle; (d) a rachis; and finally (e) a peduncle. The branch to which the peduncle is attached is not shown.

When the individual flowers on a grape inflorescence open, the cap separates from the base of the flower and falls off, thus exposing the stigma and anthers. The anthers release pollen onto the stigma. Given the proximity of adjacent flowers and their rich multitude, pollinators are usually not required, and the nectary remains unused (although as many as half the flowers may fail to successfully self-pollinate). Germination of a pollen grain results in a pollen tube growing down the pistil to the ovary and entering the ovule, where a sperm unites with an egg and an embryo forms in the embryo sac (Figure 12.18). Fertilization has occurred, and "fruit set" is the result. The berry develops from the ovary. The ovule together with its enclosed embryo develops into the seed which thus contains the genetic material derived from stigma and anther. Most grape berries develop from self-fertilizing hermaphroditic flowers and thus retain approximately the same genome with which they started, maintaining the grape variety.

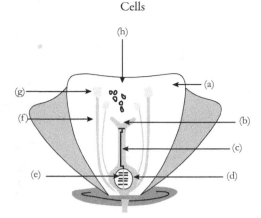

FIGURE 12.18 A cartoon of an idealized view of pollination and fertilization: (a) flower petal; (b) stigma; (c) pollen tube through which the sperm from the pollen grains pass on the way to the (d) embryo sac (ovary); (e) ovule (i.e., the small eggs, each consisting of an outer layer [integument], the nucellus, and the gametophyte); (f) the filament holding the anther (g) from which the pollen grains (h) bearing the sperm fall to the stigma (b).

The Oxford English Dictionary (OED) defines "fruit" as "the edible product of a plant or tree, consisting of the seed and its envelope, esp. the latter when it is of a juicy pulpy nature"

In the ovary of the grape there are four ovules per flower. Thus, there is a maximum potential of four seeds per fruit.

The size of the fruit is apparently related to the number of seeds. More seeds leading to larger berries. However, not all ovules successfully lead to seeds. Furthermore, fruit size can also be affected by the availability of water and other environmental conditions. Adverse conditions can also result in failure of the cluster to form (called "shatter" or "coulure") during the time (usually 1 to 3 weeks) of blooming flowers (also called "anthesis").

Not every flower on the vine gets fertilized. Depending upon the location of the vine, entomophily (fertilization by insect pollinator) can obtain, or under the right conditions anemophily (fertilization by the wind) occurs. Interestingly, in this regard, it has often been noted that the nectaries (the odor glands at the base of the flowering *V. vinifera*) do not produce nectar, and thus insects which may be around the plants and could serve as pollinators are generally not involved in pollen transfer.[18] So, wind-driven autogamy (self-fertilization) or allogamy (cross-fertilization) by adjacent blooms commonly occurs. Some unfertilized flowers will be present, and they will eventually fall from the rachis. It has been reported that only 30–60% of the flowers are commonly fertilized, with climate and the overall health of the vine playing critical roles (i.e., "high temperatures and water stress having the potential of severely reducing the amount flowers that get fertilized").[18]

Now, everything being in order, as the petals fall, signaling messages, as measured by examining the onset of new compounds forming and/or the effect of exogenous application of signaling agents, increase (beyond that involved in "petal fall"), and fruit set quickly (generally a matter of days) follows flowering.

The compounds (some of which are shown in Figure 12.19) involved at this early stage seem similar across the wide variety of different grape cultivars and many other fruit-bearing plants that have been studied. Their concentrations, dictated by the respective genomes of

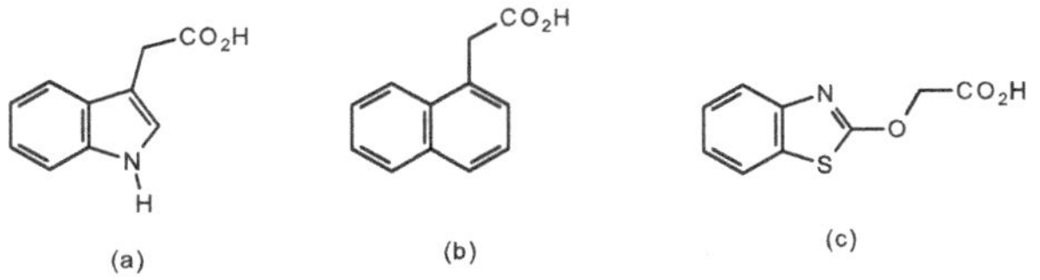

FIGURE 12.19 Representations of auxins: (a) indole-3-acetic acid; (b) naphthalene-1-acetic acid; (c) benzothiazol-2-oxyacetic acid.

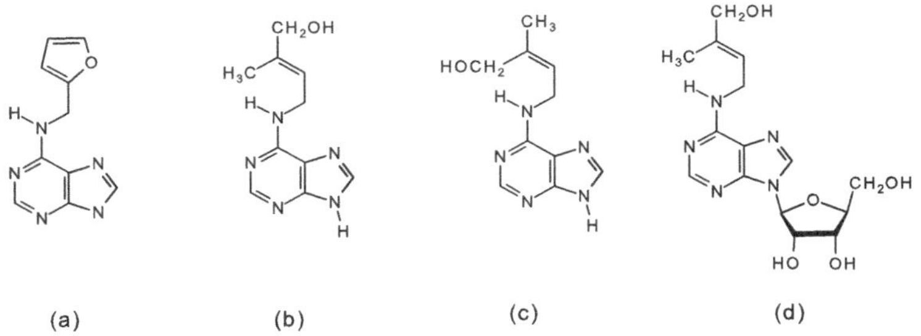

FIGURE 12.20 Representations of cytokinins (a) kinetin; (b) *trans*-zeatin; (c) *cis*-zeatin; (d) *trans*-zeatin riboside.

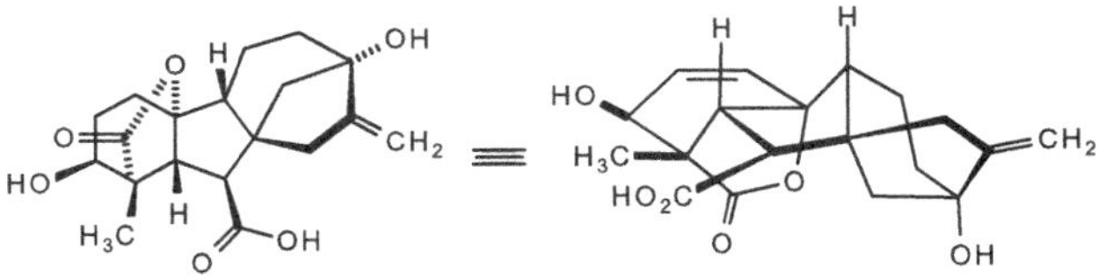

FIGURE 12.21 Representations of gibberellic acid (GA_3).

the respective cultivars, differ, and there is strong belief that not all of the compounds have been identified. And, what may be more important, very little is known about the receptors of the signals and how the signaling agents interact with them.

The materials used to transmit the signals include the naturally occurring auxin indole-3-acetic acid as well as other (unnatural) similar materials that have varying effects (e.g., naphthalene-1-acetic acid and benzothiazol-2-oxyacetic acid), the cytokinins (e.g., naturally occurring *cis*- and *trans*-zeatin, Figure 12.20), abscisic acid (Figure 6.2), gibberellins[19] (e.g., gibberellic acid, Figure 12.21), and in some cases, ethylene ($CH_2=CH_2$, Figure 6.2) (although ethylene is detected in low concentration only early in the growth cycle and grapes are generally classified as a "non-climactic" fruit whose ripening is thought to be ethylene independent).

The presence of members of the family of gibberellins is generally accounted for as arising from farnesyl diphosphate as shown in Schemes 12.1 and 12.2.

It has been pointed out[20] that oxidation of gibberellin A_{12} by cytochrome P_{450} monooxygenases and 2-oxoglutarate-dependent (oxygenases produce series of gibberellins that include the active gibberellins A_1, A_3, and A_4 (Figure 12.22).

It remains clear at this writing that while (a) the biosynthesis of the signaling agents has had much work done and is modestly well understood in a general sense in many plants and

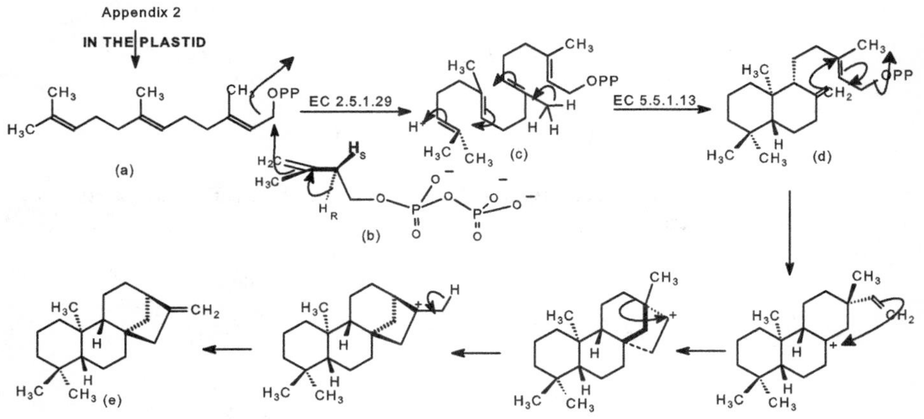

SCHEME 12.1 A brief representation of an accounting for the formation of gibberellin A_{12} from isoprenoid precursors. A discussion of the precursors and the rationale involved can be found in Appendix 2. As discussed there, in the plastid of the cell, in a reaction catalyzed by geranylgeranyl diphosphate synthase (EC 2.5.1.29), farnesyl diphosphate (a) undergoes condensation with isopentenyl diphosphate (b) with the loss of inorganic phosphate to produce geranylgeranyl diphosphate (c). In the presence of *ent*-copalyl diphosphate synthase (EC 5.5.1.13) cyclization to *ent*-copalyldiphosphate (d) takes place, and then a series of carbocation transformations leads to *ent*-kaurene (e).

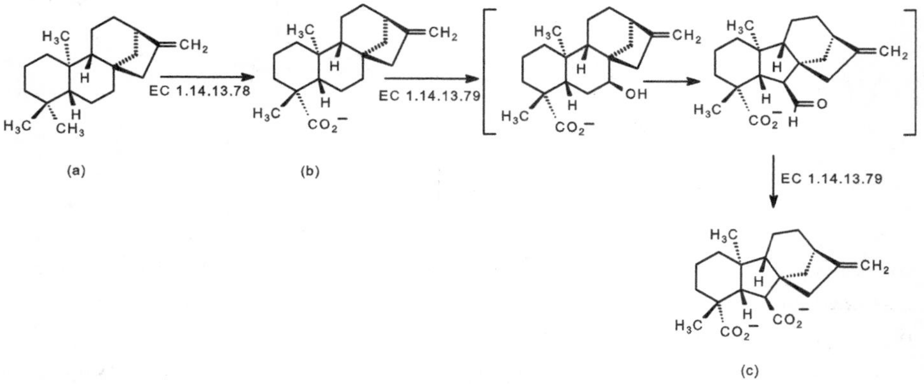

SCHEME 12.2 A brief representation of the conversion of *ent*-kaurene (a) under the influence of *ent*-kaurene oxidase (EC 1.14.13.78) using oxygen, a cytochrome P_{450} oxidase, and NAD(P)H to *ent*-kaurenoic acid (b), water, and NAD(P)$^+$. This is followed by further oxidation requiring a similar *ent*-kaurene oxidase (EC 1.14.13.79), again using oxygen, a cytochrome P_{450} oxidase, and NAD(P)H to produce gibberellin A_{12} (c).

better understood in others, and (b) the detection, by isolation from the plant or plants, of some of the signaling agents has improved and continues to do as technology improves, and (c) exogenous application of the signaling agents can be shown to have an effect, very little is known about the receptors of the signals. Indeed, it has recently been noted (in *Arabidopsis thaliana*) that, as part of biphasic reproduction, gibberellin (GA$_4$) levels are not constant and that "gibberellin acts positively and then negatively to control onset of flower formation."[21]

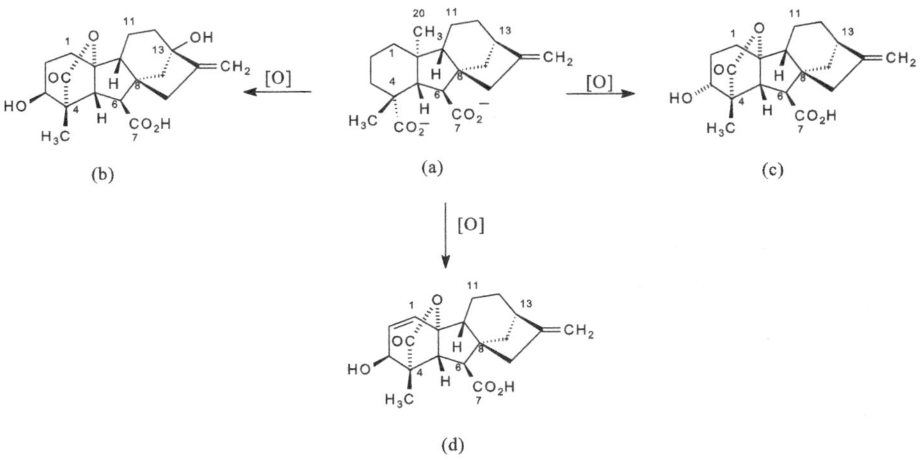

FIGURE 12.22 A representation of several gibberellins linked to oxidation of (a) gibberellin A_{12}. The details for the formation of gibberellin A_1 (b), gibberellin A_4 (c), and gibberellin A_3 (gibberellic acid) (d) are not well known. The abbreviation [O] refers to a nonspecific (i.e., details not yet known) oxidation pathway.

Nonetheless, signals having been received, the maturing fruit fills the chamber within the ovary (the locule) by cell expansion—accumulation of water and many organic compounds—all the way to the pericarp (that part of the fruit formed from the wall of the ripened ovary).

In general the filling of the locule in grapes follows a common growth pattern of many fruits. While there appears to be little cell division during anthesis, once the fruit begins to set active cell division begins.

NOTES AND REFERENCES

1. The process of removing suckers is referred to as "suckering" and is also called, simply, shoot thinning. Generally the thinning needs to be undertaken before the new shoots harden too far from where the previous season's pruning had been made.

2. The flower biology of the grapevine has been reviewed. See Meneghetti, S.; Gardiman, M.; Calò, A. *Adv. Hort. Sci.* **2006**, *20*, 317.

3. Mullins, M. G.; Bouquet, A.; Williams, L. E. *Biology of the Grapevine;* Cambridge University Press: Cambridge, UK, 1992, p 112.

4. The stem or branch that holds a group of pedicels is called a peduncle, and the main stem holding the flowers or more branches within the inflorescence is called the rachis.

5. On Friday, May 3, 2013, Courtney wrote for http://tabletograve.com the story of the Vineyard "Aphrodisiac Perfume" which leads to many additional references, such as the blog of Alice Feiring (http://www.alicefeiring.com/blog; accessed Apr 5, 2015) and many other paeans to the perfume found in the spring vineyard.

6. Buchbauer, G.; Jirovetz, L.; Wasicky, M.; Nikiforov, A. *J. Essent. Oil Res.* **1994**, *6*, 311.

7. Buchbauer, G.; Jirovetz, L.; Wasicky, M.; Nikiforov, A. *Z. Lebensmittel-Untersuchung und-Forschung* **1995**, *200*, 443.

8. Headspace can be somewhat variable as it is generally agreed to be a volume in which gas phase compounds are in equilibrium with the nonvolatile (or less volatile) sample from which they arise.

9. The claim was made by Bushdid, C; Magnasco, M. O.; Vosshall, L. B.; Keller, A. *Science* **2014**, *343,* 1370. The analytical and mathematical model was subsequently questioned by a number of investigators, including (a) Meister, M. *eLife* **2015**, e07865. DOI: 10.7554/eLife.07865; (b) Gerkin, R. C.; Castro, J. B. *eLife* **2015**, e08127. DOI: 10.7554/eLife.08127; (c) Magnasco, M.; Keller, A.; Vosshall, L. bioRxiv Preprint first posted online Jul. 6, 2015. http://dx.doi.org/10.1101/022103 (accessed Apr 5, 2017).

10. Miyasaka, N.; Arganda-Carreras, I.; Wakisaka, N.; Masuda, M.; Sümbül, U.; Seung, H. S.; Yoshihara, Y. *Nat. Comm.* **2014**, *5,* 3639. DOI: 10.1038/ncomms4639.

11. See Harkema, J. R.; Carey, S. A.; Wagner, J. G. *Toxicol. Pathol.* **2006**, *34*, 252. The exciting adventure that deals with how the signal (the odoriferous compound[s]) whose vapor is detected at the detector is handsomely described by Linda Buch (Nobel Lecture, December 8, 2004) and her prize co-recipient Richard Axel (Nobel Lecture, December 8, 2004). The story was subsequently continued and widened with details on G-protein coupled receptors, the detector(s). The latter was described handsomely by Robert Lefkowitz (Nobel Lecture, December 8, 2012) and Brian Kobilka (Nobel Lecture, December 8, 2012). For additional detectors (MS4As are not GPCRs), please see Day, S.; Stowers, L. *Cell* **2016**, *165,* 1566.

12. The G-protein coupled receptor, rhodopsin, the transmembrane region of which is held in place by a retinal isomer, is important in determining the wavelength of maximum absorption human eyes can detect.

13. The ester that gives a banana its smell is called isoamyl acetate (3-methylbut-1-yl ethanoate), and the formula for it is $CH_3COOC_5H_{11}$. The primary smell of an orange comes from octyl acetate, or $CH_3COOC_8H_{17}$. A table of ester odorants is found in Appendix 3.

14. The initial work appears as a note by Dey, S.; Stowers, L. *Cell* **2016**, *165,* 1566. (http://dx.doi.org/10.1016/j.cell.2016.06.006) referring to Greer, P. L.; Bear, D. M.; Lassance, J.-M.; Bloom, M. L.; Tsukhara, T.; Pashkovski, S. L.; Masude, F. K.; Nowlan, A. C.; Kirchner, R.; Hoekstra, H. E.; Datta, S. R. *Cell* **2016**, *165,* 1734. (http://dx.doi.org/10.1016/j.cell.2016.05.001)

15. See, e.g., http://www.newtonproject.sussex.ac.uk/prism.php?id=1 and MS "Of Colours" Add. 3975, pp. 1–22, Cambridge University Library, Cambridge, UK.

16. Clearly it is possible—and it must occur—for the absorbed energy to be reemitted. Generally, in the systems we consider here, the energy is reemitted as heat. Were it reemitted as light, either fluorescence or phosphorescence would be the result.

17. Singhal, A.; Ostermaier, M. K.; Vishnivetskiy, S. A.; Panneels, V.; Homan, K. T.; Tesmer, J. J.; Veprintsev, D.; Deupi, X.; Gurevich, V. V.; Schertler, G. F.; Standfuss, J. *E.M.B.O. Rep.* **2013**, *14,* 520.

18. Dokoozlian, N. K. in *Raisin Production Manual*, Univ. of Calif., Agricultural and Natural Resources, Publication 3393, Oakland CA, 2000.

19. The work in Appendix 2 provides a path to sesquiterpenes (C_{15} compounds). Gibberellic acid and the gibberellins are diterpenes (or derived from diterpenes) and thus C_{20} compounds. So, it is necessary, as shown in Scheme 12.1, to add yet another isopentenyl (or dimethylallyl) diphosphate to the C_{15} farnesol to make geranylgeranyl diphosphate, the C_{20} diterpene precursor. The work on gibberellin metabolism has been recently reviewed (Yamaguchi, S. *Ann. Rev. Plant Biol.* **2008**, *59,* 225).

20. Yamaguchi, S. *Ann. Rev. Plant Biol.* **2008,** *59,* 225.

21. Yamaguchi, N.; Winter, C. M.; Wu, M.-F.; Kanno, Y.; Yamaguchi, A.; Seo, M.; Wagner, D. *Science* **2014,** *344,* 638. This is noted here not because the same has been found in *V. vinifera*, but rather to indicate that simple on–off switches of growth modifiers are not always possible to determine, not the least as it may not be the case.

SECTION III
Berries

13

The Grape Berry

PART A. FRUIT SET

Beginning with fruit set (generally the grape berry is now between 1.5 and 3.0 mm, i.e., less than 1/8 of an inch in diameter) the grape berry growth is divided into three stages. Stages I and III correspond to periods of rapid growth, and the intervening slow growth phase is called Stage II. Generally the slow growth stage (Stage II) corresponds to the slowing of Stage I and the acceleration of Stage III, but it is clear that different grape cultivars have stages of different lengths even under ostensibly identical conditions.[1]

In the first stage of fruit set (also called "nouaison") the actual development of the flower ovary into the grape berry begins. The seeds in the two seed cavities (the locules) and the flesh (the pericarp) begin to take form. The pericarp separates into the exocarp (the skin with its cuticle—a thin wax coating) and the mesocarp.

The mesocarp, as it grows and divides, will eventually (by the end of Stage III) account for more than 90% of the grape's weight.

The exocarp, significantly thinner than the mesocarp, may be only five or six cells thick, and the cuticle only several layers of lipids (waxy, fatty acid esters, and compounds similar to those of cell walls and the chloroplast envelope, see pages 30 and 31).

It is in this stage that the as yet undeveloped berries are green and hard (it has been suggested that this is because chlorophyll is present and photosynthesis in the berry—as well as in leaves—is occurring). The berries are low in sugar (sucrose) but high in carboxylic acids, predominately malic acid and tartaric acid along with, generally, a lesser amount of ascorbic acid (vitamin C), hydroxycinnamic acid, and some acidic tannins (Figures 13.1 and 13.2).[2]

The grape berry structure is generally divided into three types of tissue: skin, flesh, and seed (Figure 13.3). The first, skin, as already mentioned is also known as exocarp. There is a waxy (fatty acid) cuticle external to the skin which both protects the skin from damage and serves as an additional barrier (along with the skin) to water loss as the berry ripens. Avoiding water loss is both beneficial in the ripening process (e.g., to hold carbohydrates and protect

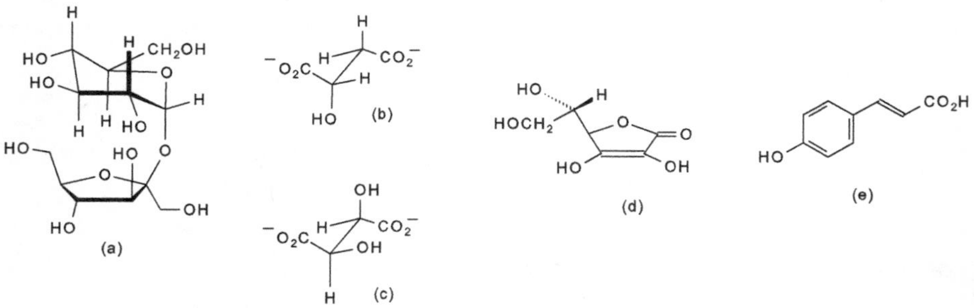

FIGURE 13.1 Representations of early berry constituents: (a) sucrose (α-D-glucopyranose bonded via an α1→β2 bond to β-D-fructofuranose); (b) the anion of L-(−)-malic acid [L-(−)-malate]; (c) the anion of L-(+)-tartaric acid [L-(+)-tartrate]; (d) ascorbic acid (vitamin C); and (e) *para*-coumaric acid, i.e., *para*-hydroxycinnamic acid.

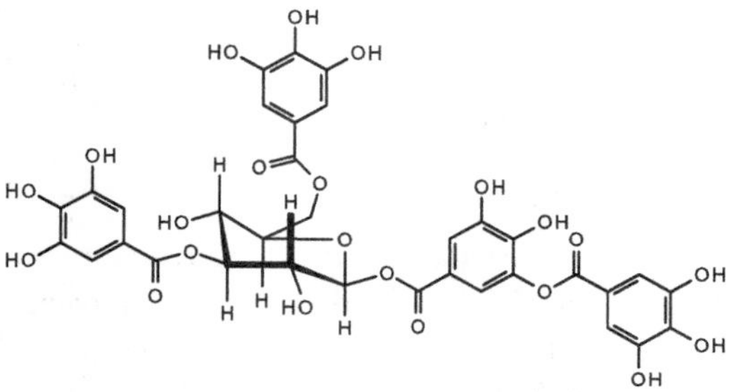

FIGURE 13.2 A representation of a tannic acid. Four (4) gallic acid units are involved: three (3) directly bonded as esters to a glucopyranoside, while the fourth is bonded (on the right of the drawing) to the phenolic unit of a gallic acid as an ester.

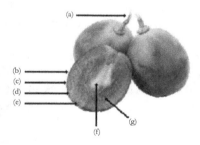

FIGURE 13.3 A representation of grape berries. Regardless of skin color, yellow, green, red, purple, etc. all berries have (a) the pedicel that remains; (b) a waxy cuticle, external to the skin; (c) the skin (exocarp); (d) the layer immediately beneath the skin, the hypodermis; (e) vascular tissue along the periphery feeding skin, etc. growth; (f) the seed, and; (g) the flesh of the berry. Retouched photo by the author.

the seed) and, occasionally, deleterious, since water loss through hindered evaporation cannot ameliorate the effect of rising temperatures, and the fruit can split.

The hypodermis lies just below the skin—at the top of the flesh. The flat cells of the hypodermis accumulate the phenols, some acids, higher-molecular-weight alcohols, terpenes, tannins, etc. as the berry ages. These products will eventually be released when the grapes are crushed and they can be extracted into the juices. Compounds are elaborated in the peripheral vascular tissue that lend to the astringent notes of the "mouthfeel" of both the grape and the final product. Other components found here in the hypodermis include anthocyanins and some terpenes, sesquiterpenes, diterpenes, and esters responsible for the flavors and fragrances of the fruit.

The cells beneath the peripheral vascular tissue, the flesh of the berry, tend to be more full than those closer to the surface and it is in these cells that water and sucrose accumulate, the latter being broken down after transport into glucose and fructose. Tartaric and malic acids along with citric and succinic acids also accumulate here. In many berries tartaric acid (Figure 13.1) eventually predominates.

The vascular tissue is connected to the vascular system of the plant through fine capillary lines. The phloem cells in the vascular system make up the transport entity for moving sugars, other organic compounds, and water into and out of the vascular tissue. The xylem cells are used for moving salts (magnesium [Mg^{2+}], calcium [Ca^{2+}], potassium [K^+], etc. cations and phosphate [PO_4^{3-}], sulfate [SO_4^{2-}], etc. anions), other minerals, and water. The vascular tissue is going to swell and provide the sweet juice of the grape and sustenance to the animals feeding on the grapes as well as the food for the yeast that will digest the sugar to produce the ethanol.

Interestingly it appears that the phloem and xylem do not work equally during the ripening process. Thus, early in ripening the phloem, transporting nutrients from the canopy is more involved than the xylem. The roles reverse later with the xylem bringing nutrients from the roots.[3]

As shown in Figure 13.1, sucrose is a dimer composed of the hexose isomers glucose and fructose. The formation of fructose and glucose from smaller fragments has been thoroughly studied and is summarized as in the term gluconeogenesis.

Gluconeogenesis is the opposite of glycolysis, the metabolic consumption of glucose. The yeast that converts glucose to ethanol is effecting glycolysis, and while the result is different, digestion of glucose to smaller fragments that are subsequently used to other purposes is also glycolysis.

The pathways of glycolysis and gluconeogenesis have been summarized numerous times. A good overall picture is available in the Kyoto Encyclopedia of Genes and Genomes (www.genome.jp/kegg; map 00010; glycolysis and gluconeogenesis), and some of what is known is summarized below.

The carboxylation of pyruvate to oxaloacetate (pyruvate carboxylase, EC 6.4.1.1) is commonly considered as a reasonable beginning for gluconeogenesis, but it is inappropriate here because that pathway, requiring biotinylated proteins (i.e., proteins to which biotin is attached), is common to yeast and animal tissues, but not plants. For plants, in the presence of rubisco (ribulose-1,5-bisphosphate carboxylase/oxygenase, EC 4.1.1.39), it is reasonable to begin gluconeogenesis with reaction of carbon dioxide (CO_2) with D-ribulose-1,5-bisphosphate to produce 3-phosphoglycerate as it occurs in the Calvin cycle (Appendix 2).

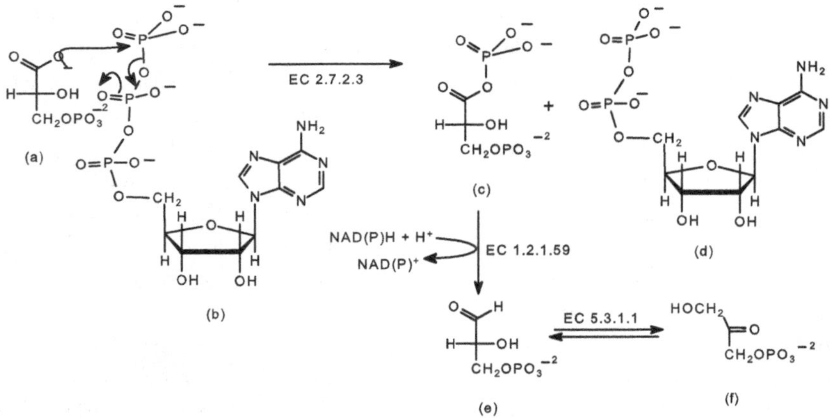

SCHEME 13.1 An accounting for the conversion of 3-phosphoglycerate (a) on reaction with adenosine triphosphate (b) in the presence of phosphoglycerate kinase (EC 2.7.2.3) to yield the mixed anhydride of glyceric acid with phosphoric acid, 1,3-diphosphoglycerate (c), and adenosine diphosphate (ADP) (d). In a separate step, the 1,3-diphosphoglycerate (c) is shown to be reduced to glyceraldehyde 3-phosphate (e), while reduced nicotinamide, NAD(P)H is oxidized to NAD(P)$^+$ in a reaction catalyzed by glyceraldehyde 3-phosphate dehydrogenase (EC 1.2.1.59). Also shown is the product of the isomerization of glyceraldehyde 3-phosphate (e) to dihydroxyacetone monophosphate (f) as catalyzed by triosphosphate isomerase, EC 5.3.1.1.

As shown in the representations in Scheme 13.1, 3-phosphoglycerate, on reaction at the carboxylate terminus with adenosine triphosphate (ATP) in the presence of phosphoglycerate kinase (EC 2.7.2.), yields an anhydride of phosphoric acid and adenosine diphosphate (ADP). Enzyme-catalyzed reduction (glyceraldehyde 3-phosphate dehydrogenase, EC 1.2.1.59) at the phosphorylated carboxyl with loss of phosphate produces glyceraldehyde 3-phosphate.

Isomerization of 3-phosphoglyceraldehyde in the presence of the enzyme triosphosphate isomerase (EC 5.3.1.1), presumably occurring through the common enol (Scheme 13.2), produces the three-carbon isomer dihydroxyacetone monophosphate. These two three-carbon fragments, 3-phosphoglyceraldehyde and dihydroxyacetone monophosphate, when added together, meet the requirement for formation of a six-carbon hexose product (3 + 3 = 6).

That process, shown as an aldol-type condensation catalyzed by fructose bisphosphate aldolase (EC 4.1.2.13) between the two fragments results in the formation of D-fructose-1,6-bisphosphate.

A cartoon representation, of a potential pathway for the condensation to D-fructose-1,6-bisphosphate followed by loss of phosphate (fructose bisphosphate, EC 3.1.3.11) and then isomerization of the fructose 6-phosphate to glucose 6-phosphate (EC 5.3.1.9) via the common enol is shown as Scheme 13.3.

Finally, coupling between glucose and fructose (sucrose phosphorylase, EC 2.4.1.7) produces sucrose (Scheme 13.4).

It has been pointed out that "Contrarily to tartaric acid, malic acid levels vary greatly as berries develop and mature."[2] L-Malate is a dicarboxylic acid member of the tricarboxylic

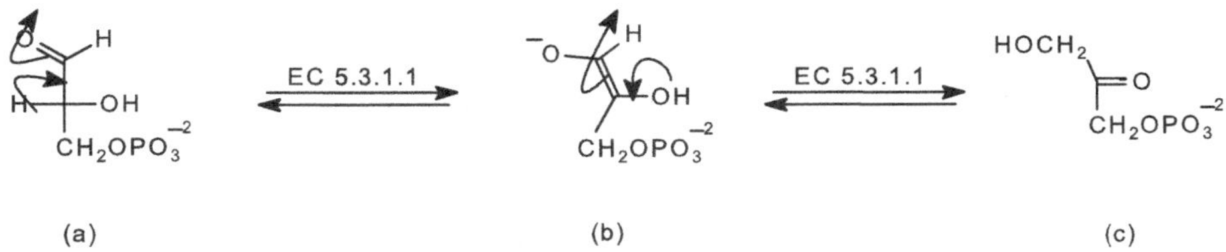

SCHEME 13.2 A cartoon representation, using curved arrows, accounting for the isomerization of 3-phosphoglyceraldehyde (a) to dihydroxyacetone monophosphate (c) via the presumed common enol (b). The presence of the enzyme triose phosphate isomerase (EC 5.3.1.1) is required.

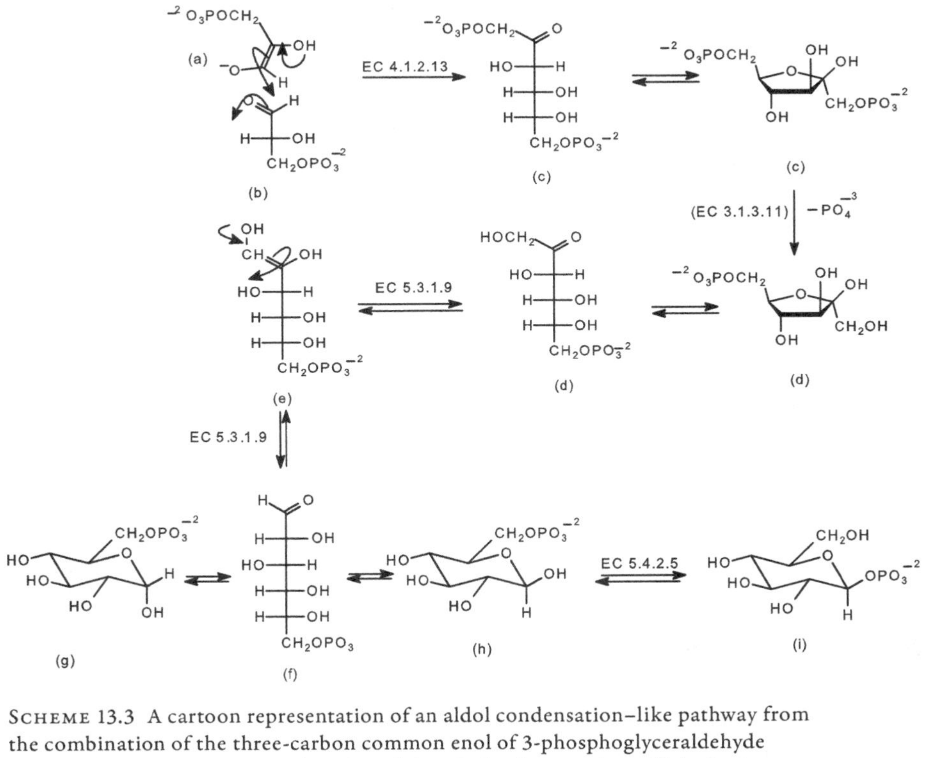

SCHEME 13.3 A cartoon representation of an aldol condensation–like pathway from the combination of the three-carbon common enol of 3-phosphoglyceraldehyde and dihydroxyacetone mono-phosphate (a) with the three-carbon aldehyde, 3-phosphoglyceraldehyde (b), to produce D-fructose-1,6-bisphosphate (c), a six-carbon compound. The enzyme fructose bisphosphate aldolase (EC 4.1.2.13) is required. D-fructose-1,6-bisphosphate (c) is shown in cyclic and acyclic forms. Loss of phosphate from D-fructose-1,6-bisphosphate (c) (fructose-bisphosphatase, EC 3.1.3.11) yields β-D-fructose-6-phosphate (d) also shown in cyclic and acyclic forms. In the presence of glucose 6-phosphate isomerase, EC 5.3.1.9, and perhaps through the common enol (e), D-fructose-6-phosphate undergoes isomerization to glucose 6-phosphate (f). The open chain (f) as well as the cyclic anomeric isomers α-D-glucopyranose 6-phosphate (g) and β-D-glucopyranose 6-phosphate (h) and the isomeric β-D-glucopyranose 1-phosphate (i) are formed from (h) on the action of phosphoglucomutase, EC 5.4.2.5.

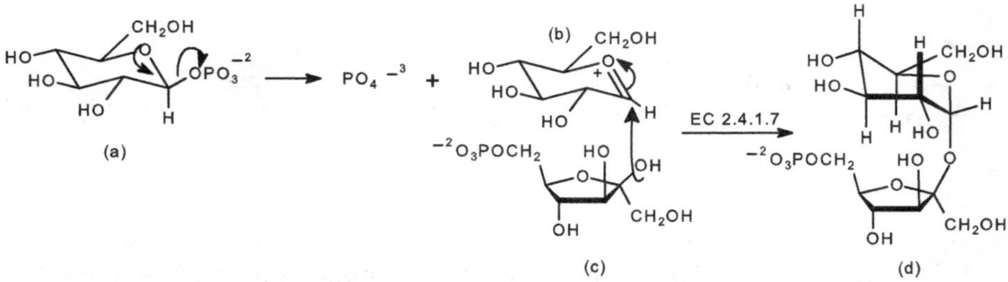

SCHEME 13.4 A representation of the conversion of β-D-glucopyranose-1-phosphate (1) to an oxonium ion (b) by elimination of phosphate and the subsequent reaction of that ion with β-D-fructose-6-phosphate (c) with the axial hydroxyl at the anomeric carbon of the latter. The reaction, in the presence of sucrose phosphorylase (EC 2.4.1.7) leads to sucrose (d) (with an α1→β2 linkage as shown).

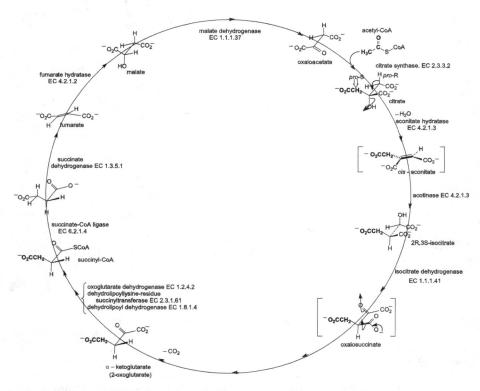

FIGURE 13.4 A representation of the tricarboxylic acid (Krebs) cycle.

acid (Krebs) cycle and, as with other members of other cycles, can be utilized by withdrawing some of it from the cycle while permitting enough to remain to allow the cycle to turn (Figure 13.4).

Additionally, L-malate can be generated (Scheme 13.5) by carboxylation of phosphoenolpyruvate (phosphoenolpyruvate carboxykinase, EC 4.1.1.49) to produce oxaloacetate and subsequent reduction (malate dehydrogenase, EC 1.1.1.37) of the latter. A representation of the tricarboxylic acid cycle (Krebs cycle) is shown in Figure 13.4.

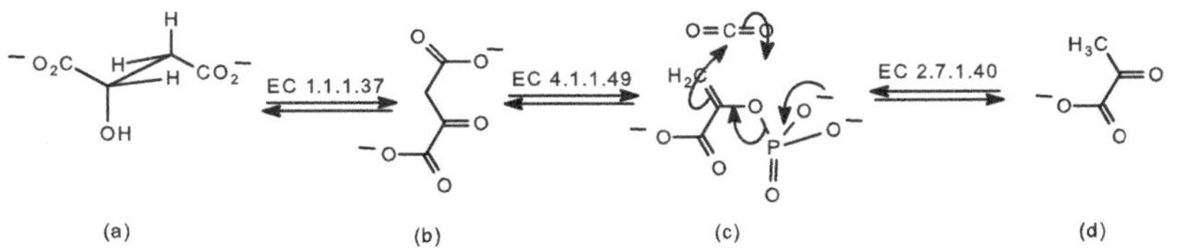

SCHEME 13.5 A representation of a pathway to L-malate in addition to that shown in the tricarboxylic acid (Krebs) cycle. Thus, in addition to hydration of fumarate (fumarate hydratase EC 4.2.1.2) as shown in Figure 13.4, L-malate (a) also results on reduction of oxaloacetate (b) (malate dehydrogenase, EC 1.1.1.37). And since oxaloacetate (b) can be formed from carboxylation of phosphoenol pyruvate (PEP) (c) (phosphoenolpyruvate carboxykinase, EC 4.1.1.49) as well as by oxidation of L-malate (a), the regulation of L-malate is not completely dependent upon the tricarboxylic acid (Krebs) cycle. Phosphoenol pyruvate (PEP) results from phosphorylation (ATP → ADP) of pyruvate (d) (pyruvate kinase, EC 2.7.1.40) (See Appendix 2).

In response to the reasonable question as to what happens to the malate to allow it to disappear, the same reference[2] notes "cytosolic and mitochondrial isoforms of malate dehydrogenase would respectively participate in malate synthesis and catabolism in response to metabolic changes occurring during grape development."

Tartaric acid, once formed, does not apparently disappear rapidly through degradation although derivatives at both the hydroxyl and carboxyl groups may be formed from it.

Interestingly, the main path to L-tartaric acid appears to begin with ascorbic acid (vitamin C), itself derived from glucose.

To form the ascorbic acid, it has been generally agreed that the Smirnoff-Wheeler pathway[4] is followed even though the quantities isolated are small (perhaps because it is further metabolized to tartaric acid). Nonetheless, it currently appears that tartaric acid lies at the end of a series of reactions beginning with D-glucose and passing through ascorbic acid.

Briefly, D-glucose is phosphorylated (glucokinase, EC2.7.1.2) to D-glucose 6-phosphate, and the latter is presumed to undergo tautomerization (as above) to D-fructose 6-phosphate (phosphoglucose isomerase, EC 5.3.1.9). Reopening of the hemiketal and, again through the common enol, tautomerization in the presence of phosphomannose isomerase (EC 5.3.1.8) leads to mannose 6-phosphate (Scheme 13.6).

Mannose 6-phosphate isomerizes to mannose 1-phosphate (phosphomannose mutase, EC 5.4.2.8) which, as shown in Scheme 13.7, is then converted to the guanosine diphosphate derivative (GDP-mannose pyrophosphorylase, EC 2.7.7.13).

It is likely that, as with other phosphohexomutases, the catalysis for the intramolecular phosphoryl transfer from C-6 to C-1 involves a phosphoserine residue on the enzyme and requires Mg^{2+} for full activity. The process is believed to involve transfer of a phosphoryl group from the phosphoserine residue to C-1 to produce a bisphosphorylated intermediate. Then, after the initial transfer of the phosphoryl group from the enzyme, the reaction is believed to proceed via diffusional reorientation of the bis-phosphorylated intermediate in the active site, followed by a second phosphoryl transfer to regenerate the phosphorylated form of the enzyme using the phosphoryl transfer from C-6 to the serine residue. The mannose 1,6-bisphosphate is required to maintain the enzyme in the phosphorylated state.[5] Finally, the conversion to the guanosine diphosphate derivative involves the enzyme GDP-mannose pyrophosphorylase (EC 2.7.7.13) and the conversion of GTP to the GDP derivative and inorganic phosphate.[6]

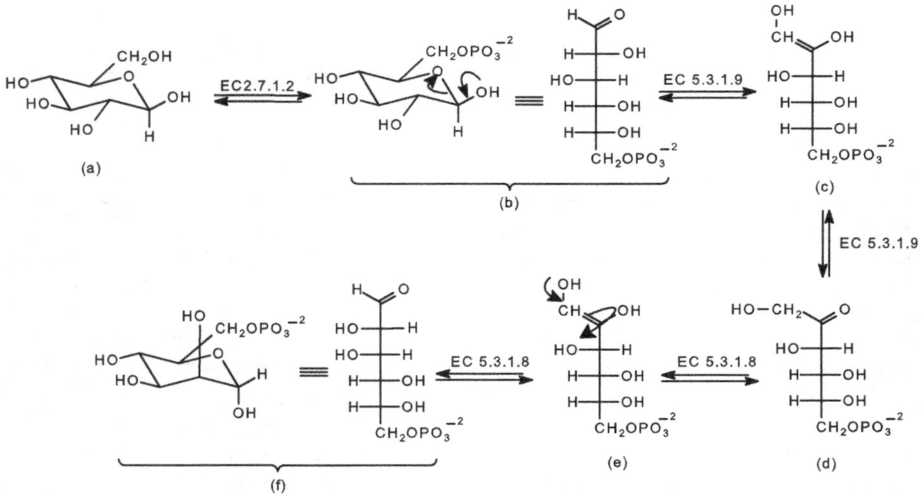

SCHEME 13.6 A representation of the conversion of β-D-glucopyranose (a) as the 6-phosphate (b) to β-D-manopyranose 6-phosphate. Thus, β-D-glucopyranose (a) undergoes phosphorylation (glucokinase, EC 2.7.1.2) to the corresponding β-D-glucopyranose 6-phosphate (b). Isomerization to the common enol (c) (glucose 6-phosphate isomerase, EC 5.3.1.9) results in the formation of D-fructose 6-phosphate (d) from which the common enol (e), now in the presence of the enzyme phosphomannose isomerase (EC 5.3.1.8) results. Picking up the proton at C2 in the presence of this isomerase (EC 5.3.1.8) now results in the formation of D-mannose 6-phosphate (f).

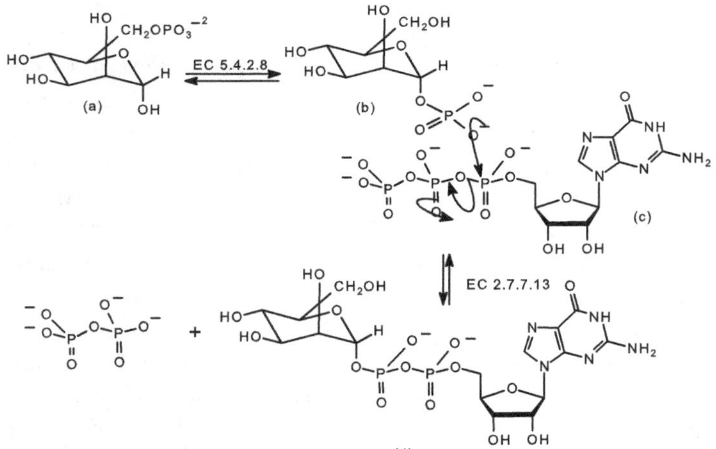

SCHEME 13.7 A representation of the conversion of β-D-mannopyranose 6-phosphate (a) to the corresponding β-D-mannopyranose 1-phosphate (b) with phosphomannose mutase (EC 5.4.2.8) participation followed by reaction of the latter with guanosine triphosphate, GTP, (c) in the presence of mannose-1-phosphate guanyltransferase (EC 2.7.7.13) to yield GDP-mannopyranose (d).

The isomerization of D-mannose to L-galactose has been shown to involve oxidations and reductions, and a brief outline of a proposed pathway is presented in Scheme 13.8.[7]

In order for the conversion of D-glucose into L-ascorbic acid and thence to L-(+)-tartrate to continue, the obligatory conversion of GDP-1-L-galactopyranose to L-galactose 1-phosphate is required.[8] The conversion was shown (in *Arabidopsis thaliana*) to involve the protein encoded by a gene called VTC2 (**V**ascular **T**ransport **C**haperone) which has homologs in "plants, invertebrates, and vertebrates."[8]

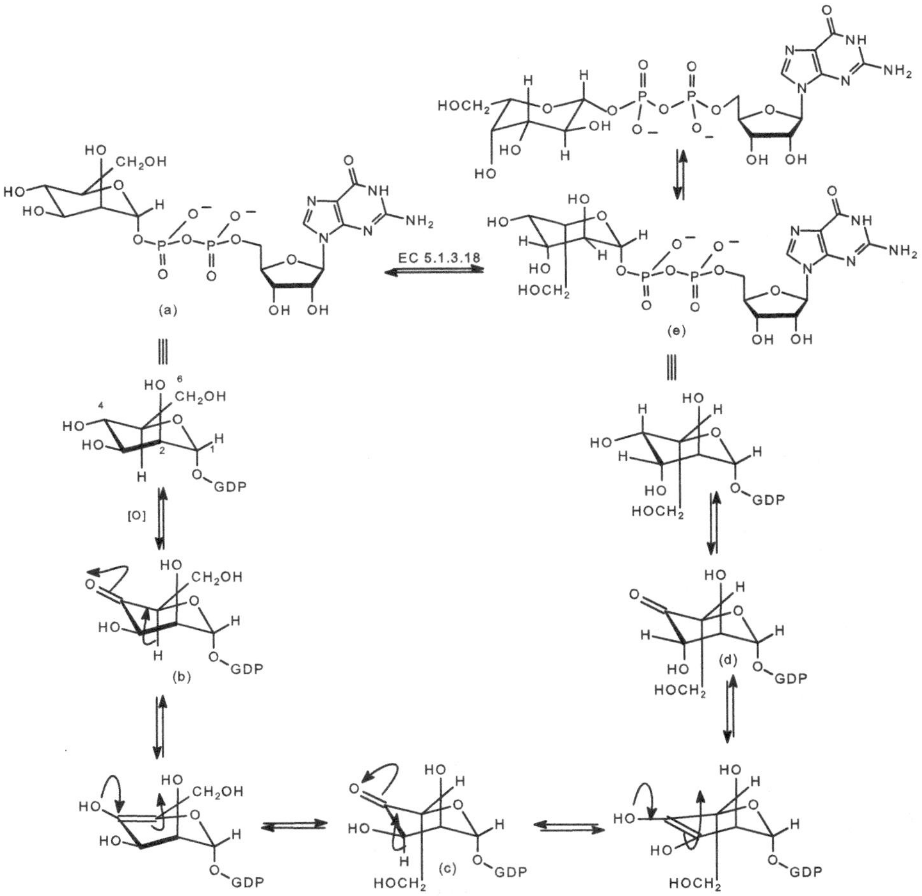

SCHEME 13.8 A representation of a potential pathway for the epimerization of D-mannose to L-galactose using GDP-mannose-3,5-epimerase (EC 5.1.3.18). Beginning with GDP-1-α-D-mannopyranose (a) oxidation at C-4 to the corresponding ketone (b), followed by enolization with loss of the proton at C-5 and reprotonation from the opposite face, leads to epimerization at C-5 with placement of the carbinol unit axial as shown in (c). Now, proton loss again, but this time at C-3 of the ketone (c), allows for reprotonation and epimerization at C-3 and placement of the hydroxyl at C-3 axial to yield (d). Reduction of the carbonyl back to the equatorial alcohol produces GDP-1-L-galactopyranose (e).

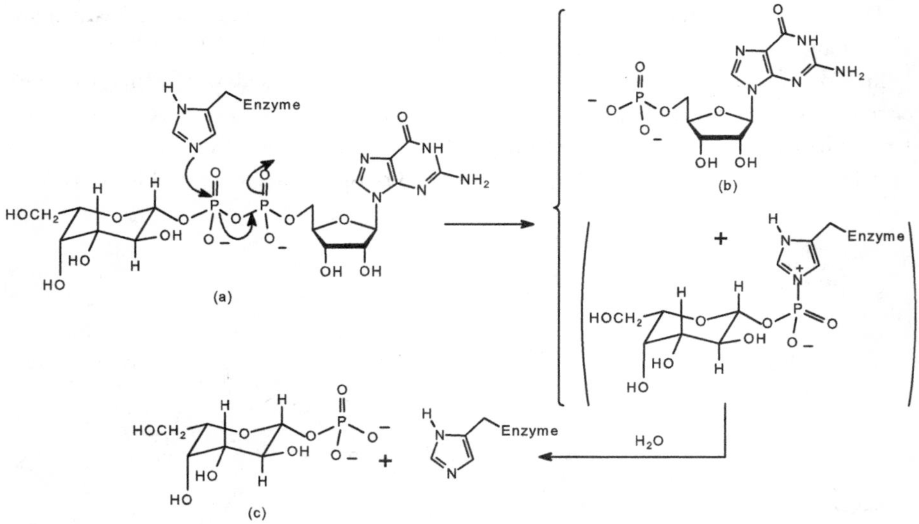

SCHEME 13.9 A representation of a potential pathway, using an unnamed histidine-bearing protein from a vascular transport protein gene for the conversion of GDP-1-L-α-galactopyranose (a) into guanosine monophosphate (GMP) (b) for recycling and also obtaining L-α-galacto-pyranose 1-phosphate (c) for oxidation to the galactonolactone needed for ascorbic acid (vitamin C) (Scheme 13.10).

The product of the VTC2 gene is characterized as a member of the histidine triad superfamily of enzymes, which consists of monophosphates with various substituents and leaving groups attached to them.[9] An example is shown in Scheme 13.9.

In a similar way then, the phosphate is removed from the hydroxyl at C-1 of the GDP-1-L-galactopyranose (the product of the VTC4 gene) to produce L-α-galactopyranose.

Oxidation of L-galactose (L-galactose 1-dehydrogenase, EC 1.1.1.316) results in formation of L-galactono-1,4-lactone. Presumably, either opening of the pyranose to the aldehyde, oxidation to the corresponding carboxylic acid, and recyclization to the five-membered lactone or oxidation without opening ($NAD^+ \to NADH + H^+$) attack on the carbonyl of the 6-membered lactone by the axial hydroxyl at C-4 occurs (Scheme 13.10).

With the L-galactono-1,4-lactone in hand, the oxidation to L-ascorbic acid can be effected (Equation 13.1). The structure of the enzyme required, L-galactono-1,4-lactone dehydrogenase (EC 1.3.2.3) has not yet been entered into the Enzyme Database (www.enzyme-database.org) if, indeed, it has been solved. However, as noted in the Enzyme Database, the conversion involves the reduction of a ferricytochrome c to ferrocytochrome c + H^+ (i.e., reduction of ferric iron [Fe^{3+} or Fe(III)] to ferrous iron [Fe^{2+} or Fe(II)]) to accompany the oxidative introduction of the double bond. Details are not known, but a picture has been drawn in the KEGG database of likely iron-porphyrin systems (see www.genome.jp/Fig/reaction/R00640.gif).

L-Tartaric acid is derived from L-ascorbic acid (Scheme 13.11).[10,11]

The proposals for the pathway suggest that opening of the lactone and formation of the ketone from the enol of ascorbic acid produces 2-keto-L-gulonic acid. Reduction of the

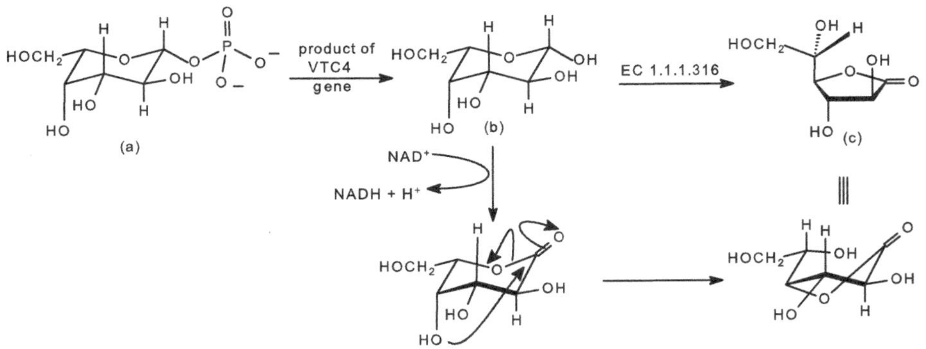

SCHEME 13.10 A representation of a potential pathway for the conversion of L-α-galacto-pyranose 1-phosphate (a) to galactose (b) utilizing the product of the VTC4 gene for de-phosphorylation of the initial phosphate ester (a) and from there to the corresponding L-galactono-1,4-lactone (c) using the NAD^+ cofactor of the L-galactose 1-dehydrogenase enzyme, EC 1.1.1.316, to effect the oxidation.

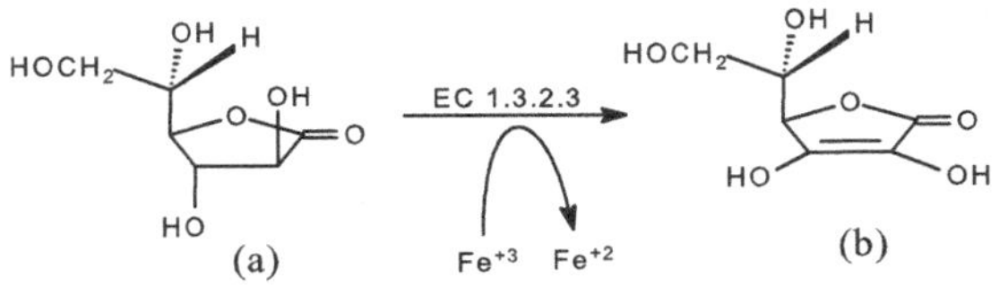

EQUATION 13.1 A representation of the oxidation of L-galactono-1,4-lactone (a) to L-ascorbic acid (vitamin C) (b) with the enzyme L-galactono-1,4-lactone dehydrogenase (EC 1.3.2.3).

carbonyl group (presumably $NADH^+ \rightarrow NAD + H^+$ although the evidence is sparse) then generates L-idonic acid. Oxidation at C-5 to generate the corresponding ketone produces 5-keto D-gluconic acid (!). Cleavage by a presumed transketolase then produces glycoaldehyde and L-*threo*-tetruronate. The latter is readily oxidized to L-tartaric acid.

Interestingly, although it is clear that ascorbic acid serves as a precursor to tartaric acid, the obvious question as to whether 5-keto-D-gluconic acid might be generated in an alternate way from D-glucose or, indeed, from another species present in the Calvin cycle, has received only cursory attention.[12]

Recently, the role of ascorbic acid and its conversion to tartaric acid as well as to oxalic acid has been reconsidered. It was pointed out "whether the use of ascorbic acid for biosynthetic purposes is regulated by developmental, environmental or tissue-specific cues remains unknown."[13]

The last two types of compounds listed as commonly present in the as yet undeveloped berries, are the coumaric acid derivative *para*-hydroxycinnamic acid and the glucose gallic acid ester, tannic acid. They are both derived from dehydroshikimic acid (see Appendix 4). For the latter (Scheme 13.12), the derivation is only in part since the glucose which serves as the esterifying alcohol of the tannic acid can be derived directly from the Calvin cycle, while 3-dehydroshikimic acid serves as the precursor to the esterifying acid (gallic acid). Similarly, 3-dehydroshikimate on reduction of the keto carbonyl to the corresponding

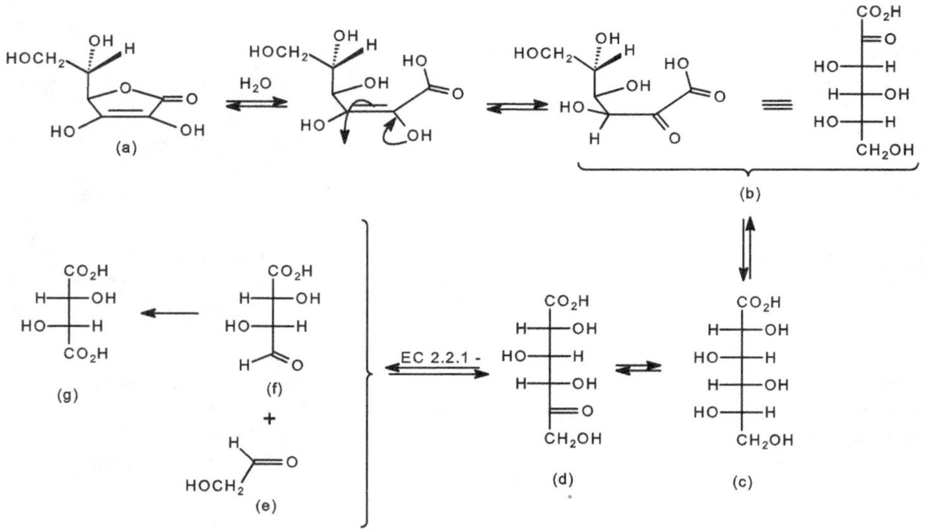

SCHEME 13.11 A representation of the path from L-ascorbic acid (vitamin C) (a) *via* hydrolysis to 2-keto-L-gulonic acid (b) and the subsequent reduction of the keto group of 2-keto-L-gulonic acid (b) at C-2 to yield L-idonic acid (c). Oxidation of L-idonic acid at C-5, to a carbonyl group, destroying the chirality, produces the same ketocarboxylic acid that would be obtained from the corresponding gluconic acid, viz., 5-keto-D-gluconic acid (d). Then a transketolase from the group EC 2.2.1. results in a retroaldol reaction to yield glycoaldehyde (e) and L-*threo*-tetruronate (threuronic acid) (f). Oxidation of the aldehyde group of L-*threo*-tetruronate yields L-tartaric acid (g).

alcohol, followed by dehydration and reaction with pyruvate (Appendix 4) yields chorismate which is the nonaromatic precursor to both tyrosine (Tyr, Y) and phenylalanine (Phe, F). Coumaric acid is derived from those amino acids (Scheme 13.13).

Finally, with regard to undeveloped berries, it is not surprising that they also contain minerals, other amino acids [in addition to the phenylalanine (Phe, F) and tyrosine (Tyr, Y) noted above], traces of metals, vitamins, halides, and numerous micronutrients found in the growing environment or added by those caring for the vineyard. That is, the small variations in the concentrations of these and related compounds are in part a function of the "*terroir*" as well as manipulation of the soil and other growing conditions. Indeed, it is for these reasons, as well as use of the techniques passed from generation to generation in established vineyards, that the same cuttings transposed to new environments will produce grapes that are capable of being distinguished from the original and, eventually, wines that differ in quality.

It is hoped it is clear that, while only a little is truly known about the larger changes, those more subtle and yet vital to the production of suitable beverage remain to be found.

Now, with rapid growth Stage I having been completed in the berry, the slow growth phase Stage II that allows the plants to prepare (gathering strength, as it were) for the acceleration of Stage III begins. As noted earlier, different grape cultivars have stages of different lengths even under ostensibly identical conditions, and the period of slow growth might last for as little as a week or as long as several weeks (depending upon the cultivar and the *terroir*). When the lag stage has ended, Stage III, the "last" stage, of berry growth and development

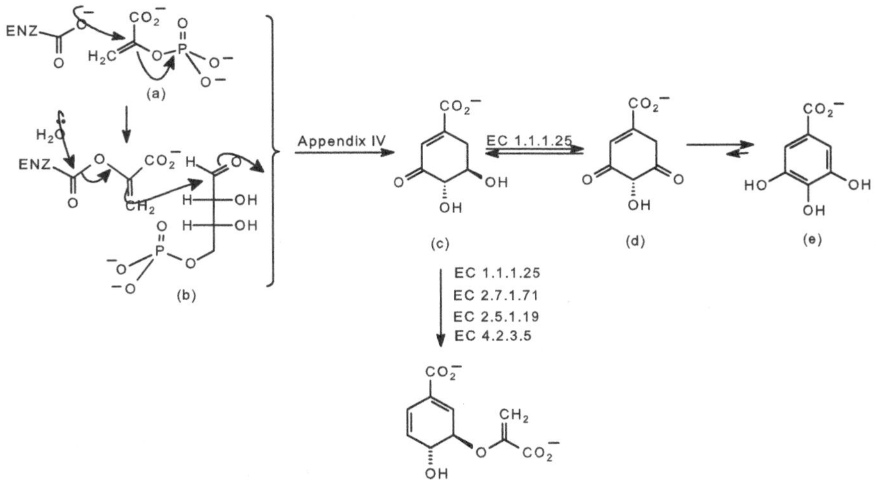

SCHEME 13.12 An abbreviated cartoon version (for more detail see Appendix 4) of the conversion of phosphoenol pyruvate (a) and erythrose 4-phosphate (b) to 3-dehydroshikimate (c). Oxidation of 3-dehydroshikimate (c) can generate 3,5-didehydroshikimate (d) a proton tautomer of gallic acid (e). Alternatively, reduction of 3-dehydroshikimate (c) (shikimate dehydrogenase, EC 1.1.1.25) followed by a series of enzymatic transformations involving phosphorylation (shikimate kinase, EC 2.7.1.71), reaction with phosphoenol pyruvate (a) (3-phosphoshikimate 1-carboxyvinyltransferase, EC 2.5.1.19), and elimination of phosphoric acid (chorismate synthase, EC 4.2.3.5) yields chorismate (f).

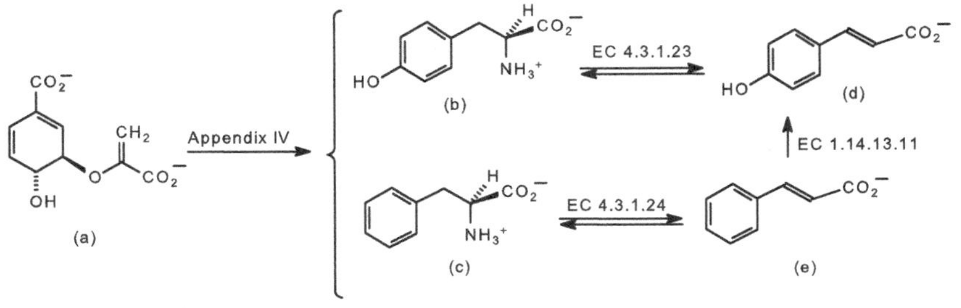

SCHEME 13.13 A representation of chorismate's (a) conversion to both tyrosine (Tyr, Y) (b) and, separately, to phenylalanine (Phe, F) (c) and the subsequent conversion of each of them to coumaric acid (d). Details, including intermediates that have been identified and the specific enzymes involved in the formation of the amino acids from chorismate are found in Appendix 4. Tyrosine (Tyr, Y) (b), on reversible loss of ammonia (NH_3) and in the presence of tyrosine ammonia lyase (EC 4.3.2.23), yields coumaric acid (d). Phenylalanine (Phe, F) (c), on reversible loss of ammonia (NH_3) and in the presence of phenylalanine ammonia lyase (EC4.3.1.24), yields *trans*-cinnamate (e), which subsequently in the presence of *trans*-cinnamate 4-monooxygenase (EC 1.14.13.11), also yields coumaric acid (d).

and the onset of berry ripening and softening (véraison) finally sets in. Exciting recent high-resolution electron microscopy results show morphological changes as growth phases change (albeit in another plant).[14]

PART B. VÉRAISON

At the beginning of véraison (ripening and softening) the grapes have very little glucose and fructose (sucrose) but they are rich in malic and tartaric acids as well as other simple acids (e.g., mevalonic, quinic, shikimic, prephenic) that serve as progenitors of the more fully elaborated terminal carboxylic acids (e.g., coumaric, ascorbic, gallic, syringic, and other phenolic and phenylpropenoic acids) (Figure 13.5).

As water (H_2O) and some minerals with mono- and divalent cations such as potassium (K^+), magnesium (Mg^{2+}), calcium (Ca^{2+}), and sodium (Na^+) and anions such as sulfate (SO_4^{2-}), phosphate (PO_4^{3-}), chloride (Cl^-), and nitrate (NO_3^-) move through the woody xylem, and the aqueous sucrose solution and related water-soluble carbohydrates move through the phloem, the grapes begin to grow. It can now be appreciated how hard the root system must work to bring in sufficient water for the grapes while, in

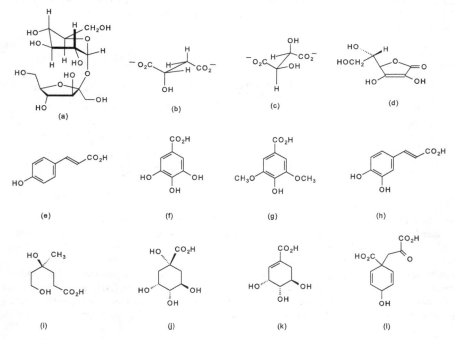

FIGURE 13.5 A representation of some of the components known to present in most grape varieties at the beginning of véraison. These include: (a) sucrose; (b) L-malate; (c) L-tartrate; (d) ascorbic acid; (e) para-coumaric acid; (f) gallic acid; (g) syringic acid; (h) caffeic acid; (i) mevalonic acid; (j) quinic acid; (k) shikimic acid; and (l) prephenic acid.

the warm sunshine, the leaves lose water through transpiration (water loss from plants through evaporation) and guttation (water loss from plants as a consequence of pruning and damage). Since growth of the fruit is a measure of growth of the component cells, it can be argued that a valid approach to describing fruit growth is to quantify water balance (net growth = flow in − flow out).[15] Of course such an attempt needs to take into account the exchange of water between all of the plant constituents serviced by the phloem and xylem tissues, not only the tissues of the fruit and its pedicel. Inflow of water to the fruit and leaves as well as other elements must come from the parent plant and, in principle, might be easily measured. Outflow via transpiration or guttation is more difficult to measure.[15]

In short, water balance is critical because, lacking sufficient water, the grapes will not swell, sucrose, minerals, and growth factors will fail to be transported, and the crop will die. Experimentally, berries deprived of water leads to smaller grapes and, presumably, to wine of lesser quality since the berry will have higher concentrations of the solutes present before minerals, sucrose, other carbohydrates, and related materials are added.

It is at this time that intensive field work in regard to the canopy is also taking place. For the most part, this work will take place well before the harvest and during the nearly 2 months (July and August north of the equator and January and February in the south) of ripening after fruit set. Additionally, depending upon the variety of grape, the density of berries on the rachis (which may vary from year-to-year), and the skill of the grower, the canopy care will vary.

As noted earlier (page 38), differences in the concentration of some of the constituents of the grape (as subsequently measured in the wine itself) have been measured as a function of base and head leaf trimming.

It is also during this time that aggressive agricultural pest management may be deployed. Indeed, it is common to utilize a fumigant (generally, an easily vaporized material or a material that can be dissolved in a suitable solvent—such as water—for spraying as a vapor and used as a pesticide and/or disinfectant) by driving a spraying machine or using hand-operated sprayer(s) throughout the vineyard. Although potent fumigants such as the sodium salt of methyldithiocarbamate (metam-sodium, Figure 13.6) are used to kill pests (as well as fungi), it is necessary to take great care to avoid poisoning workers engaged in their application.

Of course the advantage of a material such as sodium methyldithiocarbamate (metam-sodium), toxic though it may be, remains its solubility and thus ready removal during washing of the grapes before processing or by rain during growing.

Many other pest management tools, including herbicides, to manage plants intruding into the space wanted for growing vines are also widely employed. A list of more than one hundred and sixty (160) such materials used at various times throughout the growing season (and before and after) are provided through EXTOXNET the (Extension Toxicology

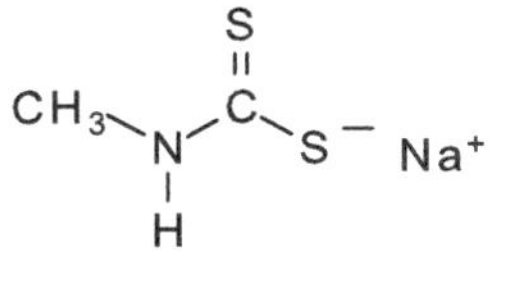

FIGURE 13.6 A representation of the structure of the dithiocarbamate (metam-sodium).

Network; http://extoxnet.orst.edu). Other lists of materials used to control fungi and other pests (e.g., rodents and birds) can be obtained from various state agricultural extension services (e.g., http://www.ipm.ucdavis.edu/PMG/selectnewpest.grapes.html) as well as the National Pesticide Information Center (http://npic.orst.edu).

With regard to weed control in the vineyard, it has often been found that, while it is relatively easy to deal with (mowing or plowing under) grasses and weeds in the middle of the rows, it is much more difficult to control them in the area beneath the vines themselves. It is currently common to use herbicides ([e.g., Roundup®, N-(phosphonomethyl)glycine, $HO_2CCH_2NHCH_2PO_3H_2$] or to attempt to use small mechanical weeding devices or weeding by hand for the more difficult-to-reach growth. Other, older and perhaps more widely used herbicides include paraquat (1,1'-dimethyl-4,4'-bipyridinium dichloride), deadly to a broad spectrum of growing plants, and 2,4-D (2,4-dichlorophenoxyacetic acid), a systemic herbicide for broadleaf plants (Figure 13.7). However, as noted above, other herbicides are also widely used, and other materials, such as powdered sulfur, are occasionally brought to bear to fight fungi. The armamentarium of chemical warfare against animal and vegetable life forms that might intrude on the produce of the vineyard seems to know no bounds. Interestingly, the extent to which the added chemicals are incorporated into the product, where it has been examined, appears minimal or absent.

During the ripening (engustment) and depending upon the variety, the colors of the grape begin to appear. The sugar free anthocyanidins, exemplified by the flavylium cation (Figure 13.8 and Appendix 4) with a variety of anions, are produced in the red wine grapes and they begin to appear red (but see Figure 12.12, page 78), and the color may deepen as time goes on, the pH of the berry changes, and more anthocyanidins form by hydrolysis of the anthocyanins. Carotenoids, exemplified by the isoprenoid β-carotene (Figure 13.8 and Appendix 2), begin to appear in low concentration as yellow, with the color deepening to yellow-green.

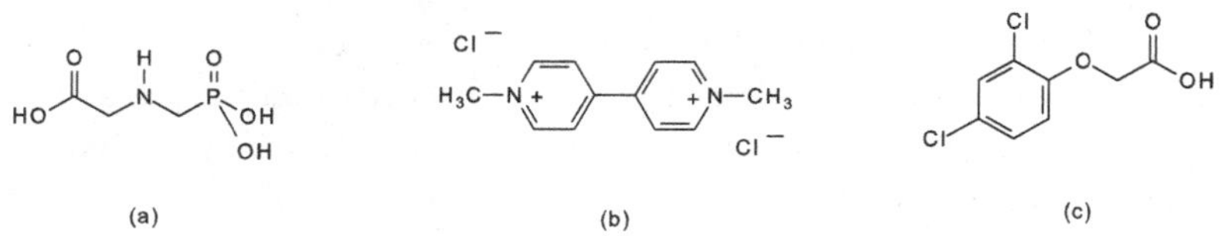

FIGURE 13.7 Representations of: (a) N-(phosphonomethyl)glycine (Roundup®); (b) 1,1'-dimethyl-4,4'-bipyridinium dichloride (paraquat); and (c) 2,4-dichlorophenoxyacetic acid (2,4-D).

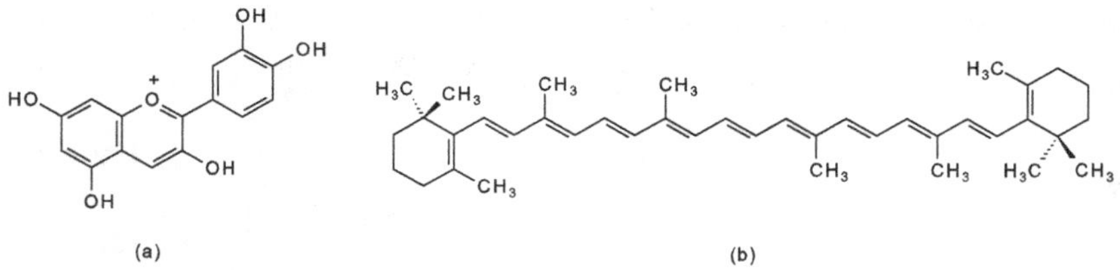

FIGURE 13.8 Representations of species that impart color to vegetable matter: a flavylium cation, viz. cyanidin at low pH (a) and the highly conjugated carotenoid β-carotene (b).

So, in summary, the berries are also beginning to swell and to soften, water fills the cells, the concentrations of the sugars (carbohydrates including sucrose, which is transported from the leaves and then hydrolyzed to glucose and fructose) begins to rise, and the acid levels begin to fall even more rapidly as the tartaric acid is diluted and the malic acid is consumed.

It has recently become clear[16] that, at about the same time, anthocyanins and anthocyanin-related species, which are also highly conjugated and may be colored as well as glycosylated, are on the move from their sites of synthesis (usually the endoplasmic reticulum) in the cytosol to deposits called anthocyanin vacuolar inclusions where they are then held. And it appears that it is in these same vacuolar inclusions that some hydrolysis of the anthocyanins to free sugars and anthocyanidins may also occur.

Canopy work is very important at this time.

Clusters of berries that are most exposed to the warmth of the sun because they are on the outer portions of the grape vine canopy ripen more rapidly than those shaded on the inside. Usually a high fruit-to-leaf ratio is desired so that the energy expended by the plant goes into fruit production and enhancement rather than leaf production. Indeed, it is anthropomorphically argued that the plant needs the seeds (in the fruit) to prepare its progeny to survive. So, while the vines have been growing, caretakers in the vineyard have been monitoring their orderly progress. Vines have been encouraged (removing "suckers" as noted earlier) to stay close to the trellis, and leaves near the interior of the trellis are removed so that clusters of berries in the interior are open to air flow and sunlight. Clearly, enough canopy, but not too much (since photosynthesis must continue) as dictated by experience, should be removed.

With rapid growth, it is anticipated that the fruit size will more than double before it is ready to be harvested. Some of the compounds that were present in the berries before this period remain (e.g., tartaric acid, sucrose), but their relative concentrations change. Other compounds are used up. For example, malic acid, a normal constituent of the Krebs cycle, *vide supra*, is fed back into the cycle—undergoing oxidation to oxaloacetate (malate dehydrogenase, EC 1.1.1.37) or, by decarboxylation, converted to phosphoenolpyruvate (PEP) on the way to carbohydrates and aromatic amino acids. It has been suggested[2] that the gallic acid derivatives of carbohydrates (i.e., tannins), as well as cinnamic acid derivatives, Figure 13.9

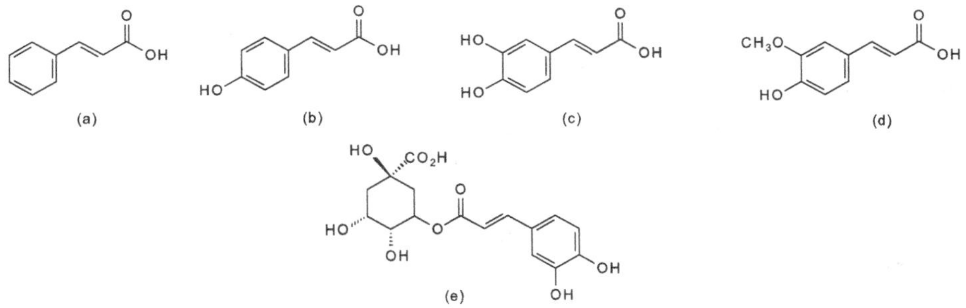

FIGURE 13.9 Representations of compounds considered "derivatives" of (a) cinnamic acid. While the compounds themselves may not actually be derived from cinnamic acid (a) oxidation pathways for introduction of hydroxyl groups onto aromatic rings are well known (e.g., phenylalanine 4-monooxygenase, EC 1.14.16.1). The "derivatives" include (b) coumaric acid; (c) caffeic acid; (d) ferulic acid; and (e) the quinic acid ester of caffeic acid (chlorogenic acid).

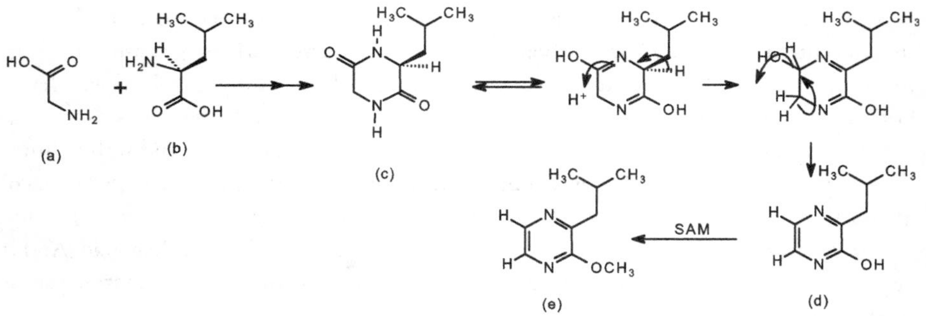

SCHEME 13.14 A representation of a potential pathway from common amino acid constituents of grapes to 2-methoxy-3-isobutylpyrazine, which is found in some wines. Reaction of the amino group of glycine (Gly, G) (a) with the acid function of leucine (Leu, L) (b) and vice versa to produce a diketopiperazine (c) followed by tautomerization and dehydration to the hydroxy-substituted pyrazine (d) and eventual methylation from S-adenosylmethionine (SAM) are all possible steps to 2-methoxy-3-isobutylpyrazine (e).

also begin to decline on a "per berry basis." Interestingly, some of the nitrogenous materials also begin to disappear. These nitrogen-containing compounds are generally considered derived from amino acid fragments and are often found in green vegetables such as asparagus, peas, and peppers and many are synthesized by beetles, molds, and fungi.

For example, it has been pointed out[17] that the presence of the pyrazine, 2-methoxy-3-isobutylpyrazine, in grapes is a function of the genotype of the grape (Scheme 13.14).

Little is known about the genesis of such secondary metabolites although it is clear that the methyl group attached to the oxygen is derived from S-adenosylmethionine (SAM), and it is likely that amino acids are utilized.[18] A cartoon of a potential pathway is shown as Scheme 13.14.

While hormones wax and wane during this rapid growth and ripening process, the most striking change appears to be the large increase in the transport and build-up of sucrose and its hydrolysis to its glucose and fructose fragments (Figure 13.10).

Again, it has been pointed out specifically[2] that hormones play an important role during the period from petal fall to *l'époque de la nouaison* (the time of fruit set). These hormones apparently fall into the same groups of auxins, cytokinins, and gibberellins seen before (i.e., those that promote cell division and cell expansion) (Figure 13.11). And, to the extent that they have been measured, compounds in these groups appear to reach a maximum concentration just before véraison and then decrease sharply along with ripening of the berries.

While it is very common for many fruits (called "climacteric" fruits) during their ripening (engustment) to produce ethylene gas ($CH_2=CH_2$) which acts as a hormone, it is clear that for grape crops the signal is missing. Interestingly, while major changes are occurring in the grape, it has been reported that ethylene ($H_2C=CH_2$) production is low and apparently more or less constant even before véraison begins. Indeed, since relatively little is known about the ethylene receptor in this or any fruit, it appears that, generally, for grapes, the presence of ethylene and ripening are coincidental.

A rise in abscisic acid (Figure 13.12) and other putative hormones, including the brassinosteroids (*vide infra*), has been reported. Exogenous application of the latter is also reported to have hastened ripening.

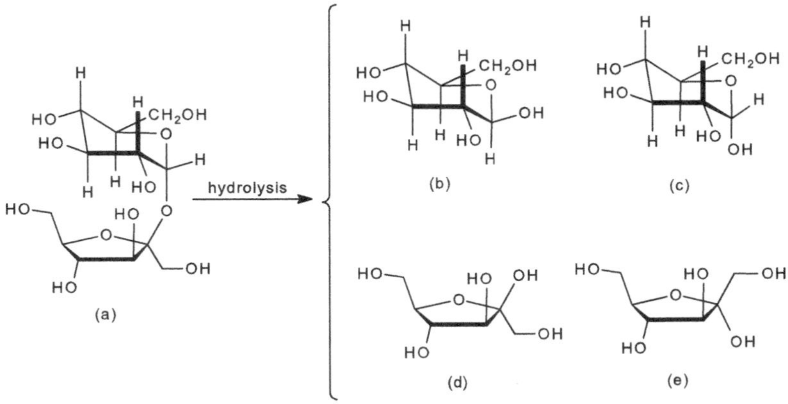

FIGURE 13.10 A representation of the hydrolysis of sucrose (a) into the anomeric glucopyranoses β-D-glucopyranose (b) and α-D-glucopyranose (c), as well as β-D-fructofuranose (d) and α-D-fructofuranose (e). It is presumed that on hydrolysis epimerization at the anomeric carbon might occur.

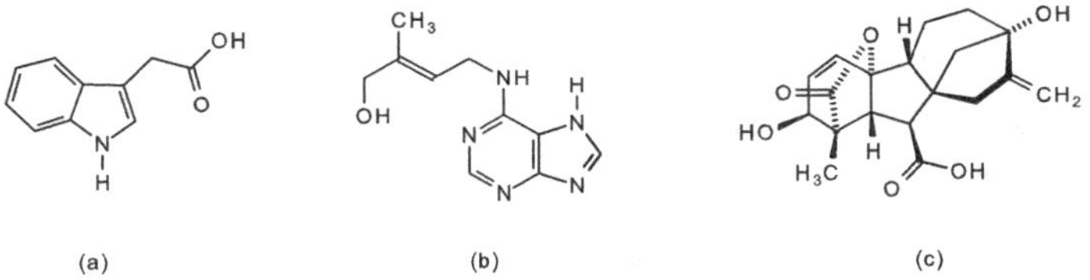

FIGURE 13.11 A representation of the same group of auxins such as indole-3-acetic acid (a), cytokinins, such as zeatin (b) and gibberellins such as gibberellic acid (c), seen before and implicated now in fruit set (*nouaison*).

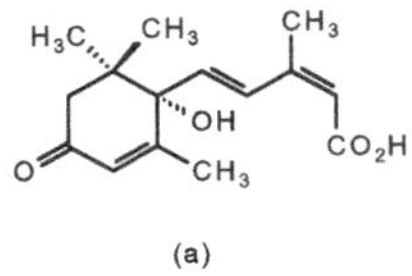

FIGURE 13.12 A representation of abscisic acid (a).

With regard to the brassinosteroids and as discussed in Appendix 2, it is in the cell cytoplasm[19] (rather than in plastids) that isopentenyl diphosphate and dimethylallyl diphosphate combine to, eventually, produce farnesol and its isomers. "Head-to-tail" dimerization of farnesol equivalents (i.e., $2 \times C_{15} = C_{30}$) produces the parent thirty-carbon hydrocarbon squalene. Squalene oxide, with only a few modest changes, undergoes the remarkable cyclization to the parent plant sterol lanosterol. Modification of lanosterol produces other plant sterols (steroids), and it should be clear that only additional oxidative modifications of lanosterol are required to produce the brassonosteroid, epi-brassinolide. It is the latter which has been demonstrated to have a dramatic effect on grape ripening.[20]

An abbreviated cartoon representation showing the relationship between farnesol and squalene and thence to lanosterol, cholesterol, and epi-brassinolide is provided as Scheme 13.15. Steps have been omitted, and a somewhat more elaborate exposition is provided in Appendix 2.

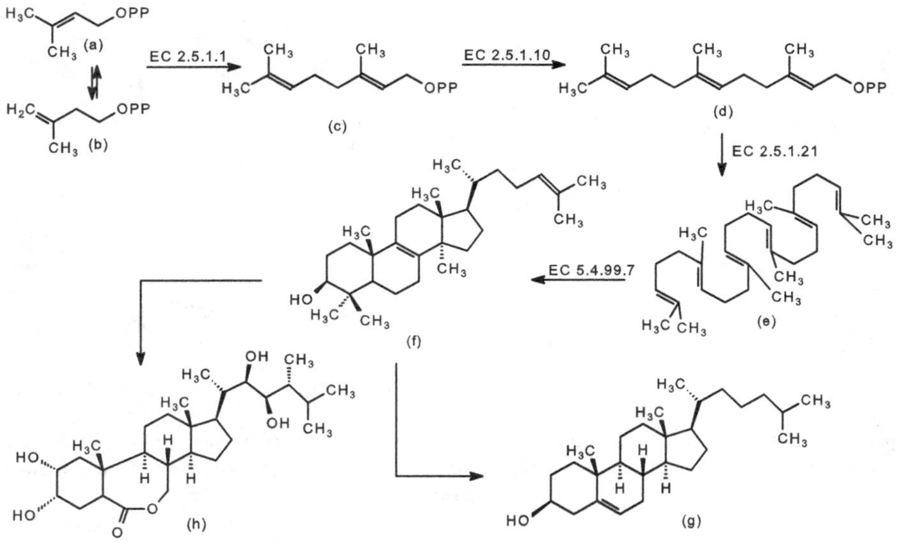

SCHEME 13.15 A heavily abbreviated cartoon representation of selected steroid formation. The abbreviation "PP" is for diphosphate. Beginning at the upper left, dimerization between the C-5 isomers dimethylallyl diphosphate (a) and isopentenyl diphosphate (b) in the presence of geranyl diphosphate synthase (EC 2.5.1.1) produces the ten-carbon terpene derivative, geranyl diphosphate (c) and inorganic phosphate (Pi). Then, with another isopentenyl diphosphate (b) and in the presence of (2E,6E)-farnesyl-diphosphate synthase (EC 2.5.1.10), geranyl diphosphate (c) goes on to the C-15 compound farnesyl diphosphate (d). Two equivalents of farnesyl diphosphate (d) combine under the influence of squalene synthase (EC 2.5.1.21) to produce the C-30 hexadiene, squalene (e). Oxidation, cyclization, and rearrangement of squalene (e) in the presence of lanosterol synthase (EC 5.4.99.7) generates lanosterol (f). Approximately twenty additional steps are required to convert lanosterol (f) to cholesterol (g) and a similar number to epi-brassinolide (h), which may or may not have cholesterol (g) as an obligatory precursor.

Finally, although it is traditionally held that chemical processes occurring in wine occur at a relatively slow pace, the same is probably not true in the late ripening stages of the grape.

Thus, a recent study[21] has demonstrated the existence of daily (!) changes in plant transcriptome oscillating in response to changes in light and temperature as well as in the circadian clock components to which all living systems respond.

PART C. BEGINNING TO HARVEST

The completion of ripening is the beginning of harvest.

Different grapes complete ripening at different times. Not only does the genome dictate when the grape (bearing the seed for the next generation of grape) is ripe, but the *terroir* and the viticulturist play critical roles.

Presumably, it is clear how the genome would play a role in the timing of ripening and the signaling of completion as the pattern for production of all of the components of the grape is set by the biology of the plant. It is probably also clear that the *terroir* is important since it should now be appreciated that cuttings from the same vine on similar root stock but in a different environment (soil, sun, water, etc.) will not produce equally successful crops. Small differences in the complex symphony of a good wine can be significant. However, it may not be clear what role the viticulturist plays.

The viticulturist must make timely and rapid decisions regarding the amount of sugar (the "must weight"), the quantity and identity of acids as well as the pH (see Appendix 1) of the grapes, and the amount and kinds of tannins, phenols, terpenes (and sesquiterpenes), esters, and other coloring and flavoring components present *in the grape*!

Since time is of the essence, the viticulturist or surrogate, while in the vineyard, normally crushes a few grapes and allows the liquid to flow onto the surface of a refractometer, Figure 13.13. The refractometer is a device for measuring the refractive index of the liquid, a measure of the propagation of light through the liquid. The refractive index can be correlated with the amount of sugar present in the colorless, clear sample. The amount of sugar is traditionally measured in "brix" (one degree "brix" = 1 gram of sucrose in 100 grams of solution). The viticulturist, knowing the grape and the strain of yeast that will be used and the way the fermentation will be carried out, can correlate the amount of sugar with the quantity of alcohol that can be obtained from the grape.

It is currently possible (although it has not always been so) to utilize only a few grapes and to attempt to determine a variety of other constituents in real time and on a small scale. While it is not generally possible (yet) to perform these measurements in the field, many vineyards have access to spectroscopic and spectrometric tools (see Appendix 1) that will permit analysis of tannins, phenols, anthocyanins, amino acids, steroids, carotenoids, sesquiterpenes, proteins, pyrazines, and esters, as well as their acids and alcohols (not yet including much if any ethanol), as well as other primary and secondary metabolites and other compounds giving rise to the symphony of components producing the aroma, color, and flavor of the wine.

FIGURE 13.13 A photograph of an Extech hand-held refractometer. Photo courtesy of Extech Instruments.

However, despite the plethora of tools available, what the viticulturist decides is "ripe" is a function of the grape being grown under the specific conditions found in the vineyard and, of course, the opinion of the vintner.[22]

Among the considerations for the specific grape being grown will be features of the specific season. Thus, in any given year, a hot summer prior to harvest may well mean that harvest must be undertaken as soon as a month after ripening has begun,. However, should the summer be cool, harvest might be postponed for two months (or more) after véraison unless heavy rains are in the forecast. Too much water, absorbed by the vines can cause the berries to swell and crack. Indeed, too much water will promote mildew which can rapidly spread across the vineyard. The effects of too little water were noted earlier, and as commented there, ripening can be significantly delayed as a consequence.

While it could be argued that as long as sucrose is being moved into the grape it might be best to allow harvest to be postponed, the argument is faulty. The fault does not lie in sugar production. Although the sugar is fermented by yeast into ethanol, and although it might be supposed that the higher the concentration of sugars in the grape, the greater the potential alcohol yield, it falls out that most strains of yeast die when the alcohol solution reaches between 14 and 17% (on a volume-to-volume basis). This might leave some sugar unreacted if there is an excess in the grape juice, and so it clearly changes the "sweetness" of the wine. As will be seen, some wines are either left with this additional sugar or even have sugar added, but in general, a viticulturist attuned to the vineyard and keeping close watch will note when the grapes should be harvested so that neither too much nor too little sucrose is present.

Indeed, as véraison progresses and as the harvest draws ever closer, vineyard workers will test individual berries picked from clusters across the vineyard in ever shorter intervals. The berries near the middle of clusters, but (generally) not on vines near the ends of rows, are chosen as typical. The test (for which the refractometer might be used) is for the amount of sugar, and the result is plotted as a function of time so that the degree of "ripeness" can be judged. The "appropriate" must weight depends on the grape and, of course, is a judgment based on experience that the viticulturist or owner of the vineyard chooses to make. If a specific yeast whose result is known with the specific grape is to be used, it is generally possible to predict the eventual concentration of alcohol that will be obtained from the measured amount of sugar.

As noted earlier, the crop near or at harvest will have been affected by appropriate managing of the canopy. Some foliage is necessary to provide the energy to move the carbohydrates and secondary metabolites used for the life of the grape as well as the growth of the grape and the seeds housed within. Too much foliage, while supporting the vine, will shade the grape clusters, deny them access to the warmth of the sun and encourage the development of mildew and rot. Additionally, a healthy vine with heavy leaf growth will be competing for the carbohydrate resources needed to produce the sweet juice. Secondarily, without (and sometimes even with) proper management, all the grapes on a given bunch will not ripen to the same extent at the same time. It has been suggested that this might result from a lack of soil nutrients which resulted in inhomogeneous fertilization during flowering, with some flowers being fertilized significantly later than others.[22]

Grapes that have been left on the vine too long may become overripe and dehydrated. Further, with rising sugar levels the relative acid (malate, citrate, tartrate, etc.) levels fall, and the fine balance required of the juice of the grape destined to become a good wine needs to be carefully judged. This is because it has been found by vintners and it is generally agreed by consumers that some degree of acidity is needed for what enthusiasts refer to as good "mouth feel." It is also found that the appropriate level of acidity will aid in the aversion of spoilage and promote interaction with the oak casks in which the wine will eventually be aged.

In principle, acid (e.g., citric, tartaric, malic, succinic, cinnamic) can be added to adjust acidity (see Figure 13.14), and there is some evidence that such emendation can be successful. However, as will be discussed subsequently, the total acidity can also be modified after sulfur dioxide (SO_2) or salts of metabisulfite ($S_2O_5^{2-}$) or bisulfite (HSO_3^-), commonly added (*vide infra*) to manage the growth of (undesirable) bacteria during processing, have been used.

The acidity is either expressed using the pH scale (Appendix 1), where a value lying between 3 and 4 is common, or as a function of the amount of tartaric acid (the major carboxylic acid present) in the juice. A value between about 0.6–0.8% per 100 mL is wanted.[22] So, as the pH levels rise (the solution becomes less acidic), bacteria and undesirable adventitious yeasts common to gardens proliferate. Phenols, such as resveratrol and the anthocyanin malvidin-3-glucoside (Figure 13.14) and similar compounds are acidic enough so that, if the pH is too high they will suffer proton loss, and subsequent changes leading to unwanted flavors and colors will occur. The measurement of the pH can be done with a variety of indicators (Appendix 1) or with carefully calibrated pH meters suitably adjusted for the medium in which the pH is being measured. To the extent that the vintner bases the decision to harvest on the pH of the grape juice, the accurate measurement is, clearly, important.

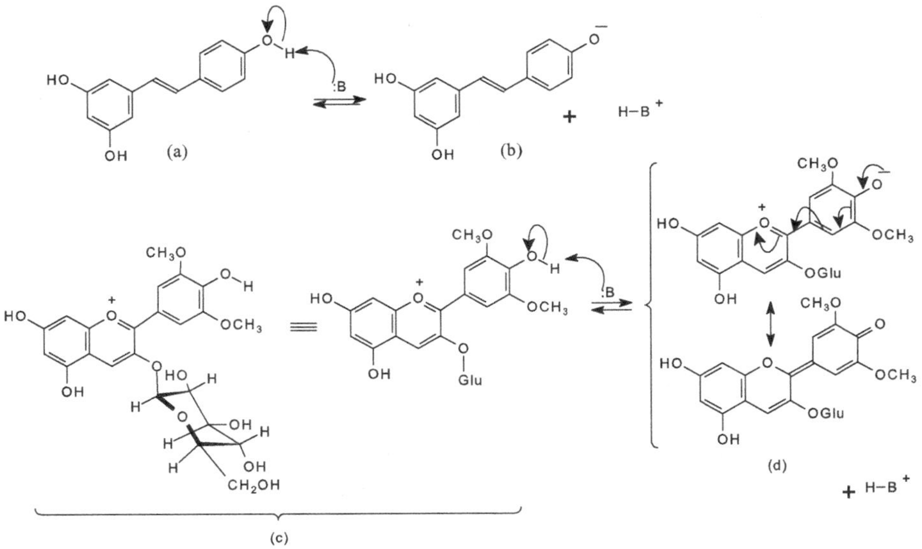

FIGURE 13.14 A representation of potential changes wrought by increasing the pH of solutions containing (a) resveratrol and its conversion to one of the possible phenoxide anions (b) in the presence of a base "B." Additionally, a representation of the anthocyanin malvidin-3-glucoside (c) and the formation of the corresponding cross-conjugated ketone (d) for which two resonance (Appendix 1) structures are drawn, as one of the possible responses to treatment with base B are also shown.

Nonetheless, while the pH is important, the quantity of sugar and the specific carboxylic acids present also contribute to the overall balance of the wine. In the end it is the decision of the vintner.

There is, among experts in the art, an overall evaluation of ripeness which, while usually including the above mentioned criteria of sweetness, acidity, and development of tannins and other phenolics, involves a thorough understanding of the specific grape variety. Thus, the "physiological ripeness" ("engustment") of the grape is detected using all of the senses of the expert: the look of the berry (i.e., visually); the smell of the grape and the vine (i.e., aroma); and the feel of the skin, the hardening of the stem, and the texture of the berry (i.e., tactile). Specific compounds might be examined. Thus, as seen (Figure 13.14), malvidin (as a typical example of a phenol) is found as its glucoside (i.e., a glucosylated phenol) rather than as a free phenol. The overall extent of oxidation and methylation of phenols might also be measured. For example, the oxidation of coumaric acid to caffeic acid and methylation of caffeic acid to ferulic acid (*vide supra*, Figure 13.9) are all "late stage" events in the grape ripening process. Therefore, even though the sugar/acid ratio might be appropriate for the novice to have judged ripeness, the flavors necessary for the production of a robust wine have not yet matured, and the experience that an expert vintner brings to the harvest is needed for the fullness of flavor expected from the grape to produce quality vintage.

Finally, pre-harvest, there are reports of analytical techniques being used in attempts to quantitate (perform quantitative analysis) and quantify (express the quantity) various aspects of the ripening process. Leaf analysis has been reported with photoacoustic spectroscopy (PAS),[23] berry analysis with high-performance liquid chromatography coupled with tandem mass spectrometry (HPLC-MS/MS) and protein analysis with isobaric (same mass) tags for relative and absolute quantitation (iTRAQ) (incorporation of stable isotopes to study proteome changes).[24]

The more pedestrian infrared (IR), ultraviolet (UV), visible (VIS), and nuclear magnetic resonance (NMR)[25] spectroscopies, as well as routine mass spectrometry (MS), are all widely used, but moving them into the field for real-time analysis remains a problem.

PART D. THE HARVEST

As might be anticipated from what has already been written, the exact time and method of harvest is a function of a number of variables.

Both mechanical and manual grape harvesting methods are currently employed.

Depending upon the topography, mechanical harvesting may be prohibited or may be preferred. Vineyards on hillsides (often preferred for drainage reasons), widely spaced or valuable vines (vine density), low- or high-growing grape bunches (vine training), and concern for a really good vintage all generally require more traditional (i.e., manual) harvesting methods. Mechanical methods often damage the grapes as the bunches are struck or slapped from the vines. The grape damage can require rapid remediation (depending upon the specific variety), since some grapes are torn open and adventitious yeasts—commonly found in vineyards but not desired for the fermentation of the grape juice—begin to work on the juice, and tannins and other components associated with the grape skins are undergoing extraction into the juice before the vintner is ready. Additionally, any pesticides or other materials present on the grapes, if washing was being considered, will already have been incorporated into the juices.

However, these disadvantages may be overcome by the need to bring in the harvest because of heat or cold, rain or drought, or even lack of sufficient manpower to manage the vineyard. Thus, while skilled workers might cover as little as one or as much as 10 hectares (1 hectare = 10,000 m² ≈ 2.5 acres) per day (depending on the size of the harvest) a mechanical harvester might accomplish the task in a few hours/hectare. And, as is becoming more common, skilled harvest workers are not always available. The harvesting workforce, normally composed of itinerant workers, follows varieties of crops as they ripen. Labor laws, as well as the desire to treat a workforce with dignity and respect, may be insufficient to attract sufficient numbers of people to attend the harvest. Interestingly, in many countries, it was common to have a workforce that celebrated the harvest with the vineyard owners. Some small vineyards (often European and/or manned by family groups) continue to enjoy those benefits. However, commercialization of the industry has introduced changes, and itinerant workers have become necessary participants.

For manual harvest, individual workers, using a specialized knife (Figure 13.15) cut through the peduncle and drop the grape bunch into a basket carried on the back of a coworker or coworkers. When the basket is full the worker walks to a wagon into which the basket is emptied, and (s)he then returns for more grapes to the next available harvest worker. The advantage of such work is that only those grapes ready to be harvested (as decided upon by the vintner) are taken. The disadvantage is that when it is time to harvest such care may be unwarranted, as the remainder of the crop could be lost.

When the harvest wagon is full it is either removed to the winery or to a field pressing station as appropriate, where destemming (égrappage), along with crushing occurs. It is generally the case that at the winery or at the field pressing station, the grapes, if they are to be washed to remove residues of pesticides, insects, bird droppings, etc. before entering the crusher, will be so treated. Then the grapes, washed or unwashed, are pushed through a destemmer by a worm gear. The separated stems are ready for recycling (as mulch), and the grapes now ready for crushing. Since the stems are composed of polymerized carbohydrates

FIGURE 13.15 Representations of some manual harvest grapes and implements. (a) A grape harvest knife; (b) a grape bunch cut at the peduncle; (c) a harvest basket for carrying on one's back.

(cellulose) and phenols and thus contain tannins specific to the stems, crushing or breaking them along with the grapes will cause increase in the tannin concentration. Further, as already noted, not all tannins are the same. For example, the amount of gallic acid and extent of crosslinking is not the same within the grape as the stem, and thus the wine will not accurately reflect the grape.

PART E. THE DECISION

The grapes having been separated from the stems are ready to be processed. Now, a decision is wanted concerning the "color" of the product. White wines (actually slightly yellowish) can be produced from all grapes except teinturier varieties where pigments are found in the juice of the grape.[26] Of the large number of modern cultivars from which most wines are derived (whether or not derived from North American root stock as a consequence of Phylox infestation), most are the tight skinned *Vitis vinifera* with clear and largely colorless juice.[27]

If the skins are kept with the juices, the extraction of coloring matter begins. For white (light green/yellow) varieties, the coloring matter (largely composed of long-chain unsaturated hydrocarbons, e.g., carotene-like species) is not particularly soluble in water. The rapid separation and removal of the skins is not as critical as it is with more deeply colored skins. However, for the more deeply colored skins, extraction of anthocyanins, their corresponding anthocyanidins, and related phenolics begins almost at once, and the more heavily hydroxylated and thus more water-soluble species are extracted first. The longer the skins are with the juice, the more deeply colored the extract. So, if a "white" wine is desired from a red grape, the separation of skins and juice is effected quickly.

There are a multitude of varieties of white and red grape cultivars, although some are more generally available than others, and lists of common and not so common cultivars are available.[28] However, despite the large variety of cultivars it appears that a relatively small number of varieties of grapes have taken over the marketplace. These have become known as the "international variety."[29]

Some of the characteristics of an expanded international or classic variety are discussed in the material that follows, but the chemistry of what is involved does not differ much from one variety to another. The subtle differences between good wines and wines that are excellent lies in the trace flavor quantities detected in the nose and palate of the consumer.

In that vein, Appendix 5 provides several lists of compounds (and their structures) that are associated with the international and other groups of varieties of grapes. These lists are of recent vintage, as the technology that allows for the determination of specific compounds is recent. Most of the identification work is currently done with mass spectrometric tools (briefly introduced in Appendix 1). However, once the structure of the particular compound has been identified and it has been independently prepared, the odors and/or tastes associated with that material and the wines containing it are best noted as subjective. Additional compounds can be shown to be associated with yeasts, with oak barrels, etc. as the juice of the grape is treated on its way to the bottle. Some of these compounds are also shown in Appendix 5.[30]

Nonetheless it is of overarching importance to recognize that the same grape variety grown (terroir), harvested (i.e., the opinion of the viticulturist), and treated (e.g., a different yeast clone, oak cask, etc. as directed by the vintner) under different conditions will not produce the same wine.

As recently pointed out[24] "The dynamic interaction among diverse factors including the environment, the grapevine plant and the imposed viticultural techniques means that the wine produced in a given terroir is unique." Most recently,[31] berries from a single clone of *V. vinifera* cv. Corvina grown in different vineyards, cultivated in different macrozones (i.e., environmental areas), have been examined. It was found that each vineyard could be characterized by the unique profile of specific metabolites [metabolites are small molecule intermediates and products of (plant) metabolism].

In conjunction with earlier work[32] it is now clear that the authors[31] believe that the specificity of terroir may extend not only to the observed products (and their concentration differences) as noted above regarding the metabolites, but also to the transcriptome (the producer of the metabolites) as dictated by the genome of the plant. Thus, while it appears true that the transcriptome can vary with environmental conditions (seen in the expression of messenger RNAs [mRNAs] of genes), it turns out that actual examination of any of the set of RNAs (mRNA, rRNA, tRNA, etc.) under any conditions has not yet been reported. The experiments are wanting.

However, it has been suggested "that grape-associated microbial biogeography is nonrandomly associated with regional, varietal and climatic factors across multiscale viticultural zones."[33] Indeed, it appears possible that "the mix of microorganisms associated with a grapevine can influence the basic characteristics of the resulting wine." Of course, a very large number of these organisms are simply associated with the soil and the details surrounding them, and their activities are only now beginning to be recognized.[34,35]

In the same vein, it has been pointed out that members of the "microbial community ... were significantly different between plant parts and soils ... (but) above ground samples shared more operational taxonomic units (OTUs) with below ground environments than they did with each other, suggesting that soil is a major microbial reservoir."[36]

NOTES AND REFERENCES

1. Coombe, B. G. *Ann. Rev. Plant Physiol.* **1976**, *27*, 507.

2. Conde, C.; Silva, P.; Fontes, N.; Dias, A. C. P.; Tavaris, R. M.; Sousa, M. J.; Agasse, A.; Delrot, S.; Gerós, H. *Food,* **2007**, *1*, 1.

3. Jackson, R. S. *Wine Science,* Chapter 3, 3rd ed.; Academic Press: Burlington, MA, 2008; pp 50–98.

4. Wheeler, G. L.; Jones, M. A.; Smirnoff, N. *Nature* **1998**, *393*, 365.

5. Regni, C.; Tipton, P. A.; Beamer, L. J. *Structure* **2002**, *10*, 269.

6. Preiss, J.; Wood, E. *J. Biol. Chem.* **1964**, *239*, 3119.

7. Major, L. L.; Wolucka, B. A.; Naismith, J. H. *J. Amer. Chem. Soc.* **2005**, *127*, 18309.

8. Linster, C. L.; Gomez, T. A.; Christensen, K. C.; Adler, L. N.; Young, B. D.; Brenner, C.; Clarke, S. G. *J. Biol. Chem.* **2007**, *282*, 18879.

9. Brenner, C. *eLS* (Wiley Online Library), Histidine Triad (HIT) Superfamily), **2014.** DOI: 10.1002/9780470015902.a0020545.

10. DeBolt, S.; Cook, D. R.; Ford, C. M. *Proc. Nat. Acad. Sci. (USA)* **2006**, *103*, 5608.

11. Salusjärvi. T.; Povelainen, M.; Hvorslev, N.; Eneyskaya, E. V.; Kulminskaya, A. A.; Shabalin, K. A.; Neustroev, K. N.; Kalkkinen, N,; Miasnikov, A. N. *Appl. Microbiol. Biotechnol.* **2004,** *65,* 306.

12. Saito, K.; Kasi, Z. *Plant Plysiol.* **1984,** *76,* 170 and Saito, K.; Morita, S.; Kasai, Z. *Plant Cell Physiol.* **1984,** *25,* 1223.

13. Debolt, S.; Melina, V.; Ford, C. M. *Annals Bot.* **2007,** *99,* 3.

14. Toyooka, K.; Sato, M.; Kutsuna, N.; Higaki, T.; Sawaki, F.; Wakazaki, M. Goto, Y.; Hasezawa, S.; Nagata, N.; Matsuoka, K. *Plant Cell Physiology*, **2014.** DOI: 10.1093/pcp/pcu084.

15. Matthews, M. A.: Shackel, K. A. In *Vascular Transports in Plants;* Holbrook, N. M., Zwieniecki, M., Eds.; Elsevier: New York, 2005; p 181.

16. Chanoca, A.; Kovinich, N.; Burkel, B.; Stecha, S.; Bohorquez-Restrepo, A.; Ueda, T.; Eliceiri, K. W.; Grotewold, E.; Otegui, M. S. *Plant Cell* **2015,** *27,* 254; and Bassham, D. C. *Nature* **2015,** *526,* 644.

17. Koch, A.; Doyle, C. L.; Matthews, M. A.; Williams, L. E.; Ebeler, S. E. *Phytochem.* **2010,** *71,* 2190.

18. Vallarino, J. G.; López-Cortés, X. A.; Dunlevy, J. D.; Boss, P. K.; González-Nilo, F. D.; Moreno, Y. M. *J. Agric. Food Chem.* **2011,** *59,* 7310. DOI: 10.1021/jf200542w.

19. Lichtenthaler, H. K. *Ann. Rev. Plant Physiol. Plant Mol. Biol.* **1999,** *50,* 47.

20. Symons, G. M.; Davies, C.; Shavrukov, Y.; Dry, I. B.; Reid, J. B.; Thomas, M. R. *Plant Physiol.* **2006,** *140,* 150.

21. Carbonell-Bejerano, P; Rodríguez, V.; Royo, C.; Hernáiz, S.; Moro-González, L. C.; Torres-Viñals, M.; Martínez-Zapater, J. M. *BMC Plant Bio.* **2014,** *14,* 78.

22. (a) Robinson, J. *The Oxford Companion to Wine*, 3rd Ed.; Oxford University. Press: New York, 2006; p 688; (b) Jackson, R. S. *Wine Science*, 3rd Ed.; Academic Press: New York, 2008.

23. Nery, J. W.; Pessoa, O.; Vargas, H.; Reis, F. de A. M.; Gabrelli, A. C.; Miranda, L. C. M.; Vinha, C. A. *Analyst* **1987,** *112,* 1487.

24. Martínez-Esteso, M. J.; Casado-Vela, J.; Sellés-Marchart, S.; Elortza, F.; Pedreño, M. A.; Bru-Martínez, R. *Mol. BioSyst.* **2011,** *7,* 749.

25. For some very handsome early NMR work, please see Ramos, A.; Santos, H. *Ann. Rep. NMR Spectroscopy* **1999,** *37,* 179–202.

26. The flesh and juice of the grape variety *teinturier* is actually red in color. It has been suggested that anthocyanins appear to have accumulated in the pulp of the grape itself. See also note 27.

27. It is hoped that the principles hitherto discussed are sufficient to remark now that any sugar-containing fruit can be utilized in fermentation. The details of the fermentation of a specific cultivar will be similar, while the taste is a function of the multitude of subtle minor constituents and the receptor cells of the taster. So, it is not a surprise that less common use is made of the "slip-skin" *Vitus labrusca* which is, largely, used for grapes and jellies—but which can, of course, be used for wine too.

28. These include but are not limited to: (1) Robinson, J.; Harding, J.; Vouillamoz, J. *Wine Grapes—A Complete Guide to 1,368 Vine Varieties, Including Their Origins and Flavours;* Allen Lane (Penguin): London, 2012. (2) A statement indicating there are "10,000 varieties" and a somewhat shorter list is to be found at http://en.wikipedia.org/wiki/Wine_Grapes (accessed Apr 5, 2017). (3) Malenica, N.; Šimon, S.; Besendorfer, V.; Maletić, E.; Kontić, J. K.; Pejić, I. *Naturwissenschaften* **2011,** *98,* 763. DOI: 10.1007/s00114-011-0826-8.

29. MacNeil, K. *The Wine Bible;* Workman: New York, 2001; p 48.

30. Note: In addition to the personal opinion of the author, comments regarding characteristics of the grapes and the wine produced therefrom to be discussed are taken from: (a) Robinson, J., Ed.; *Oxford Companion to Wine*, 3rd Ed.; Oxford University Press: New York, 2006; (b) Robinson, J.; Harding, J.; Vouillamoz, J. *Wine Grapes—A Complete Guide to 1,368 Vine Varieties, Including Their Origins and Flavours*, Allen Lane (Penguin): London, 2012; and (c) MacNeil, K. *The Wine Bible;* Workman: New York, 2001.

31. Anesi, A.; Stocchero, M.; Dal Santo, S.; Commisso, M.; Zenoni, S.; Ceoldo, S.; Tornielli, G. B.; Siebert, T. E.; Herderich, M.; Pezzotti, M.; Guzzo, F. *BMC Plant Bio.* **2015**, *15,* 191. DOI: 10.1186/s12870-015-0584-4.

32. Dal Santo, S.; Tornielli, G. B.; Zenoni, S.; Fasoli, M.; Farina, L; Anesi, A.; Guzzo, F.; Delledonne, M.; Pezzotti, M. *Genome Biol.* **2013**, *14,* R54.

33. Bukulich, N. A.; Thorngate, J. H.; Richardson, P. N.; Mills, D. A. *PNAS* **2013**, E139–E148. DOI: 10.1073/pnas.1317377110.

34. Gruber, K. *Nat. Plants*, **2015**, *1*. DOI: 10.1038/nplants.2015.194.

35. Roesch, L. F. W.; Fulthorpe, R. R.; Riva, A.; Casella, G.; Hadwin, A. K. M.; Kent, A. D.; Daroub, S. H.; Camargo, F. A. O.; Farmerie, W. G.; Triplett, E. W. *ISME J.* **2007**, *1,* 283.

36. Zarraonaindia, I.; Owens, S. M.; Weisenhorn, P.; West, K.; Hampton-Marcell, J.; Lax, S.; Bokulich, N. A.; Mills, D. A.; Martin, G.; Taghavi, S.; van der Lelie, D.; Gilbert, J. A.; *mBio*, **2015**, *6(2),* e02527. DOI: 10.1128/mBio.02527-14.

SECTION IV
A Sample of Grape Varieties

14

A Selection of Grapes

AIRÉN: A WHITE GRAPE.

This undistinguished, productive, drought resistant, vigorous white grape, Airén, from the La Mancha region of Spain, was said to be the most widely planted grape in the world.[1] In part the justification for this claim relies upon the observation that it is planted at a very low density!

Except for its use in blending to make other wines "lighter," it has not found wide acceptance. In part, it appears that its lack of popularity is the result of what is reported to be a mild, neutral flavor, and advertising has not pushed wines produced from it to the fore.

Although it is now common to attempt to analyze the headspace (or ullage)[2] in bottled wine (as well as the wine itself) by chromatographic and mass spectrometric techniques it is less common to find that the grapes (skin, must,[2] and seeds) are also subjected to such analysis. Nonetheless, the phenolic composition of *V. vinifera* var Airén was subjected to just such analysis during ripening from véraison to "technological" maturity (i.e., maturity which might actually be earlier than harvest, the latter being the decision of the viticulturist and vintner).[3] The analysis of the ethyl ether extract of macerated skins, seeds, and accumulated solids (the *pomace*)[2] was undertaken. Procyanidins and anthocyanins which would (the authors claim) interfere with subsequent analysis would not move into the ether phase. It was also found (using controls) that other highly polar materials (e.g., carboxylic acids) were only poorly extracted from the macerated skins and seeds.

The isolated compounds and some information about their sources are provided in Figures 14.1 and 14.2.

The analysis of the seeds, skin, and must did lead to the conclusion that "the maximum concentrations of benzoic and cinnamic acids and aldehydes and flavonol aglycones and glycosides at the end of the ripening period did not coincide with the minimum concentrations of the flavan-3-ols and hydroxycinnamic tartaric esters."[3] Depending upon what was sought, this information might thus affect decisions concerning the harvest date. Along similar lines, examination of the must from crushing Airén grapes during the ripening period between the beginning of August through the beginning of October was undertaken.[4] During that time

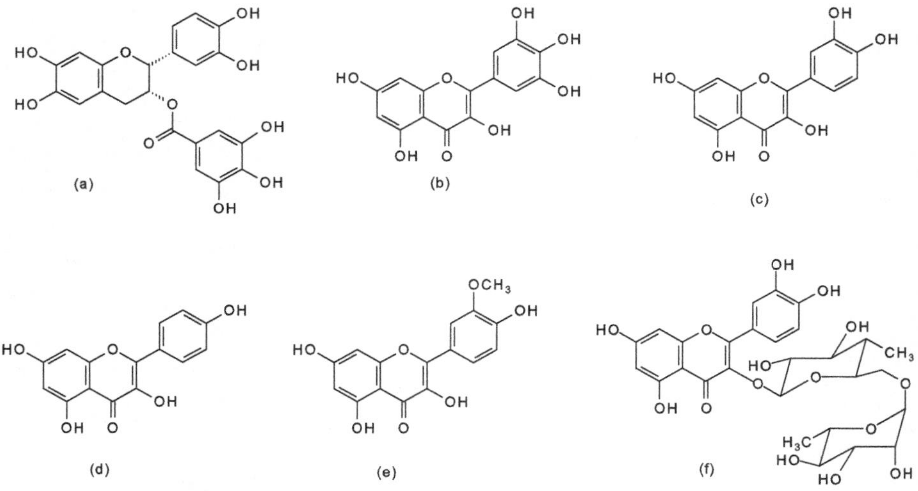

FIGURE 14.1 Representations of some of the components found in the skins, seeds, and *must* of the Airén grape. These include (a) epicatechingallate, found in seeds; (b) myricetin, found in must; (c) quercetin, found in skin and must; (d) kaempferol, found in skin; (e) isorhamnetin, found in skin; and (f) rutin, also found in skin and must. Not shown are some additional carbohydrate-linked derivatives found in skin and must which include quercetin-3-galactoside (hyperoside), quercetin-3-glucoside, quercetin-3-rhamnoside (quercitrin), kampferol-3-rhamnoglucoside, and myricetin-3-rhamnoside.

frame, the amount of sugar (measured in degrees Baumé[5]) increased from 5.81 to 11.20, the total acidity (measured in grams/L for tartaric acid) decreased from 13.30 to 3.78, the pH increased from 3.05 to 3.44, and the amount of malic acid (grams/L) decreased from 7.12 to 0.45. A variety of measurements of volatile components (Figure 14.3) over the same time frame were also made. Interestingly, it was found that C-6 aldehydes and alcohols (e.g., *trans*-2-hexenal and *trans*-3-hexen-1-ol) as well as terpenes (e.g., citronellol[6] and geranic acid) reached a maximum at 6–8 Baumé in Airén grapes and that these values might be considered during the ripening period by the vintner.

Cabernet Sauvignon: A red grape.

The cultivar Cabernet Sauvignon has been described as "the world's most renowned ... red wine grape."[7] This widely grown cultivar has been shown to most likely be the progeny of two other Bordeaux cultivars, *viz.*, Cabernet franc and Sauvignon blanc, all three of which are derived from root stock in the Bordeaux region of France.[8] Of course it is important to note again, that, while the cuttings that give rise to the specific cultivar can be transported to a variety of regions, those regions leave their imprint on the result. Despite this, Robinson[7] argues that it is "Cabernet Sauvignon's remarkable concentration of phenolics that really sets it apart from most other widely grown vine varieties." In part as a consequence of this difference and the importance of phenols (e.g., anthocyanins and other polyphenolics) in this wine, it has been chosen for studies that compare methods of maximizing extraction of must (*vide infra*) as fermentation continues.[9]

However, the issue of terroir remains important, and within that concept it appears that, while a variety of regions will accept the cultivar and produce good to excellent wines, the

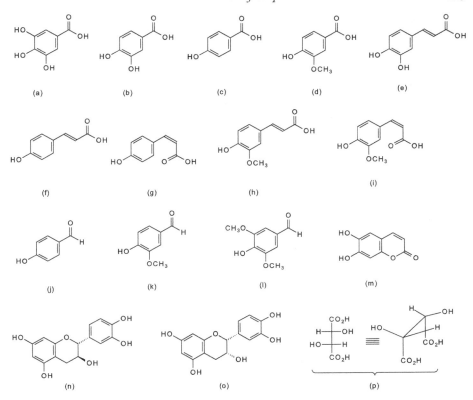

FIGURE 14.2 Representations of some of the components found in the skins, seeds, and *must* of the Airén grape. All of the compounds are reported to be found in the skin and must unless otherwise specified. These include: (a) gallic acid, found in the skin and seeds; (b) protocatachuic acid; (c) *para*-hydroxybenzoic acid; (d) vanillic acid; (e) *trans*-caffeic acid; (f) *trans-para*-coumeric acid; (g) *cis-para*-coumeric acid; (h) *trans*-ferulic acid; (i) *cis*-ferulic acid; (j) *para*-hydroxybenzaldehyde; (k) vanillin; (l) syringaldehyde; (m) 6,7-dihydroxycoumarin (aesculetin); (n) (+)-catechin, found in the seeds; (o) (−)-epicatechin, found in the seeds; and (p) esters at one of the two hydroxyl (−OH) groups of (L)-(+)-tartaric acid with most of the above cited carboxylic acids.

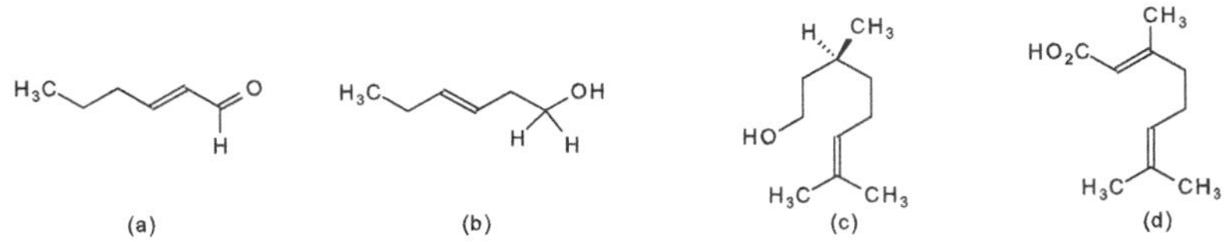

FIGURE 14.3 Representations of some early volatile components reported found in the Airén grape: (a) *trans*-2-hexenal; (b) *trans*-3-hexen-1-ol; (c) (−)-citronellol; and (d) geranic acid.

climate of the growing season is critical, because Cabernet Sauvignon is one of the varieties to bud and ripen "late." Thus bright sunshine and long warm days provide a better outcome. Interestingly, the history of derivation of the Sauvignon blanc component may account for the presence of pyrazines[10] (which some find undesirable) in the cultivars found in cooler climates. The pyrazines are apparently destroyed with fuller ripening in the sunshine. Therefore, while the grape remains the same, the kind and number of phenolics that affect the essence of this grape will vary as a function of where the grape is grown. Vineyards with this cultivar are found growing in soils as diverse as the rich soils of Bordeaux and the Loire Valley, France, as well as in the United States in northern and southern California, in the Piedmont and Tuscan regions of Italy, as well as regions within many other countries. Thus, it bears the reputation as a truly cosmopolitan "aristocrat" grape which is, nonetheless, different from region to region. And yet it is so generous in its bounty that it is often mixed with the wine from other grapes to produce desirable flavors.

The subsequent treatment of the harvest will differ as dictated by custom, the aging process in oak (not all oak being the same either—as will be discussed).

In the Castilla-La Mancha region of Spain, the rich phenolic nature of Cabernet Sauvignon grapes (and other *V. vinifera* cultivars) was investigated.[11] The phenolic compounds found in the red grape skin of Cabernet Sauvignon included those shown in Figure 14.4. Some of these phenols were also found in the seeds.

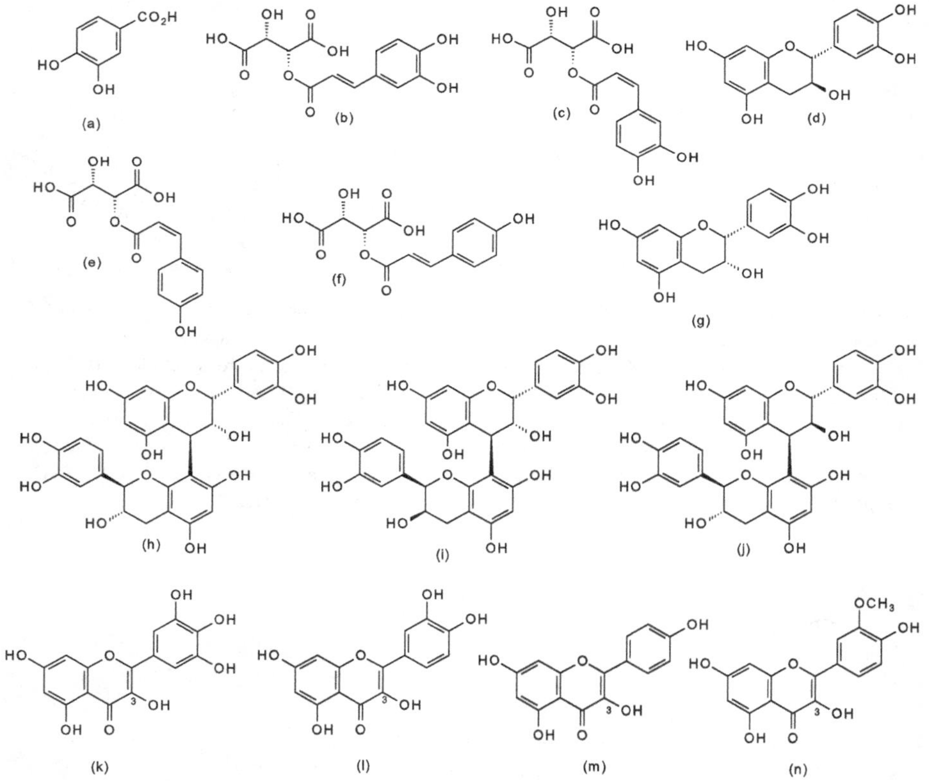

FIGURE 14.4 Representations of phenolics reported to be found in the skin of Cabernet Sauvignon. The more extensively conjugated phenols lend color to the skin. Those isolated include: (a) protocatechuic acid; (b) *trans*-caftaric acid; (c) *cis*-caftaric acid; (d) (+)-catechin; (e) *cis*-coutaric acid;, (f) *trans*-coutaric acid; and (g) (−)-epicatechin. The dimers of catechin and epicatechin are also present as: (h) procyanidin B1; (i) procyanidin B2; (j) procyanidin B3. The compounds: (k) myricetin; (l) quercetin; (m) kaempferol; and (n) isorhamnetin were reported to be found as the 3-*O*-glucosides, glucuronides, glucosylxylosides, etc.

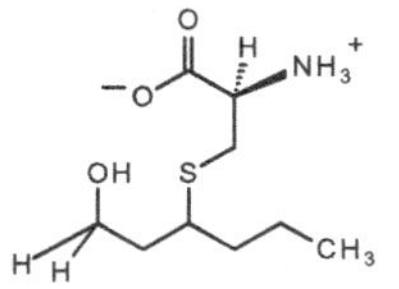

A Selection of Grapes

FIGURE 14.5 A representation of 3-mercaptohexan-1-ol-L-cysteine, reported to found in the skin of Cabernet Sauvignon.

The grapes themselves have also been examined. In one case, the major volatile compounds from Cabernet Sauvignon grapes from fruit-set to harvest were isolated and compared to those from Riesling grapes (*vide infra*). The Cabernet Sauvignon grapes were generally found richer in variety of components.[12]

The pre-véraison volatile compounds found in the Cabernet Sauvignon grapes examined are shown in Figure 14.6, and while the concentrations changed during and post-véraison, only a few additional compounds (not shown here, but known, e.g., benzaldehyde, benzyl alcohol, etc.) appeared. There is also some evidence that at least one (as one is listed but not identified) C_{13} norisoprenoid was also present during and after véraison. A full list is to be found in Kalua, C. M.; Boss, P. K. *J. Agric. Food Chem.* **2009,** *57,* 3818. Finally, it has been reported that the skins of the grape contain small quantities of a sulfur-containing (cysteine) derivative of hexan-1-ol (i.e., 3-mercaptohexan-1-ol-L-cysteine (Figure 14.5).[13]

Chardonnay: A white grape.

The dry, clean, and crisp white wine from Burgundy generated from the light green-to-yellow grape called Chardonnay has found its way into vineyards around the world. As is expected, numerous cultivars have come into being largely as a function of planned changes in growing season as well as the terroir which, as expected, remains critical to the production of suitable wines. Indeed while Chardonnay can adapt to almost all vineyard soils, those rich in calcium (Ca^{2+}), magnesium (Mg^{2+}), and potassium (K^+) such as chalk, clay, and limestone, seem particularly favored.

In addition to being grown in the Burgundy region, Chardonnay is also grown in the Champagne region, where it is popular for the production of Champagne. In part, this is because the Chardonnay grape can be used not only to make dry, still wines but also sparkling wines, sweet wines, and even botrytized wine (wine from grapes infected with a fungus that, under suitable conditions, will produce sweet dessert wines, *vide infra*). Topics to be discussed subsequently, include botrytized wine, the temperature of the fermentation, the time spent on the lees (dead yeast and other solids deposited during fermentation), the options of using malolactic fermentation (malic acid conversion to other products by a so called "secondary fermentation"), the conversion of the still wine to the sparkling wine, and when, how, and to what extent oak barrels are used to age the fermented product.

Interestingly, the same sets of compounds found in the Airén grape (*vide supra*) were also found in Chardonnay and are considered typical of "white grapes."[4]

However, even before the isolation of comparable compounds, it was pointed out that "Chardonnay juices seldom have a distinct aroma," although the wines of this variety can usually be readily recognized.[14] Implicitly then, Chardonnay juices contain constituents which, although odorless, are capable of producing characteristic aroma

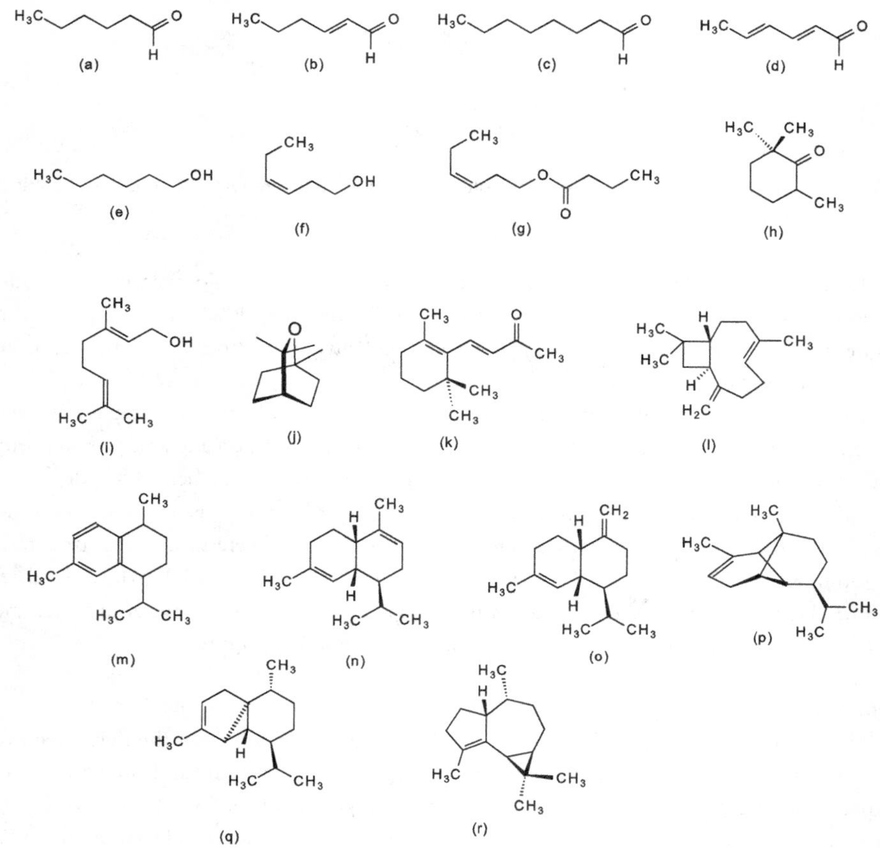

FIGURE 14.6 A representation of some pre-véraison volatile compounds reported to be found in the Cabernet Sauvignon grape. As shown, the compounds detected are a mixture of aldehydes, alcohols, esters, ketones, a C-13 norterpene and a variety of terpenes, diterpenes and sesquiterpenes and their derivatives. Those shown are: (a) hexanal; (b) (*E*)-2-hexenal; (c) octanal; (d) 2,4-hexadienal; (e) 1-hexanol; (f) (*Z*)-3-hexen-1-ol; (g) (*Z*)-3-hexenyl butanoate; (h) 2,2,6-trimethylcyclohexanone; (i) geraniol; (j) eucalyptol; (k) β-ionone; (l) caryophyllene; (m) calamenene; (n) α-muurolene; (o) γ-muurolene; (p) (−)-α-copaene; (q) (−)-α-cubebene; and (r) α-gurjunene.

compounds in the finished wine.[15] The authors[14] identified one hundred eighty compounds, of which twenty-eight were not previously reported as grape components. Interestingly, more than 70% of the total concentration of volatile secondary metabolites was comprised of thirteen-carbon (C_{13}) norisoprenoids apparently derived from carotene or carotene-like species (the genesis of which are discussed below, *vide infra*, under the heading "Nebbiolo"). Indeed, in that vein, it has recently been pointed out that "the physiological functions of carotenoids in plants go beyond their traditional roles as accessory light-harvesting pigments, natural colorants, and quenchers of triplet chlorophyll and singlet oxygen ($^1O_2$)."[16] Indeed, in the process of quenching singlet oxygen, the C_{13} compounds form, and many of them and their sugar conjugates are unique to specific grapes. Thus, while esters, alcohols, ketones, and terpenes that are expected

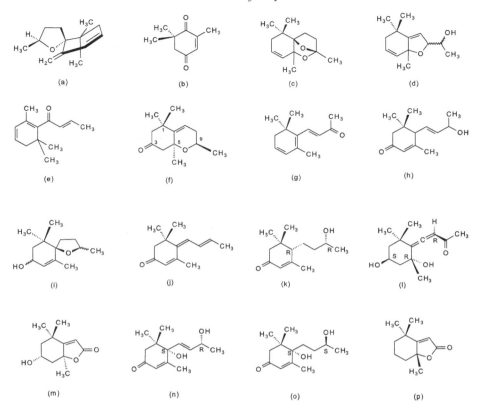

FIGURE 14.7 A representation of the rich array of C-13 compounds, derived by degradation of terpenoids and found in Chardonnay. Those shown here are: (a) vitispirane; (b) 2,6,6-trimethyl-cyclohex-2-ene-1,4-dione; (c) Riesling acetal; (d) actinidol (of uncertain stereochemistry); (e) β-damascenone; (f) (5S,9R)-3,4-dihydro-3-oxoedulan; (g) *trans*-3-dehydro-β-ionone; (h) 3-oxo-α-ionol; (i) 8-hydroxytheaspirane; (j) (6Z,8E)-megastigma-4,6,8-trien-3-one; (k) blumenol C; (l) grasshopper ketone; (m) loliolide; (n) vomifoliol (or blumenol A); (o) blumenol B; and (p) dihydroactinidiolide.

are also present (Appendices 2–5), the unique C_{13} compounds shown in Figure 14.7 deserve special note as terpene degradation products.

As noted earlier, the Chardonnay grape (and the wine produced from it) has been examined[17] as a function of leaf treatment. Both basal leaf plucking and head trimming were employed to study the effect on synthesis of aromatic precursors and the impact on wine aroma. Chardonnay was least affected!

Finally, it has been reported[13] that the skins of the grape contain small quantities of sulfur-containing derivatives of hexan-1-ol (i.e., 3-mercaptohexan-1-ol-L-cysteine and 3-mercapto-hexane-1-ol-L-glutathione) (Figure 14.8).

Chenin blanc: A white grape.

In the Loire valley region of France, Chenin blanc is known as an easily grown white grape variety. The juice of this wine is, generally, acidic, and both sparkling wines and dessert wines

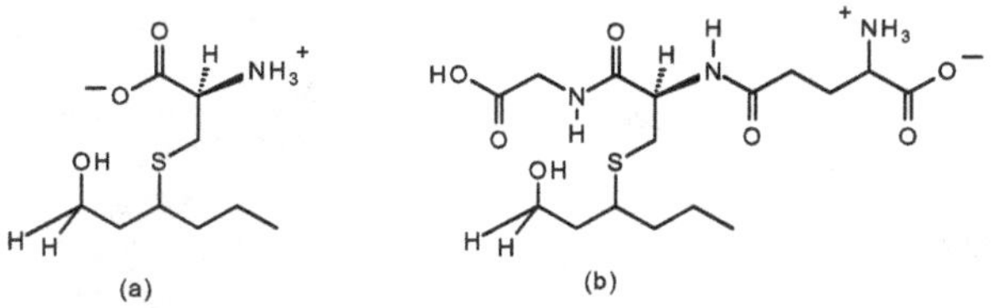

FIGURE 14.8 Representations of sulfur-containing compounds reported to be found in the skin of Chardonnay grapes: (a) 3-metcaptohexan-1-ol-L-cysteine and (b) 3-mercaptohexan-1-ol-L-glutathione.

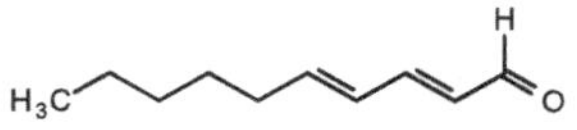

FIGURE 14.9 A representation of the structure of one of the 2,4-decadienals, *viz.*, (2*E*,4*E*)-2,4-decadienal.

have been produced from this grape. However, if the vintner's treatment is chosen appropriately, a neutral, bland wine can be generated that is good for mixing with other varieties. Indeed, since this grape can provide a neutral palate, the expression of added sugars, acids, and other flavoring notes from oak barrels and other sources allows for a variety of flavors all of which would be called Chenin blanc. It also appears that cuttings taken from a growing vine may have enough variability to allow the Chenin blanc clone to introduce minor changes in acidity and sweetness (as a function of *terroir*) to overcome its normal bland nature, and perhaps for that reason, the grape has been referred to as "the world's most versatile grape."[18]

In the early 1980s[19] grapes from different localities and at various stages of maturity were harvested, extracted with a halogenated solvent, and examined gas chromatographically with a mass spectrometer as the detector (GC-MS). Interestingly, "no measureable amounts of terpenoid components were detected." Two isomers of 2,4-decadienal (Figure 14.9) were found which, at low concentration, are known to have the odor of citrus fruit.

Subsequently, it was reported[20] that leaf volatile compounds isolated from steam distillation of Chenin blanc leaves included the six-carbon (C_6) compounds: *cis*-2-hexenal, *trans*-2-hexen-1-ol, *cis*-3-hexen-1-ol, 1-hexanol, and *cis*- or *trans*-hexa-2,4-dienal. In addition, terpene and terpenoid compounds including myrcene, linalool, α-terpineol, isopulegone, citral, geraniol, and ionone along with the sulfur containing compounds tetrathiocane and *cis*- or *trans*-3,5-dimethyl-1,2,4,-trithiolane were identified and are shown in Figure 14.10.

Recently, work was undertaken[21] to examine Chenin blanc during attack on the grape by *Botrytis cinerea*[22] in an attempt to provide objective identification of levels of botrytisation.

While most of the effort lay in physical examination of the grape and properties of its skin, phenolic composition utilizing isolation and mass spectrometry did deal with some of the chemistry of the grape. It was pointed out that uninfected grapes had levels of gluconic acid (a botrytisation marker) below detection. It was also found that the total level of (unidentified) polyphenols decreased as the infection progressed, as did the amounts of caftaric acid, coutaric acid, quercetin-3 glucoside, kaempferol-3 glucoside, astilbin, and resveratrol.

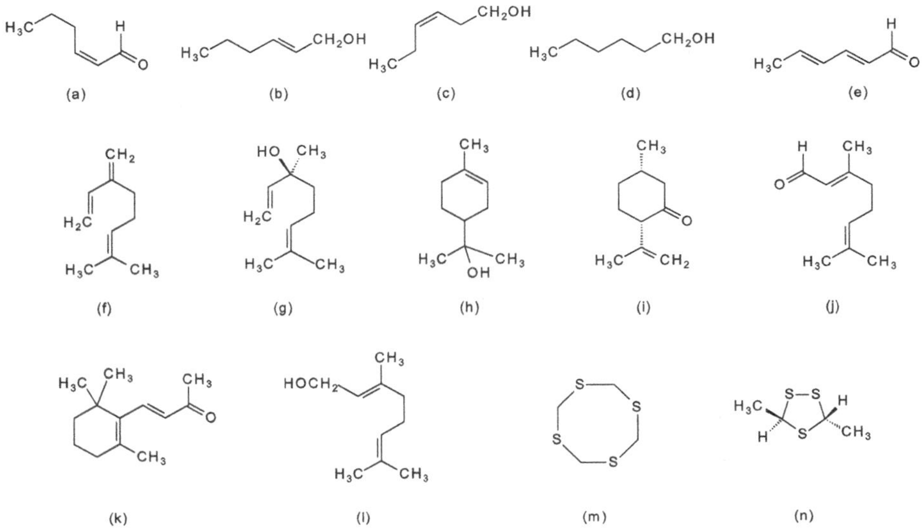

FIGURE 14.10 Representations of the structures of some of the compounds found in the leaves of the Chenin blanc vine. There is no evidence that these compounds are found in the grape itself, but the data remains incomplete. Thus, the plant does have: (a) *cis*-2-hexenal; (b) *trans*-2-hexenal; (c) *cis*-3-hexen-1-ol; (d) 1-hexanol; (e) (*E,E*)-hexa-2,4-dienal; (f) myrcene; (g) linalool; (h) α-terpineol; (i) *cis*-isopulegone; (j) citral (geranial); (k) β-ionone; (l) geraniol; (m) 1,3,5,7-tetrathiocane; and (n) *trans*-3,5-dimethyl-1,2,4-trithiolane available for biosynthetic utilization.

However, the concentrations of catechin, epicatechin, and epicatechin gallate all increased. The compounds are shown in Figure 14.11.

Gewürztraminer: A white grape.

It has been argued that this "Perfumed Traminer" originally came from the Alsace region of France and that the grape variety is a mutation of Sauvignon blanc (and also found under the name "Traminer" *viz*, "*vin blanc issu de ce cépage*" or "white wine made from the vine").[23]

Thus, it has recently been reported "that 75% of the vinifera cultivars in the USDA grape germplasm collection are related to at least one other cultivar by a first-degree relationship ... (and that) ... the genetic structure of vinifera can be largely understood as one large complex pedigree." Further, it was "propose(d) that this pedigree structure is the result of a limited number of crosses made among elite cultivars that were immortalized and sometimes vegetatively propagated for centuries." And finally that "... the most highly connected wine grape in the present sample is Traminer ... believed to be an ancient cultivar widely used during the history of grape breeding."[24] A representation of the connectivity network is provided as Figure 3 in that reference and seen below as Figure 14.12

At some point, either a Traminer or Sauvignon blanc mutated into a form with pink-skinned berries which are common today. In addition, it has been held that a general genetic instability led to a further mutation finally yielding the extra-aromatic Gewürztraminer. Thus, as noted above, the "Traminers" are apparently all related clones, and of course, the very large number of such vigorous early budding grapes, needing dry and warm summers, are widely found in many grape-growing regions.

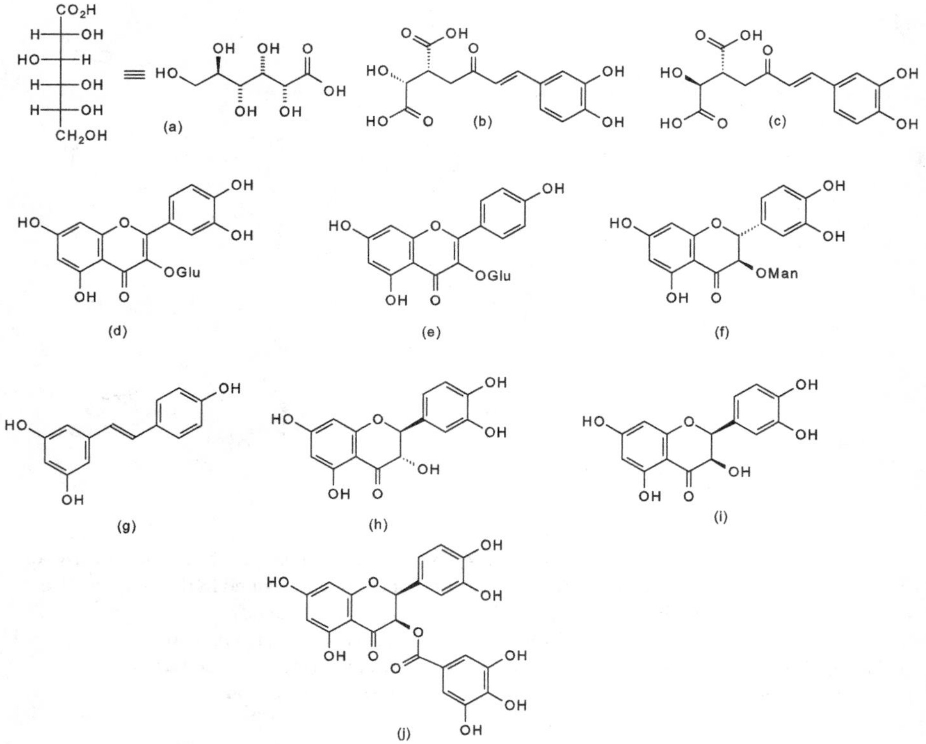

FIGURE 14.11 A representation of constituents reported present and changing on infection with the fungus *Botrytis cinerea*. These include the absence of: (a) gluconic acid and decrease in the concentration of (b) caftaric acid; (c) coutaric acid; (d) quercetin-3 glucoside; (e) kaempferol-3 glucoside; (f) astilbin; and (g) resveratrol. The concentrations of (h) catechin; (i) epicatechin; and (j) the gallic acid ester of epicatechin (epicatechin gallate) all increased.

Interestingly, although the grapes are white to pink (as a consequence of traces of anthocyanins), the color is often lost during fermentation as the phenolic materials lead to polymers that can be removed and, should that fail, those traces can be removed after fermentation has been completed with treatment by appropriate enzymes.[25]

Curiously, there appears to be striking similarities in the aroma of a number of Gewürztraminer variety wines and lychee (*Litchi chinesis Sonn.*) fruit.[26]

Although a few of the compounds (Figure 14.13) found common to both lychee and the Gewürztraminer variety examined were not specifically identified despite the gas chromatographic analysis (e.g., the specific stereoisomer of linalool and 2-methylbutanoate and the ethyl hexanoate/ethyl isohexanoate ratio), it was clear, nonetheless, that they were comparable. So it seems that while the character of this full-bodied aromatic white wine is clearly distinguishable among white wines, almost regardless of the place and name under which it is known, the compounds responsible are also used elsewhere and are also distinguishable.

As noted earlier, this grape (and the wine produced from it) has been examined[17] as a function of leaf treatment. Both basal leaf plucking and head trimming were employed to

FIGURE 14.12 Figure 3 from Myles, S.; Boyko, A. R.; Owens, C. L. Brown, P. J.; Grassi, F.; Aradhya, M. K.; Prins, B.; Reynolds, A.; Chai, J.-M.; Ware, D.; Bustamante, C. D.; Buckler, E. S. *Proc. Nat. Acad. Sci. (US)* **2011**, *108*, 3530. Used with permission.

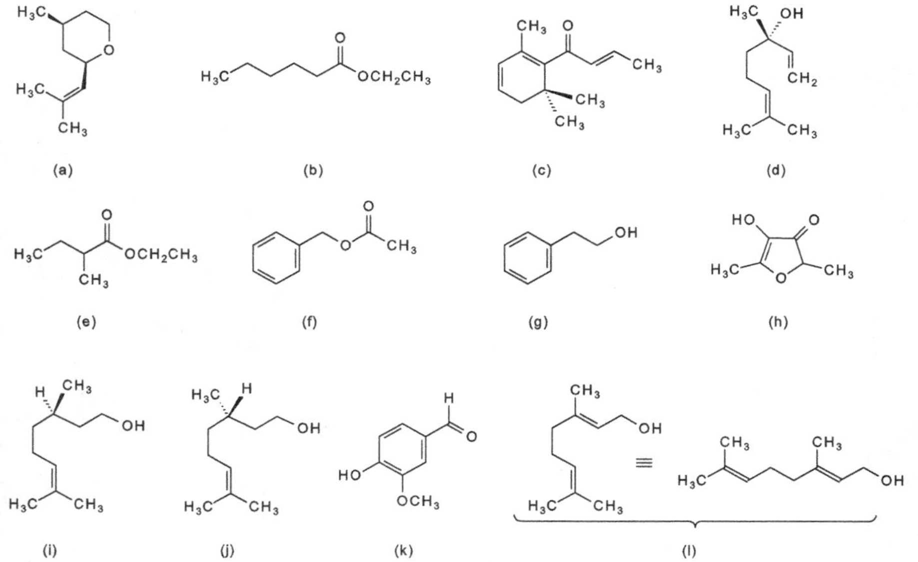

FIGURE 14.13 Representations of the compounds found in both Gewürztraminer variety wines and lychee (*Litchi chinesis Sonn.*) fruit. The compounds shown are: (a) rose oxide; (b) ethyl hexanoate; (c) β-damascenone; (d) (S)-(+)-linalool; (e) ethyl 2-methylbutanoate; (f) benzyl acetate; (g) 2-phenylethanol; (h) furaneol; (i) (−)-citronellol; (j) (+)-citronellol; (k) vanillin; and (l) geraniol.

study the effect on synthesis of aromatic precursors and the impact on wine aroma. The effect on Gewürztraminer was clear.

It has been reported[13] that the skins of the grape contain small quantities of sulfur derivatives of hexan-1-ol (i.e., 3-mercaptohexan-1-ol-L-cysteine and 3-mercaptohexane-1-ol-L-glutathione) and 4-mercapto-4-methyl-2-pentanone (i.e., 4-mercapto-4-methyl-2-pentanone-L-cysteine and 4-mercapto-4-methyl-2-pentanone-L-glutathione) (Figure 14.14).

A GC-MS analysis of volatile compounds found in Gewürztraminer wines in addition to that referenced above[17] has recently been made available, and structures assigned to the compounds in the vapor are shown in Figures 14.15a–14.15e.[27] These were chosen since the contrast between the *relative concentrations* of some of the volatile constituents found in a Gewürztraminer wine produced from normally harvested grapes and ice-harvest grapes are substantive and serve to demonstrate that, overall, flavor is, at least in part, a function of relative concentrations of the same constituents.

The contrast can be found in the section of this work under the heading *Ice Wine*.

It is important to note that the compounds shown in Figures 14.15a–14.15e are from the vapor above a wine prepared from Gewürztraminer grape. The constituents of the wine, as will become clear in the following pages, are both more and less than what might have been (or actually was) present in the grape. Some of the constituents of the grape may remain in the wine. However, it is clear that, along with sucrose, which has been modified through fermentation by the added yeast (presumably a strain of *Saccharomyces cerevisiae* was used), other factors resulting from the treatment of the crushed grape may also change what was originally present through deletions by reaction or through rearrangement to new products.

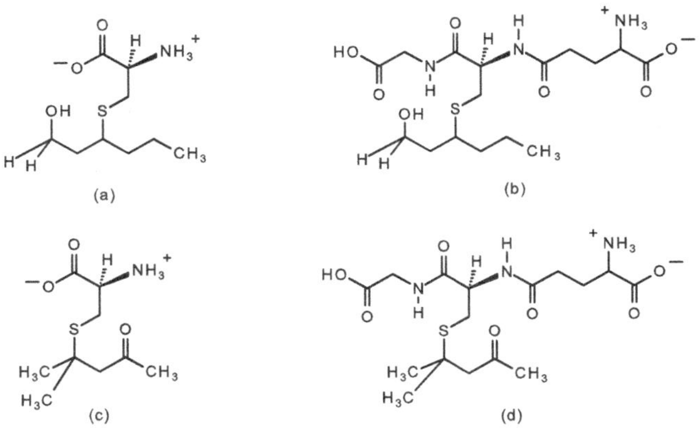

A Selection of Grapes 135

FIGURE 14.14 A representation of the sulfur containing compounds reported to be found in the skins of Gewürztraminer grape. Some of them are also found in the skins of grapes yielding white wines. Those shown are: (a) 3-metcaptohexan-1-ol-L-cysteine; (b) 3-mercaptohexan-1-ol-L-glutathione; (c) 4-mercapto-4-methyl-1-pentanone-L-cysteine; (d) 4-mercapto-4-methyl-2-pentanone-L-glutathione.

So the many compounds existing in the fraction of the liquid that is neither water nor ethanol, and which has been further fractionated (as a consequence of having sufficient vapor pressure at room temperature) into the vapor, is truly complex. And, finally, it is important to keep in mind that the respective quantities of the vapor components vary as a function of age, some falling below levels of detection.

Merlot: A red grape.

"Merlot has traditionally been Bordeaux's 'other' red grape."[28]

The relationship between Merlot and Traminer has been noted above. Many of the same compounds are also found in Merlot, and while Cabernet Sauvignon has been the main variety in Bordeaux, it has been argued that television made a difference in the popularity experienced by Merlot and lost by Cabernet Sauvignon. The virtue of drinking red wine was expounded upon during a television show in 1991 (the subject of a *60 Minutes* program on the so called "French Paradox"—discussed more fully in Chapter 22) and that exposition subsequently sufficed to convinced wine consumers in the United States to imbibe. The soft mouthfeel (Chapter 22) of Merlot was appealing, and the fact that the early vintages came from Bordeaux, where it had been mixed with Cabernet Sauvignon to soften its rough edge, made it desirable.

There is another interesting story regarding Merlot. It has been reported[29] in a study regarding the origin of grape cultivars, that in 1851, Don Silvestre Ochagavia imported a French Merlot cultivar into Chile. Chilean soil and terroir in general proved conducive to its growth. The Phylloxera infection (Chapter 3), beginning in about 1860, ravaged the European vineyards. Recalling that the rescue of the vineyards of Europe involved grafting scions onto the rootstock of a resistant American native species and with the new knowledge[30] that there may be exchange of genetic material between cells in the plant tissue grafts and that may lead to gene transfer,[31–33] it was of some interest to compare the genome of current French and Chilean Merlot cultivars.

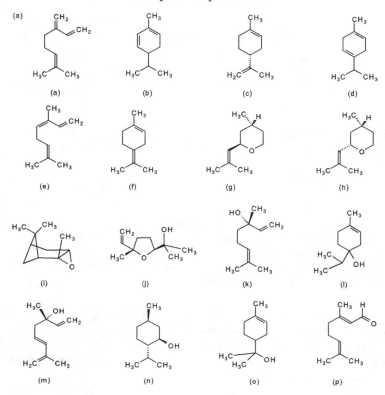

FIGURE 14.15a Representations of some of the terpene-derived compounds found in the vapor above wine made from the Gewürztraminer grape. They include: (a) β-myrcene; (b) α-phellandrene; (c) (R)-(+)-limonene; (d) γ-terpinene; (e) trans-β-ocimene; (f) terpinolene; (g) trans-rose oxide; (h) cis-rose oxide; (i) α-pinene oxide; (j) trans-linalool oxide; (k) (S)-(+)-linalool; (l) 4-terpineol; (m) hotrienol; (n) L-(−)-menthol;(o) α-terpineol; and (p) geranial. Where stereochemistry is not shown, it was not defined.

In the event[29] examination of the genome of the Merlot cultivar from Chile (presumably unaffected by Phylloxera) with the genome of current French cultivar resulted in the observation that "a different DNA fingerprint (was found) when the cultivar Merlot cultivated in Chilean vineyards was compare(d) to its French ancestor stock . . . (and) a 64% similarity, which is a remarkably low similarity (was observed) . . ."[29] In the same vein, "it is unusual for a grape to become so fashionable without there being any real consensus about how it should be grown and made."[34] Clearly, such clones both in the New World and Old demonstrate the role of climate and terroir along with the malleability of the genome and yet the retention of the fundamental characteristics so necessary to retain the name. Table 2 in Appendix 5 provides a comparison of the compounds found in New and Old World wines, and it is clear that there are both similarities and differences.

In the same vein, it has been pointed out[35] that Merlot wines (from a group of different vintages and varieties) when compared for color and astringency (which included other similar attributes) as well as phenolics, tartaric esters, flavonols, and anthocyanins among others, were found to "encompass a wide range of characteristics without exhibiting high levels of

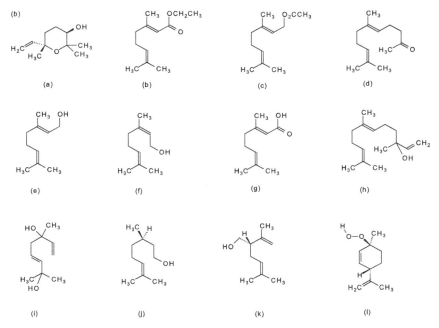

FIGURE 14.15b Representations of some of the terpene-derived compounds found in the vapor above wine made from the Gewürztraminer grape. They include: (a) (3R,6S)-*trans*-linalool oxide (pyran form); (b) ethyl geranate; (c) geranyl acetate; (d) geranylacetone; (e) geraniol; (f) nerol; (g) geranic acid; (h) *trans*-nerolidol; (i) diendiol-1; (j) (R)-(+)-citronellol; (k) (S)-lavandulol; and (l) (1S,4R)-1-methyl-4-(1-methylethenyl)-2-cyclohexane-1-hydroperoxide.

any one attribute." Merlot grapes have been studied under a variety of conditions, and since the wine from those grapes enjoy a modicum of popularity, such research has apparently proved useful. Thus, a serious attempt, using the Merlot grape, has been made to measure the effect of water deficiency on the concentrations of tannins and anthocyanins across a number of growing seasons.[36] The result has shown that although anthocyanins and non-hydrolysable tannins have similar initial biosynthetic pathways (i.e., beginning with phenylalanine and acetyl coenzyme A to form naringenin chalcone as shown in Figure 14.16), "vine water deficit during ripening is a much more effective tool to increase anthocyanins than tannins in Merlot grapes."[36]

It was pointed out in 2009[37] that abscisic acid (Figure 14.17), one of the plant hormones involved in signaling during grape development and ripening, varied during véraison (the onset of ripening). Interestingly, although the pulp and skin of pre-véraison berries contained moderate amounts of abscisic acid, they also found high concentrations of dihydrophaseic acid, which is (logically) considered to be a product arising from further oxidation and rearrangement of abscisic acid (Figure 14.17) So, it was concluded that abscisic acid had been present earlier in the berry development and had been oxidized and trapped as dihydrophaseic acid. Further metabolites, of which some had been studied before, were not reported.[38] Additionally, there were two peak periods during which larger concentrations of abscisic acid were found in the pulp and skin of the berry: the first was two weeks pre-véraison and the second at late véraison. When abscisic acid was applied to the berries at

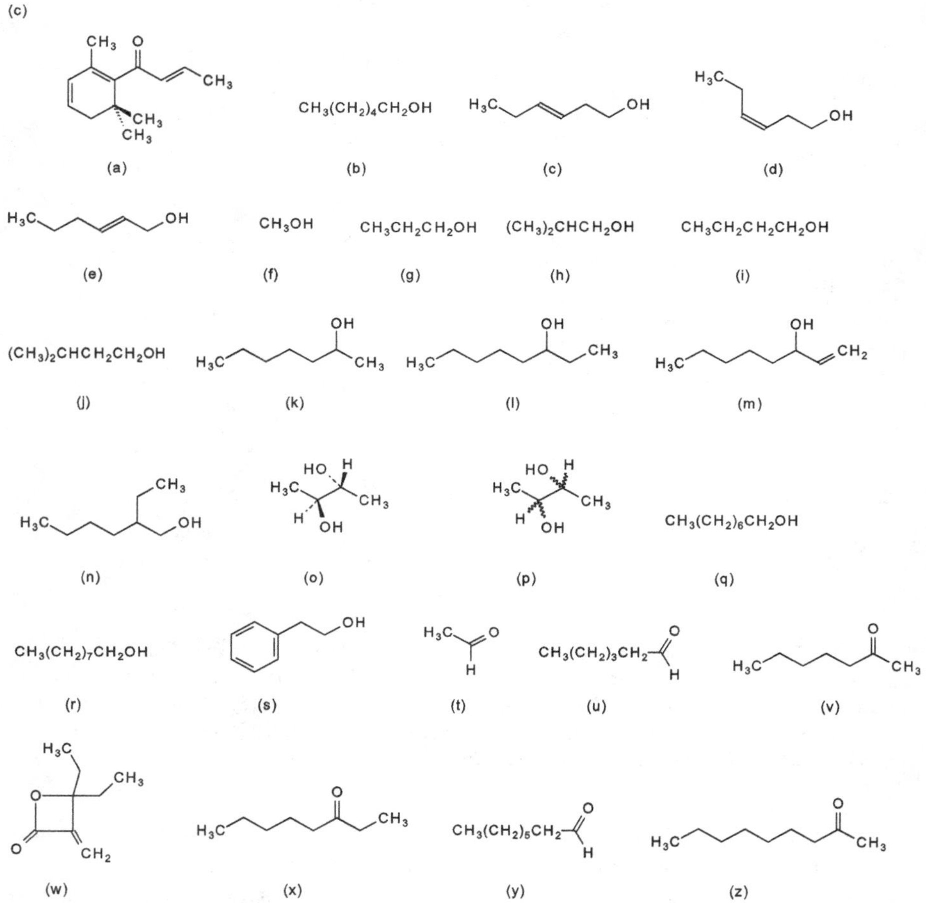

FIGURE 14.15c Representations of some of the aldehydes, alcohols, and ketones found in the vapor above wine made from the Gewürztraminer grape. They include: (a) (*E*)-β-damascenone; (b) 1-hexanol; (c) (*E*)-3-hexen-1-ol; (d) (*Z*)-3-hexen-1-ol; (e) (*E*)-2-hexene-1-ol; (f) methanol; (g) 1-propanol; (h) 2-methyl-1-propanol (isobutanol); (i) 1-butanol; (j) 3-methyl-1-butanol (isoamyl alcohol); (k) 2-heptanol; (l) 3-octanol; (m) 1-octen-3-ol; (n) 2-ethyl-1-hexanol; (o) (2*R*,3*S*)-2,3-butanediol; (p) (2*R*,3*R*)-2,3-butanediol and (2*S*,3*S*)-2,3-butanediol; (q) 1-octanol; (r) 1-nonanol; (s) 2-phenylethanol; (t) acetaldehyde; (u) hexanal; (v) 2-heptanone; (w) 4,4-diethyl-3-methylene-1-oxetan-2-one; (x) 3-octanone; (y) octanal; and (z) 2-nonanone.

véraison, a large increase in the concentration of dihydrophaseic acid was detected, but there seemed to be no effect on berry maturation rate.

As noted earlier, this grape (and the wine produced from it) has been examined as a function of leaf treatment. Both basal leaf plucking and head trimming were employed to study the effect on synthesis of aromatic precursors and the impact on wine aroma. The effect on Merlot was clear as noted in Chapter 7, with more terpenes and related products produced from trimmed vines.[17]

Finally, for this grape, it has been reported[13] that the skins of the grape contain small quantities of a sulfur-containing (cysteine) derivative of hexan-1-ol (i.e., 3-mercaptohexan-1-ol-L-cysteine) (Figure 14.18).

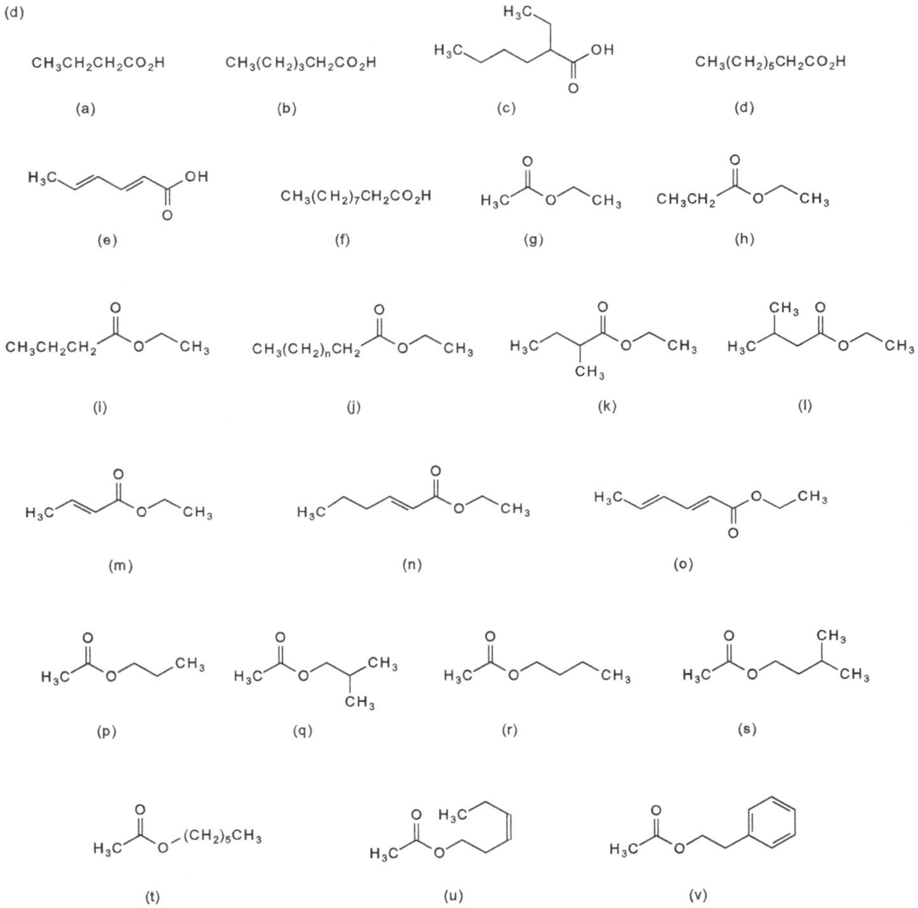

FIGURE 14.15d Representations of some of the acids and esters found in the vapor above wine made from the Gewürztraminer grape. They include (a) butanoic acid; (b) hexanoic acid; (c) 2-ethylhexanoic acid; (d) octanoic acid; (e) sorbic acid; (f) decanoic acid; (g) ethyl acetate; (h) ethyl propionate; (i) ethyl butanoate; (j) ethyl esters of n=2, pentanoic; n=3, hexanoic; n=4, heptanoic; n=5, octanoic; n=6, nonanoic; n=7, decanoic; n=9, dodecanoic; n=11, tetradecanoic acids; (k) ethyl 2-methylbutanoate, (l) ethyl 3-methylbutanoate; (m) ethyl (*E*)-2-butenoate; (n) ethyl (*E*)-2-hexenoate; (o) ethyl sorbate; (p) propyl acetate; (q) 2-methylpropyl acetate; (r) butyl acetate; (s) 3-methylbutyl acetate; (t) hexyl acetate; (u) (*Z*)-3-hexenyl acetate; and (v) 2-phenethyl acetate.

Muscat: A white grape.

A suggestion for the relationship of **Muscat** to **Traminer** is noted graphically above under the heading of the latter (*vide supra*). However, the Muscat family of grapes includes a wide variety of related, characteristically aromatic (terpene-based "musk" -like odorants) grapes.

Muscat Blanc à Petits Grains is a white (*blanc*) grape with characteristically small (*petite*) berries and tight clusters. While the small size and tightness is generally maintained, there is some evidence that from year to year its color may vary. Indeed, Muscat Giallo is characterized by yellow grapes, Moscato rose del Trentino by pink grapes, and both Muscat Hamburg

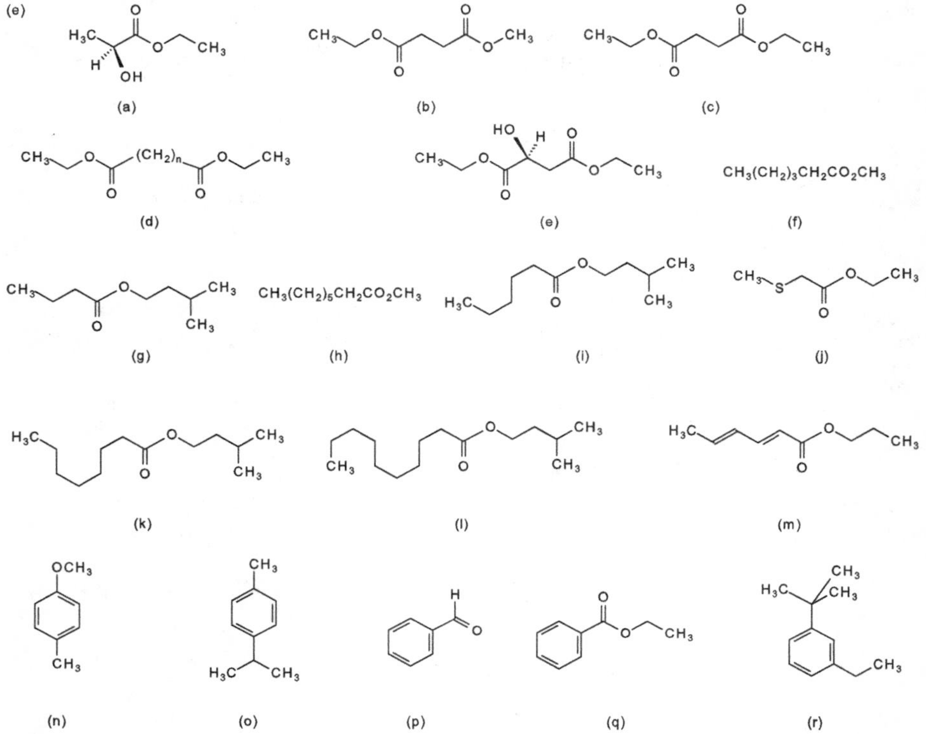

FIGURE 14.15e Representations of some of the esters found in the vapor above wine made from the Gewürztraminer grape. They include (a) (−)-ethyl (S)-lactate; (b) ethyl methyl succinate; (c) diethyl succinate; (d) diethyl esters of n = 3, glutaric; n = 4, adipic; n = 5, pimelic; n = 6 subaric acids; (e) (S)-diethylmalate; (f) methyl hexanoate; (g) 3-methylbutyl butanoate; (h) methyl octanoate; (i) 3-methylbutyl hexanoate; (j) ethyl 3-(methylthio)propanoate; (k) 3-methylbutyl octanoate; (l) 3-methylbutyl decanoate; (m) propyl sorbate; (n) 4-methylanisol; (o) *para*-cymene; (p) benzaldehyde; (q) ethyl benzoate; and (r) 1-ethyl-3-(2-methyl-2-propanyl)benzene.

and Muscat Norway by dark, almost black grapes. Moscato Asti, an Italian white grape, is used in the production of the sparkling Asti wine. The pleasure associated with a dark Muscat is described vividly by Oz Clarke.[39]

Using a version of gas chromatography (GC) with a mass spectrometer (MS) as a detector, the terpene alcohols (and other compounds) found in the headspace above Sicilian Muscat wine cultivars "Moscato di Siracusa," "Moscato di Noto," and "Moscato di Pantelleria" were recently examined.[40] While there was some variation in the quantities of the various compounds detected as a function of cultivar, the main conclusion was that the quantity of terpenes decreased with aging. The reason for this was not mooted, and indeed, it is not clear if this is because of their volatility or because they were undergoing reactions with other components in the mixture. Some of the compounds reported are shown in Figure 14.19. Where more than one optical isomer (each with a different aroma) is known, only the most common is shown. The GC-MS did not, of course, distinguish between the optical isomers.

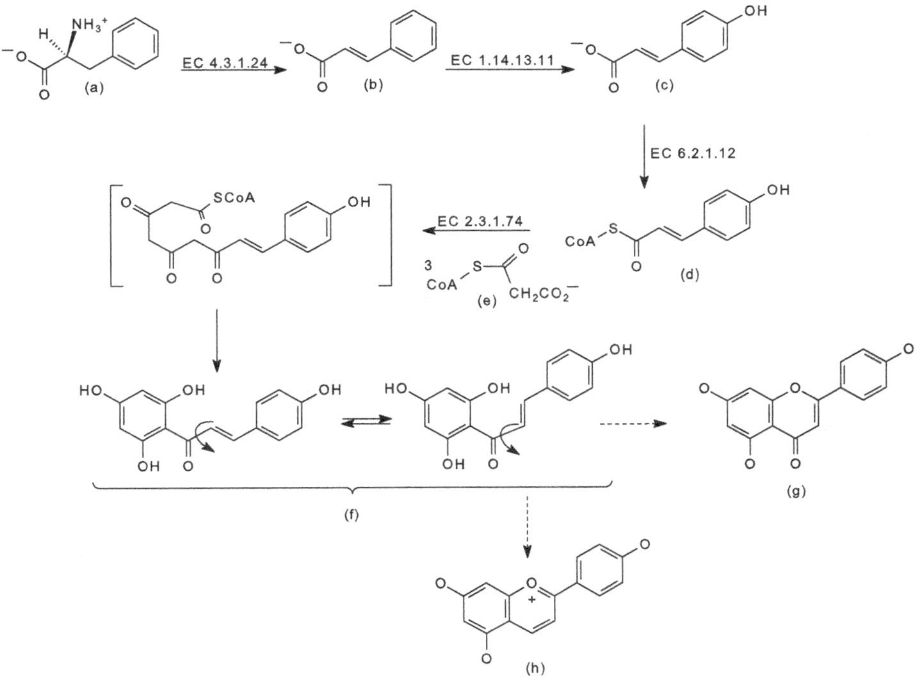

FIGURE 14.16 A representation showing the relationship between phenylalanine (a) and the flavone (g) and anthocyanidin (h) ring systems. Thus phenylalanine (a), through the action of phenylalanine ammonia lyase, EC 4.3.1.24, yields *trans*-cinnamate (b). Then, oxidation of *trans*-cinnamate by *trans*-cinnamate 4-monooxygenase, EC 1.14.13.11, generates 4-coumarate (*para*-coumarate) (c). Ligation of the *para*-coumarate with acetyl-CoA to (d) and subsequent elaboration with malonyl-CoA (e) in the presence of naringenin-chalcone synthase (EC 2.3.1.74) leads to naringenin (f) with the oxygen substitution pattern (whether –OH, –OCH$_3$, etc.) specified. The dashed lines show, broadly, that naringenin (f) leads to flavone (g) and anthocyanidin (h) systems.

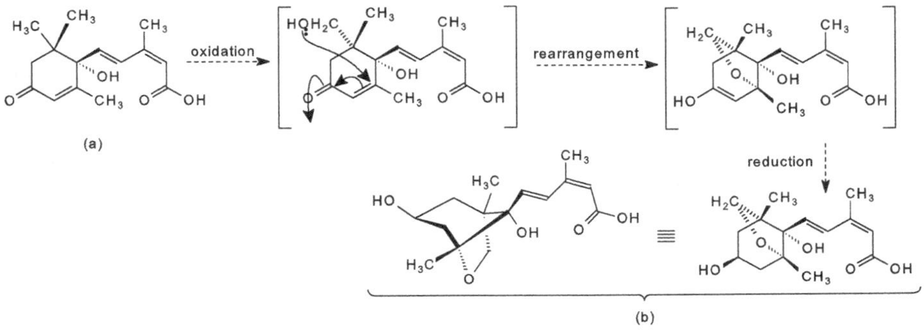

FIGURE 14.17 A representation of a potential pathway from abscisic acid (a) to dihydrophaseic acid (b) *via* presumed oxidation, rearrangement, and subsequent reduction.

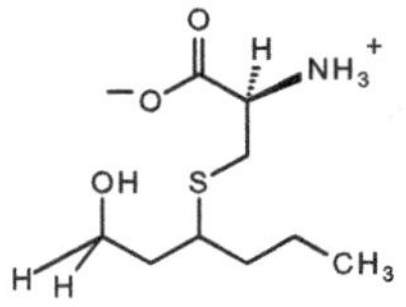

FIGURE 14.18 A representation of 3-mercaptohexan-1-ol-L-cysteine.

Most of the terpenes have unique but "flowery" aromas. The sweeter and "unique" (e.g., as unique to pineapple, strawberry, apple, etc.) aromas in fruits are generally esters (Appendices 1 and 3). Esters reported[40] in Muscat are shown in Figure 14.20. Some are already present in the grape before it is converted to wine, and others form during vinification.

Finally, for this grape, there is a group of simple alcohols that, in addition to ethanol, are present and shown in Figure 14.21.

Although the details can be found in the original work (*vide supra*) it is clear, in summary, that differences in the concentration of compounds rather than differences in the kinds of compounds present in the grapes are to be held responsible for the subtle (and not so subtle) differences in aromas and tastes. Presuming relatively minor changes, it was argued that analyses such as that performed might serve to "characterize Muscat wines on the basis of geographic origin."[40]

Mourvèdre: A red grape.

The dark purple-to-black heavy, phenol-rich juice of the earthy product of the Mourvèdre grape is said to have Spanish heritage (where, today, the grape is called "Monastrell"). Interestingly, although it is argued that the grape, under a variety of names,[41] was very popular (indeed, once "the 4th most planted red wine grape")[42] and did well as long as it was warm and sunny, grafting onto American root stock after the Phylloxera infestation was not easy. So, although the small, sweet, thick-skinned grapes routinely produced a flavorful (if somewhat gamey) high-alcohol and high-tannin beverage, more fruitful varieties replaced it in many vineyards. However, the grape remains the principle grape variety of the Bandol AOC (*appellation d'origine contrôlée*) official French label. The wines of Bandol are characterized by their need for warmth and sunshine likely to be found in Provence (Bandol itself lies on the Mediterranean coast between Marseille and Toulon).

An extensive characterization of Mourvèdre of this "capricious and delicate" variety was carried out and published in 1993.[43] Much of what follows is derived from what was learned in that study.

The bouquet of the wine called Mourvèdre is generally described as "resembling ripe berries (raspberries . . .) . . . with spicy . . . and . . . animal like (musk) notes."[43] The structure[44] of the major component of musk, muscone, as isolated from the (protected) musk deer (*Moschus Moschidae*) is known,[43] and there is no evidence that it (or similar compounds) are actually present in the wine. It is not clear exactly which component or combination of components excites the neuron or sets of neurons that elicit the same response as muscone.

The same authors note that the extent of polymerization of the anthocyanins (glucosides of anthocyanidins) (Table 14.1) to form tanins (Figure 14.22) continues long after the wine is bottled. The consequence is that the taste appears to "mellow" with time.

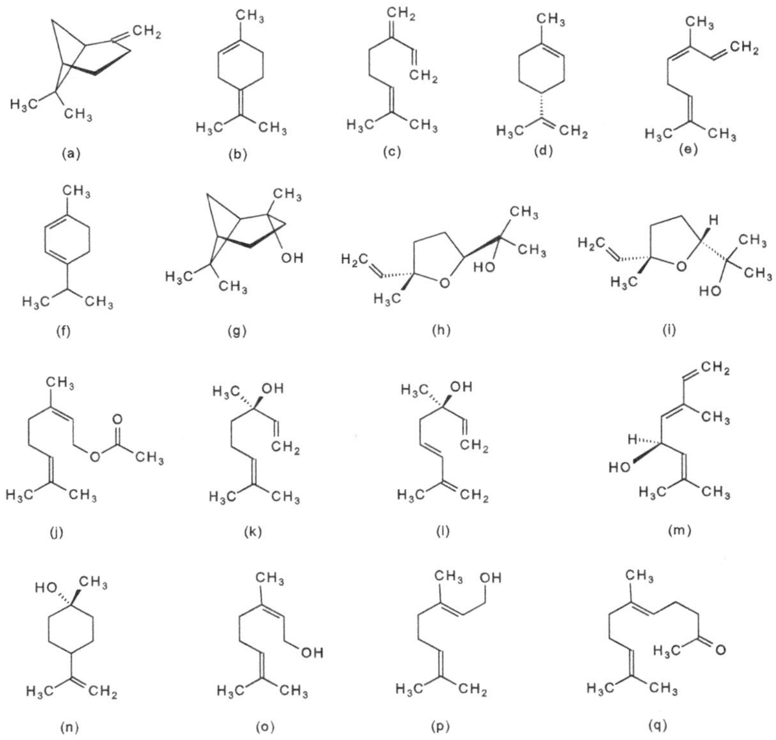

FIGURE 14.19 Representations of terpenoid compounds found in the headspace above Sicilian Muscat wine cultivars and presumed present, perhaps at other oxidation states, in the Moscato grapes examined. The compounds shown are: (a) β-pinene; (b) terpinolene; (c) myrcene; (d) (R)-limonene; (e) α-ocimene; (f) α-terpinene; (g) cis-2-pinanol; (h) trans-linalool oxide; (i) cis-linalool oxide; (j) neryl acetate; (k) (R)-(−)-linalool; (l) (R)-(−)-hotrienol; (m) trans-ocimenol; (n) β-terpineol; (o) nerol; (p) geraniol; and (q) geranyl acetone. See Appendix 2 regarding terpenoids.

The presence of these polyphenolic polymers, as well as polyphenolic tannic acid itself (Figure 14.23), serves to allow the more fragile components (e.g., alcohols) to resist oxidation. The oxidation of the phenols and subsequent further polymerization will often occur preferentially (depending upon the alcohol and the phenol components).

The major acid, as expected, is tartaric acid, although both malic and lactic acids are also present in lesser quantities (Figure 14.24)

Although a large number of alkanes and alkenes (as terpenes, etc.) have been isolated (as expected), it appears that there are many alcohols and esters with significantly long hydrocarbon chains also found (to add to the fruity aroma). The results of examination of the gas chromatographic analysis showing this are in Figure 14.25 (below) from that publication.[43]

Nebbiolo; a red grape.

The dark red-to-blue fog-like (*nebbia*) milky veil covering the berries as they reach maturity might be the origin of the name of this Italian grape, although the grape itself was, presumably, referred to by Pliny the Elder (aka Gaius Plinius Secundus, 23–79 AD) in his seminal work on wine, *Naturalis Historia* (77 AD), Volume 3, Book 14, Chapters 1–23.[45] The grape itself is

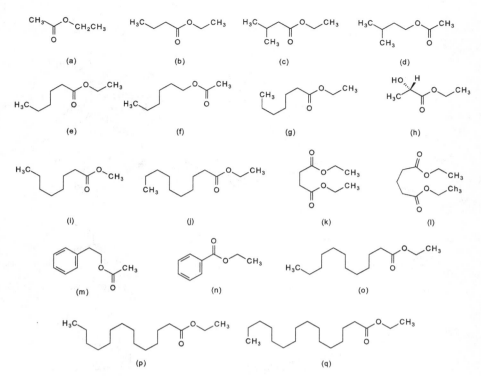

FIGURE 14.20 Representations of esters found in Muscat grape–derived wines. Some esters are also found in the grape prior to fermentation. Those shown here are: (a) ethyl acetate; (b) ethyl butanoate; (c) ethyl 3-methylbutanoate; (d) 3-methyl-1-butyl acetate; (e) ethyl hexanoate; (f) hexyl acetate; (g) ethyl heptanoate; (h) ethyl (R)-(+)-lactate; (i) methyl octanoate; (j) ethyl decanoate; (k) diethyl succinate; (l) diethyl pentanedioate; (m) 2-phenylethyl acetate; (n) ethyl benzoate; (o) ethyl dodecanoate; (p) ethyl tetradecanoate; and (q) ethyl hexadeconate.

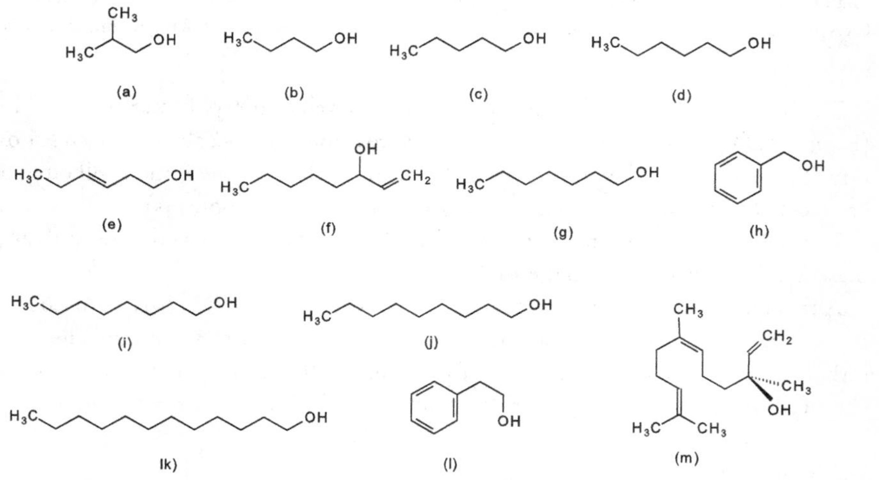

FIGURE 14.21 Representations of alcohols that are found in the Muscat wine, some of which are likely to have been present in the grape. They include: (a) 2-methyl-1-propanol; (b) 1-butanol; (c) 1-pentanol; (d) 1-hexanol; (e) 3-hexen-1-ol; (f) 1-octen-3-ol; (g) 1-heptanol; (h) benzyl alcohol; (i) 1-octanol; (j) 1-nonanol; (k) 1-dodecanol; (l) 2-phenylethanol; and (m) (S)-(+)-(Z)-nerolidol.

A Selection of Grapes

TABLE 14.1

A representation of some of the known and more common anthocyanidins.

Fundamental Structure	Anthocyanidin	R_1	R_2	R_3	R_4	R_5	R_6
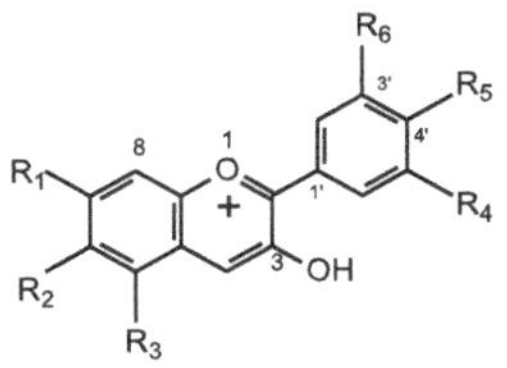	aurantinidin	OH	OH	OH	H	OH	H
	cyanidin	OH	H	OH	H	OH	OH
	delphinidin	OH	H	OH	OH	OH	OH
	europinidin	OH	H	OCH_3	OH	OH	OCH_3
	malvidin	OH	H	OH	OCH_3	OH	OCH_3
	pelargonidin	OH	H	OH	H	OH	H
	peonidin	OH	H	OH	H	OH	OCH_3
	petunidin	OH	H	OH	OCH_3	OH	OCH_3
	rosinidin	OCH_3	H	OH	H	OH	OCH_3

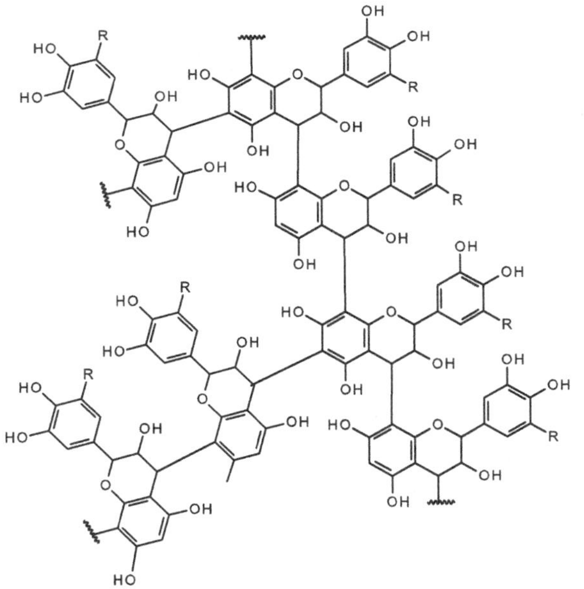

FIGURE 14.22 A representation of a portion of a polymeric tannin.

associated with the Piedmont region of northern Italy. The bright, ruby-red, young wine from the Nebbiolo is rich in tannins, and there is some evidence that the geography plays an important role it its taste. Thus, examination of the elemental composition (twenty [20] elements were checked) of leaves, grapes, and derived wines from eight (8) sites by atomic emission spectroscopy clearly showed that differences in origin could be easily detected.[46]

However, of somewhat more general interest was the observation that the concentration and profile of the various volatiles, as well as the timing of their relative appearances, were

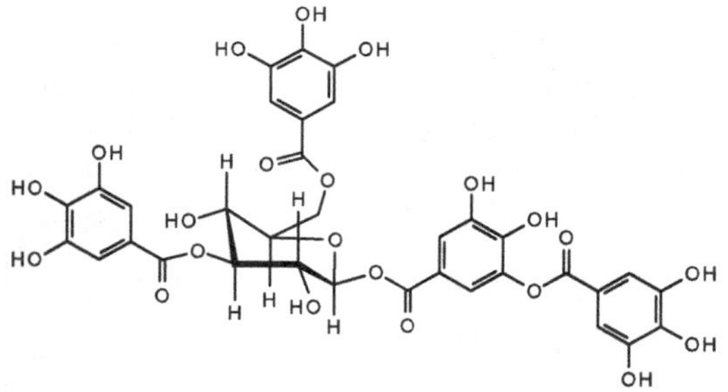

FIGURE 14.23 A representation of a tannic acid. Four (4) gallic acid units are involved; three (3) directly bonded as esters to a glucopyranoside, while the fourth is bonded (on the right of the drawing) to the phenolic unit of a gallic acid as an ester.

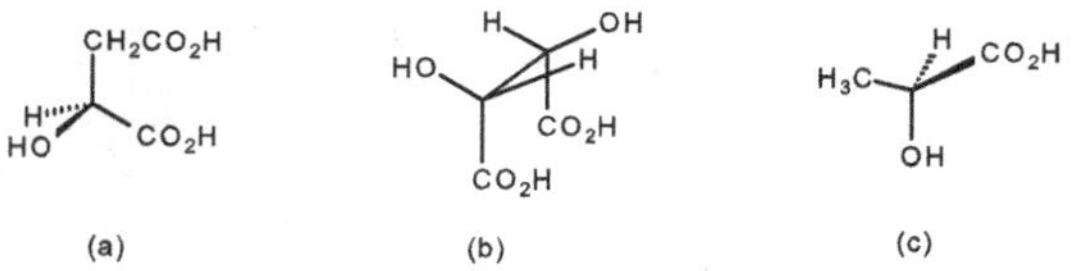

FIGURE 14.24 A representation of some of the acids reported found in Mourvèdre. It is likely that all three of these acids were initially present in the grape but in different amounts than ultimately found in the wine from the grape. The acids are: (a) (S)-(−)-malic acid; (b) (2R,3R)-(+)-tartaric acid; and (c) (S)-(+)-lactic acid.

correlated with the separate growing areas. Additionally there is the observation that while the list of pre-fermentative volatiles is one filled with the expected aldehydes, alcohols, terpenes and esters (as found elsewhere and shown above for other grapes) the relatively unusual β-ionone and 1,1,6-trimethyl-1,2-dihydronaphthalene (TDN), β-carotene degradation products (!) are present.[47–49]

Figures 14.26a, b, and c, outline expected pathways for these compounds.

Although β-ionone and 1,1,6-trimethyl-1,2-dihydronaphthalene (TDN) are, as shown, both found, the cartoon showing the formation of the latter from the former may not apply, as many pathways are potentially available and evidence is lacking. Additional "norisoprenoid" compounds derived in the same fashion are also found in the wine produced from the grape.

Pinot blanc: A white grape.

Pinot blanc[52] is a genetic mutation of Pinot noir. It has been suggested that this widely grown grape is "not a star...a genuine Cinderella" and a precursor to a "mild...drinkable beverage."[50]

As noted above for Nebbiolo, current investigations lend themselves to profiling various characteristics as a function of separate growing areas. Thus, it was reported that "statistically reliable differences between vineyard sites were found" and that, "for this grape, "locations between 400 and 500 m (about 1,300 to 1,500 ft) above sea level on easily warmed, fertile, loamy, slightly alkaline soils represent favorable conditions."[51]

Although the Pinot blanc grape is widely planted in France and elsewhere (where its name is occasionally changed), regulations permit the wine from the grape to be mixed with wine from Pinot noir and Pinot gris when these are, in the process of vinification, kept away from

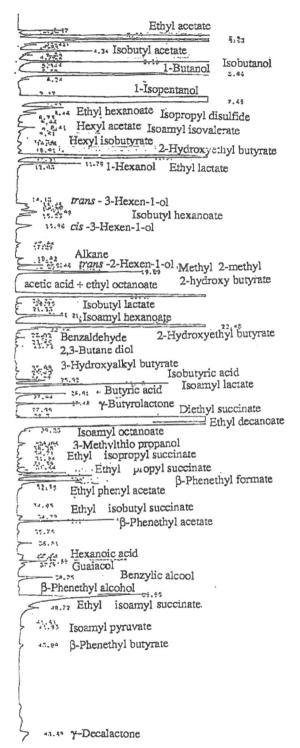

FIGURE 14.25 This representation of the gas chromatographic output from the Mourvèdre examined was published as Figure 5 in Vernin, G.; Pascal-Moussellard, H.; Metzger, J.; Párkányi, C. *Developments in Food Science, 33* (*Shelf Life Studies of Foods and Beverages: Chemical, Biological, Physical and Nutritional Aspects*; Charalambous, G., Ed.), 1993, 945 ff. Used with permission. Copyright Elsevier Limited, 1993.

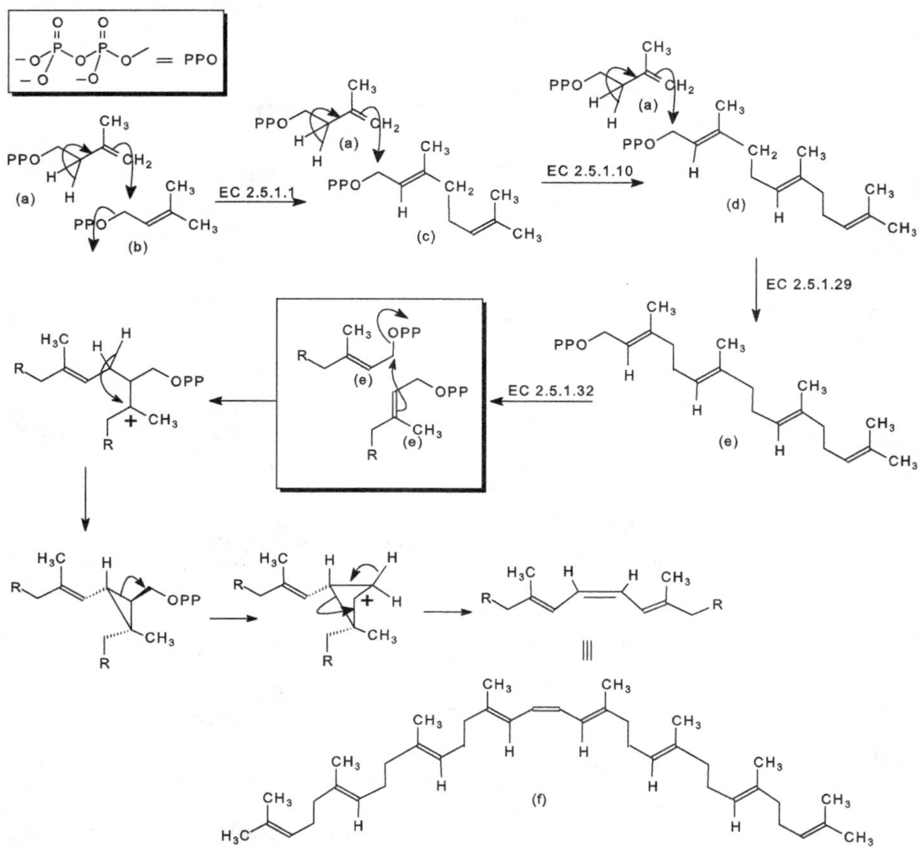

FIGURE 14.26a A representation of a pathway from isopentenyl diphosphate (a) and dimethylallyl diphosphate (b) to geranyl diphosphate (c) (dimethylallyl*trans*transferase, EC 2.5.1.1) and then, *via* farnesyl diphosphate (d) (2*E*,6*E*)-farnesyl diphosphate synthase, EC 2.5.1.10), to geranylgeranyl diphosphate (geranylgeranyl diphosphate synthase, EC 2.5.1.29) (e) a C-20 terpenoid. Dimerization of geranylgeranyl diphosphate (15-*cis*-phytoene synthase, EC 2.5.1.32) then yields the C-40, 15-*cis*-phytoene (f), which is the β-carotene precursor.

skin contact. This mixing is permitted in order to produce what are believed to be more favorable characteristics despite the observation that the genomic differences are, while small, nonetheless observable.

Indeed, partial genomic analysis is among the most interesting recent developments with regard to Pinot blanc and its relationship with Pinot gris and Pinot noir. Thus, it appears that both Pinot blanc and Pinot gris (*vide infra*) are independent somatic mutation of Pinot noir. That is, Pinot blanc arises from a mutation of the Pinot noir genome as does Pinot gris. However, the mutations giving rise to the two mutants are different from each other.[52,53]

In that vein, it has also been found that a somatic mutation occurs[54] and that two Myb-related (i.e., myeoloblastosis-related) transcription factor genes, VvMybA1 (i.e., Vitus vinifera myeoloblastosis A1) and VvMybA2, regulate anthocyanin biosynthesis and thus

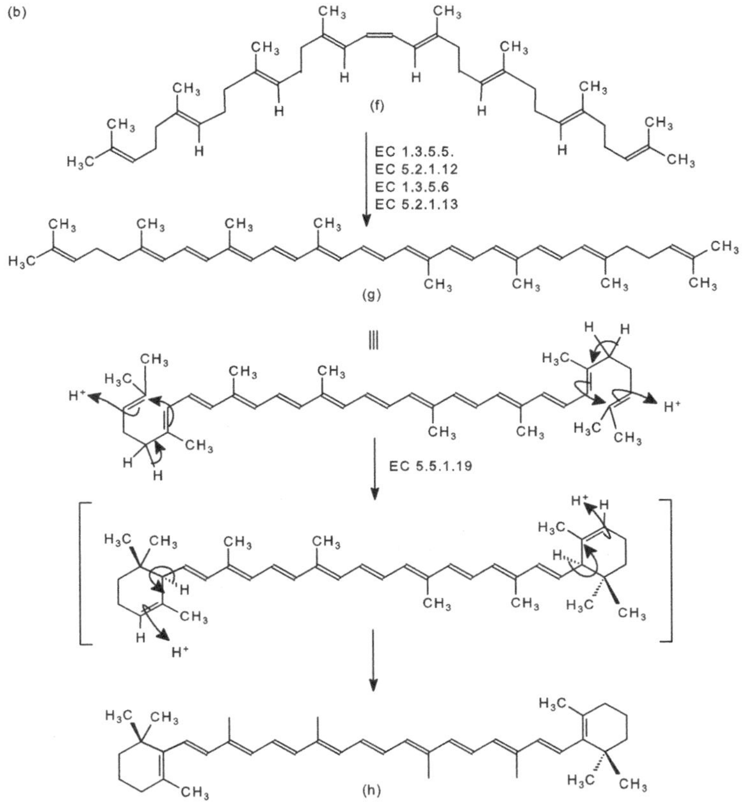

FIGURE 14.26b An abbreviated representation of the pathway from the C-40 compound 15-*cis*-phytoene (f) to its isomer all-*trans*-lycopene (g) involving the enzymes 15-*cis*-phytoene desaturase, EC 1.3.5.5 and ζ-carotene isomerase, EC 5.2.1.12 followed by 9,9′-*dicis*-ζ-carotene desaturase, EC 1.3.5.6 and polycopene isomerase, EC 5.2.1.13. With all-*trans*-lycopene in hand isomerization (lycopene β-cyclase, EC 5.5.1.19) can be effected and β-carotene (h) generated. It appears that β-carotene (h) serves as the precursor to many of the C-13 "norisoprenoid" components found in wines produced from the Nebbiolo and some other grapes.

the color of the grape. Inactivation of these two functional genes, through the insertion of a retrotransposon[54] in the VvMybA1 promoter and through a non-synonymous single nucleotide polymorphism (SNP) present in theVvMybA2 coding region, gives rise to the white berry, Pinot blanc, phenotype. If the transcription factor genes are unmodified, Pinot noir results.

So, since anthocyanins (the carbohydrate derivatives of anthocyanidins) and anthocyanidins found in red-skinned and darker cultivars are responsible for the red wine color, and since the same compounds result in the condensed tannins that lead to the astringency found in the red wine, their loss results in generation of the "mild . . . drinkable beverage" rather than its more robust progenitor.

Finally, it is of some interest that similar retrotransposons are found to control the accumulation of anthocyanins found in oranges and other fruits.[55] In this vein, it was as early as 1967[56] when it was found that the skin and seeds of Pinot blanc grapes that were used as a

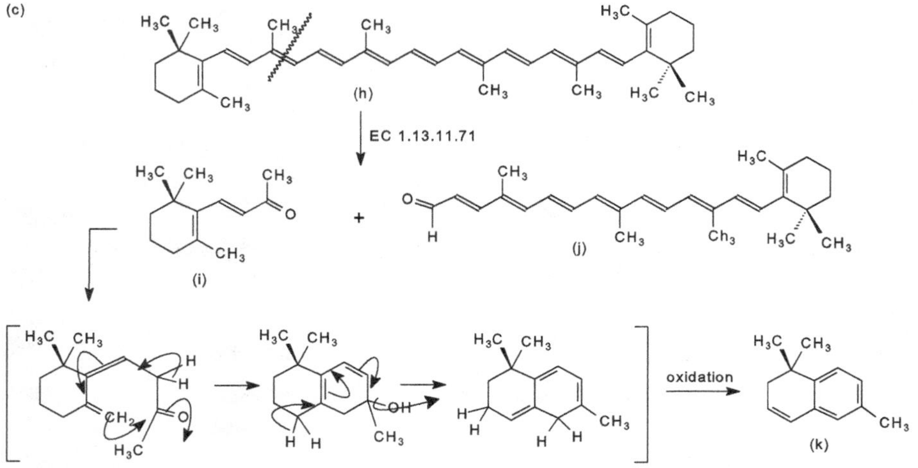

FIGURE 14.26c A representation of the conversion of β-carotene (h) in the presence of carotenoid-9′,10′-cleaving dioxygenase, EC 1.13.11.71 into the C-13 "norisoprenoid" β-ionone (i) and the remaining C-27 fragment, 10′-apo-β-carotenal (j). Additional arrows are shown in an attempt to account for the conversion of β-ionone (i) into 1,1,6-trimethyl-1,2-dihydronaphthalene (TDN) (k).

source of different proanthocyanidins yielded cyanidin upon heating with n-butyl alcohol (n-butanol) and hydrochloric acid (HCl). More than two anthocyanidins can combine to form proanthocyanidins and many-bodied (i.e., polymeric) mixtures can occur. A potential path to a dimer is shown as Figure 14.27. Repeating the same reaction at either end with additional cyanidins being added one-by-one will lead to polymeric proanthocyanidins.

Pinot gris: A white grape.

The "gray" version of this Pinot noir mutant produces the expected "cone" shaped bunches of grapes whose coloration varies through shades of gray (hence the name) as well as to grayish brown, pink, and almost white. The clone (i.e., its genetic copy) of this grape when grown in Italy is known as Pinot grigio.

As was noted for Pinot blanc, minor genetic changes account for the coloration of these grapes but, even beyond this, some of those modifications result in eventual production of wines with neither the soft flavor of the Pinot blanc nor the astringency of their progenitor. So, the low acidity of the Pinot blanc remains, but more robust and fruity flavors, in part a function of growing locale or "terroir," result.

Also, as was noted above for Pinot blanc, it appears that Pinot gris is an independent somatic mutation of Pinot noir. That is, it arises from a mutation of the Pinot noir genome as does Pinot blanc—but it is a different mutation![53]

A recent study[57] using certified clones of Pinot gris (from the collection of the Institut National de la Recherché Agronomique [INRA, Colmar, France]) was undertaken because of the known diversification of the *V. vinifera* Pinot clones. The study was begun in order to examine the chimeric structures of the clones (a chimera is a single organism with genetically distinct cells such as the cells in the meristem, Chapter 1, that give rise to different tissues; whereas the genetic information of a clone is identical to that of the parent). Interestingly, it was found on analysis of the leaf tissue DNA that the polymorphism (i.e., the structurally

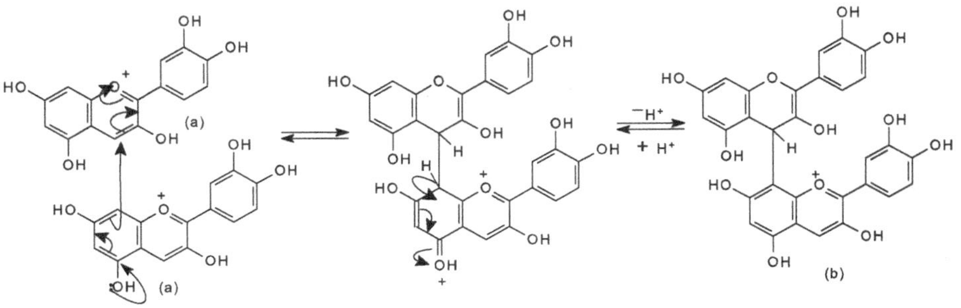

FIGURE 14.27 A cartoon showing a possible path and its reverse from the anthocyanidin named cyanidin (a) to a proanthocyanidin (b). The process serves to demonstrate how polymeric products might both form and decompose on standing.

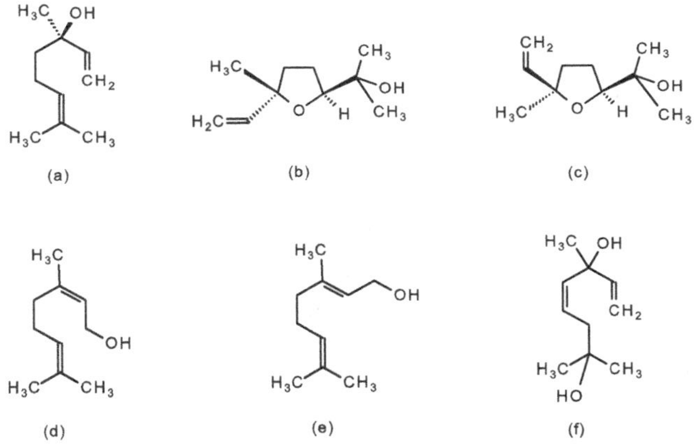

FIGURE 14.28 A representation of some of the monoterpenes reported to be found in Pinot gris. Those shown are: (a) (R)-(−)-linalool; (b) (2S,5S)-*trans*-linalool oxide; (c) (2R,5S)-*cis*-linalool oxide; (d) nerol; (e) geraniol; and (f) (Z)-3,7-dimethyl-1,4-octadiene-3,7-diol.

different types within the same organism) "mainly resulted from the appearance of a third allele when two ... were expected." *V. vinifera* is diploid.

Earlier work[58] had demonstrated that Pinot gris (Ruländer) and similar grape varieties could be distinguished from others based on the analytical chemistry of "12 monoterpene compounds" as determined by gas chromatography. Interestingly, however, the author chose to provide only six examples of those compounds, shown here as Figure 14.28.

The stereochemistry of the isolates was not provided in the reference.[58] So, in some of the structures provided in Figure 14.28, where more than one configurational or stereochemical isomer is known, the wrong configuration or enantiomer (or diasterioisomer) may be drawn. Nonetheless, it was argued that the "terpene profiles ... (showed that) ... clear differences exist between the grape varieties ..." Further, because in the Pinot gris (Ruländer) and other "neutral grape varieties" where the concentration of monoterpenes is low, glycosidically bound aroma substances (which include monoterpenes, aromatic alcohols, and norisoprenoids, i.e., C_{13} compounds) can help characterize those varieties. Again, no further specific identification of the compounds was provided.

A patent application (US2008/0276339 P1) for a clone of this life form "Grape plant named Pinot grigio/Pinot gris (Tehachapi clone)" was filed on 4 May 2007. According to the application, this is a distinct variety "characterized by producing small round white berries," and further this variety is distinguished "by ripening earlier and having a larger, more prominently winged cluster." It is not clear that the application succeeded.

Finally, it has been reported[13] that the skins of the grape contain small quantities of sulfur-containing derivatives of hexan-1-ol (i.e., 3-mercaptohexan-1-ol-L-cysteine and 3-mercaptohexane-1-ol-L-glutathione), the same compounds already reported as being found in the skin of Chardonnay grape and already shown in Figure 14.8.

Pinot noir: A red grape.

Although the grape Pinot noir ("pine cone shaped" and "black") is very cosmopolitan, it remains largely associated with the Burgundy region of France. Despite (or perhaps because of) its relationship with Burgundy and the oft noted observation that it is difficult to cultivate, it is said to produce some of the finest wines both young and well-aged. Further, it has served to foster serious genetic experimentation. Indeed, "Pinot noir provides a great pool of clonal phenotypes displaying plasticity in canopy growth, cluster architecture, fruit yield, and maturity."[59] The authors were able to distinguish between various clones using the appropriate analytical technique (methylation-sensitive amplified polymorphism).

To investigate canopy involvement, vine canopy of Pinot noir growing in the Vipava Valley, Slovenia (close to the border of Italy), was manipulated[60]. Grapes were allowed to grow under different canopies (heavy leaf, thin leaf, etc.). Then, over 10-day intervals from June to September 2010 (vineyard planted in 2004), some grapes were harvested, and methanolic extracts of their carefully removed skins were examined using high pressure liquid chromatographic (HPLC) separation and mass spectrometry (MS). A variety of different phenolic compounds were found, with those resulting from complex interactions between smaller fragments (i.e., dimers, trimers, etc.) more common in the late harvest. Typical compounds, many blue and red, are shown in the Figures 14.29a and b.

The compounds in Figures 14.29a and b impart color to grape skin and, if the skins are left with the must, color to the wine. In addition, they will doubtlessly impart some flavor. There are also some volatile materials that are present in the grape as well as in the wine, subsequently produced.[61] Some of these materials are shown in Figure 14.30.

Interestingly, it appears that the wines produced from the grape are somewhat variable. As Oz Clarke has pointed out[62] "stop arguing as to whether it tastes like it should. There is no 'should' with Pinot."

Pinotage; a red grape.

Pinotage is reported to "arouse(es)...fierce disagreement."[62] The source of the disagreement appears to be attributed to the observation that the crossing of Pinot noir with Cinsaut (called "Hermitage" where crossed so... "Pino" + "tage") was effected in South Africa (in 1925).[62]

Apparently, the consequence of this crossing can produce high quality but low quantity beverage which is, as expected, neither Pinot noir nor Cinsaut. Interestingly, the grapes themselves, unlike others already discussed so far here, do not, at this writing, appear to have been examined for constituents even though they are intensely colored (even by mid-vintage), arguing for a significant concentration of anthocyanins and/or anthocyanidins.[63]

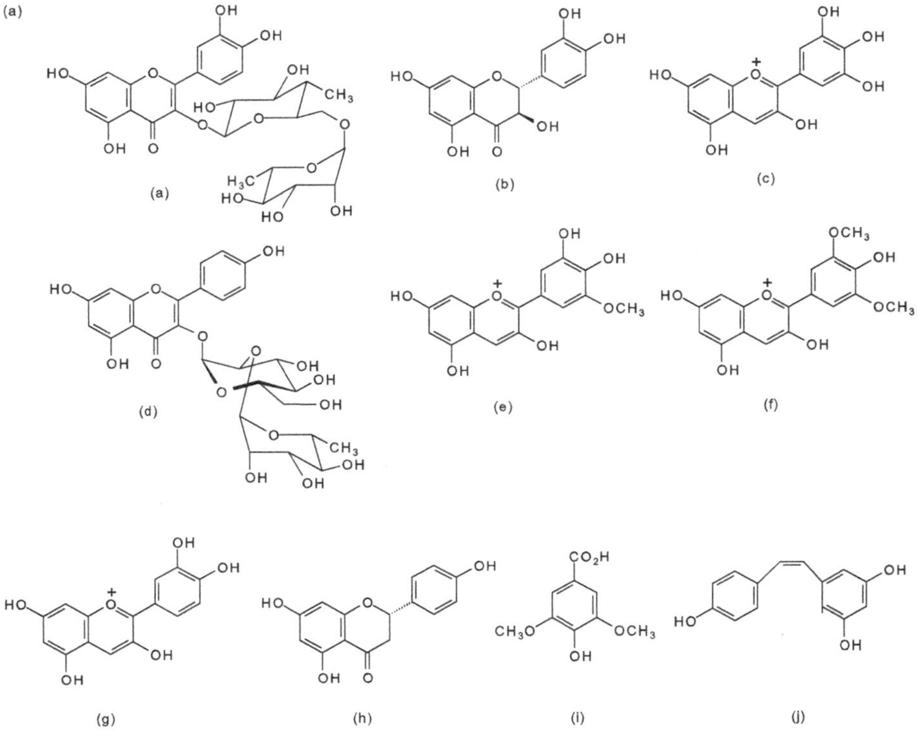

FIGURE 14.29a Representations of compounds, many with deep (pH dependent) colors reported found in the skin of a Pinot noir. The compounds include: (a) rutin (with a rhamnose and glucose attached at C-3 as is typical of anthocyanins); (b) taxifolin (a flavonoid); (c) delphinidin (an anthocyanidin); (d) kaempferol 3-O-rutinoside (an anthocyanin); (e) petunidin; (f) malvidin; (g) cyanidin; (h) naringenin; (i) syringic acid, and (j) cis-resveratrol.

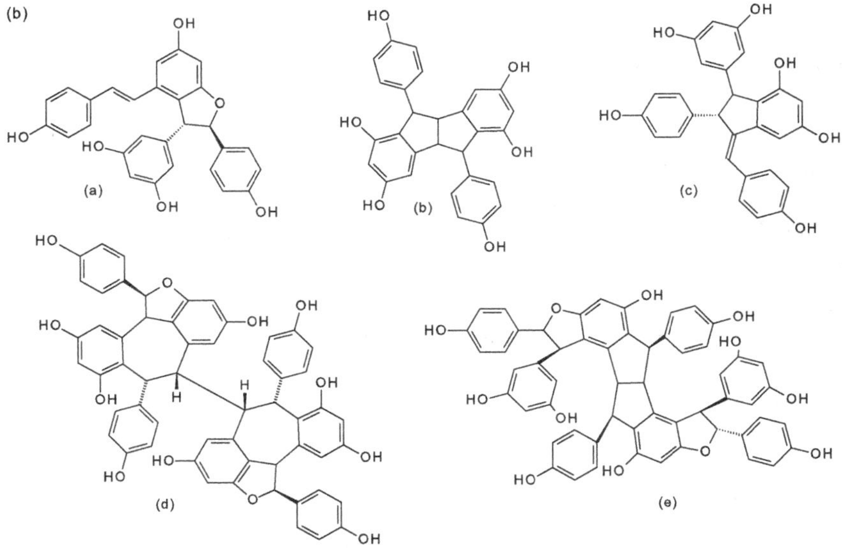

FIGURE 14.29b Representations of dimeric and higher polymeric derivatives of resveratrol reported to be found in the skin of Pinot noir. The dimers include: (a) ε-viniferin; (b) pallidol; (c) ampelopsin-D; while (d) isohopeaphenol; and (e) ampelopsin-H are tetramers. These lignan-type compounds, derived from phenylalanine and tyrosine, are discussed subsequently.

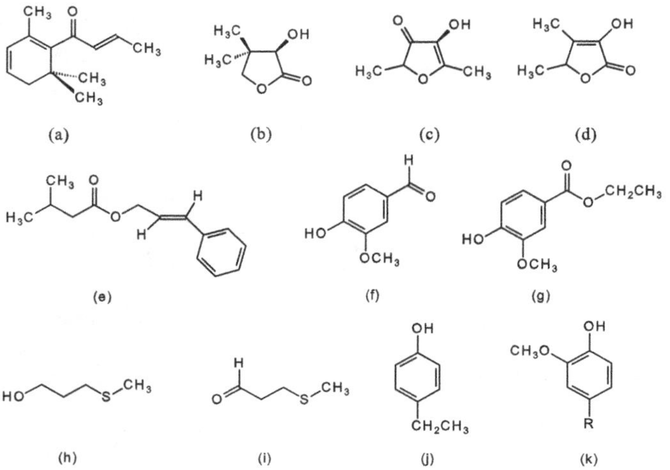

FIGURE 14.30 A representation of some of the volatiles reported to be found in the skin of Pinot noir. They include: (a) β-damsacenone; (b) pantolactone; (c) furaneol; (d) sotolon; (e) *trans*-cinnamyl isovalerate; (f) vanillin; (g) ethyl vanillate; (h) methionol; (i) methional; (j) *para*-ethylphenol; and (k) R=H, guaiacol; R=CH$_2$CH$_3$, 4-ethylguaiacol; and R=CH=CH$_2$, 4-vinylguaiacol.

Curiously, although the grapes (in contrast to the wine derived from the grapes) have not been examined in the same way as those previously discussed here, much more significant work has been done on its genome.

In 2012[64] the genes and enzymes of the carotenoid metabolic pathway in *Vitis vinifera* L. were examined, and forty-two (42) genes on sixteen (16) chromosomes that are putatively involved in carotenoid biosynthesis/catabolism in grapevine were mapped.

Cloned deoxyribonucleic acid (cDNA) copies of eleven corresponding genes from *Vitis vinifera* L. cv. Pinotage were characterized, and four were shown to be functional. Additional work on three separate berry developmental stages in Sauvignon blanc were also carried out.

Using these two varieties and comparing the carotenoid metabolic pathway found, it was concluded that the "carotenoid pathway genes of the grapevine showed that they share a high degree of similarity with other eudicots." (Eudicots are considered "true" dicots [or dicotyledons], and they, along with monocots [or monocotyledons] make up the majority of the hundreds of thousands of known flowering plants.) The work, both graphically and written, concluded that "these data ... form a baseline ... (for a) central metabolic pathway and provide insights into the evolution of ... compounds that also serve as substrates for quality impact factors (i.e., β-ionone, β-damascenone and vitispirane) and regulating phytohormones (i.e., abscisic acid and strigolactone)" which, along with some other terpenoid type lactones (see Figure 14.31), are catabolic products of carotenoids. Indeed, taken along with the earlier observation[65] that monoterpene (e.g., linalool, nerol, geraniol, etc.) content in grapevine is associated with 1-deoxy-D-xylulose 5-phosphate synthase, provides a major step in analysis of the details of the *V. vinifera* genome.

Riesling: A white grape.

"... (T)he wine experts' favorite grape."[66]

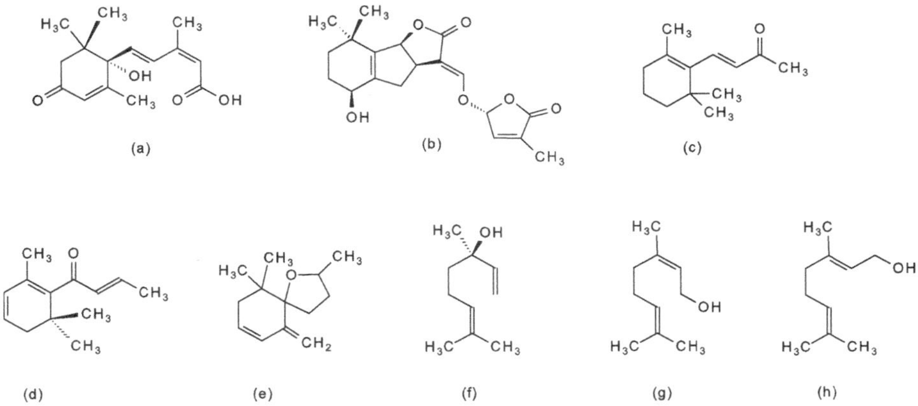

FIGURE 14.31 Representations of terpene and C-13-terpene-derived volatiles derived from the Pinotage grape. The structures shown represent: (a) abscisic acid; (b) (+)-strigol; (c) β-ionone; (d) β-damascenone; (e) vitispirane; (f) (R)-(−)-linalool; (g) nerol; and (h) geraniol.

Regrettably, it appears that most people are not as enthusiastic about the wine derived from the Riesling grape. Of course, there are those who appreciate the variety that the various clones grown in various soils and under varying climate conditions can provide. Indeed, as Clarke[66] notes, "It reflects the vineyard more transparently than almost any other grape." In this vein, the report of a study of sensory and flavor analysis of different terroirs of German Riesling is clearly telling.[67] As demonstrated there, sensory information using a subset of forty-nine (49) aroma compounds (not specified in the report) found in wines taken from a standardized wine estate vinification set, varied among the types of bedrock in which the grapes were grown. "... (T)he concentration of aroma compounds present in the head space allowed discrimination between wines grown on the bedrock types greywacke (a grey, mixed rock base), basalt, and slate. The wines from rotliegend (German "underlying red") were well separated from those from sandstone, however they showed minor overlapping with the limestone cluster."[67]

The grapes themselves have also been examined. In one case, the major volatile compounds found in Riesling grapes from fruit-set to harvest were isolated and compared to those from Cabernet Sauvignon grapes. The latter were found richer in variety of components.[68] In a second work, C_{13} norisoprenoids and benzenoids were followed during ripening.[69]

The pre-véraison volatile compounds found in the Riesling grapes examined are shown in Figure 14.32. While the concentrations changed during and post-véraison, only a few additional compounds (not shown here, but known, e.g., benzaldehyde, benzyl alcohol) appeared. There is also some evidence that at least one (as one is listed but not identified) C_{13} norisoprenoid was also present during and after véraison.

Following the concentrations of C-13 norisoprenoid alcohol and the benzenoids present as phenols in the grapes proved to be somewhat more complicated, since it was claimed that these compounds would be found as glycosides and their isolation made more difficult as a consequence. The work[69] not only describes in detail the collection and handling of the grapes and the subsequent extraction of glycosides but also the details of the hydrolysis to produce the aglycones, representations of whose structures are shown in Figures 14.33a–14.33c.

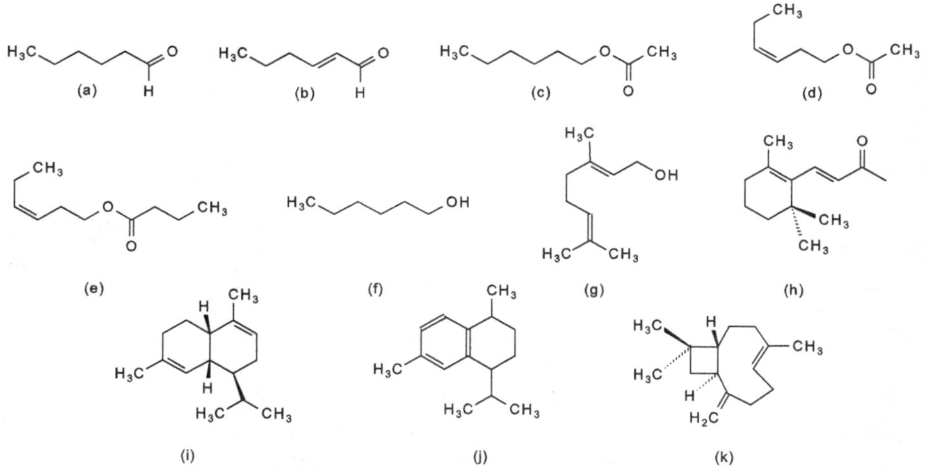

FIGURE 14.32 Representation of some of the pre-véraison volatiles from Riesling grapes which were also found to be present (albeit in changed concentrations) post-véraison. The compounds shown are the aldehydes (a) hexanal and (b) (E)-2-hexenal; the esters (c) hexyl acetate; (d) (Z)-3-hexenenyl acetate; and (e) (Z)-3-hexenyl butanoate; the alcohols (f) 1-hexanol and (g) geraniol; the ketone (h) β-ionone; and the sesquiterpenes (i) α-muurolene; (j) calamenene; and (k) caryophyllene.

The method did not allow clear separation of materials carried along with the extractions (i.e., those not removed whose characteristics matched those of the aglycones). Secondly, as the hydrolysis was effected by an acidic solution, some of the products isolated might have resulted from rearrangements of the initially produced compounds, and thus the ethers and hydrocarbons shown, which could not have linked (as they lack the wanted hydroxyl, –OH, group) to the carbohydrate, are presumed to arise by subsequent processes. The list, nonetheless, is interesting. Finally, with regard to the compounds shown, some of the simpler, ubiquitous aromatic compounds (e.g., benzyl alcohol, 2-phenylethanol) are not shown.

Interestingly, it has been reported that fermentation by native yeasts, as well as by *S. cerevisiae* of the glycosidically bound components of the terpenes and aromatic alcohols, produced differences that were below the sensory threshold.[70]

Finally, it has been reported[13] that the skins of the Riesling grape contain small quantities of sulfur-containing derivatives of hexan-1-ol (Figure 14.34) (i.e., S-3-(hexan-1-ol)-L-cysteine and S-3-(hexane-1-ol)-L-glutathione) and S-4-mercapto-4-methyl-2-pentanone (i.e., S-4-(4-methyl-2-pentanone)-L-cysteine and S-4-(4-methyl-2-pentanone)-L-glutathione). These were also seen in the skins of the Gewürztraminer grape.

Sangiovese: A red grape.

Sangiovese, ("*sanguis Jovis*," "the blood of Jove") remains largely appreciated (although not usually singled out) as a major component of Chianti. Common though it may be, its origins are debated. Thus, it has been commented that "Sangiovese probably had its origins in a wild *Vitis silvestris* vine ...,"[71] and while that may be true, the origins of the vine, based on examination of its genome, are currently also debated!

Indeed, there are two very different hypotheses for the parentage of this "most important Italian wine grape." Thus, Vouillamoz and coworkers[72] argue that Sangiovese is a progeny of

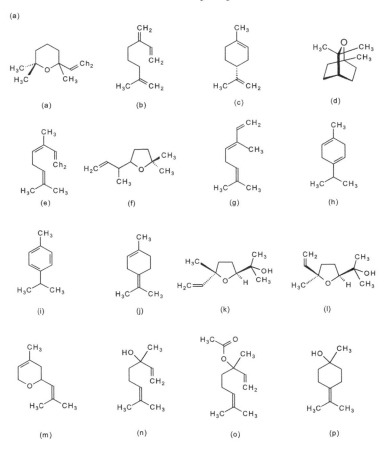

FIGURE 14.33a Representations of some of the terpenes and their derivatives found as both pre- and post-véraison volatiles from Riesling grapes. The compounds include: (a) limetol; (b) α-myrcene; (c) limonene; (d) eucalyptol; (e) *cis*-ocimene; (f) tetrahydro-2,2-dimethyl-5-(1-methyl-1-propenyl)furan (ocimene quintoxide); (g) *trans*-ocimene; (h) γ-terpinene; (i) *para*-cymene; (j) terpinolene; (k) (2S,5S)-*trans*-linalool oxide; (l) (2R,5R)-*cis*-linalool oxide; (m) neroloxide; (n) linalool; (o) linalool acetate; and (p) γ-terpineol.

an ancient Tuscan "Ciliegiolo" and an obscure grapevine from Campania, Italy, "Calabrese di Montenuovo," while Bergamini and colleagues[73] report that an ancient "Vegrodolce," believed lost, is the most likely parent.

Interestingly, Clarke and Rand[71] note that Sangiovese can produce "light, juicy wines or big, complex ones according to where it is grown and how it is cultivated" thus again emphasizing the terroir aspect of the production of wines and the grapes from which they come. They point out that the soils of Tuscany, while varied, produce the most desirable grapes, and they grow best in a soil known as "galestro." Although it is generally recognized that galestro is a metamorphic schist type of soil, even these vary in detail with regard to the minerals present. All of them contain potassium (K^+), sodium (Na^+), and calcium aluminum silicates (e.g., anorthite, $CaAl_2Si_2O_8$). They also contain quartz (silicon dioxide, SiO_2) and graphite

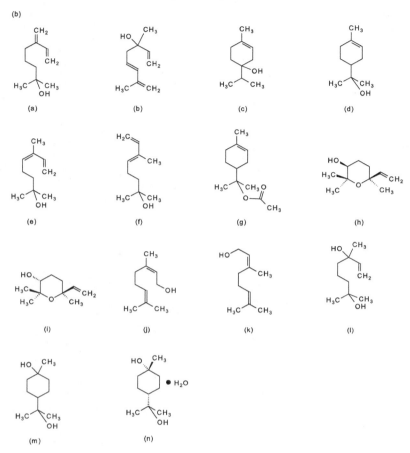

FIGURE 14.33b Representations of some of the terpenols and their derivatives found as both pre- and post-véraison volatiles from Riesling grapes. They include: (a) myrcenol; (b) hotrienol; (d) 4-terpinenol; (e) 3,7-dimethyl-1,3-octadien-7-ol; (f) cis-β-ocimenol; (f) trans-β-ocimenol; (g) α-terpinyl acetate; (h) linalool oxide, Z-pyranoid; (i) linalool oxide, (E)-pyranoid; (j) nerol; (k) geraniol; (l) 6,7-dihydo-7-hydroxylinalool; (m) 1,8-terpin; and (n) 1,8-terpin hydrate.

(carbon), as well as many other elements. Indeed, it is their variability which makes them interesting and difficult to copy from place to place.

For some years, the Sangiovese vineyards have been troubled by the European grapevine moth, *Lobesia botrana,* and in 2009 it was reported[74] that the volatiles emitted by inflorescences of Sangiovese attracted the pest. The air in the headspace above the flowers was collected, and compounds found there were examined, identified, and quantified. Six major components (Figure 14.35) active to *Lobesia botrana* female antennae were reported to be limonene, (*E*)-4,8-dimethyl-1,3,7-nonatriene, (±)-linalool, (*E*)-caryophyllene, (*E,E*)-α-farnesene, and methyl salicylate. Interestingly, "depending upon the dosage, the synthetic lure either attracted or repelled oviposition," but it was found that the lure could be tuned to the right mixture, and that it might be useful in "monitoring female activity in the field."

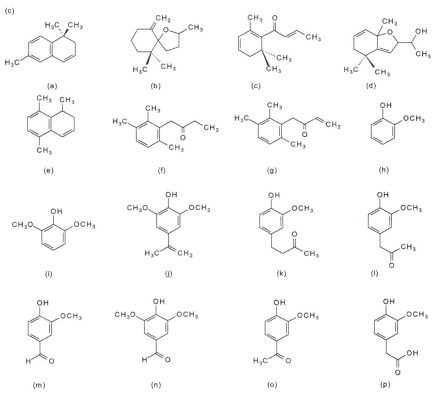

FIGURE 14.33C Representations of some of terpene, norterpene and aromatic compounds found as both pre- and post-véraison volatiles from Riesling grapes. They include; (a) 1,1,6-trimethyl-1,2-dihydronaphthalene (TDN); (b) vitispirane; (c) β-damsacenone; (d) actinidol; (e) 1,5,8-trimethyl-1,2-dihydronapthalene; (f) 1-(2,3,6-trimethylphenyl)-2-butanone; (g) 1-(2,3,6-tri-methylphenyl)-3-buten-2-one; (h) guaiacol; (i) syringol; (j) 2,6-dimethoxy-4-(2-propenyl)phenol; (k) gingerone; (l) methyl vanillyl ketone; (m) vanillin; (n) syringaldehyde; (o) acetovanillone; and (p) homovanillic acid.

More recently, efforts have been made to determine the amount and specific structures of anthocyanins that are present in various Sangiovese wines (grapes harvested at different vineyards). Apparently this was undertaken with the understanding that "the anthocyanic pattern recognitions, genetically controlled by plant variety, was (*sic*) shown to be inherited."[75] The work involved using a rare variety whose anthocyanin patterns consisted mainly of variations of *para*-coumaric acid ester condensation products. The compounds were extracted from the skins of the grapes with methanol, and that extract was subsequently manipulated before analysis by high pressure liquid chromatography (HPLC) using a mass spectrometer (MS) as the detector. Structural assignments for some of the pigments isolated were provided and are shown in Figure 14.36. Regrettably, there is a problem with the structures for vitisin A and vitisin B given in the text of the reference, and there is insufficient information in the publication itself to account for the structural assignments. The difficulty is that the structures provided (for the names given) do not match those in the Chemical Abstracts database (SciFinder). The derivatives of malvidin shown below do not suffer from the same problem,

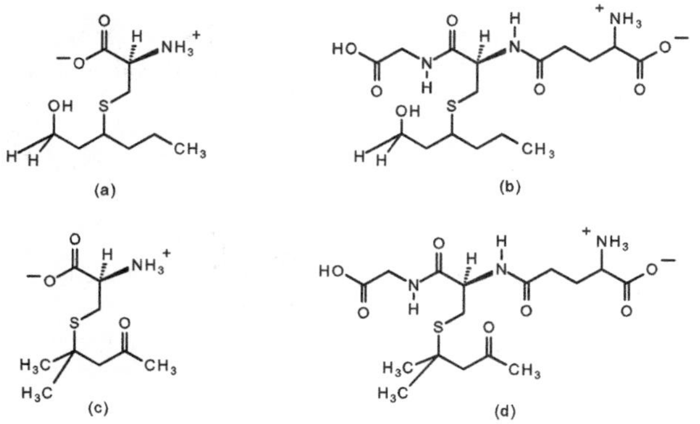

FIGURE 14.34 A representation of the sulfur-containing compounds reported to be found in the skins of Riesling grapes. These are the same as previously seen (Figure 14.14) and at that time were noted as present in the Gewürztraminer grape. Those shown are (a) 3-metcaptohexan-1-ol-L-cysteine; (b) 3-mercaptohexan-1-ol-L-glutathione; (c) 4-mercapto-4-methyl-1-pentanone-L-cysteine; (d) 4-mercapto-4-methyl-2-pentanone-L-glutathione.

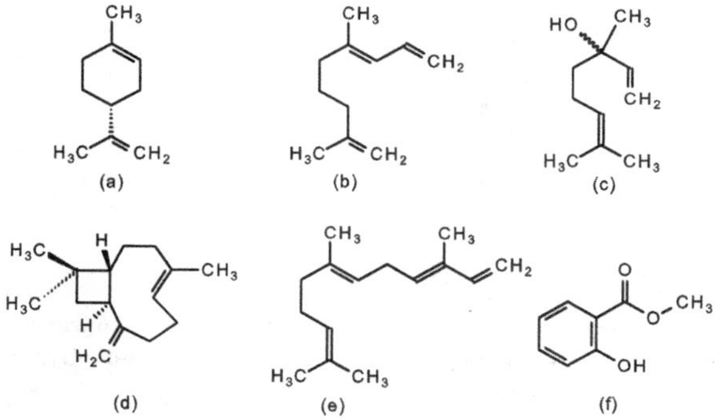

FIGURE 14.35 Representations of volatiles associated with inflorescences of the Sangiovese grape that were tested with regard to antennae of the female of the moth *Lobesia botrana*, a pest of the grape. The compounds identified are: (a) limonene; (b) (*E*)-4,8-dimethyl-1,3,7-nonatriene; (c) linalool; (d) caryophyllene; (e) (*E*,*E*)-α-farnescene; and (f) methyl salicylate.

as they do match those provided in the database, but again, the data that would confirm the structural assignments is absent (Figure 14.37).

Most recently, and ignoring structures completely, it has been pointed out that since anthocyanins (and presumably anthocyanidins since there is no chromophore in the carbohydrate) all absorb at about 540 nm, it should be possible to assess their concentration in grapes nondestructively.[76–78] The work is actually done by assessing the chlorophyll fluorescence excitation spectrum which responds because, as the anthocyanin concentration increases, less excitation light is transmitted to chlorophyll. So, by excitation at 540 nm (where anthocyanin absorption occurs) and at 635 nm (where no absorption by anthocyanin occurs but absorption by chlorophyll does occur) a good correlation with the concentration of anthocyanins is obtained.

Finally, with regard to Sangiovese, it has been pointed out[79] that while it is well known that grape berries are regarded as a non-climacteric fruit (i.e., ripening is not affected by

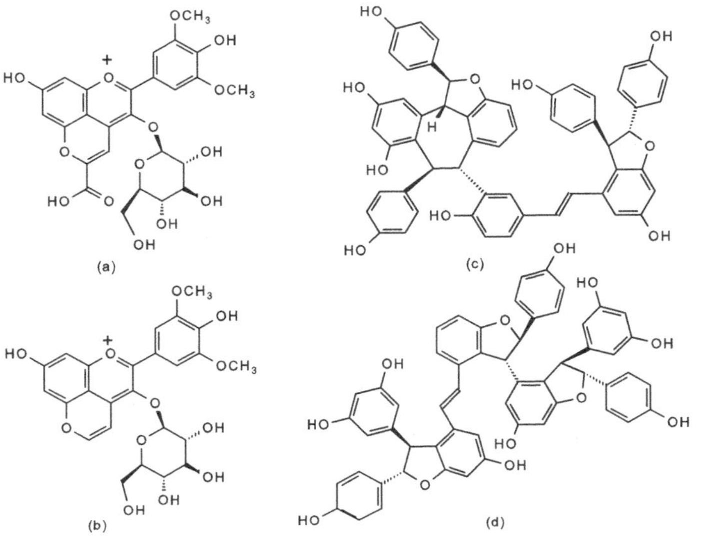

FIGURE 14.36 Representations of vitisin A and vitisin B. On the left, the vitisin A (a) and vitisin B (b) structural representations as provided in Arapitsas, P.; Perenzoni, D.; Nicolini, G.; Mattivi, F. *J. Agric. Food Chem.* **2012,** *60,* 10461. On the right, the corresponding vitisin A (c) (CA 142449-89-6) and vitisin B (d) (CA 142449-90-9) as provided in the Chemical Abstracts (CA) database.

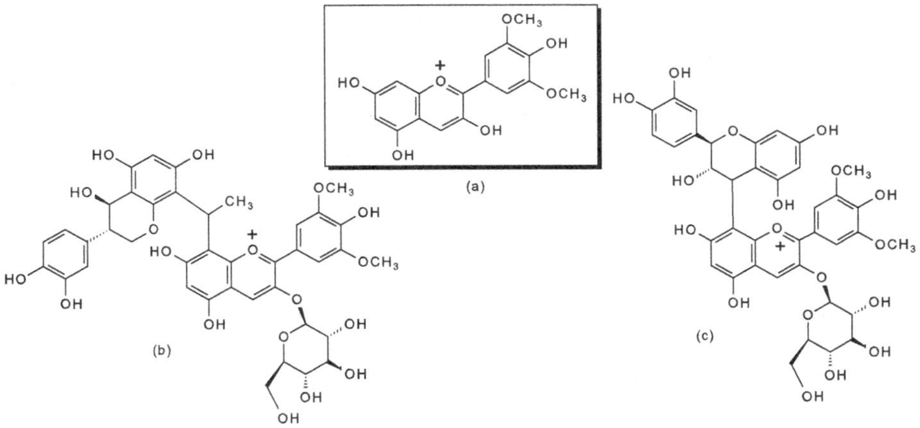

FIGURE 14.37 Representations of (a) malvidin and derivatives (b) malvidin-3-glucoside-(8,8′-ethylepicatechin and (c) malvidin-3-glucoside-(4,8′-catechin) reported in the Sangiovese grape.

the presence of the hormone ethylene [ethene, $CH_2=CH_2$]) exogenous ethylene addition, performed in the field (at véraison) or after harvest, does apparently have an effect on skin composition and, in particular, anthocyanins found therein.

Sauvignon blanc: A white grape.

As recently pointed out "... *Vitis vinifera* L. cv. Sauvignon blanc wines have become increasingly popular, since it is a cultivar that can be influenced in the vineyard and the cellar

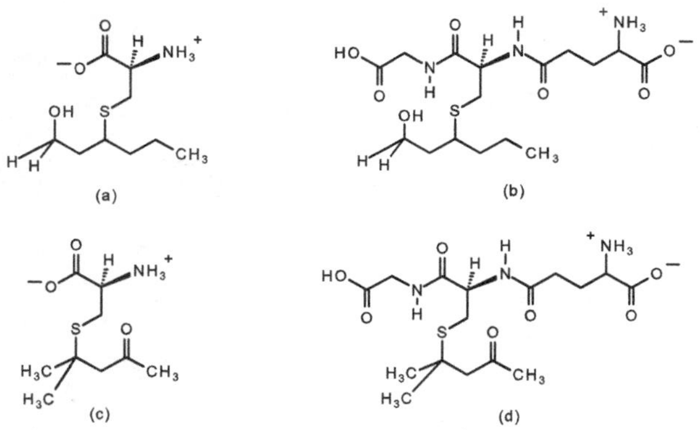

FIGURE 14.38 A representation of the sulfur-containing compounds reported to be found in the skins of Sauvignon blanc grapes. These are the same as previously seen (Figures 14.14 and 14.34) and at that time were noted as present in the Gewürztraminer and Riesling grapes, respectively. Those shown are: (a) 3-metcaptohexan-1-ol-L-cysteine; (b) 3-mercaptohexan-1-ol-L-glutathione; (c) 4-mercapto-4-methyl-1-pentanone-L-cysteine; and (d) 4-mercapto-4-methyl-2-pentanone-L-glutathione.

to produce a range of wine styles. Although originally from France, Sauvignon blanc is now widely cultivated in the wine growing regions of the world. These wines are usually given aroma descriptors such as green pepper, grassy, asparagus; while other more tropical aromas include passion fruit and guava."[80]

However, the odors associated with wines (as distinct from grapes) usually develop gradually after manipulations of the vintner. Some specific odors, as noted above for other grapes, are commonly associated with the grapes themselves. Generally these odors are the result of small compounds with high vapor pressures, and so, for example, while low-molecular-weight alcohols might be found in the vapor, their carbohydrate conjugates would have vapor pressures that are too low.

Some of the simple sulfur-containing compounds that are reported to be present (Figure 14.38) are 4-mercapto-4-methyl-2-pentanone, 3-mercapto-1-hexanol, and the acetate ester of the latter. However, it has also been reported[13] that the skins of the grape contain small quantities of more elaborate derivatives of those simpler compounds, viz., 3-mercaptohexan-1-ol-L-cysteine and 3-mercaptohexane-1-ol-L-glutathione, as well as 4-mercapto-4-methyl-2-pentanone-L-cysteine and 4-mercapto-4-methyl-2-pentanone- L-glutathione. As expected, the different optical isomers of the compounds are reported to have different odors.

Pyrazines (Chapter 13) were also earlier noted as being present in Cabernet Sauvignon along with the expected terpenes and terpene derivatives. They are also found as flavorings in the Sauvignon blanc grapes (Figure 14.39).

Earlier, over 100 volatile components (including those shown in Figure 14.40) were isolated and identified from Sauvignon blanc skins, leaving about an equal amount yet to be determined.[81] Most of them were the low-molecular-weight esters and hydrocarbons expected on the basis of similarity to other systems.

Semillon: A white grape.

The Semillon grape is used to make the most famous sweet wines (Sauternes) in France. The distinctive beverage is due to the action of *Botrytis cinerea* (Noble Rot) a fungus that

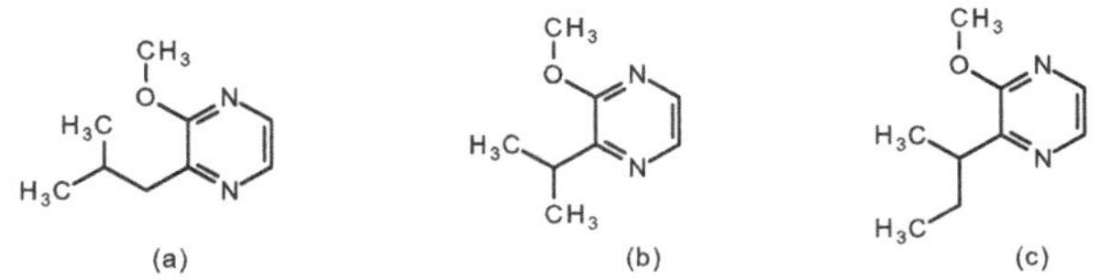

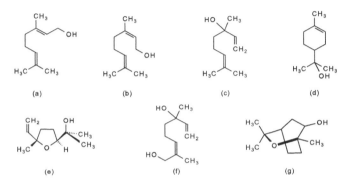

FIGURE 14.39 Representations of pyrazines reported to be found in Sauvignon blanc grapes. As shown they are: (a) 2-methoxy-3-(2-methylpropyl)pyrazine; (b) 2-methoxy-3-(1-methylethyl)-pyrazine; and (c) 2-methoxy-3-(1-methylpropyl)pyrazine.

FIGURE 14.40 Representatives of terpenols found in the skins of Sauvignon blanc grapes. Those shown are: (a) geraniol; (b) nerol; (c) linalool; (d) α-terpineol; (e) *trans*-linalool oxide; (f) *cis*-8-hydroxylinalool; and (g) 2-hydroxy-1,8-cineole.

removes water from the grapes and leaves behind a sugar-enriched residue which, with difficulty (the fungus appears to produce compounds that interfere with yeast metabolism), can then be subjected to fermentation.

A discussion of "Noble Rot" is deferred until this brief description of some of the compounds present in this and other classic grapes is completed. It can be found in "Specialized Wines" (Chapter 21, Part B). Interestingly, a report of the draft genome sequence of *B. cinerea* BcDW1, a strain isolated from Semillon grapes in Napa Valley, California, in 1992 and used with the intent to induce Noble Rot for botrytized wine production has appeared. Presumably, this was done in order to understand better how some of the unique compounds that result from action of the fungus came about.[82] Representations of some of the unusual compounds found in the skins are shown in Figure 14.41.

Secondary metabolites that have been isolated from *Vitis vinifera* grape cv. Semillon have been reported.[83] Some of the norisoprenoids obtained from Semillon grapes on glycoside hydrolysis are shown (Figure 14.42a) followed by terpenoids (Figure 14.42b) obtained from the same source.

Note that magnetic resonance technology (i.e., proton magnetic resonance, ^{1}H NMR) has recently been applied to the study of the *Vitis vinifera* L. cv. Semillon at seven different stages of berry development. The study was undertaken on berries from four weeks post-anthesis to over-ripe berries. In the study, the water that is present in the tissue cells in the berries was imaged using diffusion tensor and transverse relaxation MRI acquisition protocols. Then, the variations in diffusive motion of cellular water in the various stages of berry

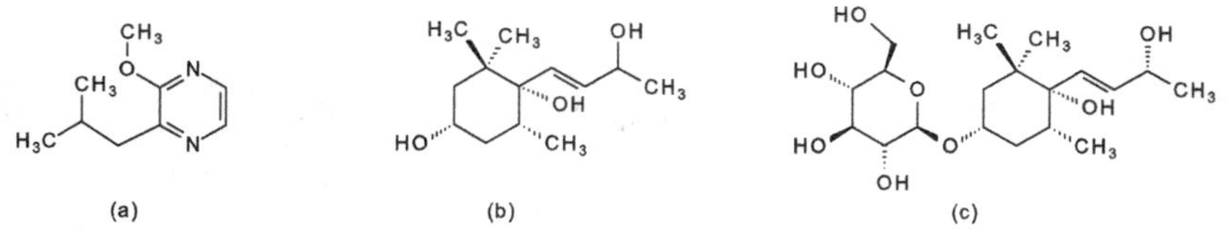

FIGURE 14.41 Representations of several unusual compounds found in the skins of Semillon grapes. The atypical compounds are: (a) 1-methoxy-3-(2-methylpropyl)pyrazine; (b) (3S,5R,6S, 9)-megastigma-7-ene-3,6,9-triol; and (c) dendranthemoside A.

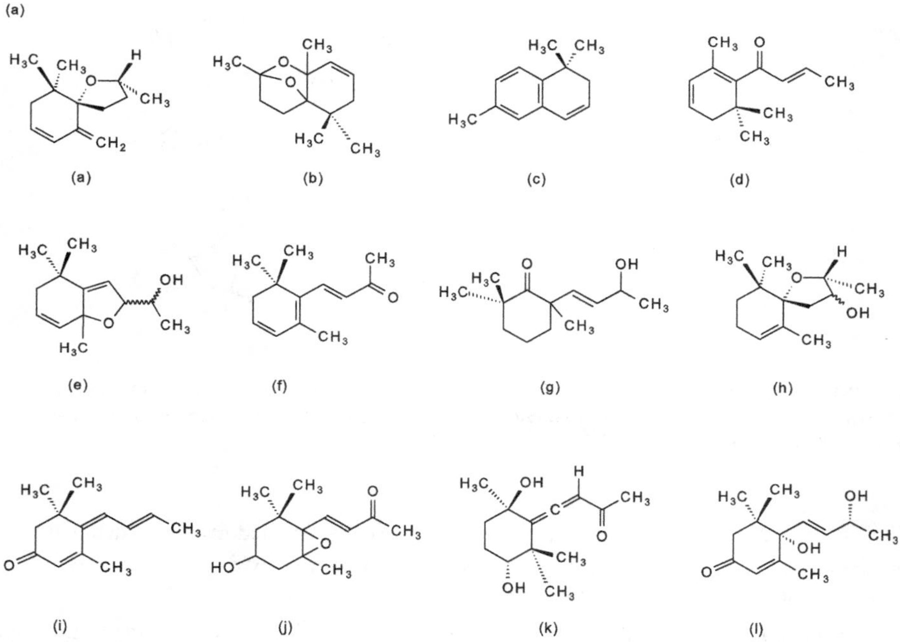

FIGURE 14.42a Representations of C-13 norisoprenoids reported found in the Semillon grape. They include: (a) vitispirane; (b) Riesling acetal; (c) 1,1,6-trimethyl-1,2-dihydronaphthalene (TDN); (d) damsacenone; (e) actinidols; (f) *trans*-3-dehydro-β-ionone; (g) 2-(3-hydroxybut-1-enyl)-2,6,6-trimethylcyclohex-3-en-1-one; (h) 3-hydroxytheaspirane; (i) (6Z,8E)-megastigma-4,6,8-trien-3-one; (j) 5,6-epoxy-3-hydroxymegastigma-7-en-9-one; (k) grasshopper ketone; and (l) vomifoliol.

development were linked to known events in the morphological development of the berries themselves. A strong association between patterns of diffusion within grape berries and the underlying tissue structures was found.[84]

Syrah/Shiraz; a red grape.

"It is somehow appropriate that a grape so startling perfumed should come laden with legend . . . or did it simply originate in the northern Rhône and stay there. Sadly for romantics it looks as though the last . . . is correct."[85]

Prior to the destruction of European vines by the pest Phylloxera (Chapter 3), cuttings of a number of different vines were taken from Europe by James Busby and brought to Australia

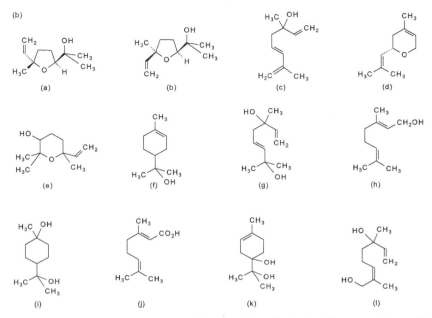

FIGURE 14.42b Representations of isoprenoids reported found in the Semillon grape. They include: (a) *trans*-linalool oxide; (b) *cis*-linalool oxide; (c) hotrienol; (d) nerol oxide; (e) linalool oxide (pyran form); (f) α-terpineol; (g) (*E*)-2,6-dimethyl-3,7-octadiene-2,6-diol; (h) geraniol; (i) 1,8-terpin; (j) geranic acid; (k) menth-1-ene-4,8-diol; and (l) *cis*-8-hydroxylinalool.

(http://en.wikipedia.org/wiki/James_Busby; February 2015). Syrah, renamed Shiraz, did well. Indeed, as recently as 2012, it was argued that "Shiraz is Australia's most important red grape variety... but little is known about the aroma compounds that are key..."[86]

Indeed, the Australian rich, warm-climate grape has been the subject of significant study since it has been argued that the aroma compounds that are present are major contributors to its success (Figure 14.43).[87] To that end, after early work on the extraction of sesquiterpenes and the tentative identification of α-ylangene rather than its isomer α-copaene[88] as the compound providing the unique, strong, spicy peppercorn aroma (which was, interestingly enough, subsequently shown to be due to rotundone)[89] RNA transcript analysis has been undertaken to provide insight into gene expression.[90]

A few sesquiterpenes (Figure 14.44) were isolated by extraction of the grape skins reported in the earlier work of Vernin[87] It is likely that all of them contribute to the overall flavor of the wine.

Earlier, over 100 volatile components (including those shown in Figure 14.44) were isolated and identified from Syrah skins, leaving about an equal amount yet to be determined.[91] Most of them were the low-molecular-weight esters and hydrocarbons expected on the basis of similarity to other systems.

Tempranillo: A red grape.

As appears to be the case for so many other grapes, this thick-skinned, black, early ripening grape of (apparently) Spanish origin is found under many other names too. However it is called, it seems that commercial wines formed using it are the result of blending.

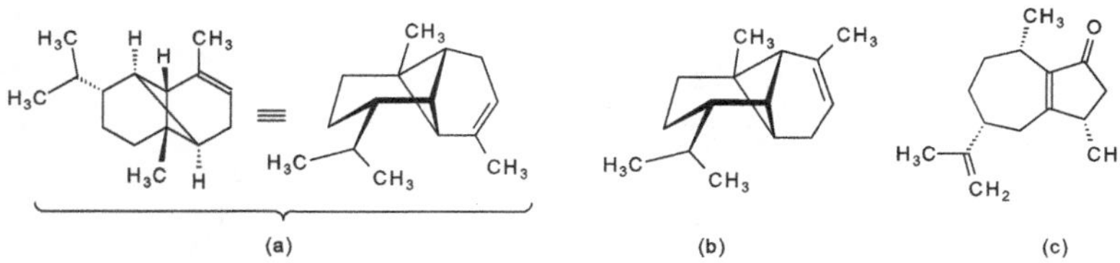

FIGURE 14.43 Representations of the terpenes that were initially identified as providing a strong aroma to Shiraz/Syrah and the currently terpenone now held responsible. As shown they are: (a) α-ylangene; its isomer (b) α-copaene; and, finally, the ketone (c) rotundone.

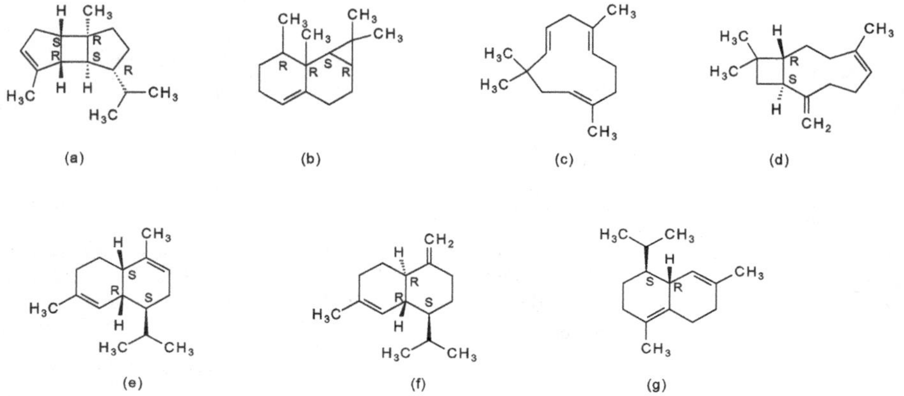

FIGURE 14.44 Representations of a few sesquiterpenes, isolated by extraction of the grape skins of Shiraz/Syrah. Those shown are: (a) α-bourbonene; (b) calarene; (c) α-humulene; (d) β-caryophyllene; (e) α-muurolene; (f) γ-cadinene; and (g) δ-cadinene.

Profiting from recent Analytical Chemistry advances, it is clear that the Tempranillo grape contains the usual set of terpenoid, norterpenoid, phenolic, and aldehydic compounds, some of which are shown in Figures 14.45 and 14.46.[17]

In addition to the phenolic compounds (Figure 14.45), anthocyanins are reported to be found in Tempranillo grapes. Structural representations are provided in Table 14.2.[92,93]

In a very interesting study, it was pointed out that, while fungicides are widely used to treat grape fungal diseases, and while it appears that their proper used does not have "adverse effects for public and environmental health," there might be a risk, and so it is necessary to develop appropriate analytical techniques to evaluate their potential presence in wine.[94]

The fungicides (Figure 14.47) were applied to grapes of the Tempranillo variety since it appears that Tempranillo grapes are more susceptible to attack by adventitious fungi. The specific compounds examined were: (a) N-{[Dichloro(fluoro)methyl]sulfanyl}-N',N'-dimethyl-N-phenylsulfuric diamide, the trade name for which is Dichlofluanid®; (b) 2-[(2,6-dimethylphenyl)-(2-methoxy-1-oxoethyl)-amino]propanoic acid methyl ester, the trade name for which is Metalaxyl® (marketed for control of fungi but which apparently failed to protect potatoes); (c) 4,6-dimethyl-N-phenylpyrimidin-2-amine (Pyrimethanil®);

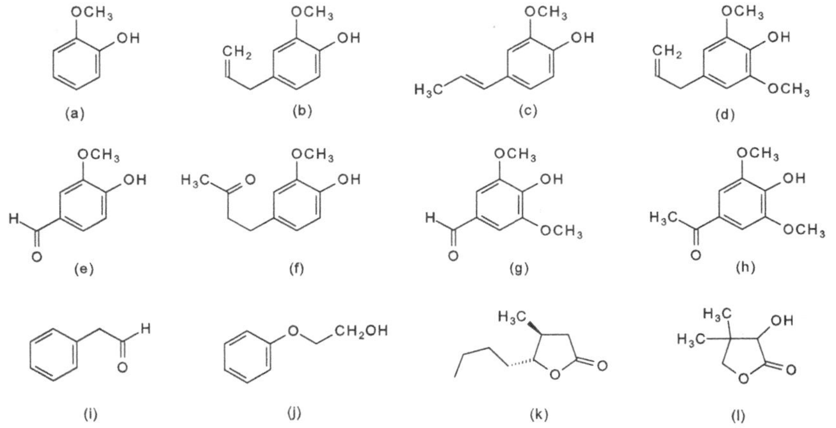

FIGURE 14.45 Representations of some aromatic compounds and two lactones reported to be present in the skin of the Tempranillo grape. Those shown include: (a) guaiacol; (b) eugenol; (c) (*E*)-isoeugenol; (d) 4-allyl-2,6-dimethoxyphenol; (e) vanillin; (f) zingerone; (g) syringaldehyde; (h) acetosyringone; (i) 2-phenylacetaldehyde; (j) 2-phenoxyethanol; (k) whiskey lactone; and (l) pantolactone.

and (d) 1-[2-(2,4-dichlorophenyl)-pentyl]-1*H*-1,2,4-triazole the trade name of which is Penconazole®.

Based upon an experimental protocol using standard isolation techniques and gas chromatography (GC) with a mass spectrometer (MS) as a detector, it was concluded that, while the highest concentrations of residues were found in the skin, penetration into the pulp occurred using all of the fungicides. Nonetheless, "in grapes collected within the safety period" . . . (of treatment and standing) . . . "the levels found for the fungicide residues were below maximum residues (*sic*) levels."

Viognier: A white grape.

"There are exceptions to every rule, but the general rule for Viognier is that it does not improve with age."[95]

This is an interesting grape, since although it is only modestly popular, it has contributed to popular fiction.[96] Mysteriously, there is a report[97] that its DNA has been sequenced and that the sequence shows a relationship to Syrah/Shiraz. At this writing (2016) the sequence has not appeared in the open literature and cannot be verified. The same source[97] notes ". . . Viognier could truly be said to be the hedonist's white grape variety, even if it is often the vintner's headache—and the drinker's headache too, come to that—for it has to be left on the vine for a very long time before its characteristic heady aroma fully develops . . ."

There are additional interesting aspects of this grape as compared to others also discussed above. That is, there is very little information about the components and about testing to determine what terpenes, esters, phenols, etc. are to be found there. Thus, for example, there is a conference report concerning the volatile terpenes before harvest as a function of fruit-zone leaf removal at "critical stages of berry development." After conversion of the grapes to wine, there was a "sensory analysis . . . four months after harvest" but no information about what the terpenes were and how they were detected.[98]

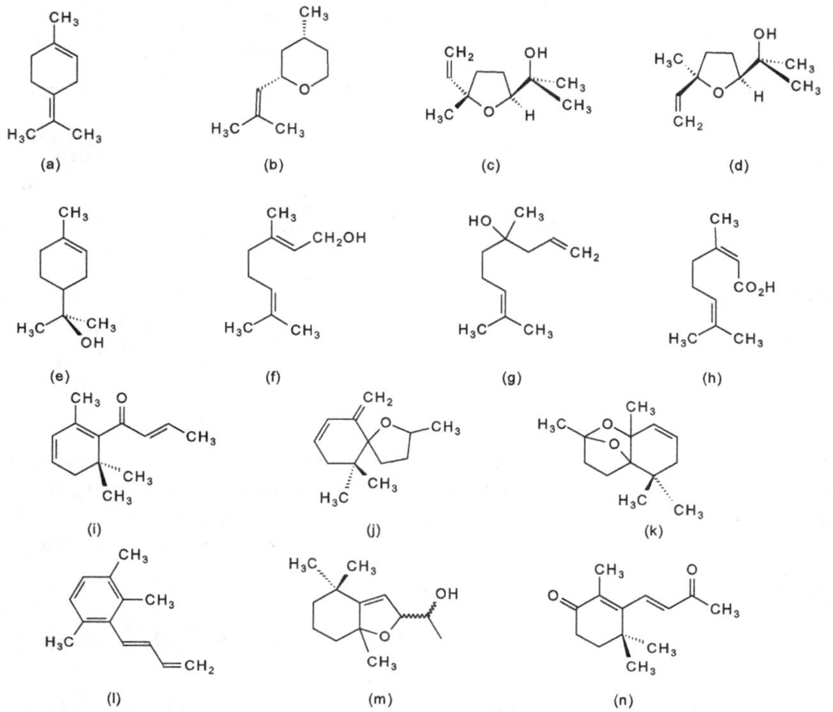

FIGURE 14.46 Representations of terpenes and norterpenes reported to be present in the skin of the Tempranillo grape. Those shown include: (a) α-terpinolene; (b) *cis*-Rose oxide; (c) *trans*-linalool oxide; (d) *cis*-linalool oxide; (e) α-terpineol; (f) geraniol; (g) linalool; (h) neric acid; (i) β-damsacenone; (j) vitispirane; (k) Riesling acetal; (l) (*E*)-1-(2,3,6-trimethylphenyl)-but-1,3-diene; (m) actinidols; and (n) 3-oxo-β-ionone.

TABLE 14.2

Representations of some of the anthocyanins reported to be found in Tempranillo grapes.

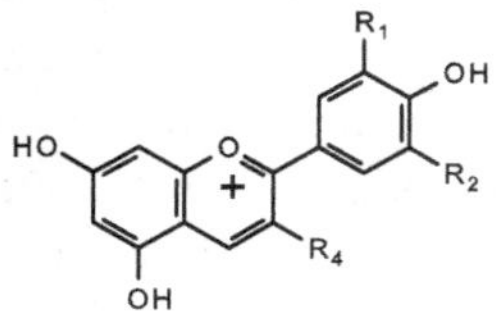

	R_1	R_2	R_3
delphinidin-3-O-glucoside	OH	OH	O-glucose
cyanidin-3-O-glucoside	OH	H	O-glucose
petunidin-3-O-glucoside	OCH_3	OH	O-glucose
peonidin-3-O-glucoside	OCH_3	H	O-glucose
malvidin-3-O-glucoside	OCH_3	OCH_3	O-glucose
malvidin-3-O-(6″-acetyl)glucoside	OCH_3	OCH_3	O-(6″acetyl)glucose
malvidin-3-O-(6″-*p*-coumaroyl)glucoside	OCH_3	OCH_3	O-(6″*p*-coumaroyl)glucose

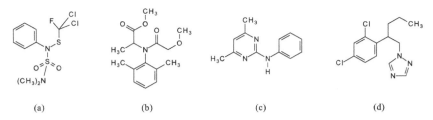

FIGURE 14.47 Representations of fungicides reported to be used on Tempranillo grapes during their growth and subsequently tested for in the wine. As shown, these were: (a) Dichlofluanid®; (b) Metalaxyl®; (c) Pyrimethanil®; and (d) Penconazole®. Their IUPAC systematic names are provided in the text.

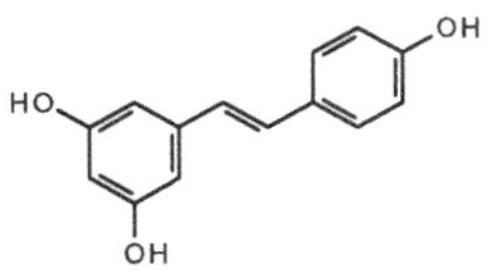

FIGURE 14.48 A representation of *trans*-resveratrol [(*E*)-5-(2-(4-hydroxyphenyl)ethenyl)-1,3-benzenediol].

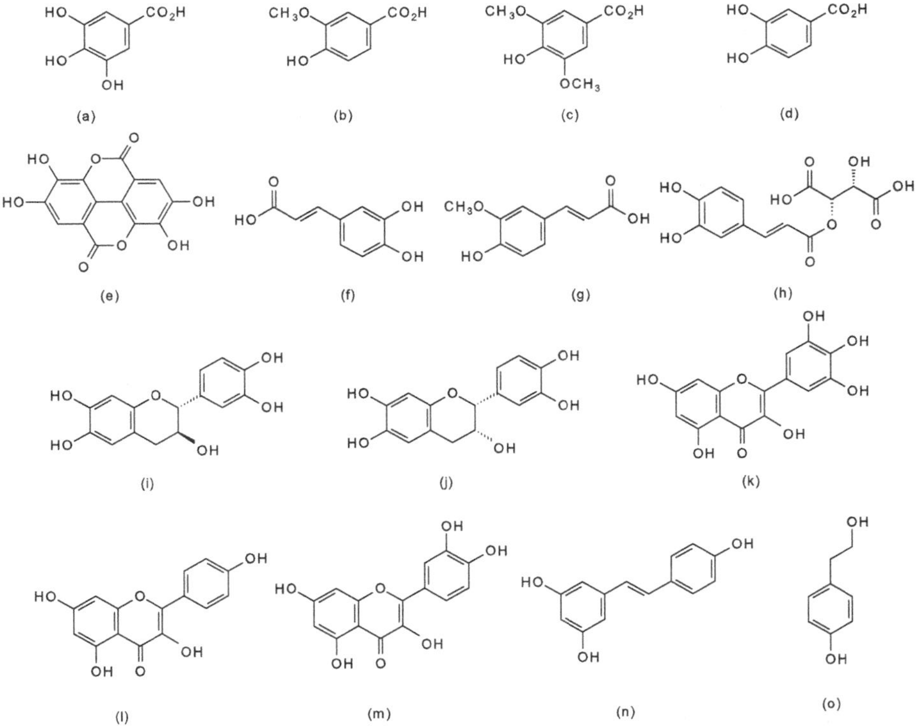

FIGURE 14.49. Representations of phenolic compounds generally found in white grapes and compared to those also found in the Viognier grape. The compounds are: (a) gallic acid; (b) vanillic acid; (c) syringic acid; (d) protocatechuic acid (below detection limit or not found in Viognier grapes examined); (e) ellagic acid; (f) caffeic acid; (g) (*E*)-ferulic acid; (h) caftaric acid; (i) catechin; (j) epicatechin; (k) myricetin; (l) kaempferol (below detection limit or not found in Viognier grapes examined); (m) quercetin (below detection limit or not found in Viognier grapes examined); (n) *trans*-resveratrol; and (o) tyrosol.

Finally, in this vein, there is one other report which includes Viognier grapes—along with other white and red grapes (five other *V. vinifera* and five *V. labrusca* varieties) in a comparison of extraction methods for phenolic "bioactive compounds."[99]

Before coming to the interesting phenolic compounds themselves it is worthwhile to note that (a) the best extraction method (best being defined carefully) was liquid–liquid extraction and (b) Concord red grapes (a *V. labrusca* variety) contained more *trans*-resveratrol [(E)-5-(2-(4-hydroxyphenyl)ethenyl)-1,3-benzenediol, Figure 14.48] than any of the *V. vinifera*, and only the Bordo grape (Cabernet franc) which is also a *V. labrusca* variety had more!

The phenolic compounds identified as present are shown in Figure 14.49. Protocatechuic acid, kaempferol, and quercetin, while present in red grapes, were missing (or below detectable levels) from Viognier and other white grapes.

NOTES AND REFERENCES

1. Robinson, J., Ed.; *Oxford Companion to Wine*, 3rd Ed.; Oxford University Press: New York, 2006; p 8.

2. The headspace (or *ullage*) above the wine is that volume lying between to surface of the liquid and the bottom of the cork (or stopper) in bottled wine. The *must* is the juice of the freshly crushed berries before separation of the skins and seeds (and stems if destemming has not taken place). The solids are, together, called the *pomace*.

3. de Simón, B. F.; Hernández, T.; Estrella, I. *Food Chem.* **1993**, *47*, 47.

4. Garcia, E.; Chacón, J. L.; Martinez, J.; Izquierdo, P. M. *Food Sci. Technol. Internat.* **2003**, *9*, 33 (DOI: 10.1177/1082013203009001006).

5. Degrees *Baumé* refers to a system for measuring the sugar content of a solution as a function of the density of the solution. Each degree Baumé is equal to about 1.75% sugar. The Baumé scale should not be confused with the Brix scale, where each degree is equivalent to 1 g sugar in 100 g of solution.

6. The specific enantiomer was not described. The (−)-isomer is shown in Figure 14.3.

7. Robinson, J., Ed.; *Oxford Companion to Wine*, 3rd Ed.; Oxford University Press: New York, 2006; p 119.

8. Bowers, J. E.; Meredith, C. P. *Nat. Genet.* **1997**, *16*, 84.

9. Bai, B.; He, F.; Yang, L.; Chen, F.; Reeves, M. G.; Li., J. *Food Chem.* **2013**, *141*, 3984.

10. A number of intriguing nitrogen and sulfur compounds have been isolated from the Sauvignon blanc grape and are discussed more fully under that heading.

11. Montealegre, R. R.; Peces, R. R.; Vozmediano, J. L. C.; Gascueña, J. M.; Romero, E. G. *J. Food Comp. Anal.* **2006**, *19*, 687.

12. Kalua, C. M.; Boss, P. K. *J. Agric. Food Chem.* **2009**, *57*, 3818 and Kalua, C. M.; Boss, P. K. *Aust. J. Grape and Wine Res.* **2010**, *16*, 337.

13. Peña-Gallego, A.; Hernández-Orte, P.; Cacho, J.; Ferreira, V. *Food Chem.* **2012**, *131*, 1.

14. Sefton, M. A.; Francis, I. L.; Williams, P. J. *Am. J. Enol. Vitic.* **1993**, *44*, 359.

15. It is important to note that the authors (Sefton, M. A.; Francis, I. L.; Williams, P. J. *Am. J. Enol. Vitic.* **1993**, *44*, 359) were clearly aware that wrenching the compounds out of the cellular matrix and subsequently subjecting them to hydrolytic enzymes as well as the isolation techniques may have produced materials not originally present. That is, some of the initially present

compounds may have rearranged into multiple isolable materials, whereas only a single precursor was actually present originally.

16. Shumbe, L.; Bott, R.; Havaux, M. *Molec. Plant* **2014**, *7*, 1248.

17. Hernandez-Orte, P.; Boncerjero, B.; Astrain, J.; Lacau, B.; Cacho, J.; Ferreira, V. *J. Sci. Food Agric.* **2015**, *95*, 688 provides examples of hundreds of isolated compounds from wines produced from this and other grapes.

18. Robinson, J., Ed.; *Oxford Companion to Wine*, 3rd Ed.; Oxford University Press: New York, 2006; p 160.

19. Augustyn, O. P. H.; Rapp, A. *S. Afr. J. Enol. Vitic.* **1982**, *3*, 47.

20. Wildenradt, H. L.; Christensen, E. N.; Stackler, B.;Caputi, Jr., A.; Slinkard, K.; Scutt, K. *Am. J. Enol. Vitic.* **1975**, *26*, 148.

21. Carbajal-Ida. D.; Maury, C.; Salas, E.; Siret, R.; Mehinagic, E. *Eur. Food Res. Technol.* **2016**, *242*, 117. DOI: 10.1007/s00217-015-2523-x.

22. The subject of the Noble Rot caused by this fungal infection *Botrytis cinerea* can be found beginning in Chapter 21.

23. Wilson, S. *Understanding, Choosing, and Enjoying Wine*; Hermes House: London, 1996; p. 88.

24. Myles, S.; Boyko, A. R.; Owens, C. L. Brown, P. J.; Grassi, F.; Aradhya, M. K.; Prins, B.; Reynolds, A.; Chai, J.-M.; Ware, D.; Bustamante, C. D.; Buckler, E. S. *Proc. Nat. Acad. Sci. (US)* **2011**, *108*, 3530.

25. Huang, H. T. *Agric. Food Chem.* **1955**, *3*, 141.

26. Ong, P. K. C.; Acree, T. E. *J. Agric. Food Chem.* **1999**, *47*, 665.

27. Lukić, I.; Radeka, S.; Grozaj, N.; Staver, M.; Peršurić, D. *Food Chem.* **2016**, *196*, 1048.

28. Clarke, O.; Rand, M. *Oz Clarke's Encyclopedia of Grapes*; Harcourt: New York, 2001; p 127.

29. Herrera, R.; Cares, B.; Wilkinson, M. J.; Caligari, P. D. S. *Euphytica* **2002**, *124*, 139 and Ochagavia Wines. http://www.ochagaviawines.com/historia-silvestre-en.html (accessed Apr 6, 2017).

30. Stegemann, S.; Bock, R. *Science* **2009**, *324*, 649.

31. As noted in the accompanying references 32 and 33, it was found that signal exchange processes between scion and root stock as soon as 2–3 days after grafting are seen. Most recently, the entire nuclear genome across graft junctions has been defined.

32. Yin, H.; Yan, B.; Sun. J.; Jia, P.; Zhang, Z.; Yan, X.; Chai. J.; Ren, Z.; Zheng, G.; Liu, H. *J. Exp. Botany* **2012**, *63*, 4219.

33. Fuentes, I.; Stegemann, S.; Golczyk, H.; Karcher, D.; Bock, R. *Nature* **2014**, *511*, 232.

34. Clarke, O.; Rand, M. *Oz Clarke's Encyclopedia of Grapes*; Harcourt: New York, 2001; p 130.

35. Cliff, M. A.; King, M. C.; Schlosser, J. *Food Res. Internat.* **2007**, *40*, 92.

36. Bucchetti, B.; Matthews, M. A.; Falginella, L.; Peterlunger, E.; Castellarin, S. D. *Scientia Hort.* (Amsterdam, Netherlands) **2011**, *128*, 297.

37. Owen, S. J.; Lafond, M. D.; Bowen, P.; Bogdanoff, C.; Usher, K.; Abrams, S. R. *Amer. J. Enol. Vitic.* **2009**, *60*, 277.

38. Loveys, B. R. *New Phytol.* **1984**, *98*, 575.

39. Clarke, O.; Rand, M. *Oz Clarke's Encyclopedia of Grapes*; Harcourt: New York, 2001; p 146 ff.

40. Barbera, D.; Avellone, G.; Filizzola, F.; Monte, L. G.; Catanzaro, P.; Agozzino, P. *Nat. Prod. Res.* **2013**, *27*, 541.

41. Clarke, O.; Rand, M. *Oz Clarke's Encyclopedia of Grapes*; Harcourt: New York, 2001; p 140.

42. Robinson, J., Ed.; *Oxford Companion to Wine*, 3rd Ed.; Oxford University Press: New York, 2006; p 459 ff.

43. Vernin, G.; Pascal-Moussellard, H.; Metzger, J.; Párkányi, C. *Developments in Food Science*, **1993**, *33 (Shelf Life Studies of Foods and Beverages: Chemical, Biological, Physical and Nutritional Aspects*; Charalambous, G. Ed.), p 945 ff.

44. Interestingly (but not surprisingly), the biosynthesis of this fifteen-member cyclic ketone, (R)-3-methylcyclopentadecanone, *aka* muscone, found naturally in species of the genus *Moschus*, does not seem to have been explored. Presumably it is derived by cyclization and decarboxylation of a sixteen-carbon α,ω-dicarboxylic acid. The source of the methyl group is not known.

45. Some of the details of the work of Pliny the Elder are now readily found. See https://archive.org/details/natural_history_3_1301_librivox (accessed Apr 6, 2017).

46. Cugnetto, A.; Santagostini, L.; Rolle, L.; Guidoni, S.; Gerbi, V.; Novello, V. *Scientia Horticulturae*, **2014**, *172*, 101.

47. Ferrandino, A.; Carlomagni, A.; Baldassarre, S.; Schubert, A. *Food Chem*. **2012**, *125*, 2340.

48. Simkin, A.; Schwartz, S. H.; Auldridge, M.; Taylor, M. G.; Klee, H. J. *Plant J*. **2004**, *40*, 882 for suggestions regarding their biosynthesis from, for example, β-carotene under the action of members of the carotenoid cleavage dioxygenase (CCD) family.

49. Auldridge, M. E.; McCarty, D. R.; Klee, H. J. *Curr. Opin. Plant Biol*. **2006**, *9*, 315, regarding carotenoid cleavage oxygenases.

50. Clarke, O.; Rand, M. *Oz Clarke's Encyclopedia of Grapes*; Harcourt: New York, 2001; p 170.

51. Pedri, U.; Pertoll, G. *Mitteilungen Klosterneuberg* **2013**, *63*, 173.

52. It is argued (Oxford English Dictionary) that "Pinot" is a variant of "Pineau," the diminutive of "pin" (pine) referring to the shape of the grape cluster; "blanc" refers to the "white" color of the grape; noir to a black grape, and gris to one that is gray.

53. Vezzulli, S.; Leonardelli, L.; Malossini, U.; Stefanini, M.; Velasco, R.; Moser, C. *J. Exp. Bot*. **2012**, *63*, 6359.

54. A *somatic mutation* is an alteration in the genome that is acquired by a cell and that can be passed to the progeny of the mutated cell in the course of cell division. A *transcription factor* is a protein that binds to a specific DNA sequence and regulates the transcription of that part of the code. A *retrotransposon* is a self-amplifying genetic element, and the *phenotype* is the composite of observable characteristics.

55. Butelli, E.; Licciardello, C.; Zhang, Y.; Liu, J.; Mackay, S.; Bailey, P.; Reforgiato-recupero, C.; Martin, C. *Plant Cell* **2012**, *24*, 1242.

56. Joslyn, M. A.; Dittmar, H. F. K. *Amer. J. of Enology and Viticulture*, **1967**, *18*, 1.

57. Hocquigny, S.; Pelsy, F.; Dumas, V.; Kindt, S.; Heloir, M.-C.; Merdinoglu *Genome* **2004**, *47*, 579.

58. Rapp, A. *Nahrung* **1998**, *42*, 351.

59. Ocaña, J.; Wlater, B.; Schellenbaum, P. *Mol. Biotechnol*. **2013**, *55*, 236.

60. Lemut.; M. S.; Sivilotti, Pl; Franceschi, P.; Wehrens, R.; Vrhovsek, U. *J. Agric. Food Chem.* **2013**, *61,* 8976.

61. Yuan, F.; Qian, M. *Abstracts, 68th Northwest Regional Meeting, Amer. Chem. Soc., Corvalis, OR*; (NORM-34), 2013.

62. Clarke, O.; Rand, M. *Oz Clarke's Encyclopedia of Grapes*; Harcourt: New York, 2001; p 186.

63. Robinson, J., Ed.; *Oxford Companion to Wine*, 3rd Ed.; Oxford University Press: New York, 2006 and Jackson, R. S. *Wine Science*, 3rd Ed.; Academic Press, New York, 2008; p 528.

64. Young, P. R.; Lashbrooke, J. G.; Alexandersson, E.; Jacobson, D.; Moser, C.; Velasco, R.; Viver, M. A. *Genomics* **2012**, *13,* 243. http://www.biomedcentral.com/1471-2164/13/243 (accessed Apr 6, 2017).

65. Battilana, J.; Costantini, L. Emanuelli, F.; Sevini, F.; Segala, C.; Moser, F.; Velasco, R.; Versini, G.; Grando, M. S. *Theor. Appl. Genet.* **2009**, *118*, 653. DOI 10.1007/s00122-008-0927-8.

66. Clarke, O.; Rand, M. *Oz Clarke's Encyclopedia of Grapes*; Harcourt: New York, 2001; p 191 ff.

67. Bauer, A.; Wolz, S.; Schormann, A.; Fischer, U. *Progress in Authentication of Food and Wine*; Ebler, S., E.; Takeoka, G. R.; Winterhalter, P. , Eds.; ACS Symposium Series 1081, American Chemical Society: Washington, DC, 2011; Ch. 9, pp 131–49.

68. Kalua, C. M.; Boss, P. K. *Aust. J. Grape Wine Res.* **2010**, *16,* 337.

69. Ryona, I.; Sacks, G. L. *Carotenoid Cleavage Products*; Winterhalter, P.; Ebler, S. E.; Eds.; ACS Symposium Series, American Chemical Society1134: Washington, DC, 2013; Ch. 10, pp 109–24.

70. Zoecklein, B. W.; Marcy, J. E.; Williams, J. M.; Jasinski, Y. *J. Food Comp. Anal.* **1997**, *10,* 55.

71. Clarke, O.; Rand, M. *Oz Clarke's Encyclopedia of Grapes*; Harcourt: New York, 2001; p 211.

72. Vouillamoz, J. F.; Monaco., A.; Costantini, L.; Stefanini, M.; Scienza, A.; Grando., M. S. *Vitis* **2007**, *46,* 19.

73. Bergamini, C.; Caputo, A. R.; Gasparo, M.; Perniola, R.; Cardone, M. F.; Antonacci, D. *Mol. Biotechnol.* **2013**, *53,* 278.

74. Anafora, G.; Tasin, M.; Cristofaro, A., Ioriatti, C.; Lucchi, A. *J. Chem. Ecol.* **2009**, *35,* 1054.

75. Arapitsas, P.; Perenzoni, D.; Nicolini, G.; Mattivi, F. *J. Agric. Food Chem.* **2012**, *60,* 10461.

76. Agati, G.; Meyer, S.; Matteini, P.; Cerovic, Z. G. *J. Agric. Food Chem.* **2007**, *56,* 1053.

77. Agati, G.; D'Onofrio, C.; Ducci, E.; Cuzzzola, A.; Rrmorini, D.; Tuccio, L.; Lazzini, F.; Mattii, G. *J. Agric. Food Chem.* **2013**, *61,* 12211.

78. The work by Agati and coworkers (reference 76) was actually carried out on *Pinot sup*. In the years between 2007 and 2013 technology improved dramatically, and implementation of those changes by the same group to the earlier work could be effected. The new work was carried out with Sangiovese (reference 77).

79. Becatti, E.; Ranieri, A.; Chkaiban, L.; Tonutti, P. *Acta Hort.* **2010**, *884,* (Proceedings XIth Internat. Symp. on Plant Bioregulators in Fruit Prod., 2009, Vol.1) 223–7.

80. Coetzee, C.; du Toit, W. J. *Food Res. Internat.* **2012**, *45,* 287.

81. Ibarz, M. J.; Ferreira, V.; Hernández-Orte, P.; Loscos, N.; Cacho, J. *J. Chromatogr. A* **2006**, *1116,* 217.

82. Blanco-Ulate, B.; Allen, G.; Powell, A. L. T.; Cantu, D. *Genome Announcements*, **2013**, 1 (3).

83. Sefton, M. A.; Francis, I. L.; Williams, P. J. *Aust. J. Grape Wine Res.* **1996**, *2,* 179.

84. Dean, R. J.; Stait-Gardner, T.; Clarke, S. J.; Rogiers, S. Y.; Bobek, G.; Price, W. S. *Plant Methods* **2014,** *10,* 35.

85. Clarke, O.; Rand, M. *Oz Clarke's Encyclopedia of Grapes*; Harcourt: New York, 2001; p 247.

86. Herderich, M. J.; Siebert, T. E.; Parker, M.; Capone, D. L. Jeffery, D. S.; Osidacz, P.; Francis, I. L. *Flavor Chemistry of Wine and Other Alcoholic Beverages;* Qian, M. C.; Shellhammer, T. H.; Eds.; ACS Symposium Series, 1104, American Chemical Society, Washington, DC, 2012; p 3.

87. Vernin, G.; Boniface, C.; Metzger, J.; Fraisse, D.; Doan, D.; Alamercery, S.; in *Proc. 5th Internat. Flavor Conf.*, Porta Karras, Chalkidiki, Greece, 1987; Charalambous, G., Ed.; Elsevier: Amsterdam, 1988.

88. Parker, M.; Pollnitz, A. P.; Cozzolina, D.; Francis, I. L.; Herderich, M. J. *J. Agric. Food Chem.* **2007,** *55,* 5948.

89. Wood, C.; Siebert, T. E.; Parker, M.; Capone, D. L.; Elsey, G. M.; Pollnitz, A. P.; Eggers, M.; Meier, M.; Vossing, T.; Widder, S.; Krammer, G.; Sefton, M. A.; Herderich, M. J. *J. Agric. Food Chem.* **2008,** *56,* 3738.

90. Sweetman, C.; Wong, D. C. J.; Ford, C. M.; Drew, D. P. *BMC Genomics* **2012,** *13,* 691.

91. Ibarz, M. J.; Ferreira, V.; Hernández-Orte, P.; Loscos, N.; Cacho, J. *J. Chromat. A.* **2006,** *1116,* 217.

92. Hernández-Hierro, J. M.; Quijada-Morín. N.; Rivas-Gonzalo, J. C.; Escribano-Bailón, M. T. *Anal. Chim. Acta* **2012,** *732,* 26.

93. Ryan, J.-M.; Revilla, E. *J. Agric. Food Chem.* **2003,** *51,* 3372.

94. Vaquero-Fernández, L.; Sanz-Asensio, J.; López-Alonso, M; Martínez-Soria, M. T. *Food Addit. Contam., Part A,* **2009,** *26,* 164. DOI:10.1080/02652030802399026.

95. Clarke, O.; Rand, M. *Oz Clarke's Encyclopedia of Grapes*; Harcourt: New York, 2001; p 282.

96. Crosby, E. *The Viognier Vendetta: A Wine Country Mystery*; Scribner: New York, 2010. ISBN 978-1-4391-6383-3.

97. Robinson, J. http://www.jancisrobinson.com/learn/grape-varieties/white/viognier (accessed Apr 6, 2017).

98. Sorokowsky, D. *Am. J. Enol. Vitic.* **2006,** *57,* 3.

99. Burin, V. M.; Ferreira-Lima, N. E.; Panceri, C. P.; Bordignon-Luiz, M. T. *Microchem. J.* **2014,** *114,* 155.

SECTION V
From the Grape to the Wine

15

General Comments

VITICULTURE, IT WILL be recalled, is the art and science of vine-growing and grape-harvesting, and it was the subject of Sections I–IV (Chapters 1–14). Enology, the subject of this Section (Chapters 15–20), is the art and science of winemaking.

Much of what follows in this brief Chapter on General Comments is expanded upon in other Chapters in this Section.

In making the wine, it appears to be generally agreed that where possible it is best to follow the traditional methods to produce the best results. However, it should be clearly understood that work is underway to engineer yeast to make it more alcohol tolerant and to use the yeast to produce specific compounds recognized as being particularly flavorful. Additionally, as the number of vintners has grown, finding the proper oak for casks is becoming ever harder. Therefore, the art of reworking old oak casks or even avoiding them altogether (e.g., by aging wine in the presence of oak chips) may be used.

In the same vein, it is widely recognized that stoppers other than cork may be used, so that the day may come when the cork stopper will be a thing of the past.

Traditionally, grapes are taken directly from the vineyard to be crushed, and it is still the case in many of the oldest and most respected vineyards that this practice will continue. However, as the use of pesticides and fungicides has increased, methods for rapid washing and then drying of grapes before crushing may be employed. The arguments against these extra steps are mainly two. First, lingering water would dilute the grape juice. Second, the adventitious yeasts that might be removed by washing or deactivated by drying are often desired for the production of the vintage. Indeed, it has been argued that unique fungi, which might be exclusive to the most prestigious vineyards, are important to the production of the best wines.

The issue of washing *versus* not washing has been investigated, and it was concluded for the case examined that only minor changes are effected by washing.[1]

With regard to the issue of wild yeasts, currently available information suggests that although they abound in most vineyards (and indeed among most fruit bearing plants) they are not as ethanol tolerant as the usually used *Saccharomyces cerevisiae* (*Saccharomyces* = Greek "sugar fungus"; *cerevisiae* from Ceres—Roman God of crops). Thus, while they might be present to provide extra substance initially (*vide infra*), they generally die off as the ethanol concentration increases and the more robust *S. cerevisiae* continues to survive and ferment the juice until its ethanol toxicity limit is reached. However, even before the role of any yeast is considered, a critical decision is required regarding grape skin.

NOTE AND REFERENCE

1. Cavazza, A.; Franciosi, E.; Pojer, M.; Mattivi, F. *International Symposium Microbiology and Food Safety of Wine;* Villafranca del Panadès, Spain, November 20, 2007.

16

More Than Skin Deep

THE GRAPE BERRY is composed of skin, flesh (pulp) and seeds.

After destemming (Chapter 13), the grapes are sent on for crushing. On crushing, the thick walls of the skin, including the waxy cuticle, are broken. Crushing the grapes (Figure 16.1) is a question of quantity. Small quantities are handled differently than large.[1]

The skins, including the contaminants thereon, as well as the majority of the materials discussed above for the individual grapes (i.e., phenols, anthocyanins, tanins, some acids, terpenes, pyrazines, and some carbohydrates including those attached to the anthocyanidins, forming anthocyanins) therein, are released.

The cells of the pulp are also broken and released into the juice on crushing. This berry cell juice is mainly water (70–80% by weight) which contains the mixture of sugars (mostly glucose and fructose, but small concentrations of many other carbohydrates are also present), carboxylic acids (mostly tartaric and malic, but additional members of the tricarboxylic acid cycle, oxalic, glucuronic, etc. are also present), complex cross-linked polysaccharides from cell walls (pectins), some phenols and proteins (as well as the peptides and simple amino acids from which they are constructed), and minerals, including oxides of iron (Fe), phosphorus (P), and sulfur (S), as well as salts of potassium (K) and sodium (Na) brought up in the xylem to the growing berry.

The seeds have their cellulose carbohydrate-based exterior coatings, which are also rich in complexed polyphenols (tannins). Additionally, amino acids, generally found as constituents of peptides, proteins, and enzymes, and their cofactors needed for all life, nucleic acids and their attached sugars needed for the next generation, are all present too.[2]

Thus, overall, the result of crushing the berries is a mixture consisting of skins, seeds, and fruit juice (the *must* = Latin *vinum mustum* = young wine).

This mixture may, if the grapes were "white," be cooled and the cap on the must—sometimes called the pomace (the solid portion of the must) removed early or late (usually between 12 and 24 hours) by the vintner. Most of the flavoring constituents are quickly extracted, and brightly colored phenols, tannins, anthocyanins, etc. are not present in any large quantity (if at all) in those grapes, as already noted in Chapter 14. However, it is important that the

FIGURE 16.1 On the left, a public domain representation in shades of gray of the colorful "Pressing Grapes," J. R. Weguelin, *ca.* 1880 (https://commons.wikimedia.org/wiki/File:John_Reinhard_Weguelin_%E2%80%93_Pressing_Grapes_(1880).jpg). On the right, a Demoisy Multi-cone Crusher, P&L Specialties, Santa Rosa, CA, USA. Used with permission. Tradition, which has its value, may be set aside in many vineyards.

vintner know when and how to remove suspended solids, since an excess can be detrimental in later stages. Usually the crushed grapes are pumped into a tank that was filled beforehand with carbon dioxide (pellets of dry ice can be used) to limit oxidation. While there may be antioxidants such as ascorbic acid (vitamin C) and glutathione (GSH) present, which will be oxidized in preference to other compounds, it is often difficult to inhibit enzymatic oxidases present in the must. Nonetheless, phenolic compounds, such as caftaric acid which might be converted to highly oxidized colored species, can be inhibited by glutathione. Indeed, it has been found[3] that the product of oxidation of caftaric acid (a quinone) reacts with glutathione, and the oxidation of the quinone stops. Further, glutathione (GSH) itself undergoes (oxidative) dimerization (unspecified other species being reduced as a consequence of that oxidation) to the corresponding disulfide (GSSG), which is then reduced back to glutathione by oxidation of vitamin C to dehydroascorbic acid. The latter, as a hydrate, is then passed to the mitochondria for processing. Figures 16.2 and 16.3 have representations of these processes.

The problem with minimizing handling is that the fresh grape juice has particulate matter suspended in it which should, generally, be removed. Thus, after a few hours (during which oxidation might be occurring as a consequence of air dissolved in the grape juice—although it is under a carbon dioxide atmosphere), the juice will separate into what appears to be two phases. Then, racking (i.e., *"soutirage"* or siphoning off an upper phase

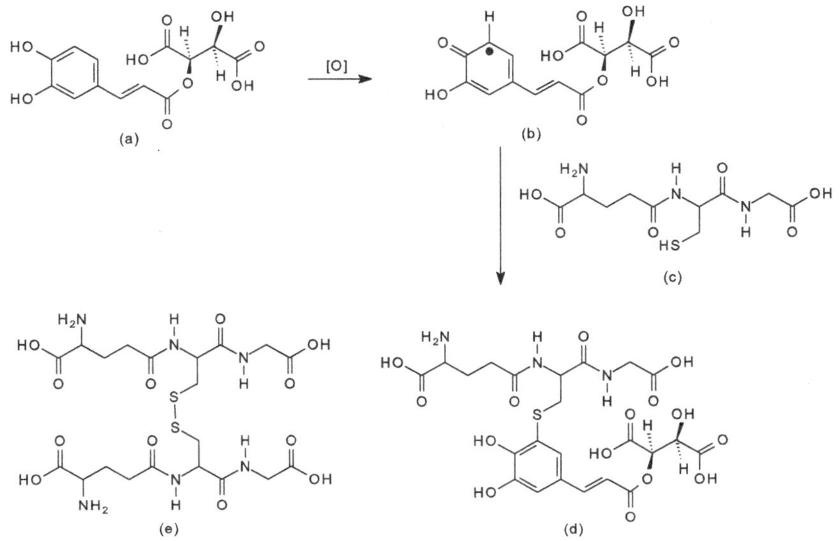

FIGURE 16.2 A representation of the oxidation of caftaric acid (a) to yield a presumed quinone type radical (b) and the trapping of that radical by glutathione (c) to form the adduct (d). Also shown is a dimer of glutathione (e).

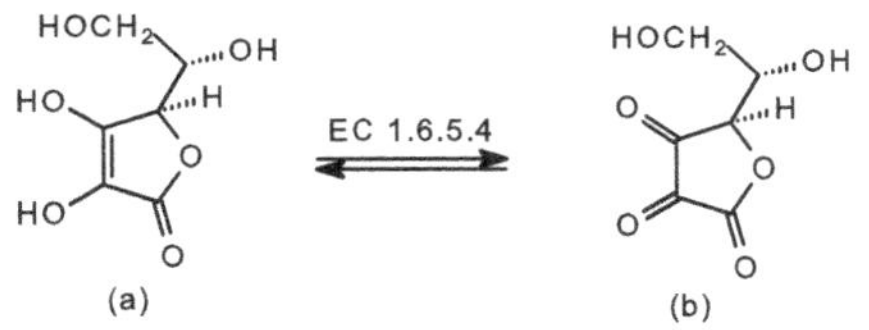

FIGURE 16.3 A representation of the oxidation/reduction equilibrium between (a) ascorbic acid (vitamin C) and (b) dehydroascorbic acid in the presence of monodehydroascorbate reductase (EC 1.6.5.4). The $NAD^+/NADH$ couple is used to transfer a hydride.

from a lower one or draining a lower from an upper phase) can be effected. Generally, the upper phase is siphoned off from the lower one by gradually lowering the siphon. The remaining solids, commonly referred to as "*lees*" (although yeast has not yet been added), are left behind.

The art of the vintner is critical during this process because all of the solids are not actually removed. Some particles will have adventitious yeasts attached as well as plant (and yeast) enzymes needed to hydrolyze glycoside linkages to free up both sugars and aglycones (which will add to flavor) and other critical elements required by the vintner. So, some particulate matter is needed. The current practice is to use a nephelometer to measure the concentration of suspended particulate matter. The nephelometer functions by measuring scattered light and is used after calibrating known solutions of solids that scatter light and comparing the result observed for the grape juice with those knowns.

The vessel used for fermentation (stainless steel is currently common, although glass carboys and oak barrels are also used) is usually filled with the partially clarified (racked) juice to about 90% capacity because carbon dioxide, formed during fermentation, results in foaming. The yeast is added. As will be discussed subsequently, most vintners utilize a strain of *Saccharomyces cerevisiae,* and some add sulfite [usually potassium metabisulfite ($K_2S_2O_5$)] early, and some late. Larger fermentation operations have used sulfur dioxide (SO_2) gas (to inhibit unwanted yeasts and spoilage causing microorganisms, Chapter 17) directly. The fermenting vessel is then fitted with an air lock (Figure 16.4) to allow carbon dioxide (CO_2) formed as fermentation progresses to escape but to prevent air from entering to spoil the process.

The fermentation process is followed by measuring the density of the fermenting juice. The density is high when glucose is present, but it steadily decreases as carbon dioxide is lost and glucose is replaced by ethanol during the fermentation. The process takes place over about two weeks. Longer than that indicates that the fermentation has stopped before all of the sugar has been consumed (i.e., there is a "stuck fermentation"[4] which will require special treatment, Chapter 18). Racking is repeated to remove the lees resulting from the death of the yeast, and a final purification through a filter such as diatomaceous earth (e.g., diatomite, celite, kieselguhr) before bottling is effected. As the density is monitored, the vessel in which the fermentation is occurring is kept under carbon dioxide and as noted above, air influx needs to be avoided.

With red grapes, the process is the same, except the cap on the must is retained. Extraction of the rich variety of materials found in the skins held in the cap, which bring color and substance to the wine, is permitted to continue. Short contact time is used for rosé wine. The extent to which the contents of the cap, which might be "punched down" from time to time

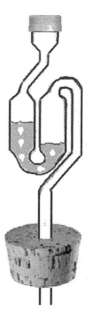

FIGURE 16.4 A cartoon representation of a typical air lock used in small-scale wine production. The liquid in the lock (usually water) permits escape of carbon dioxide (CO_2) as its pressure exceeds the pressure on the right side of liquid in the lock. Air (with oxygen, O_2) on the outside cannot pass in from the left.

to encourage mixing, is macerated (i.e., leaching the coloring and flavor compounds from the grape skins by soaking so the red wine receives its red color) is decided upon by the vintner. The decision is a function of all of the variables about which the vintner is expert, since the process removes those flavoring materials (including tannins) that will help characterize the wine.

While the more robust red grapes seem to be less sensitive to maceration processes than white grapes, it is nonetheless important to exclude air. Of course, while it is unlikely that air can be completely removed, it is important to do so to the extent possible, because in its presence it is common to find that clouding and darkening of the color of the wine (white wines too!) occurs. The process, often referred to as "*oxidasic casse*," is caused by polyphenol oxidases such as *Aureusidin synthase* (EC 2.32.3.6),[5] which has been linked to the formation of colors in flowers and is a homologue of a variety of other polyphenol synthases.

The actual extraction of different materials from the cap is a function of the variety (and maturity) of the grape, and while it might be accomplished quickly for some, others take longer. As a consequence, the cap is kept moist (e.g., by "punching down" into the must or by pumping the juice onto the cap and allowing it to drain through in a more-or-less continuous stream). Both methods often introduce air into the must, and the process might continue for as little as twelve to twenty-four hours or as long as a week. Again, the length of time is decided upon by the vintner, whose choice is a function of experience, the grape variety, the temperature of maceration, etc.

Also, during this time, yeast (in addition to whatever adventitious yeasts might have been present on the grape) may be added (usually a strain of *Saccharomyces cerevisiae*), and some sulfite [usually potassium metabisulfite ($K_2S_2O_5$)] which will (*vide infra*) serve to destroy bacteria is added too. The adventitious yeasts, even in the absence of *S. cerevisiae* will also help the overall process since, in their conversion of glucose to ethanol and as they begin the fermentation, carbon dioxide (CO_2) gas will be produced. The carbon dioxide will displace air in the juice and could help keep air above the cap from entering the fermenting juice during the "punch down." They will also contribute to a foam that often forms above the cap.

Knowing the grape and the acidity (pH) of the juice, as well as the concentration of sugars (by density) and other characteristics (e.g., potential nitrogen deficiency), specific acids present (e.g., citric acid, malic acid, tartaric acid), emendation of the must might be effected. Thus, addition of sugar, glucose, or sucrose (the process is called "*Chaptalization*")[6] and ammonium phosphate [$(NH_4)_3PO_4$] for nitrogen enhancement might be added to adjust those levels.

If the appropriate strain of *Saccharomyces cerevisiae* had not been previously added, then once the acidity, the amount of nitrogen, etc. have been optimized, the yeast is added and fermentation is begun. Generally, a time between about ten (10) days and two (or even three) weeks is used for fermentation and before the wine is racked (i.e., run off or pumped or siphoned over).

It is currently common to take the racking directly from the fermentation vat (tank, bottle, etc.) into a barrel (commonly oak, the value of which will be subsequently discussed). The pomace (or lees or residual cap) on compression produces "press wine" which may contain valuable taste constituents and might be used by the vintner for mixing with the now barreled racked wine or treated separately.

Malolactic "fermentation" (*vide infra*) is begun, and after completion small amounts of bisulfite might be added again to destroy the bacteria (Lactobacillales, i.e., lactic acid bacteria) used for this process. Additional racking may be wanted before bottling.

Rosé wines can be made by separating the cap after a day or so, pressing the juice, and then continuing the processing of the juice. Some rosé wines, it is also claimed, are produced by mixing wines already finished, although it is also claimed that such activity is prohibited by law in France.[7]

Now, returning to the processing, in the temperature-controlled (fermentation generates heat) fermenting vat (tub, barrel, tank, etc., as a function of volume to be treated), yeast and sulfur dioxide (SO_2) have been added. While complicated, the role of the sulfur dioxide (SO_2) (Chapter 17) is easier to deal with than that of the life form called "yeast" (Chapter 18), and so it will be treated first.

NOTES AND REFERENCES

1. Truly explicit and clearly well written details of the treatment of grapes to make wine and well beyond this work are to be found in Ribéreau-Gayon, P.; Dubourdieu, D.; Donèche, B.; Lonvaud, A. *The Handbook of Enology*, Vols. 1&2, 2nd Ed.; Wiley: Chichester, UK, 2006.

2. Feldmann, H. In *Yeast: Molecular and Cell Biology*, 2nd Ed.; Feldman, H., Ed.; Wiley-VCH Verlag & Co, KGaA: Weinheim, Germany, 2012; p 1.

3. Singleton, V. L.; Zaya, J.; Trousdale, E.; Salgues, M. *Vitis*, **1984**, *23*, 113 and Singleton, V. L.; Salgues, M.; Zaya, J.; Trousdale, E. *Am. J. Eno. Vitic.* **1985**, *36*, 50.

4. Christ, E.; Kowalczyk, M.; Zuchowska, M.; Claus, H.; Löwenstein, R.; Szopinska-Morawska, A.; Renaut, J.; König, H. *J. Agric. Sci.* **2015**, *7*, 18.

5. Nakayama, T.; Yonekura-Sakakibara, K.; Sato, T.; Kikuchi, S.; Fukui, Y.; Fukuchi-Mizutani, M.; Ueda, T.; Nakao, M.; Tanaka, Y.; Kusumi, T.; Nishino, T. *Science* **2000**, *290*, 1163.

6. The process of *Chapatalization* (named after French chemist Jean-Anteine-Claude Chaptal who developed the method in the 18th century) consists of adding a sugar (a variety of sources of sucrose have been reported) to grape must from grapes with low sugar content.

7. Samuel, H. *The Telegraph*, Home, Food and Drink, Wine, Mar 10, 2009. http://www.telegraph.co.uk/foodanddrink/wine/4969209/Rose-Just-mix-red-and-white-wine-says-EU.html (accessed Apr 7, 2017).

17

Adding Sulfur Dioxide (SO_2)

THE JUDICIOUS USE of sulfur dioxide (SO_2) will inhibit the growth of microorganisms (e.g., bacteria) present on the grape skins as the berries come from the vineyard. Its early use presumes the vintner has decided that the adventitious wild yeasts which might be destroyed or inhibited by sulfur dioxide will not contribute to the vintage.[1]

It appears that *Saccharomyces cerevisiae* might be less susceptible to the action of sulfur dioxide than other yeasts that may be present. So, if the particular strain of *S. cerevisiae* used can cope, it may be able to function unimpeded. Regardless, sulfur dioxide might still be used because, in addition to suppression of deleterious microorganisms, it appears to reduce oxidation of particularly fragile white wine components.

In industrial settings, both gaseous sulfur dioxide and sulfur dioxide as a liquefied gas (boiling point −10 °C [14 °F]) are used. In either form it is a dangerous tool. It is dangerous first because it is toxic and second because an excess of it will ruin the wine. In many cases, because its value is recognized as beneficial, sulfur dioxide is replaced by addition of either sodium metabisulfite ($Na_2S_2O_5$) or potassium metabisulfite ($K_2S_2O_5$) with the latter generally preferred. Indeed, while it is best to look at the MSDS. (Manufacturer's Safety Data Sheet) before use, the solubility of the two salts is the same and given as 450 grams/liter (g/L) at 68 °F (20 °C) and the pH on dissolution as between 3.5 and 4.5. The potassium (K) salt appears, at this writing, to be more readily available in food quality (as opposed to chemical quality) grade.

So, with regard to sulfur dioxide (SO_2), and as shown in Figure 17.1, its structure is much more similar to water and to ozone than it is to carbon dioxide (CO_2); sulfur lies beneath oxygen (O_2) in the periodic table (silicon, Si, lies beneath carbon). Nonetheless, sulfur dioxide (SO_2) reacts with water much the same way that carbon dioxide (CO_2) does. As shown (Figure 17.1), the reaction of SO_2 with water produces (in equilibrium with their progenitors) sulfurous acid (H_2SO_3) which, on proton loss to water, produces the bisulfite anion (HSO_3^-) and the hydronium ion H_3O^+. Similarly, although not shown, carbon dioxide (CO_2) reacts with water to produce carbonic acid (H_2CO_3) which, on proton loss to water,

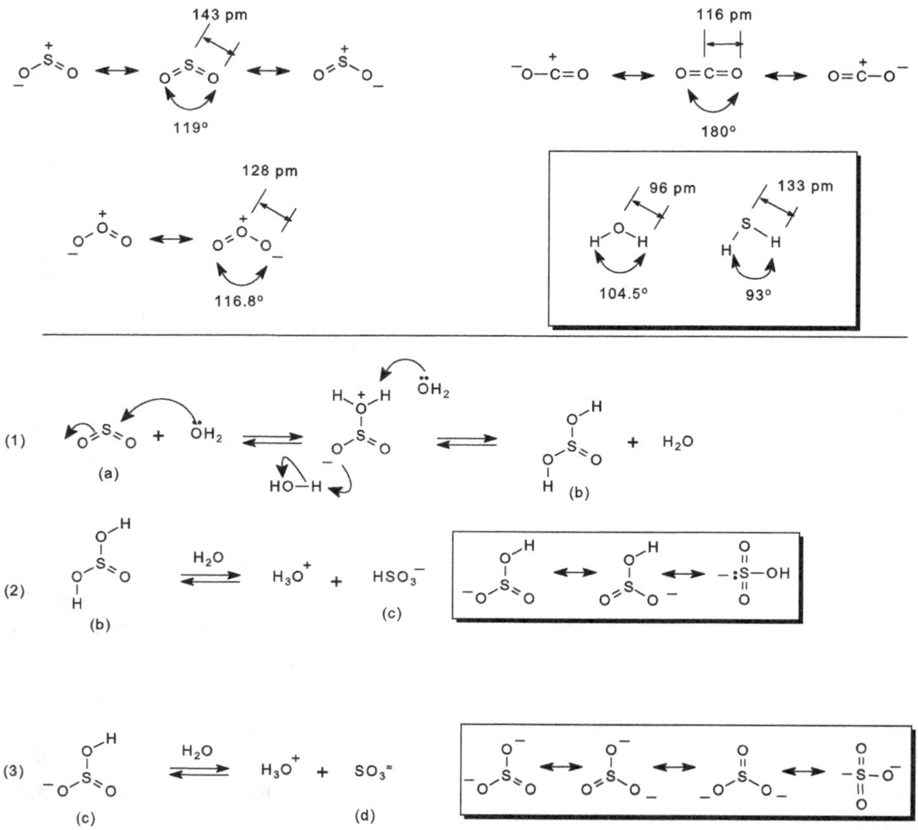

FIGURE 17.1 At the top, some properties of sulfur dioxide (SO_2), suggesting that its shape is closer to that of ozone (and water) than it is to carbon dioxide. Distances are shown in picometers (pm, 10^{-12} m). Below, cartoon representations of: (1) the reaction of water with sulfur dioxide (a), SO_2, to produce (b) H_2SO_3, sulfurous acid; (2) the ionization of sulfurous acid (b) in aqueous solution to form the bisulfite anion (c) HSO_3^- where several classical resonance forms for that anion—in particular showing the presence of a charge on sulfur (!) are provided; and (3) the ionization of the bisulfite anion in aqueous solution (c) to the sulfite anion (d) SO_3^{2-}, where again resonance forms showing the presence of a charge on sulfur are provided.

produces the bicarbonate anion (HCO_3^-) and the hydronium ion H_3O^+. Of course these similar processes proceed with different equilibrium constants.

For these equilibria, a second proton loss from bisulfite (HSO_3^-) to water or to another electron-rich species (e.g., another bisulfite anion) produces the sulfite (SO_3^{2-}) anion. Identical equilibria, but with different equilibrium constants, attend the (not shown) reaction of the bicarbonate anion (HCO_3^-) to produce the carbonate anion (CO_3^{2-}). As expected, the aqueous equilibria shown between sulfur dioxide (SO_2), the bisulfite anion (HSO_3^-) and the sulfite anion (SO_3^{2-}) are pH dependent.

The graphic in Figure 17.2 shows where the equilibria lie as a function of pH at or near room temperature. It is particularly important to note that, because the pH of the freshly crushed white grapes is generally around 3.3, while that of the red is close to 3.4, the

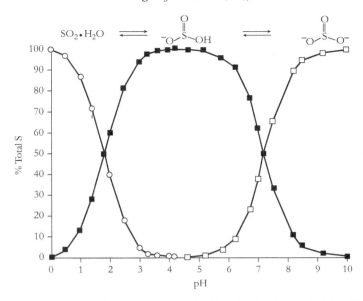

FIGURE 17.2 The relative abundance as percent total sulfur of molecular SO_2 (○), bisulfite (□), and sulfite (■) at different pH values in aqueous solution. After Fugelsang, K.C. In *Wine Microbiology*, 2nd Ed.; Fugelsang, K. C., Edwards, C. G., Eds.; Spinger: New York, 2007; p 67. DOI: 10.1007/978-0-387-33349-6-5. Used with permission.

EQUATION 17.1 A representation of the reaction between water and potassium metabisulfite (a) resulting in the cleavage of the sulfur–sulfur bond in the latter and formation of potassium bisulfite (b).

concentration of sulfur dioxide (SO_2) is relatively small, but constant, as a function of the equilibrium with bisulfite as long as the pH remains in the appropriate range.

The interaction between sulfur dioxide (SO_2) and components in the wine must, as well as its use as an antioxidant (i.e., by reaction with oxygen (O_2) itself to form sulfur trioxide (SO_3), a reaction normally run at high temperature and in the presence of metal oxide catalysts), has been reviewed, and information is available regarding its use.[2]

Most recently it has become clear that it is probably better to treat the must with potassium metabisulfite ($K_2S_2O_5$), since there is little possibility of overloading the system with an initial pulse of an excess of sulfur dioxide (SO_2). Thus, as shown in Equation 17.1, when potassium metabisulfite ($K_2S_2O_5$) is added to an aqueous solution it is proposed that water attacks the sulfur–sulfur bond, producing two equivalents of potassium bisulfite ($KHSO_3$) which then enters into the equilibria to produce sulfur dioxide (SO_2)

and sulfite (SO_3^{2-}) already described. In this way, the addition of sulfur dioxide (SO_2) itself is avoided.

In this vein, while the concentration of sulfur dioxide (SO_2) is important, it is also important to recognize that bisulfite (HSO_3^-) itself is an important and reactive species in the must and will intrude into the biochemistry of living bacteria and fungi (i.e., yeast) that may be present in the young wine.

First, aldehydes that are not hindered will react to form bisulfite adducts. The kinetics of the formation of bisulfite adducts has been studied, and the technique is widely used in organic chemistry because the adducts are often crystalline and thus allow the separation of aldehydes from other components.[3] Since the adduct is formed in equilibrium with the free aldehyde, once the adduct has been obtained, it can be reconverted to the aldehyde itself.[4] It is clear that a carbon-to-sulfur bond has formed (Equation 17.2).[5]

Although the reaction (Equation 17.2) is shown for ethanal (acetaldehyde), it occurs with many aldehydes such as glyceraldehyde and all of its carbohydrate relatives including glucose (in the acyclic form), as well as ketones such as pyruvic acid and fructose. Thus, albeit in very low concentration, these materials too will be in equilibrium with the corresponding bisulfite adducts.

Secondly (but not widely appreciated) it has been well known for many years (at least since 1868)[6] that good leaving groups attached to carbon could be displaced by sulfite. Thus, in the cofactor thiamine (vitamin B_1) the methylene carbon linking the pyrimidine ring to the positively charged nitrogen of the thiazole ring is ideally set up to use the latter as a leaving group (Equation 17.3). Indeed, it has been recognized that thiamine, crucial to the formation of acetyl-CoA from pyruvate, and thus lying at the foundation of the synthesis of many of the materials necessary for life, is destroyed by sulfite.[7]

EQUATION 17.2 A representation of the reaction between acetaldehyde (a) and the bisulfite anion (b). The adduct (c) is shown as an anion and will have as its counter ion whatever the cation of bisulfite used was. A proton shift to produce the adduct (d) is denoted by (~H^+).

EQUATION 17.3 A representation of the reaction between bisulfite (a) and thiamine (vitamin B_1) (b) showing the attack of the bisulfite anion on the methylene carbon linking the pyrimidine ring to the thiazole. It has been argued that this destruction of the vitamin B1 is deleterious to health.

In the presence of the multienzyme pyruvate dehydrogenase complex, pyruvate bound to a core of enzymes including acyl-transferring pyruvate dehydrogenase (EC 1.2.4.1), dihydrolipoyllysine-residue acetyltransferase (EC 2.3.1.12), and multiple copies of dihydrolipoyl dehydrogenase (EC 1.8.1.4) is converted to acetyl coenzyme A.

Acetyl coenzyme A (CH_3COS-CoA) is the fundamental building block for a wide variety of required metabolic products and is fundamental to life. Thus, loss of vitamin B_1 and the failure of the conversion of pyruvate to acetyl CoA strikes at the fundamental chemistry of life.

Equation 17.4 outlines a path for the overall process of the conversion of pyruvate to acetyl CoA and demonstrating the requirement for the cofactor thiamine (vitamin B_1). A more detailed representation of coenzyme A is provided in Figure 17.3, and a cartoon elaborating on Equation 17.4 is shown as Scheme 17.1.

Among the other reactions of note, it is important to recognize that sulfur–sulfur bonds are cleaved by sulfite and sulfur dioxide, destroying the high-order structure of proteins and peptides[8] and, of current interest, that bisulfite is used to measure the extent of epigenetic remodeling of cytosine (C) to 5-methylcytosine (Scheme 17.2).[9,10]

Because 5-methylcytosine is more hindered than cytosine (C) and thus reacts with bisulfite more slowly than cytosine (C), the latter will generate uracil (U) and the former will not. Thus it becomes clear which cytosines (C) in a DNA have been methylated and which not.[9,10]

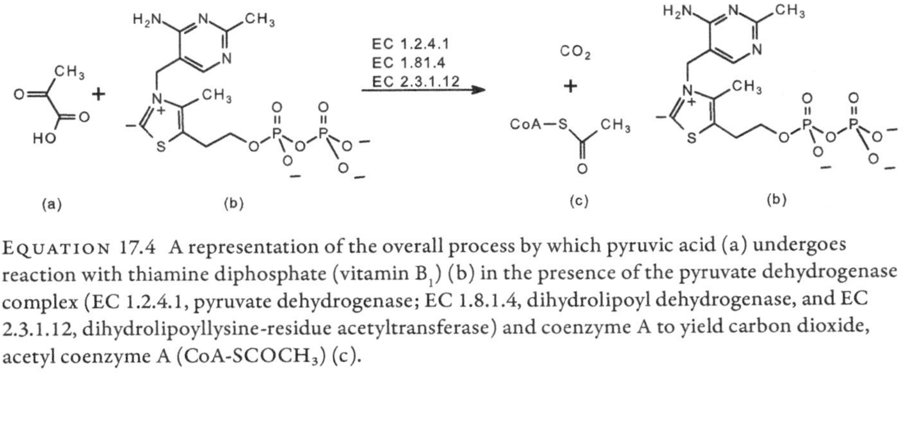

EQUATION 17.4 A representation of the overall process by which pyruvic acid (a) undergoes reaction with thiamine diphosphate (vitamin B_1) (b) in the presence of the pyruvate dehydrogenase complex (EC 1.2.4.1, pyruvate dehydrogenase; EC 1.8.1.4, dihydrolipoyl dehydrogenase, and EC 2.3.1.12, dihydrolipoyllysine-residue acetyltransferase) and coenzyme A to yield carbon dioxide, acetyl coenzyme A (CoA-SCOCH$_3$) (c).

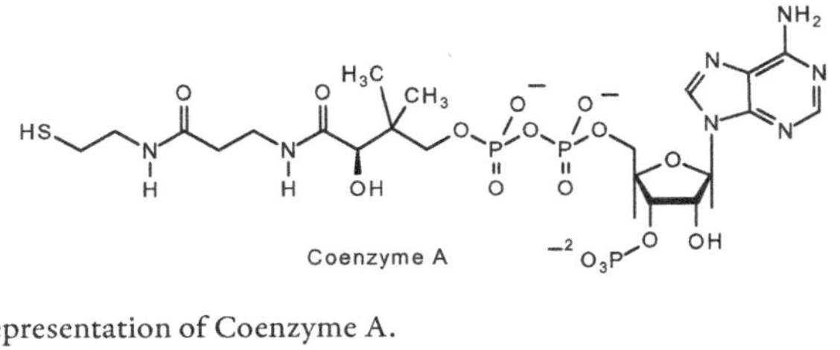

FIGURE 17.3 A representation of Coenzyme A.

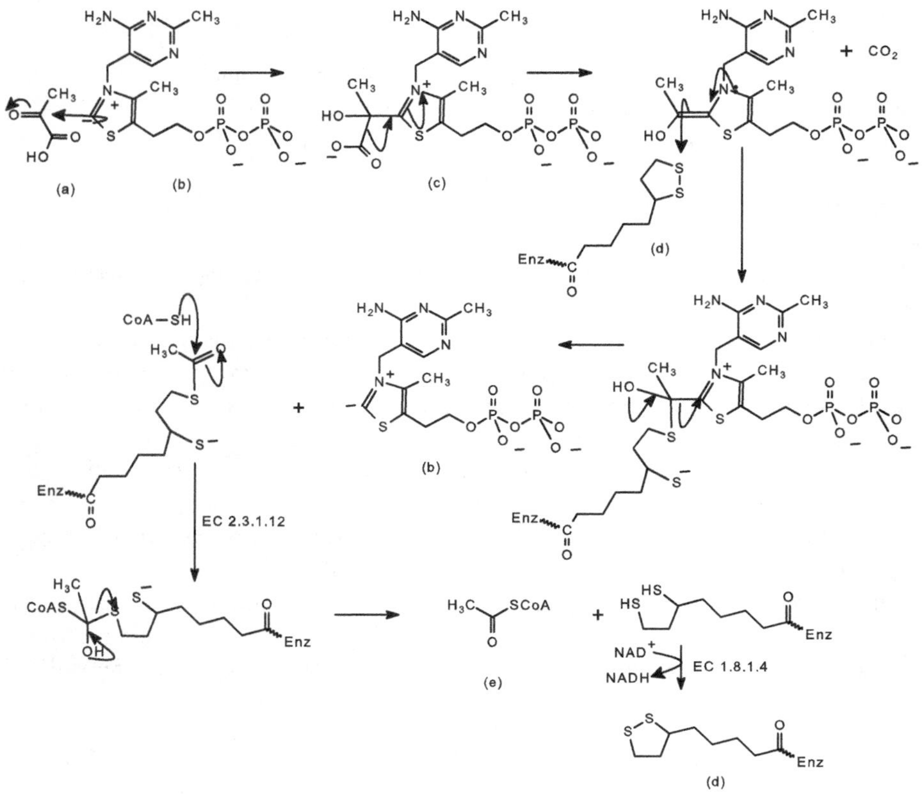

SCHEME 17.1 A more detailed cartoon type representation of the reaction of pyruvate (a) with thiamine diphosphate (b) in the presence of the pyruvate dehydrogenase complex. The adduct (c) between (a) and (b), losing carbon dioxide (CO_2), then reacts with a lipoyl group (d) (lipoic acid attached to a nitrogen in the enzyme) providing an acetyl group attached to the sulfur, breaking the sulfur–sulfur bond of the lipoyl group and liberating thiamine diphosphate. Then, in the presence of the part of the complex with dihydrolipoyllysine-residue acetyltransferase (EC 2.3.1.12), the acetyl group is transferred to coenzyme A, yielding acetyl-CoA (e), and subsequently, the sulfur-sulfur bond is reformed by the action of NADH and dihydrolipoyl dehydrogenase (EC 1.8.1.4).

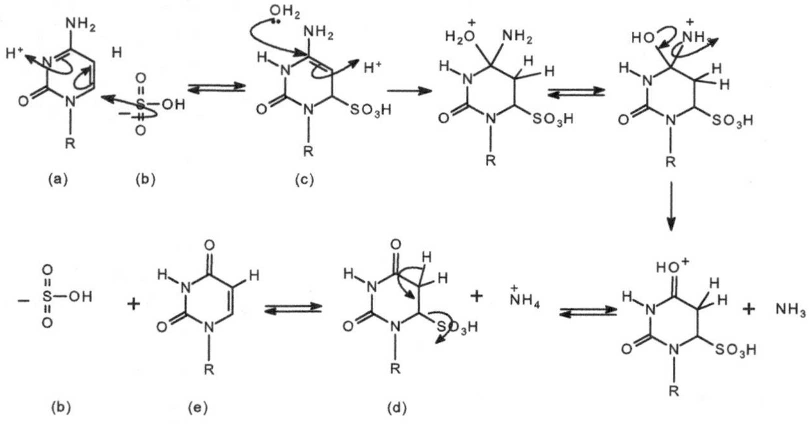

SCHEME 17.2 A representation of a likely pathway for the reaction between cytosine (a) and the bisulfite anion (b) to generate, first, cytosine sulfonate (c) which undergoes hydrolysis with loss of ammonia to produce uracil sulfonate (d). Uracil sulfonate (d), on loss of bisulfite, yields uracil (e).

NOTES AND REFERENCES

1. Pateraki, C.; Paramithiotis, S.; Doulgeraki, A. I.; Kallithraka, S.; Kotseridis, Y.; Drosinos, E. H. *Euro. Food Res. Tech.* **2014,** *239,* 1067.

2. Ribéreau-Gayon, P.; Dubourdieu, D.; Donèche, B.; Lonvaud, A. *The Microbiology of Wine and Vinifications*, Vol, 1, 2nd Ed.; Wiley: Chichester, UK, 2005. DOI: 10.1002/0470010363.ch8.

3. Lowry, T. H.; Richardson, K. S. *Mechanism and Theory in Organic Chemistry,* 3rd Ed.; Harper & Row: New York, 1987; pp 682–3.

4. Kjell, D. P.; Slattery, B. J.; Semo, M. J. *J. Org. Chem.* **1999,** *64,* 5722.

5. Shriner, R. L.; Land, A. H. *J. Org. Chem.* **1941,** *6,* 888.

6. Strecker, A. *Ann. Chem.* **1868,** *148,* 90. DOI:10.1002/jlac.18681480108.

7. Dwivedi, B. K.; Arnold, R. G. *J. Agric. Food Chem.* **1973,** *21,* 54.

8. Cecil, R.; Wake, R. G. *Biochem. J.* **1962,** *82,* 401.

9. Darst, R. P.; Pardo, C. D.; Ai, L.; Brown, K. D.; Kladde, M. P. *Curr. Protoc. Mol. Biol.* **2010** Jul; CHAPTER: Unit–7.917. DOI: 10.1002/0471142727.mb0709s91.

10. Hattori, N.; Ushijima, T. *Handbook of Epigenetics;* Tollefsbol, T., Ed.; Academic Press, Elsevier: New York, 2011; pp 125–34.

18

Yeasts

PART A. GENERAL COMMENTS ON YEASTS

The yeast, *Saccharomyces cerevisiae*, is a fungus, one of the group of eukaryotes (organisms with membrane-enclosed organelles and nuclei in their cells) that lie on that branch of the tree of life that, as shown in Figure 18.1, includes plants and animals.[1] Many years of debate preceded their notation as a separate branch on the tree while advocates forcing them into either plant or animal families battled. Thus, although the cell walls of yeast are strikingly similar to plants (save that yeasts utilize N-acetylglucosamine and related nitrogenous carbohydrate polymers [chitin-like] in place of polyphenols [lignin] for cross linking), it is clear that chloroplasts, common to plants, are missing. Similarly, while their organization and food disposition is similar to animals, the very presence of a cell wall, rather than a simple membrane, forces their exclusion from the family of animals. Of course, all life utilizes the same set of purine and pyrimidine bases bonded to a ribose or deoxyribose carbohydrate and amino acids. So while classifications are necessary, they may also be specious.

A generic eukaryotic cell and a plant cell (seen before in Figure 7.1) are shown in Figure 18.2.

Hundreds of yeasts and strains[2] of those yeasts are available for use in the wine industry for fermenting the must obtained on crushing the grapes. Some of the yeasts are referred to as "wild" and are brought in with the grapes from the vineyard. Others, originally "wild," have been isolated and maintained because it is held that their use adds value to the vintage.

Indeed, it is here that a great deal of experience is required. Generally, the vintner has a good idea of the amount of sugar (measured as glucose) in the grapes harvested.[3]

However, different strains of yeast (some 1500 yeast species, including *S. cerevisiae* are a subgroup of 700,000 or so fungi),[4] while probably processing glucose in the same way, will also process other sugars too and, in that vein, there are other issues to be faced.

First, there is the issue relating to the fermentation of glucose to ethanol. The fermentation process is very important, and it has been investigated in detail. Second, there is the issue of complexity. Clearly, it is important to distinguish between allowing yeast to act on the must

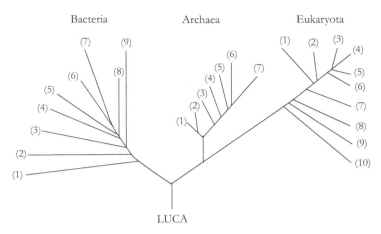

FIGURE 18.1 A representation of a more or less current phylogentic tree of life. This tree is drawn in an attempt to show currently held evolutionary relationships arising from LUCA (the Last Universal Common Ancestor). Such relationships may change as we learn more. *Bacteria* (1) Aquifex; (2) Thermotoga; (3) Bacteroides Cytophaga; (4) Planctomyces; (5) Cyanobacteria; (6) Proteobacteria; (7) Spirochetes; (8) Gram positives; (9) Green Filamentous Bacteria. *Archaea* (1) Pyrodicticum; (2) Thermoproteus; (3) Thermococcus celer; (4) Methanococcus; (5) Methanobacterium; (6) Methanosarcina; (7) Halophiles. *Eukaryota* (1) Entamoebae; (2) Slime molds; (3) Animals; (4) Fungi; (5) Plants; (6) Ciliates; (7) Flagellates; (8) Trichomonads; (9) Microsporidia; (10) Diplomonads. Modeled after Woese, C. R.; Kandler, O.; Wheelis *Proc. Natl. Acad. Sci.* **1990**, *87,* 4576 *et seq.*

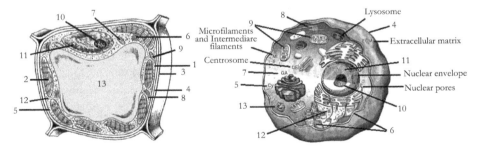

FIGURE 18.2 Representations of a typical plant cell (left) and a typical eukaryotic cell (right). Common features but with distinct functions, include items such as: (4) a membrane; (5) cytoplasm (6) endoplasmic reticulum (ER); (7) Golgi bodies; (8) microtubules; (9) mitochondria; (10) a nucleolus; (11) a nucleus; (12) ribosomes and a vacuole. The amyloplasts (1) and chloroplasts (2) of the plant are clearly missing. On the left, the representation of the plant cell is taken from Harold, F. M. *The Way of the Cell;* Oxford University Press, 2001; p 29. On the right, the representation of a generic cell that does not belong to a plant is taken from Allen, T.; Cowling, G. *The Cell*, Oxford University Press, 2011; p 6. Reprinted courtesy of Oxford University Press.

derived from the grape rather than glucose dissolved in water. So it might be important to know if the other components in the must are being acted upon by the yeast too. Third, fungi (yeast) are alive. All of the processes of living, reproducing, and dying occur in their families too.

Nutrients, in addition to those found in grapes, may be necessary to keep the yeast viable before they, having generated too much ethanol in which to live, die. What portions of these living systems remain in the wine (in solution) after their demise? Further, the yeast genome[5] has been a source of experimental tinkering in attempts to induce production of valuable commodities and, since many sequences in yeast resemble those in related animals and plants, to serve to help understand the relationship between sequence and function. Has much changed in the wine industry as a consequence of such tinkering?

PART B. GLUCOSE TO ETHANOL

The complete pathway for the conversion of glucose to pyruvate is often referred to as the Embden-Meyerhof-Parnas pathway. Pyruvate can then go on to a variety of products which include lactate, acetaldehyde (reduced to ethanol), carbon dioxide, and acetate (as in acetyl coenzyme-A). Generally, the overall process is considered to involve two major stages. In the first stage, glucose undergoes phosphorylation, rearrangement, and eventual cleavage to two equivalents of 3-phosphoglyceraldehyde. In the second stage, the phosphorylated glyceraldehyde is converted to pyruvate.

As already seen (Chapter 17), pyruvate can be converted oxidatively to acetyl coenzyme-A (lipoic acid being reduced). Acetyl coenzyme-A is needed for a variety of life preserving processes. However, under anaerobic (absence of oxygen) conditions, pyruvate can undergo reductive decarboxylation to acetaldehyde which, in turn, can be reduced to ethanol.

First, it will be recalled (Chapters 8 and 10) that sucrose is transported in the phloem and stored; thus the hydrolysis of sucrose to glucose and fructose is required (Figure 18.3).

Under neutral conditions the hydrolysis is slow. However, in the presence of the enzyme invertase (a family of invertases under the rubric EC 3.2.1.x can be invoked) which might be present in the must but is known to be present in yeast), or simply under the acidic conditions found in the must, the cleavage of the acetal linkage occurs more rapidly.

Then, the glucose that is liberated is converted (glucokinase, EC 2.7.1.2) as shown in Scheme 18.1 to the corresponding fructose-6-phosphate (glucose-6-phosphate isomerase, EC 5.3.1.9). The fructose itself can also be phosphorylated (ATP → ADP, fructokinase, EC 2.7.1.4)

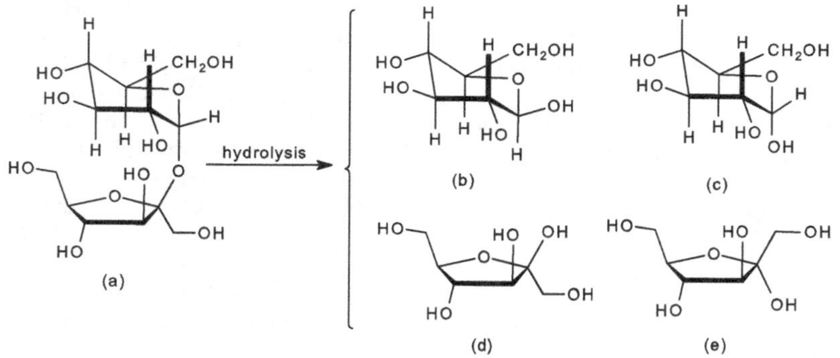

FIGURE 18.3 A representation of the hydrolysis of sucrose (a) into the anomeric glucopyranoses β-D-glucopyranose (b) and α-D-glucopyranose (c) as well as β-D-fructofuranose (d) and α-D-fructofuranose (e). On hydrolysis epimerization at the anomeric carbon may occur.

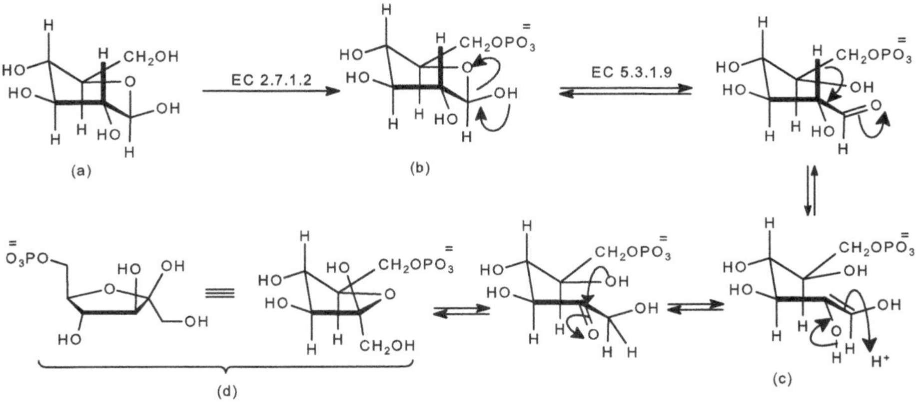

SCHEME 18.1 A cartoon representation showing the interconversion of β-D-glucopyranoside (a) via phosphorylation (glucokinase, EC 2.7.1.2) to the corresponding 6-phosphate (b) and then, isomerization (glucose-6-phosphate isomerase, EC 5.3.1.9) to the common glucose-fructose enol (c). Finally, recyclization generates β-D-fructofuranose-6-phosphate (d). The respective epimers at the anomeric carbon are expected to behave similarly.

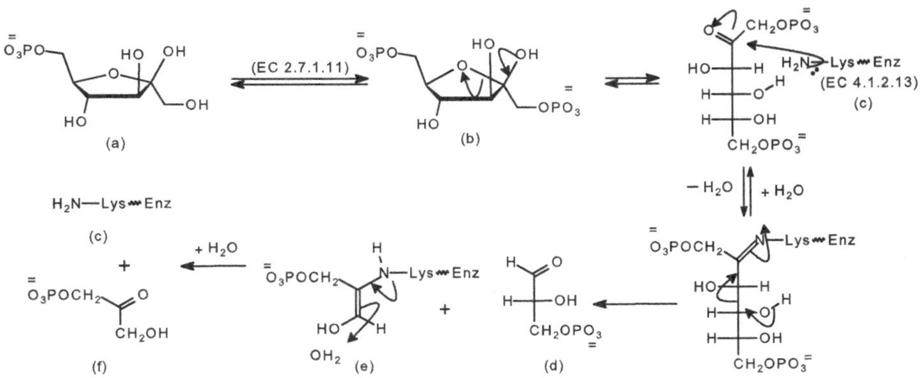

SCHEME 18.2 A cartoon representation of the process of the conversion of β-D-fructofuranose-6-phosphate (a) to the corresponding 1,6-bisphosphate (6-phosphofructokinase, EC 2.7.1.11) (b) followed by imine formation using a terminal amino group from a lysine (Lys, K) (c) of fructose-bisphosphate aldolase (EC 4.1.2.13). Enamine formation from the imine accompanied by carbon–carbon bond cleavage results in the formation of glyceraldehye-3-phosphate (d) and the three-carbon enamine (e) capable of hydrolysis to dihydroxyacetone monophosphate (f) and liberation of the enzyme (c).

After phosphorylation of β-D-fructofuranose-6-phosphate to the 1,6-bisphosphate (6-phosphofructokinase, EC 2.7.1.11), coordination of the incipient carbonyl group (shown in Scheme 18.2) to a lysine of a fructose-bisphosphate aldolase (EC 4.1.2.13) followed by fragmentation (a retroaldol-like process) generates the two three-carbon fragments, 3-phosphoglyceraldehyde (glyceraldehyde-3-phosphate) and dihydroxyacetone monophosphate (glycerone phosphate). These species, one an aldehyde and the other a ketone, are in equilibrium through the common enol which, in the presence of the enzyme triose

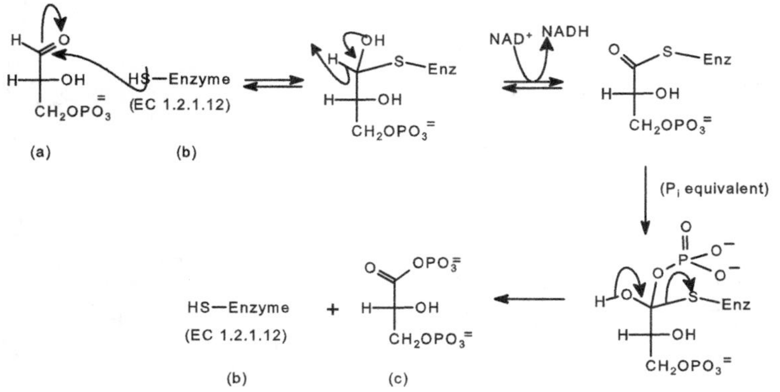

SCHEME 18.3 An accounting of the process, with curved arrows, for the conversion of glyceraldehyde 3-phosphate (a) in the presence of glyceraldehyde 3-phosphate dehydrogenase (EC 1.2.1.12) (b) and inorganic phosphate (P_i) for the formation of 3-phosphogylceroyl phosphate (c), a mixed anhydride of 3-phosphoglyceric acid and phosphoric acid.

phosphate isomerase (EC 5.3.1.1), allows their rapid interconversion. The isomerase (EC 5.3.1.1) catalyzes one of the most rapid enzymatic processes known (the rate of isomerization appears to be determined by how fast the species can diffuse into the active site of the enzyme).

The next step in the process on the way to ethanol requires the conversion of glyceraldehyde-3-phosphate to the corresponding carboxylic acid, an oxidation, thus demonstrating the need for some oxygen—usually present in the must itself. Of course too much oxygen is to be avoided since acetaldehyde, formed subsequently on the way to ethanol, yields acetic acid (CH_3CO_2H) on oxidation. Acetic acid imparts flavor to wine which is considered undesirable.

In the presence of the enzyme glyceraldehyde-3-phosphate dehydrogenase (EC 1.2.1.12) and the cofactor nicotinamide adenine dinucleotide (NAD^+) 3-phosphoglyceroyl phosphate and NADH (the cofactor having undergone reduction) are formed. The process (Scheme 18.3) requires attack on the aldehyde by a thiol group, presumably from a cysteine (Cys, C) in the active site on the enzyme and, after oxidation (hydride loss, $NAD^+ \rightarrow NADH$), substitution of the thiol by phosphate. Interestingly, this demonstrates that the normal functioning of the yeast requires phosphate (beyond, perhaps, what might normally be present). It is not uncommon for ammonium phosphate [$(NH_4^+)_3PO_4^{3-}$] to be added to the must.

Next, with the participation of phosphoglycerate mutase (EC 5.4.2.12) the phosphate at C-3 is transferred first to an active site serine (Ser, S) and then back to the substrate but at C-2, rather than C-3 (Scheme 18.4).

Now, the 2-phosphoglyceroyl phosphate anhydride undergoes hydrolysis to phosphate anion and the corresponding carboxylic acid. In the presence of phosphopyruvate hydratase (EC 4.2.1.11) [which was called "enolase" in the earlier literature],[6] the carboxylic acid undergoes loss of water. The process has been studied in some detail.[7] When

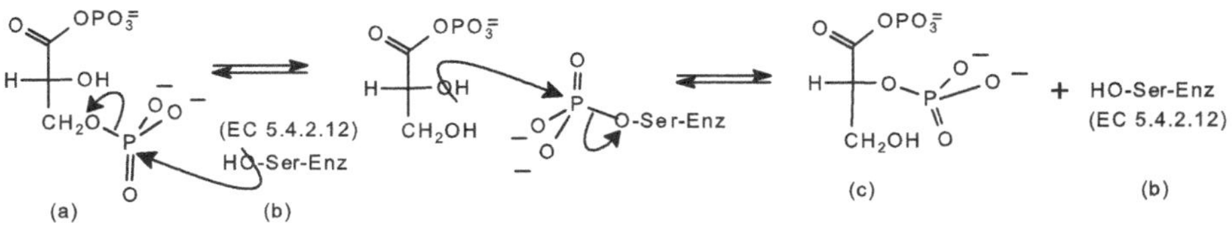

SCHEME 18.4 A cartoon construction with arrows attempting to account for the conversion of 3-phosphoglyceroyl phosphate (a) and its reaction with phosphoglycerate mutase (EC 5.4.2.12) (b) to first, remove the phosphate anion from C-3 and then to esterify the hydroxyl at C-2 with the same phosphate to form 2-phosphoglyceroyl phosphate (c).

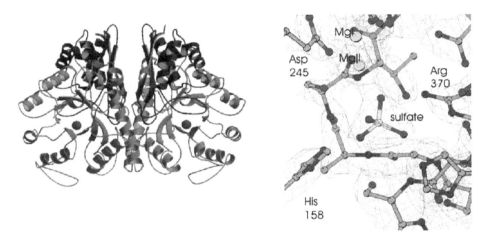

FIGURE 18.4 On the left, the enolase (EC 4.2.1.11) dimer PDB (1e9i). Mg^{2+} shown as small spheres. On the right, the active site of the enolase (EC 4.2.1.11). Sulfate introduced during crystallization. After Kühnel, K.; Luisi, B. F. *J. Mol. Biol.* **2001**, *313,* 583. DOI: 10.1006/jmbi.2001.5065. Reprinted with permission from Elsevier.

2-phosphoglycerate, binds in the active site of enolase, magnesium cations (Mg^{2+}) are coordinated with the carboxyl group. Lysine[345] (Lys^{345}, K) removes the proton *anti* to the leaving –OH group (i.e., on the opposite face of the molecule). It is argued that the carbanion formed in this way is stabilized by the carbonyl group and the hydrophobic nature of the pocket (Figure 18.4). Following the creation of the carbanion intermediate, the hydroxyl group, possibly after protonation by glutamic acid[211] (Glu^{211}, E), is lost as water to yield phosphoenol pyruvate (PEP) (Scheme 18.5).

Phosphoenolpyruvate (PEP) is converted to pyruvate through the action of pyruvate kinase (EC 2.7.1.40), which also converts ADP (adenosine diphosphate) to ATP (adenosine triphosphate) in the process (Scheme 18.6)

As shown in Chapter 17 (Scheme 17.1) pyruvate, bound into the multienzyme pyruvate dehydrogenase complex, is converted to acetyl coenzyme-A. However, when the yeast is not

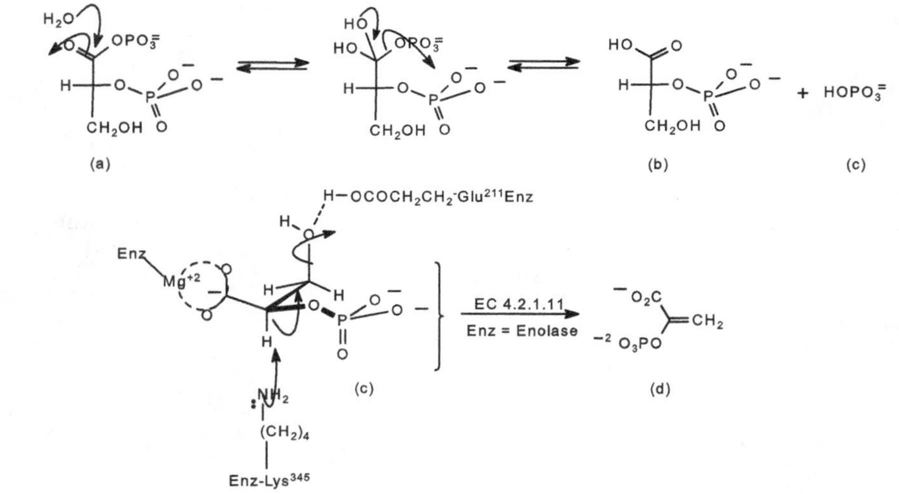

SCHEME 18.5 A cartoon attempt to account for the conversion of 2-phosphoglycerol phosphate (a) to 2-phosphoglyceric acid (b) and hydrogen phosphate (c) by hydrolysis of the former. Also shown is a rationalization of the loss of water from 2-phosphoglycerate (c) in the presence of the enzyme enolase (phosphopyruvate hydratase EC 4.2.1.11). Proton removal by lysine (Lys^{345}, K) and protonation of the leaving hydroxyl by glutamate (Glu^{211}, E) along with magnesium cation (Mg^{2+}) complexation of the substrate carboxylate, all in the catalytic site, are presumed to be involved in the formation of phosphoenol pyruvate (d).

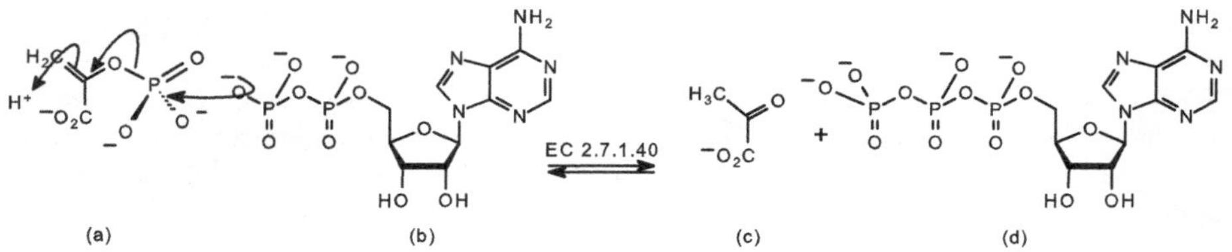

SCHEME 18.6 A representation of an accounting of the conversion of phosphoenol pyruvate (a) on reaction with adenosine diphosphate (ADP)(b) in the presence of pyruvate kinase enzyme (EC 2.7.1.40) to pyruvate (c) and adenosine triphosphate (ATP) (d).

producing acetyl CoA and is continuing to produce pyruvate,[8] then pyruvate can be converted to carbon dioxide and acetaldehyde by pyruvate decarboxylase (EC 4.1.1.1) as shown in Scheme 18.7.

Acetaldehyde is reduced to ethanol by the zinc (Zn) protein alcohol dehydrogenase (alcohol NAD$^+$ oxidoreductase, EC 1.1.1.1) (Scheme 18.8).

PART C. THE LIVING YEAST

Fungi in general, yeasts used in brewing and baking[9] and *Saccharomyces cerevisiae* in particular, have been investigated widely not only because of their practical utility but also because of their easy manipulation to ends other than in the food industry.

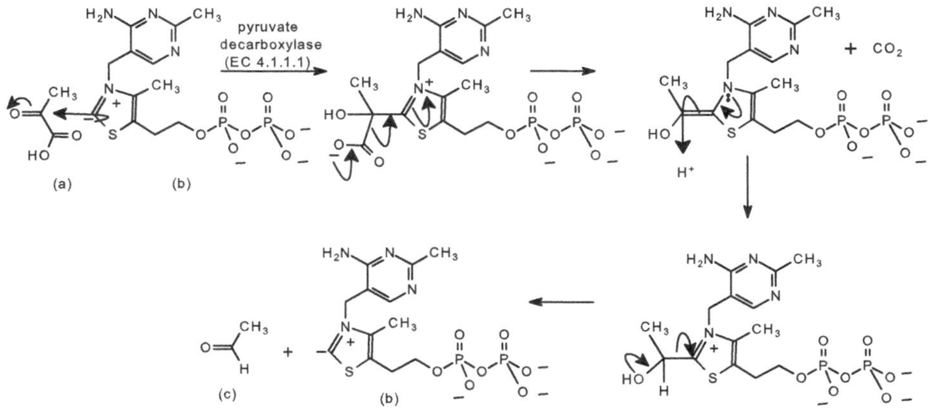

SCHEME 18.7 A cartoon pathway linking the reaction between pyruvate (a) and thiamine diphosphate (b) in the presence of pyruvate decarboxylase (EC 4.1.1.1) (which is not the same enzyme in the pyruvate dehydrogenase complex as found on the path to acetyl coenzyme-A seen in Scheme 17.1) leading to decarboxylation of pyruvate (a). Here, lipoyl group is absent, protonation occurs, and on regeneration of thiamine diphosphate (b), acetaldehyde (c) results.

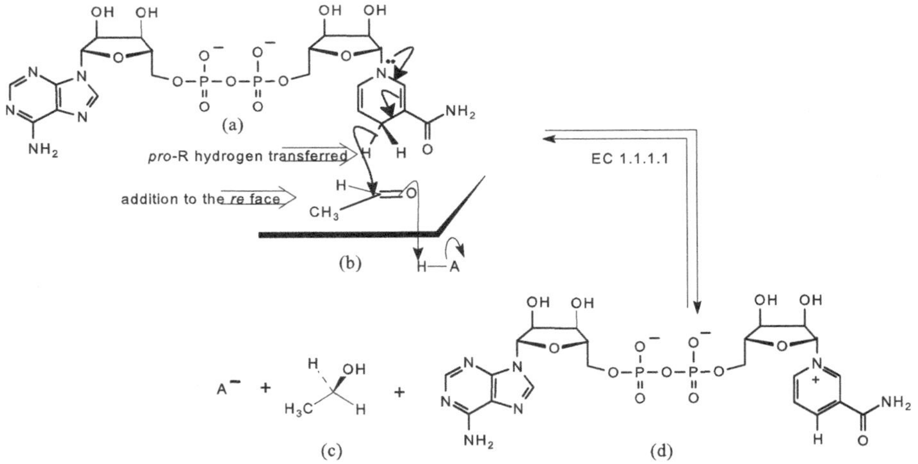

SCHEME 18.8 A cartoon representation of the use of NADH (a) in the reduction of acetaldehyde (b) to ethanol (c) while oxidation to NAD^+ (d) also occurs. The reaction utilizes the enzyme alcohol dehydrogenase, EC 1.1.1.1.

Keeping firmly in mind that yeast (see Figure 18.5), along with other fungi, are alive, their metabolic requirements need to be met just as with other life forms. A very useful synopsis of a large amount of work can be found in the volume edited by Feldman[10] as well as the continuing resource *The Yeasts* (currently in its fifth edition) edited by Kurtzman, et al.[11] Added to these, there is the journal *Yeast*[12] and the separate publication *Yeast Research* from the Federation of European Microbiological Societies (FEMS).[13] Finally, with specific reference to wine yeasts the interested reader should consult the *Handbook of Enology*.[14]

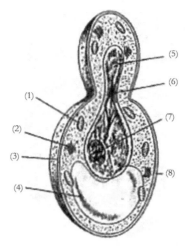

FIGURE 18.5 A labeled cartoon drawing representation of a *Saccharomyces* yeast cell. The labeled items are (1) mitochondrion; (2) Golgi body; (3) Plasma membrane; (4) Vacuole; (5) Nucleus; (6) Mitotic spindle; (7) Nucleolus; (8) Cell wall. Interested readers can profitably compare this drawing with those of the eukaryotic and plant cells in Figure 18.2 above to note the overall similarity in life forms. The representation of the yeast cell is taken from Harold, F. M. *The Way of the Cell;* Oxford University Press, 2001; p 29 and is reprinted courtesy of Oxford University Press.

The efflorescence in the active study of yeasts that has come about since the dawn of the ability to engage in genetic tinkering has led to a deeper understanding of the role of various strains[15] of yeasts which, although remaining species specific, clearly permit significant changes in the outcome of wine production. At this writing, the limits of change which can produce taste variants, both desirable and undesirable, remains experimental.

Currently, there are about[16] fifteen (15) yeast species associated with commercial wine production. Most of them apparently remain genetically uninvestigated, and in any given vineyard, adventitious yeasts may or may not be known.

What is known, however, is that some of the so-called wild yeasts (or specific strains of wild yeasts) that accompany the grapes from a vineyard into the processing plant appear to be native to that specific vineyard. These strains then serve as "starter cultures" present in the initial must. Indeed, it is widely suggested that products produced by these indigenous yeasts, on a specific grape along with the differences attributed to terroir and the skill of the vintner in determining the harvest conditions, etc. are most important in determining the distinctive nature of the wine.

At the same time, however, it is clear that there does not appear to be any consensus among vintners about the use of starter cultures. It is equally clear that it is generally the case that little or no effort is expended to "clean" the grapes before or after destemming and before crushing. Indeed, it seems that there is general agreement that whatever contributions the adventitious yeasts that accompany the grapes from the vineyard into the must may accrue, their time is short since they find the ethanol they produce toxic. Then the more ethanol-tolerant strain of *Saccharomyces cerevisiae* that the vintner has added is ready to "take over" the remainder of the fermentation process.

So, considering *Saccharomyces cerevisiae* in particular, where most published work appears to have been done, it is agreed that carbohydrates (largely but not exclusively glucose) serve as the main energy source. Other sources of energy and carbon can be found in other hexoses, some pentoses, and related compounds. Although it is often found advantageous, as noted earlier, to provide additional phosphorus and nitrogen (as ammonium phosphate [$(NH_4)_3PO_4$]), amino acids necessary for yeast metabolism can be found in grapes along with phosphorus and other elements that the grapes require for their own growth. Interestingly, however, it appears that many strains of *S. cerevisiae* are incapable of producing enough ammonium nitrogen (or its equivalent) from peptides present in the grape so that "(c)arbon-nitrogen deficiencies in the supply of assimilable nitrogenous compounds, remain the most common causes of poor fermentative performance and sluggish or stuck fermentations."[17]

As the interest in utilizing yeast for the synthesis of compounds not normally produced by yeast has grown, many materials, produced in only small quantities, have been found already present. Additionally, in that vein, *Saccharomyces cerevisiae*, along with other fungi, clearly want to survive. So, the fungi need to produce cell walls (made up of carbohydrate and similar polymers (e.g., polymers of N-acetylglucosamine, GlcNAc), fatty acids (for membranes) and products of their reduction (long-chain alcohols), proteins, and peptides to catalyze the chemistry that allow maintenance of life. They need as well to produce the purine and pyrimidine bases for DNA and RNA so that genetic information can be passed on. Since different strains of yeast are different, they will be producing different compounds as they carry out their normal functions. The determination of what some of these compounds are and how they might contribute to the wine product is an exciting adventure. The pursuit of that adventure is currently confused by the different strains of *S. cerevisiae* and other yeasts that might have been used by early workers and the often unclear determination of whether the yeast was functioning under aerobic or anaerobic conditions.

So it was that the earliest reports of the composition of the steroids, apparently beginning with Gérard[18] in 1895 and proceeding through many hands into the series of works by Barton and coworkers[19] in the 1970s, laid the foundation for the understanding that some steroids in addition to ergosterol (which had been found by early investigators)[20] could be found in *Saccharomyces cerevisiae* and other yeasts.

Indeed, Barton had gone so far as to suggest that, although it was known[21] that squalene did not have the C-24 methyl group in ergosterol (and that that methyl group came from methionine), squalene could nonetheless serve as a precursor (see Figure 18.6).[22]

The suggestion of squalene is based upon significant biochemical precedent that in living systems there are two well-defined pathways to the five-carbon isomeric isopentenyl diphosphate and dimethylallyl diphosphate (Appendix 2). That pair of prototropic isomers subsequently define "isoprenoid" compounds. Squalene, a thirty-carbon (i.e., made of six five-carbon units) relative, serves as the parent isoprenoid skeleton for steroids and is subsequently methylated and demethylated (along with other actions such as oxidation) to produce the vast family of such naturally occurring compounds. Additionally, "isoprenoid" side chains are found in ubiquinones (coenzyme Q) and other compounds present in yeast (along with other eukaryotes).

The pathway to the thirty-carbon squalene skeleton is acknowledged to pass through geranyl diphosphate (the ten-carbon putative parent to terpenes) and then farnesyl diphosphate

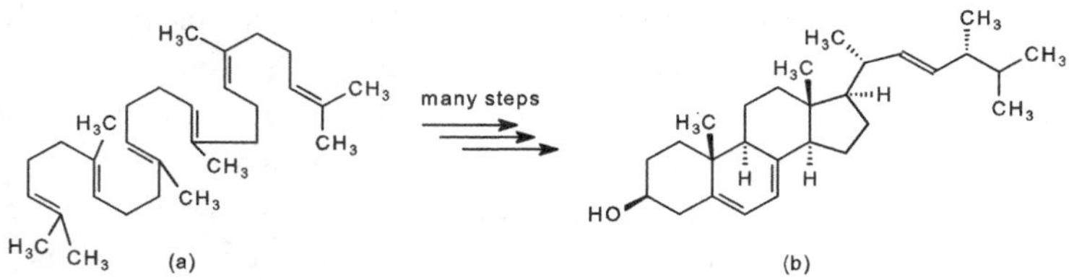

FIGURE 18.6 A broad representation of the path from squalene (a) to ergosterol (b).

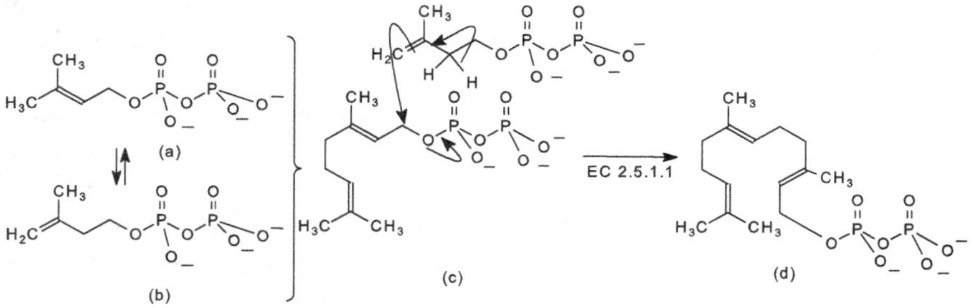

FIGURE 18.7 A broad representation of the path from dimethylallyl diphosphate (a) and isopentenyl diphosphate (b) to geranyl diphosphate (c) and farnesyl diphosphate (d) with dimethylallyl transferase, EC 2.5.1.1, participation.

(the fifteen-carbon putative parent to sesquiterpenes), which then dimerizes. Thus, it is of some surprise that little if any (only trace) amounts of geraniol, for example, are found in *S. cerevisiae*. Regardless, however, the pathway to ergosterol has been confirmed in detail (Figure 18.7).[23]

Interestingly, it has recently been pointed out[24] ... "The formation of terpenes by yeasts is limited to trace concentrations by a small number of non-*Saccharomyces* species: *Kluyveromyces lactis* ... *Torulaspora delbrueckii* ... *Kloeckera apiculate* ... *Metschnikowia pulcherrima* ... *Candida stellata* and *Ambrosiozyma Monospora*." It was also concluded that *Saccharomyces cerevisiae* only produced approximately 2 µg L^{-1} of farnesol, a hydrolysis product of the ergosterol precursor farnesyl diphosphate.[25] Having thus denied the possibility of *S. cerevisiae* producing terpenes, the same group went on to show that, in light of "intraspecific variability" in *S. cerevisiae* and perhaps because of the limitation of nitrogen availability at some growth stage, the terpenes linalool and α-terpineol were actually produced (along with farnesol) by all of the strains of *S. cerevisiae* they tested (Figure 18.8).

So, it seems clear that, at least under some circumstances, some strains of *Saccharomyces cerevisiae* produce little (if any) terpenes and sesquiterpenes. And further, that the pathway to such compounds may be frustrated by the conversion of enzyme-bound precursors (isopentenyl diphosphate and dimethylallyl diphosphate) and subsequent intermediates (e.g., geranyl diphosphate and farnesyl diphosphate) to ergosterol (and other steroids?).

Interestingly, however, genetic tinkering of *Saccharomyces cerevisiae* could not be resisted, and relatively early, noting that *S. cerevisiae* must possess the gene (or genes) for isopentenyl

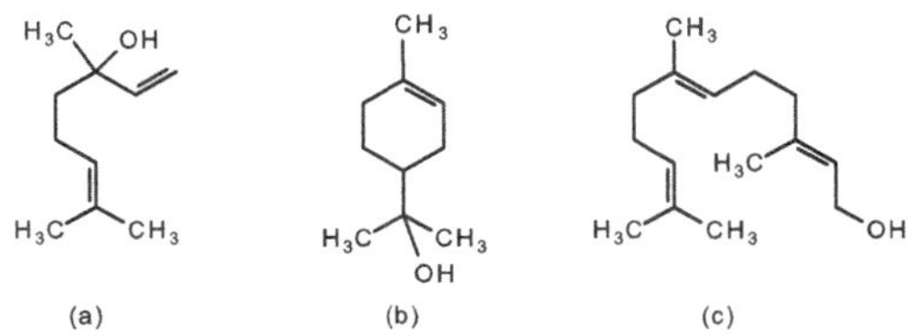

FIGURE 18.8 Isoprenoids (a) linalool, (b) α-terpineol and (c) (2*E*,6*E*)-farnesol. All three have been detected in all strains of *S. cerevisiae* that were tested for derivatives of dimethylallyl and isopentenyl diphosphates.

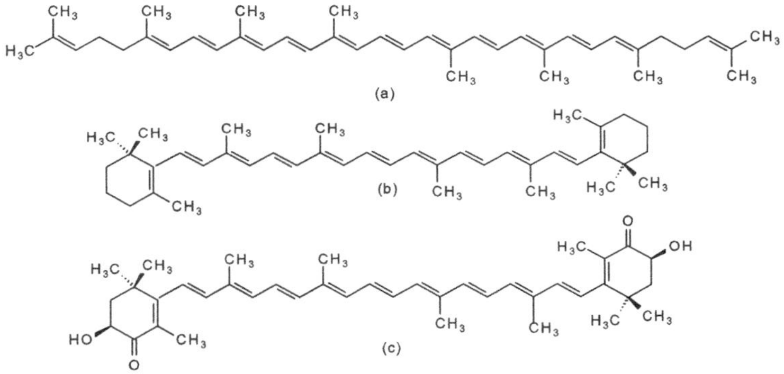

FIGURE 18.9 Representations of brightly colored isoprenoid products isolated from strains of *Saccharomyces cerevisiae* known to possess genes for isopentenyl diphosphate Δ-isomerase (EC 5.3.3.2). The compounds are (a) all-*trans*-lycopene; (b) β-carotene; and (c) astaxanthin.

diphosphate Δ-isomerase, EC 5.3.3.2, (to produce dimethylallyl diphosphate), carotenogenic genes could be inserted so that the flux for egosterol production might simply be diverted. Thus, the yeast had acquired the ability to produce carotenoids such as lycopene, β-carotene, and the brightly colored astaxanthin (see Figure 18.9) from isopentenyl diphosphate and dimethylallyl diphosphate.[26]

In the years since those early efforts, the genome of *Saccharomyces cerevisiae* has successfully been modified to produce a variety of isoprenoids.[27–32] However, it should be clear that the re-engineered strains remain uncommon, and their use in wine production has not yet (apparently) proved itself. Indeed, it is likely (as noted earlier) that the process of utilizing "wild yeasts" found in the vineyard, whose alcohol tolerance is low, and allowing them to interact with the must before *S. cerevisiae*, whose alcohol tolerance is high, is added and takes over, lends itself to the production of unique vintages.

Alternatively, use of other yeast "heritage" strains to specifically introduce unique properties to a vintage (under the guidance of the vintner) is not uncommon. As noted earlier, a wide variety of yeasts including cultures that are not common can be found through the collection at the University of California, Davis, as well as the collection held by the United State Department of Agriculture. Commercial collections also abound.[33–35]

Although it is difficult (because of unique opinions concerning flavor and the many different strains of the fungus) work has begun on the role of the flavor of wine as a function of the yeast used in the fermentation.[36,37]

In the same vein, it is of some interest to note that "yeast extract," primarily from *S. cerevisiae* is utilized by the food industry as an additive to otherwise bland foods as well as a bacterial culture medium because it is rich in nutrients and "tastes good." While the matter of taste is subjective (and will be approached again later when tasting wine is considered), the taste of yeast extract is described as "umami" (i.e., from the Japanese "pleasant savory taste"), that is distinct from sweet, sour, bitter, and salty, generally held to be the four fundamental taste sensations. It is further described as the taste derived from the monosodium salt of glutamic acid (MSG) (Figure 18.10).

The yeast extract is of some interest here as it, as well as the head space above it, might reasonably contain materials found in wine. Thus, the yeast extract is obtained by suspending the yeast cells (frequently from yeast grown on molasses or other glucose-rich media) in salt (NaCl) water. Since the ionic strength of the salty water is greater (i.e., it's *hypertonic*) than the ionic strength of the water within the wall membrane of the yeast, water flows from the yeast into the salt solution (osmosis). As a consequence, the yeast cell shrivels up, causing enzymes, programmed to lead to cell death (necrosis) to become active. The fragments resulting from this self-digestion of the cell walls are removed by filtration, and to prepare "yeast extract" for commercial purposes, water is removed from the clear aqueous solution by evaporation (e.g., by spray drying or freeze drying—lyophilization).

While several lists consisting of more than one-hundred fifty (150) volatile compounds from *S. cerevisiae,* isolated from the headspace and/or the paste before drying and detected by gas chromatography and mass spectrometry have been published,[38,39] the isolates are not identical. In part, this is because the analyses were done at different times and by different groups and with different equipment. But there is some overlap in the compounds listed even though the strain(s) of *S. cerevisiae* from which the extract was obtained were not provided and it is unlikely that they were the same. Some of the compounds that were the same are shown in Figure 18.11 and might be usefully examined.

In regard to that examination, first, there is some uncertainty in some of the structures of the compounds since the materials were not actually isolated (i.e., GC/MS comparisons under only one set of conditions may not prove definitive). Second, the "meaty" flavor of yeast extract is generally attributed to sulfur-containing compounds. Those sulfur compounds are often present in low concentrations in wine and arise as a result of sulfur metabolism of amino acids (e.g., cysteine [Cys, C] and methionine [Met, M] and their relatives such as the heterocycle[40] thiamine.

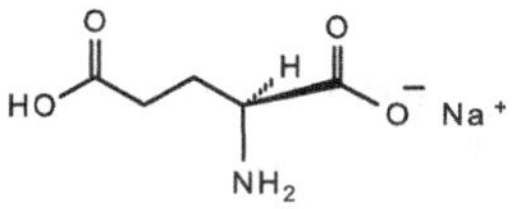

FIGURE 18.10 A representation of the structure of the monosodium salt of glutamic acid (MSG). MSG is widely used as a flavoring agent, and its taste, called "savory" or "umami," is now often considered as a separate sensation in addition to those of sweet, sour, bitter, and salty.

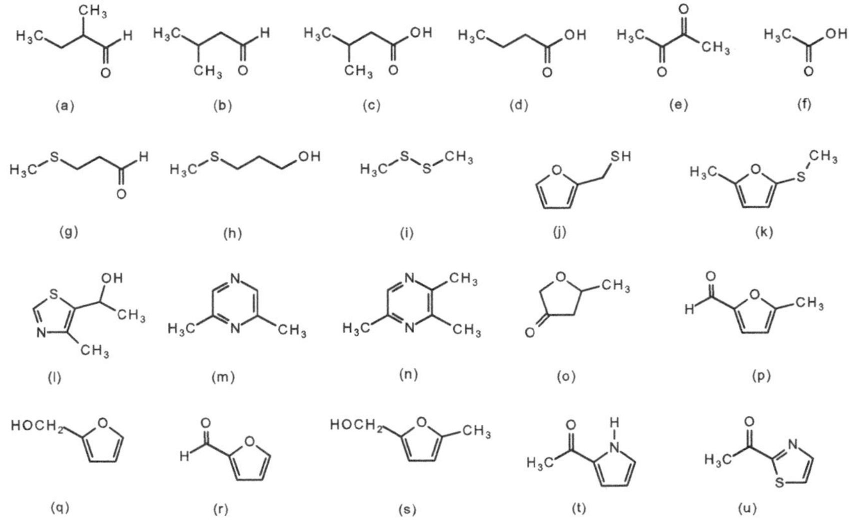

FIGURE 18.11 Representation of small acyclic and cyclic compounds isolated from *S. cerevisiae* head space and from "yeast paste." It is likely that small quantities (perhaps beneath detection limits) of these and related materials are present in wine produced. The compounds shown are: (a) 2-methylbutanal; (b) 3-methylbutanal; (c) 3-methylbutanoic acid; (d) butanoic acid; (e) 2,3-butandione (diacetyl); (f) acetic acid; (g) 3-(methylsulfanyl)propanal (methional); (h) 3-(methylsulfanyl)-1-propanol; (i) dimethyl disulfide; (j) 2-furylmethanethiol; (k) 2-methyl-5-(methyl-sulfanyl)furan; (l) 1-(4-methyl-1,3-thiazol-5-yl)ethanol; (m) 2,6-dimethylpyrazine; (n) 2,3,5-trimethylpyrazine; (o) dihydro-5-methyl-3(2*H*)-furanone; (p) 5-methylfurfural; (q) furfuranol; (r) furfural; (s) 5-methyl-2-furylmethanol; (t) 2-acetylpyrrole: and (u) 2-acetylthiazoline.

Then, too, there is a small list of aromatic, norterpene and terpenoid compounds found in the head space above yeast as well as in the yeast paste. Some of those compounds are shown in Figure 18.12.

With regard to the sulfur compounds in wine, they appear to be of lesser importance than in "yeast paste" because other wine components overshadow them in the flavors and aromas of the wine. Additionally, in this regard, it appears likely that yeast paste also incorporates some materials from the substrate on which the yeast fed. Thus, while it does not seem to have been shown, it is reasonable to argue that yeast grown on molasses will be different than yeast grown on another sucrose source such as sucrose found in grapes.

And, finally in this vein, most recently it has become possible to examine the "secretome" (i.e., the protein secretion of the living cells) of *Saccharomyces cerevisiae,* and it has been found that the concentrations of many of the hundreds of secreted proteins identified changed as a function of glucose concentration.[41] Of course these extracellular proteins or pieces of them (as a consequence of hydrolysis or other activity) are to be found in the beverage.

PART D. THE ACTION OF YEASTS ON OTHER COMPONENTS

As noted earlier, there is a wide variety of complex taste components to be found in grapes. The relatively small number of compounds whose identities are known are insufficient to

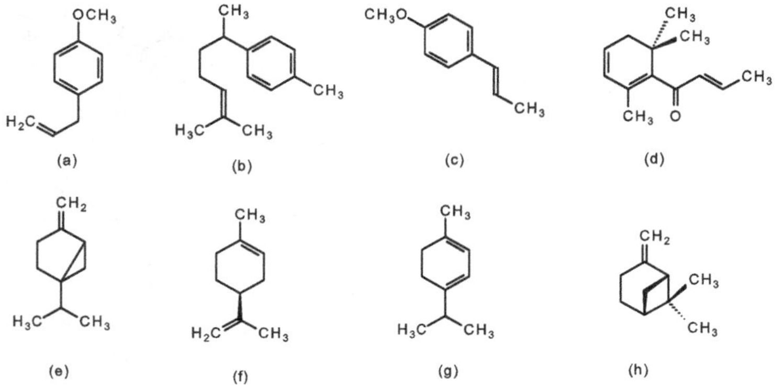

FIGURE 18.12 Representations of some aryl and norterpene compounds, as well as terpenoids reported to be found in the head space above yeast paste as well as in the yeast paste. The compounds identified are: (a) oestragole; (b) curcumene; (c) anethole; (d) *trans*-β-damsacenone; (e) sabinene; (f) (S)-(−)-limonene; (g) α-terpinene; and (h) (+)-β-pinene.

predict the actual flavor of the wine, and it is clear that the specific ratio of those that are known along with the unknown others plays a critical role. In addition, it is held that a large percentage of the compounds in the grapes are actually present bound to sugars such as glucose and mannose, and that the grape's endogenous enzymes are insufficient to effect hydrolysis. Yeast glucosidases and mannosidases can effect glycolysis, and thus phenols, terpenols, and other compounds bearing hydroxyl groups that might have been undetected originally are now set free and may engage in additional chemistry. It is supposed that various strains of "starter yeasts" will act differently than subsequently used strains of *Saccharomyces cerevisiae*, and it is argued that it is for this reason that some vineyards go to great lengths to protect their indigenous "wild" yeasts.

Second, it now appears clear that action of yeast on glucose, other sugars, and amino acids present in the grape produces, in addition to ethanol and carbon dioxide, various alcohol acetates (presumably *via* acetyl coenzyme-A [acetyl CoA]), glycerol [$HOCH_2CH(OH)CH_2OH$], fatty acid ethyl esters (esterification of cell membrane acids), and esters of short-chain straight and branched carboxylic acids (C_4–C_{10}). The latter group of esters (Appendix 3) generated from the amino acids often lead to the fruity aromas of the finished wine.

Third, while tartaric and malic acids are the major acids present in the grape and the wine and together can account for the acidity of the must, functioning of the plant and enzyme cycles to produce fatty acids along with smaller homologs including pyruvic acid, lactic acid, fumaric acid and citric acid, as well as other small acids, along with acetic acid can contribute. Phenols, also acidic, originally masked as glycones, but now liberated by yeast hydrolases, can prove detrimental to flavor, and small amounts of oxygen present can oxidize them to ameliorate the problem. These off-flavor phenols include 4-vinylphenol, 4-ethylphenol, 4-vinylguaiacol, and 4-ethylguaiacol (Figure 18.13).

There are a few additional problems. Urea is among the metabolic products of the living and functioning yeast, largely as a product of the breakdown of arginine (Arg, R), with some yeasts producing more than others (Equation 18.1). The enzyme-catalyzed reaction of urea

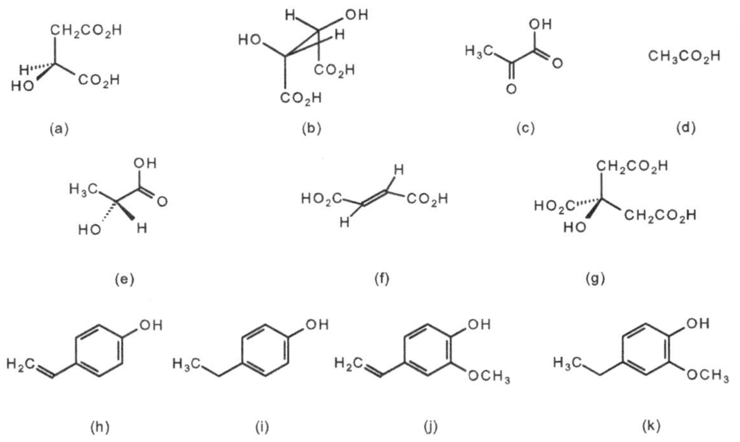

FIGURE 18.13 A representation of small molecules that have been isolated from yeast head space and/or yeast paste. They include: (a) (S)-(−)-malic acid; (b) (2R,3R)-(+)-tartaric acid; (c) pyruvic acid; (d) acetic acid; (e) (S)-(+)-lactic acid; (f) fumaric acid; (g) citric acid (h); 4-vinylphenol; (i) 4-ethylphenol; (j) 4-vinylguaiacol; and (k) 4-ethylguaiacol.

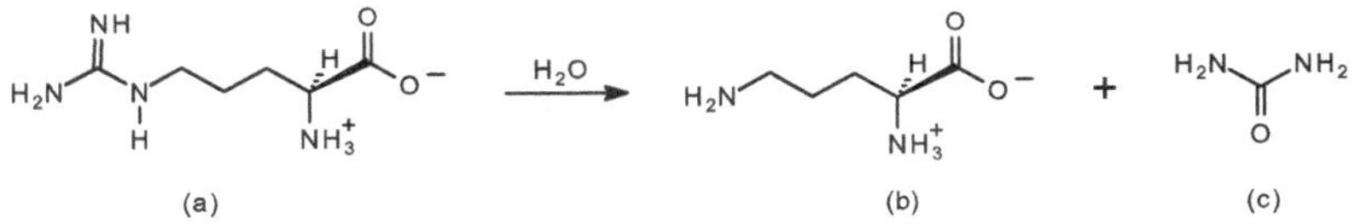

EQUATION 18.1 A representation of the hydrolytic conversion of arginine (Arg, R) (a) to ornithine (b) and urea (c).

EQUATION 18.2 A representation of the reaction between urea (a) and ethanol (b) to yield ethyl carbamate (c) and ammonia (d). This is normally a very slow reaction and unlikely to be of importance in the preparation of the beverage.

and ethanol produces the ethyl ester of carbamic acid (ethyl carbamate) (Equation 18.2), a suspected carcinogen and thus a health hazard. It has been argued that the problem is a minor one, and choosing an appropriate strain of S. cerevisiae the best solution.

In that vein, some recent work[42] has been undertaken to examine wine components after fermentation, using mixtures of different isolates of a given strain of S. cerevisiae to determine if co-culturing a grape must (Pinot Noir grapes were used) could produce new and different volatile compounds. The results, after completion of fermentation (21 days) and without aging, etc., were that while wines with unique volatile profiles were produced, the differences were more subtle than those that might be effected by temperature change. That is, in

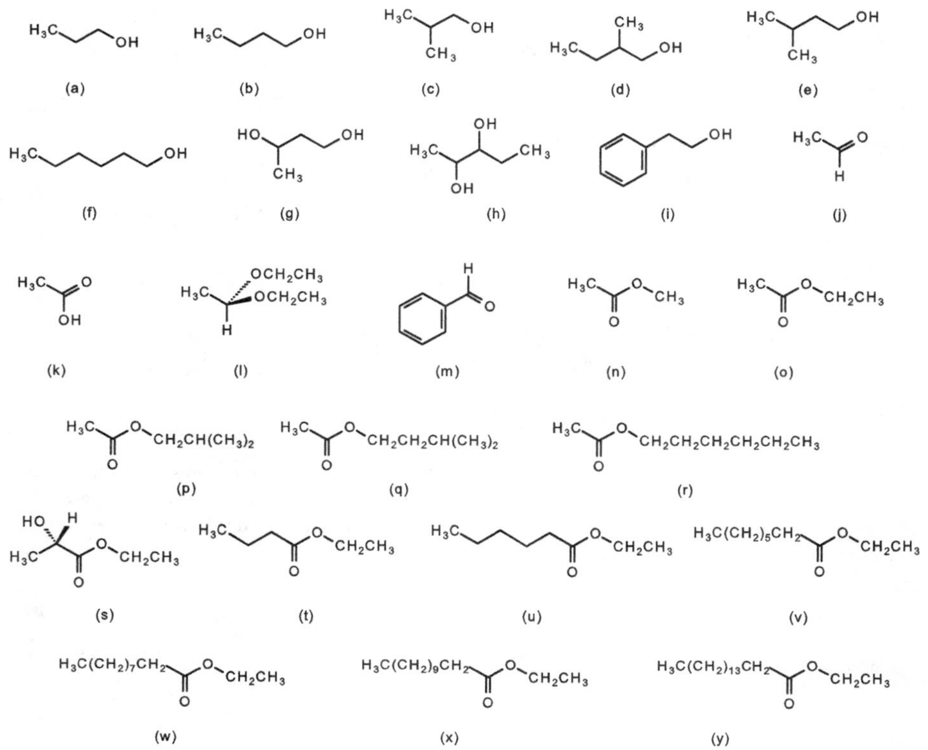

FIGURE 18.14 A representation of isolable volatiles from different isolates of the same strain of *Saccharomyces cerevisiae* and whose relative concentrations change as a function of isolate as well as temperature. The concentration changes result in different flavors. The isolates included: (a) propanol; (b) butanol; (c) 2-methylpropanol; (d) 2-methyl-1-butanol; (e) 3-methyl-1-butanol; (f) 1-hexanol; (g) 1,3-butanediol; (h) 2,3-butanediol; (i) 2-phenylethanol; (j) acetaldehyde; (k) acetic acid; (l) acetaldehyde diethyl acetal; (m) benzaldehyde; (n) methyl acetate; (o) ethyl acetate; (p) isobutyl acetate; (q) isoamyl acetate; (r) hexyl acetate; (s) ethyl lactate; and (t) ethyl butanoate; (u) ethyl hexanoate; (v) ethyl octanoate; (w) ethyl decanoate; (x) ethyl laurate; and (y) ethyl palmitate.

both ways (different isolates on the one hand and different temperatures on the other) the same volatiles were produced, but their relative concentrations were different. The work used ratios of the twenty-five volatile compounds (in addition to ethanol) shown in Figure 18.14.

Somewhat earlier than the work noted above, the issue of the action of glycosidases from plants and yeasts and other enzymes (of the group EC 3.2.1.x) on secondary metabolites in fruits that might result in an increase in volatile constituents was reviewed.[43] First, although the array of such glycosides could be large, only a small subset might give rise to detectable aromas and flavors. Second, it is known that the carbohydrate portion of the isolates are largely made up of glucosides and diglycosides, for which representations are provided in Figure 18.15. Third, gas chromatography/mass spectrometry and aroma/taste are the tools used to examine the products of hydrolysis, and not all of the products have been identified. Representations of the aglycones identified after the carbohydrates are removed are shown in Figures 18.16, 18.17, and 18.18.

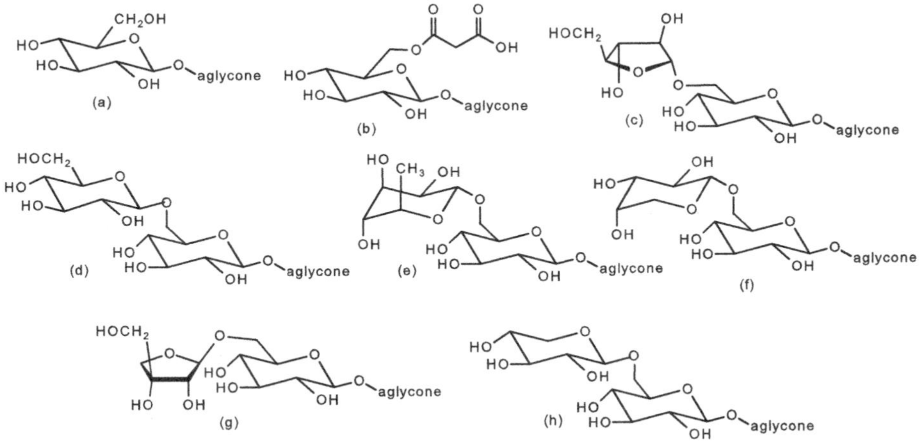

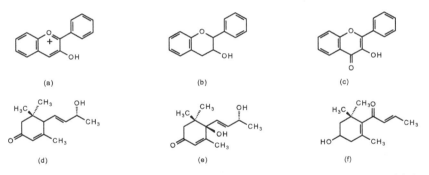

FIGURE 18.15 Representations of carbohydrates acted upon by glycosidases include: (a) β-D-glucoside; (b) malonyl-β-D-glucoside; (c) α-L-arabinofuranosyl-β-D-glucoside; (d) β-D-glucopyranosyl-β-D-glycoside (gentiobioside); (e) α-L-rhamnopyranosyl-β-D-glucoside (rutinoside); (f) α-L-arabinopyranosyl-β-D-glucoside (vicianoside); (g) β-D-apiofuranosyl-β-D-glucoside; and (h) β-D-xylopyranosyl-β-D-glucoside (primeveroside).

FIGURE 18.16 Representations of the set of aglycones that include variously substituted: (a) the anthocyanidin-3-ol parent; (b) the flavan-3-ol parent; and (c) the flavon-3-ol parent. Also: carotene or other C-40 degradation products (d) an hydroxyl-bearing C-13 norisoprenoid (3-oxo-α-ionol); (e) a dihydroxy-bearing C-13 norisoprenoid (vomifoliol, *aka* blumenol-A); and (f) an hydroxyl-bearing C-13 norisoprenoid.

The aglycones that are common are usually found to be anthocyanidins, flavanols, flavonols, carotene-derived norisoprenoids, terpenols, and hormones (including brassinosteroids, abscisic acid, and cytokinins such as zeatin, and a few others). Clearly, too, the aglycone portion of the glycoside must contain an –OH group so that it can react at the anomeric carbon of the carbohydrate (as shown in Figure 18.14). As noted above, representations of some members of the classes of compounds listed are shown in Figures 18.16, 18.17, and 18.18.

In order to determine the role of yeast in affecting the compounds in the must an experiment was undertaken using a "flavor precursor fraction" from non-floral grape varietals (Macabeo, Sauvignon blanc, Merlot, and Parraleta were used) added to the same juice.

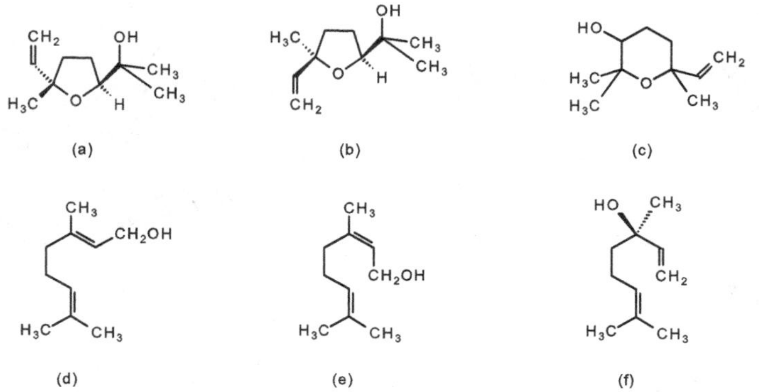

FIGURE 18.17 Representations of hydroxyl-bearing terpenoids that can serve as aglycones. These include: (a) *trans*-linalool oxide; (b) *cis*-linalool oxide; (c) linalool oxide (pyranoid form); (d) geraniol; (e) nerol; and (f) (*R*)-(−)-linalool.

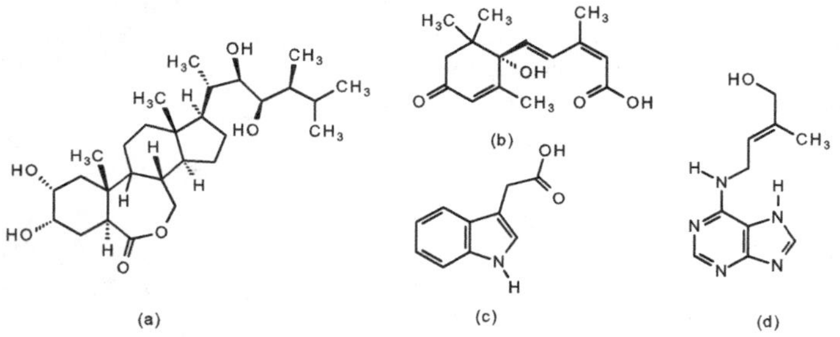

FIGURE 18.18 Representations of some −OH-bearing hormones that might serve as aglycones. These include: (a) brassinolide; (b) abscisic acid; (c) indole-3-acetic acid; and (d) zeatin.

In separate experiments the fraction was added to grape juice from *V. vinifera* var Macabeo, and different commercial strains of *Saccharomyces cerevisiae* were used for fermentation.[44]

It was reported[44] that "the addition of the precursor fraction brought about a significant increase of the wine floral notes irrespective of the yeast used" and "while the sensory effect of the addition is not extreme . . . it is strong enough to be . . . detected." The only compounds produced above threshold levels were the norisoprenoids β-damascenone and β-ionone and vinylphenols. And yet, floral notes resulted from active combinations of lactones, cinnamates, vanillins, and terpenes. Interestingly, the levels of some of them, such as β-damascenone, γ-nonalactone (apricolin), 2-methoxyphenol, 2-methoxy-4-vinylphenol, 4-allyl-2-methoxyphenol, linalool, and α-terpineol, have been found to be related to the grape variety. For example, cinnamates (ethyl cinnamate and ethyl dihydrocinnamate) have been related to Pinot noir wines. The presence of floral notes and, for some, the ability to isolate them, may be the result of acid hydrolysis of their corresponding glycosides.

Presumably, the hydrolysates in the yeasts effected the production of the compounds detected or produced precursors that could then give rise to those compounds by subsequent chemical transformations.

The representations of the compounds shown in Figure 18.19 are those identified in the groups noted.

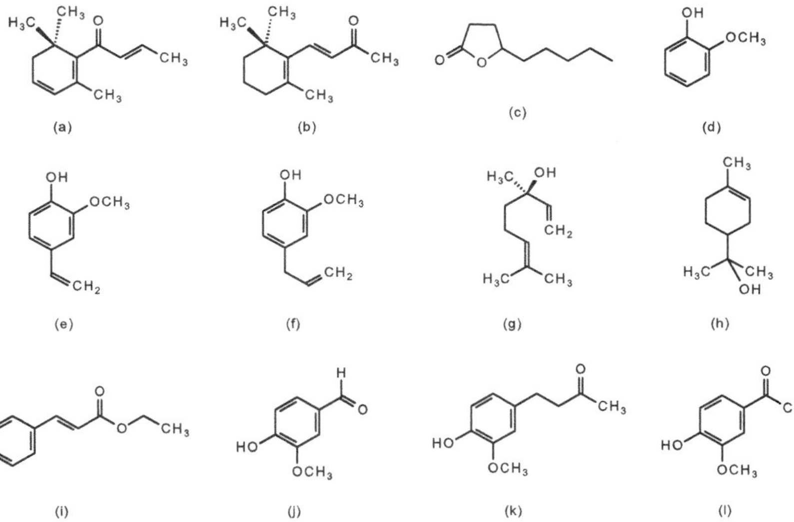

FIGURE 18.19 Representations of compounds from nonfloral grape varieties added to the same juice but subsequently treated with different *S. cerevisiae* strains. Those shown include: (a) β-damsacenone (b) β-ionone, (c) apricolin; (d) guaiacol; (e) 4-vinylguaiacol; (f) eugenol; (g) (R)-(−)-linalool; (h) α-terpineol; (i) ethyl cinnamate; (j) vanillin; (k) zingerone; and (l) acetovanillone, *aka* apocynin).

As the must derived from red grapes stands in contact with the skins, glycosides (EC 3.2.1.x) present in the yeast (as well as in the must itself) undertake their respective hydrolytic roles. During this period, it is clear that the initially colorless-to-light yellow juice begins to take on red hues which deepen and darken as time passes. In part, this color change is attributed to the presence of anthocyanidins (resulting from the hydrolysis of corresponding glycosylated anthocyanins, themselves possessed of color) where newly liberated phenolic hydroxyl groups on the flavylium cation are now capable of proton loss and participation in electron delocalization, resulting in deeper color as a function of pH (See Chapter 12, Figure 12.12 and Appendix 1).

For example, as shown in Figure 18.20, the compound cyanidin, glucosylated at C-3 and C-5 (further carbohydrate substitution on the glucose at C-3 has occurred) and with additional (not shown) substitutions on hydroxyl groups of the sugars, has recently been shown to be present as a consequence of anthocyanin biosynthetic genes in *Arabidopsis thaliana*. *A. thaliana* is the currently most thoroughly studied plant for the molecular biology of flavonoid metabolism.[45]

The chemical reactions taking place at various pH values for such flavylium cations as malvadin-3,5-glucoside have been described (Figure 18.21).[46,47] Additional reactions based upon these and related species to produce dimers, trimers, and even some low-molecular-weight polymers occur as the wine matures (*vide infra*).

PART E. OTHER YEASTS (AND A BACTERIUM)
ON THE SAME AND OTHER COMPONENTS

In general, grape and yeast proteins, functioning to establish the nature of the wine and perpetuation of the yeast, along with pectins, glucans, xylans, polysaccharides, and related

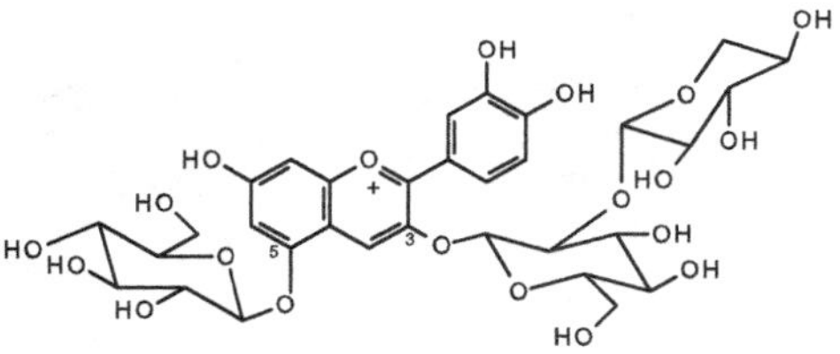

FIGURE 18.20 A representation of multiply glucosylated cyanidin.

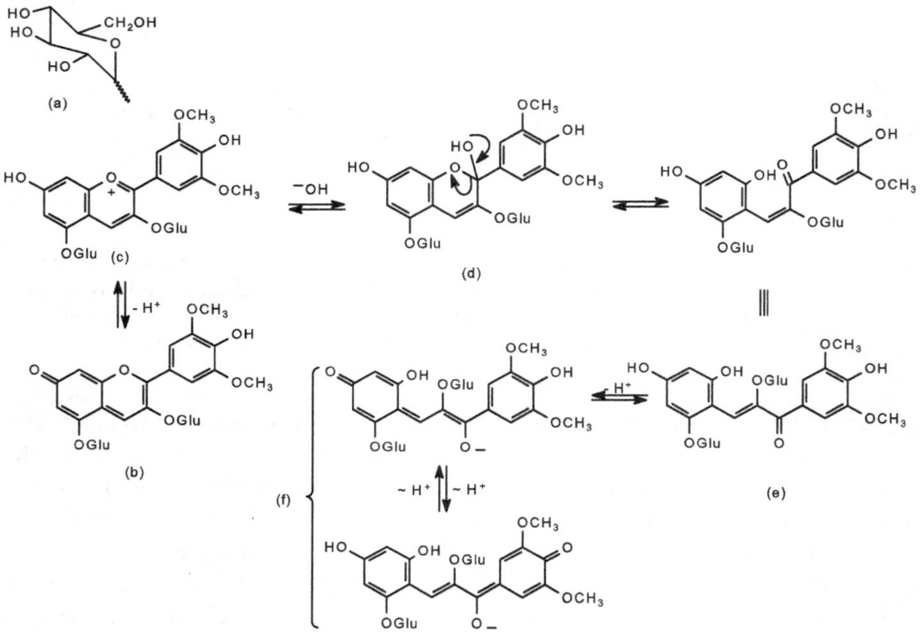

FIGURE 18.21 Representation of glucosylated malvidin changes with pH. It is noted that the glucose, Glu, (a) (the anomer is not known to be either α or β) is retained as phenolic protons are lost (b), and the flavylium ion (c) is attacked (d) and subsequently opened (e) where simple proton loss and subsequent migration ($\sim$ H$^+$) results in extended conjugation (f) (and coloration).

materials, may not be digested by endogenous hydrolyases, pectinase, glucanase, and other enzymes. Thus, there remains some detritus after their use to the living has ended. While older technologies of filtration and fining were used to remove them, newer technologies attempt to introduce additional, so called "industrial," enzymes to solve that and related problems by further reactions.[48,49] The presence of the remains of adventitious yeasts, and even the remains of some strains of *Saccharomyces cerevisiae*, have been well known since the discovery of what are called "killer yeasts." It has been found that some strains of *S. cerevisiae*, as well as other yeasts, excrete proteins that are toxic to other cells but not to themselves. This ability is, on the one hand, useful in limiting the lives of deleterious yeast strains that might be present but has also proven to be a problem since it can limit the life of the useful yeasts too.

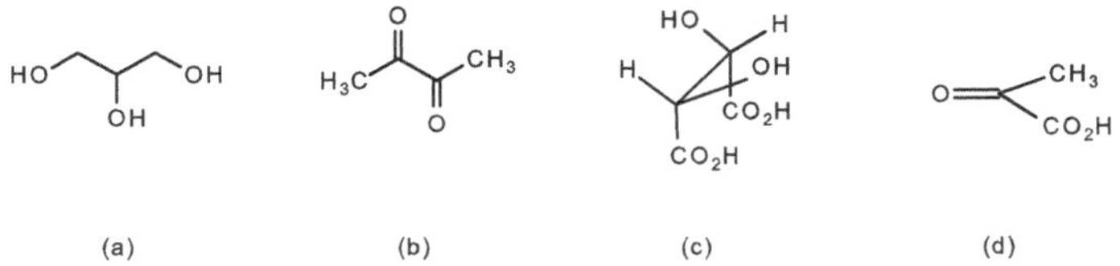

FIGURE 18.22 Representations of four small, generally occurring compounds whose respective concentrations dramatically affect flavor directly or through subsequent modification. These are: (a) glycerol; (b) diacetyl; (c) tartaric acid; and (d) pyruvic acid.

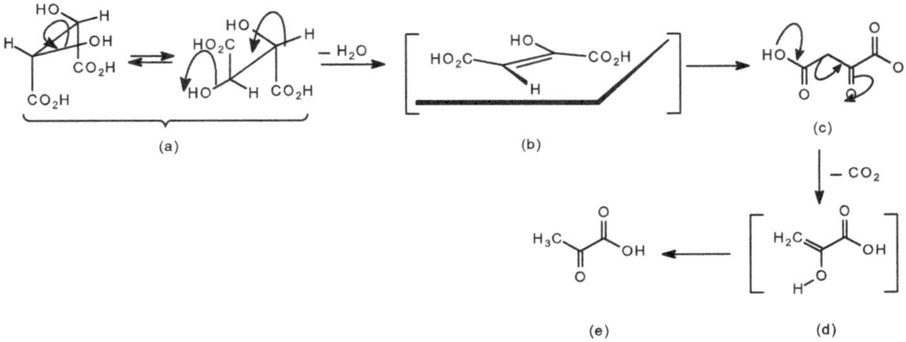

SCHEME 18.9 A representation of a path for the dehydration from either (a) D-(−)-tartrate (with the dehydratase EC 4.2.1.81) or L-(+)-tartrate (with the dehydratase EC 4.2.1.32) to (b) the enol of oxaloacetate (c). The latter, on decarboxylation, produces enolpyruvate (d) which can tautomerize to pyruvic acid (e).

Killer yeasts have proven difficult to control, although there is some evidence that pH changes can be efficacious[50] and hope that modifying the genome of the yeast might prove best.

One of the fungi infecting grapes has proven valuable (*Botrytis cinerea*) and is discussed further in Chapter 21. Briefly, the fungus often found on white grapes that have been left well past traditional harvest produces a grey rot ("Noble Rot") which appears to dehydrate the grape. Then subsequent treatment can produce exceptionally sweet white wine. In part, the flavor is the result of processing by additional enzymes, introduced by the fungus, that cause the changes. The production of new and different compounds results and the pH of the wine is altered.[51] As will be discussed, it has been argued that the smoother taste of the wine from *botrytis*-affected grapes is in no small measure due to increase in both glycerol and diacetyl concentrations and decrease in tartaric and pyruvic acid concentrations. In this vein, it may be true that changes in concentrations of these few compounds (Figure 18.22) materially affect overall changes in flavor. Indeed, adventitious yeasts could be responsible for just such alterations.

Dehydration of tartaric acid and then conversion of the resulting enol to the corresponding ketone yields oxaloacetate, the spontaneous decarboxylation of which generates pyruvate (Scheme 18.9).

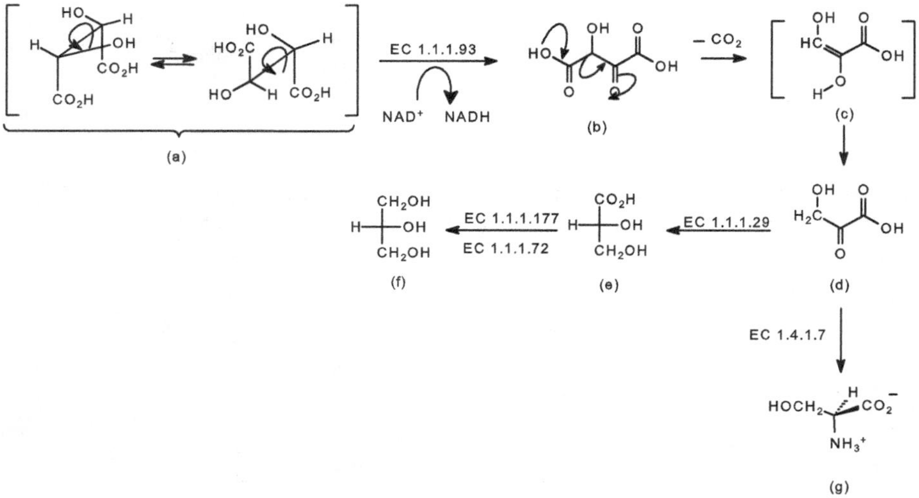

SCHEME 18.10 A representation of the pathway from tartaric acid (a) to oxaloglycolic acid (b) by oxidation (tartrate dehydrogenase, EC 1.1.1.93) and the subsequent decarboxylation of the latter to the enol of hydroxypyruvic acid (c). Tautomerization of (c) to (d) followed by reduction, first to glycerate (e) (glycerate dehydrogenase, EC 1.1.1.29) and then to glycerol (f) (via the corresponding 3-phosphate, first with glycerol 3-phosphate 1-dehydrogenase, EC 1.1.1.177, to glyceraldehyde 3-phosphate and then to the triol with glycerol dehydrogenase, EC 1.1.1.72, is expected, as well as amination of (d) to serine (Ser, S) (g) (serine 2-dehydrogenase, EC 1.4.1.7).

Alternatively, simple oxidation (Scheme 18.10) of tartaric acid (EC 1.1.1.93, tartrate dehydrogenase) produces oxaloglycolate, spontaneous decarboxylation of which generates hydroxypyruvate. The latter can be reduced (glycerate dehydrogenase, EC 1.1.1.29) to glycerate (and eventually to glycerol *via* the corresponding 3-phosphate, first with glycerol 3-phosphate 1-dehydrogenase, EC 1.1.1.177, to glyceraldehyde 3-phosphate and then to the triol with glycerol dehydrogenase, EC 1.1.1.72) or, on amination, converted to serine (Ser, S) with serine 2-dehydrogenase, EC 1.4.1.7.

Returning to pyruvate (Scheme 18.11). In the presence of acetolactate synthase (EC 2.2.1.6), pyruvate undergoes dimerization to produce acetolactate [(2S)-2-hydroxy-2-methyl-3-oxobutanoic acid], which on decarboxylation (E.C.4.1.1.5, acetolactate decarboxylase) generates acetoin [(3R)-3-hydroxybutan-2-one] and carbon dioxide. Oxidation of acetoin (EC 1.1.1.303) yields the buttery-flavored diacetyl ($CH_3COCOCH_3$ *aka* biacetyl), whilst reduction yields 2, 3-butanediol.

Indeed, smoother taste as a consequence as a decrease in acid concentration and formation of diacetyl is fairly common throughout the process of converting most must into wine. Smoother taste is also attributed, in part, to what is called "secondary" or "malolactic fermentation."

The intrusion of malolactic fermentation which, it has been pointed out,[52] is the result of participation of bacteria of the order Lactobacillales (i.e., lactic acid bacteria), rather than a fermentation arising from the use of yeast (or other fungi), generally appears to produce "more flavorful" wines.

In essence, the concept involves the conversion of malic acid (pKa 3.40) to lactic acid (pKa 3.86), thus decreasing the acidity of the must but keeping it acidic enough to encourage appropriate fermentation.

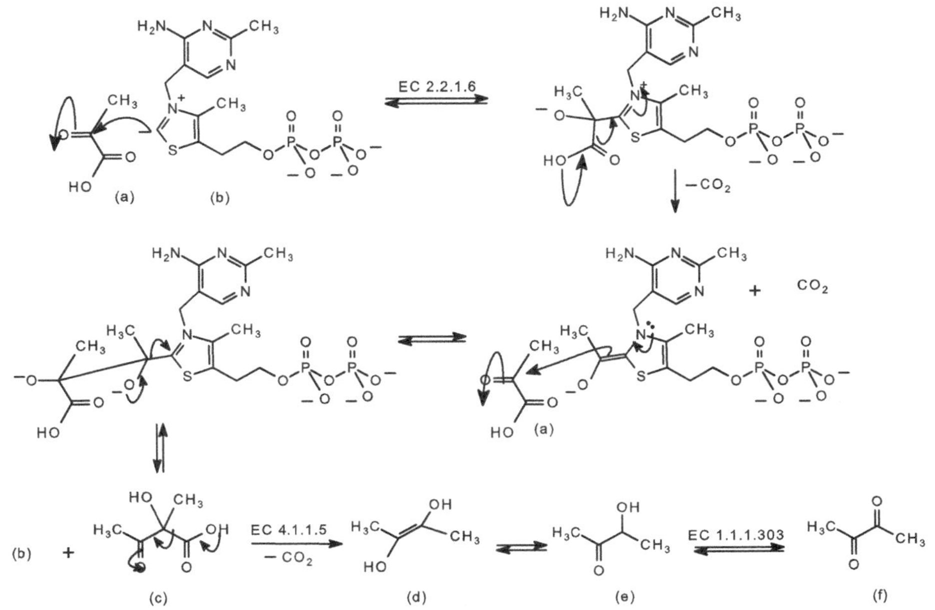

SCHEME 18.11 A brief representation of the conversion of two equivalents of pyruvate (a) with the enzyme acetolactate synthase, EC 2.2.1.6, and the cofactor thiamine diphosphate (b) to acetolactate (c). Acetolactate (c), undergoing decarboxylation (acetolactate decarboxylase, EC 4.1.1.5) to the enol of acetoin (d), which then tautomerizes to acetoin (e) which is readily oxidized to diacetyl (f) diacetyl reductase, EC 1.1.1.303).

EQUATION 18.3 An expression of the equilibrium extant between lactic acid (a) and oxaloacetic acid (b) and pyruvic acid (c) and malic acid (d) in the presence of lactate-malate *trans*-hydrogenase (EC 1.1.99.7).

Since it is common to have the bacteria present in the winery it appears that the conversion begins as soon as the grapes are pressed, concomitant with the addition of yeast. On occasion, additional quantities of the lactic acid bacteria are added when the conversion seems to have ceased or decreased to a rate insufficient to the vintner's plans.

In principle, in the presence of lactate-malate *trans*-hydrogenase (EC 1.1.99.7), which apparently has an $NAD^+/NADH$ couple fixed to the enzyme so that the oxidation and reduction reactions occur sequentially on the same enzyme, the equilibrium shown in Equation 18.3 obtains.[53]

There is other, perhaps more important, material to consider too.

Simple oxidation (EC 1.1.1.37, malate dehydrogenase) of malic acid produces oxaloacetic acid (see the Krebs cycle, Appendix 2). Then (Scheme 18.12) decarboxylation of oxaloacetate

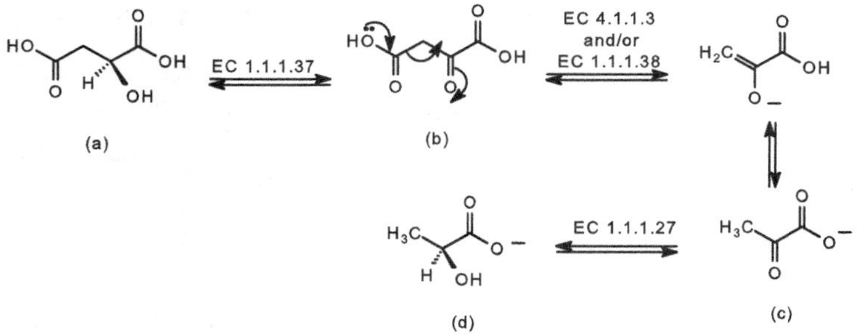

SCHEME 18.12 A representation of the conversion of S-malate (a) by malate dehydrogenase, EC 1.1.1.37 to oxaloacetic acid (b). Then, in the presence of oxaloacetate decarboxylase, EC 4.1.1.3 (or a malate dehydrogenase EC 1.1.1.38), loss of carbon dioxide (CO_2) generates the enol of pyruvate, which tautomerizes to pyruvate (c). The latter on reduction (L-lactate dehydrogenase, EC 1.1.1.27) produces S-lactate (d).

(EC 4.1.1.3, oxaloacetate decarboxylase as well as a malate dehydrogenase EC 1.1.1.38) produces pyruvate, and reduction of pyruvate (L-lactate dehydrogenase, EC 1.1.1.27) to lactate occurs.[54]

The formation of pyruvate which, as shown in Scheme 18.11, can dimerize (thiamine) and, via acetolactate, proceed to acetoin and thence to diacetyl as well as to 2,3-butanediol is also important here in lowering the acidity, as it too involves loss of carboxylate as carbon dioxide (CO_2). This pH change serves to "mellow" the final beverage. It is presumed that these mellowing flavor constituents are generated late enough in the process (i.e., after most of the active fermentation has occurred) so that they will not have been changed or, having been changed, will be somehow regenerated.

An additional point deals with the presence of citric acid (pKa 3.13) and its conversion to diacetyl.[55] As the Krebs cycle (Appendix 2) turns, malate is converted to oxaloacetate,[56] which is then carried on to citrate by addition of acetyl-CoA (citrate synthase, EC 2.3.3.2). But decarboxylation of citrate (pKa 3.13) (citrate lyase, EC 4.1.3.6) yields oxaloacetate and acetate back again. And of course, oxaloacetate leads ultimately (as noted above) to both lactate (pKa 3.86) and to the neutral α-diketone, diacetyl, thus reducing the acidity and increasing the extent to which the beverage is changed.[57,58]

So it is clear now, and noted previously, that except for the generation of hydrogen sulfide (H_2S) from sulfite (SO_3^{2-}), sulfate (SO_4^{2-}), or endogenous sulfur compounds (such as methionine, biotin, or lipoic acid) the major "aroma" compounds apparently produced by yeast during alcoholic fermentation are the familiar compounds already discussed (i.e., ethanol, acetoin, diacetyl, 2,3-butanediol, glycerol, pyruvate, acetaldehyde, acetic acid, some esters of ethanol, as well as esters of higher, amino acid–derived alcohols [and the alcohols themselves], and fatty acids). However, because of variations in strain of *S. cerevisiae* yeast used, and particular *V. vinifera* upon which it acts, only the broadest subset of compounds listed above are found throughout. The individual flavors result from the variation in the amounts of the general subset and the large number of additional compounds (also in varying quantities) produced by the different yeasts and grapes (Figure 18.23).

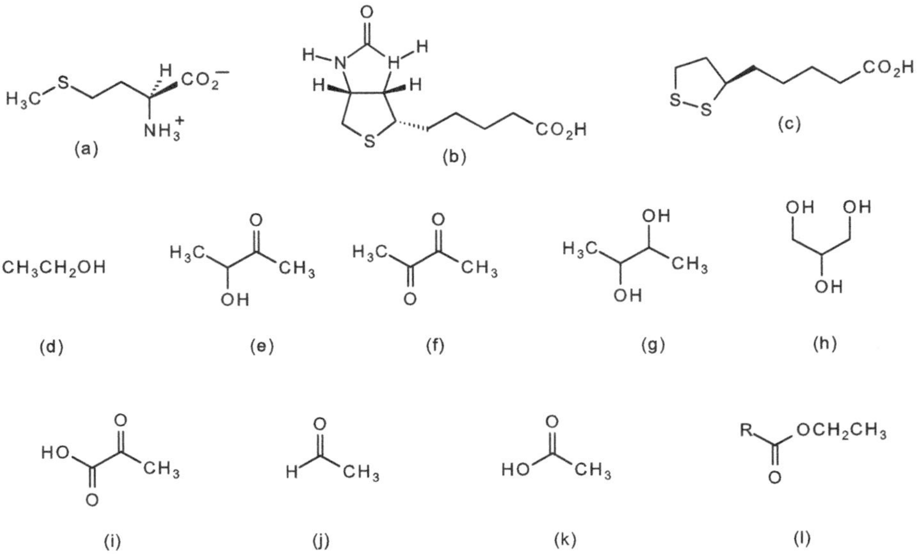

FIGURE 18.23 Representations of compounds, sometimes below the limits of detectability, generally found as a result of yeast metabolism include: (a) methionine; (b) biotin; (c) lipoic acid; (d) ethanol; (e) acetoin; (f) diacetyl; (g) 2,3-dihydroxybutane; (h) glycerol; (i) pyruvic acid; (j) acetaldehyde; (k) acetic acid; and (l) esters of ethanol where R may be large or small and will include fatty (C-12 through C-18) acids as well as small branched acids derived from amino acids.

In general, the following additional compounds have been recorded as present [59–61] in a variety of wines as a consequence of the action of a variety of strains of *S. cerevisiae* and, of course, excluding most of those compounds already found as relevant to the living system (e.g., acetic acid, pyruvic acid, lactic acid):

Acids: formic (methanoic), propanoic, butanoic, 2-methylpropanoic, 2-methylbutanoic, 3-methylbutanoic, pentanoic, hexanoic, octanoic, decanoic, *S*-butyl butanethioate (butanethioic).
Alcohols: 1-propanol, 1-butanol, 2-methyl-1-propanol, 1-pentanol, 2-methyl-1-butanol, 3-methyl-1-butanol, hexanol, tyrosol, 2-(1*H*-indol-3-yl)ethanol (i.e., indole-3-ethanol), 2-phenylethanol.
Aldehydes: acetaldehyde, benzaldehyde, butanal, propanal, 2-methylpropanal, pentanal, 3-methylbutanal, 2-methylbutanal.
Ketones: acetone, diacetyl, 2-acetyltetrahydropyridine
Esters: ethyl acetate, 2-methylpropyl acetate, 2-phenylethyl acetate, 2-methylbutyl acetate, 3-methylbutyl acetate, 2-methylpropyl acetate, hexyl acetate, ethyl 3-methylbutanoate, ethyl butanoate, ethyl 2-methylbutanoate, ethyl hexanoate, ethyl octanoate, ethyl decanoate, ethyl dodecanoate, ethyl lactate (ethyl 2-hydroxypropanoate), diethyl succinate
Phenols: 4-vinylphenol, 4-vinylguiacol (2-methoxy-4-vinylphenol), 4-ethylphenol, 4-ethylguaicol (2-methoxy-4-ethylphenol).

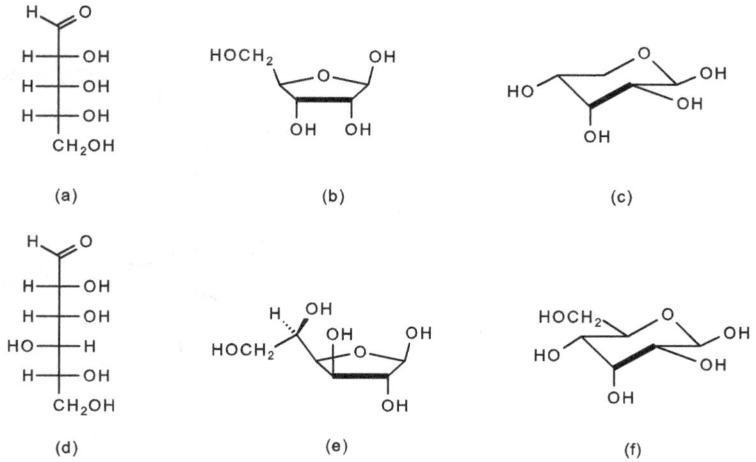

FIGURE 18.24 Representations of carbohydrate heterocycles common in both grapes and yeast. Those shown are: (a) a Fischer projection of D-(−)-ribose; (b) β-D-ribofuranose; (c) β-D-ribopyranose; (d) a Fischer projection of D-(+)-glucose; (e) β-D-glucofuranose; and (f) β-D-glucopyranose.

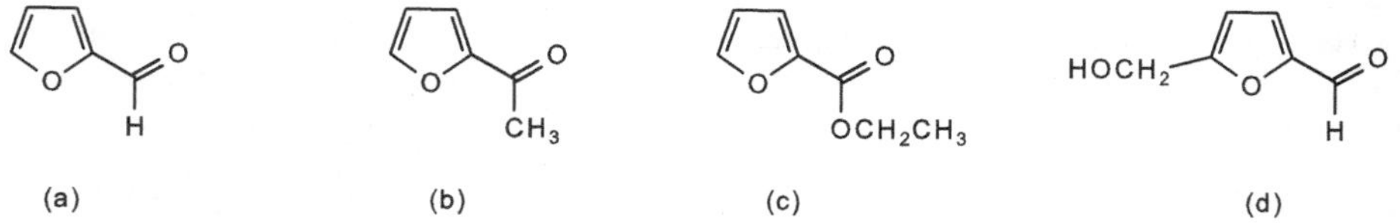

FIGURE 18.25 Representations of some derivatives of furan common to a variety of wines and thus potentially produced from carbohydrates by the action of yeasts. The drawings represent (a) furan-2-carboxaldehyde (furfural); (b) 2-acetylfuran; (c) ethyl furan-2-carboxylate; and (d) 5-hydroxymethylfurfural.

Sulfur compounds: hydrogen sulfide, dimethyl sulfide, diethyl sulfide, dimethyl disulfide, diethyl disulfide, methyl mercaptan (methyl hydrogen sulfide), ethyl mercaptan (ethanethiol), S-methyl thioacetate (S-methyl ethanethioate), 4-mercapto-4-methylpentan-2-ol (4-methyl-4-sulfanylpentan-2-ol).

Heterocycles

As already noted, there are groups of compounds containing rings of atoms and, in the context seen here, they are mostly carbocyclic (i.e., the rings consist solely of carbon atoms). However, in nature, many such cyclic systems have rings made up of carbon and other atoms too (e.g., oxygen, nitrogen, sulfur, phosphorus). Thus, five- and six-membered carbohydrates, such as ribose and glucose, are found in cyclic as well as acyclic forms (Figure 18.24).

Interestingly, furan derivatives (Figure 18.25) which can be generated chemically from carbohydrates by treatment with strong acid such as sulfuric acid (H_2SO_4), might also be produced by enzymatic action on those same carbohydrates and this is currently an area of active study. Also, in this vein, it is possible that oxidation and dehydration of intermediates,

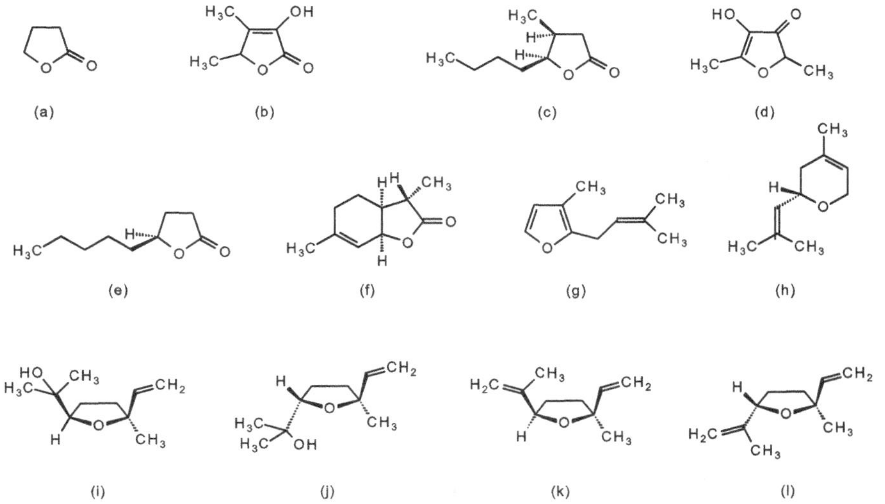

FIGURE 18.26 Representations of oxygen-containing heterocycles apparently derived from hydroxy- and or ketocarboxylic acids and/or from isoprene precursors. Those shown include: (a) butyrolactone; (b) sotolon; (c) whiskey lactone; (d) furaneol; (e) (5S)-5-pentyldihydro-2(3H)-furanone; (f) wine lactone; (g) rose furan; (h) (S)-(−)-nerol oxide; (i) cis-linalool oxide; (j) trans-linalool oxide; (k) cis-anhydrolinalool oxide; and (l) trans-anhydrolinalool oxide.

subsequent to fermentation, either in casks while aging, in bottles for distribution, or finally, glasses before consumption, might contribute to their production.

A number of such furanoid compounds are also widely found in various wines.[62] It appears that many are derived from hydroxy- or ketocarboxylic acids (as they are lactones) on the one hand or from isoprene (i.e., dimethylallyl diphosphate and isopentenyl diphosphate), on the other (Figure 18.26). As they are isolated after treatment it is not clear, for some, if they are derived from action of yeast on carbohydrates or if they are present as a consequence of enzymatic processes in the grapes themselves. It has recently been found that (at least) the precursor of wine lactone (Figure 18.26) is generated through a grapevine cytochrome P450.[63]

The so-called "norisoprenoids," from degradation of isoprene-based carotenoids, which were already discussed, in part, because some of the compounds are found in the grape "Nebbiolo" (Chapter 14), are also found in the wine itself. Indeed, these compounds are considered particularly important because they "... have very low olfactory perception thresholds and so have a high sensorial impact on wine aroma."[64]

The compounds shown earlier in the discussion of the grape Nebbiolo, viz. β-ionone and 1,1,6-trimethyl-1,3-dihydronaphthalene (TDN) are only two of the compounds also found in the wine, and as shown (Figure 18.27), neither is heterocyclic.

Interestingly, however, hydrolysis of grape juice isolates with warm acid (pH ≈ 3) provided a plethora of compounds, apparently derived by rearrangement reactions of the initial products themselves that were derived from cleavage of carotenoids. Most of these heterocyclic compounds were also subsequently isolated from wine produced from the grapes, and so it can be argued that their production occurs during and/or after fermentation (Figure 18.28).[65,66]

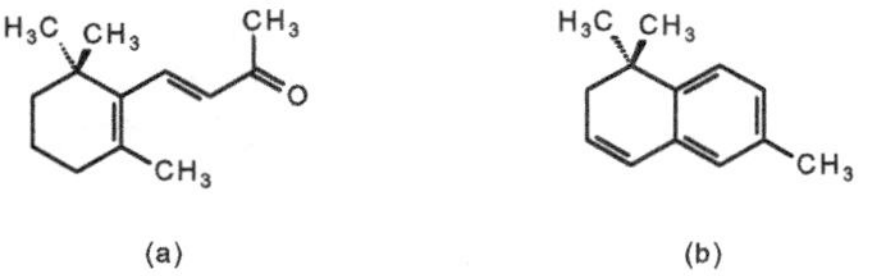

FIGURE 18.27 Representations of (a) β-ionone and (b) 1,1,6-trimethyl-1,2-dihydronaphthalene (TDN).

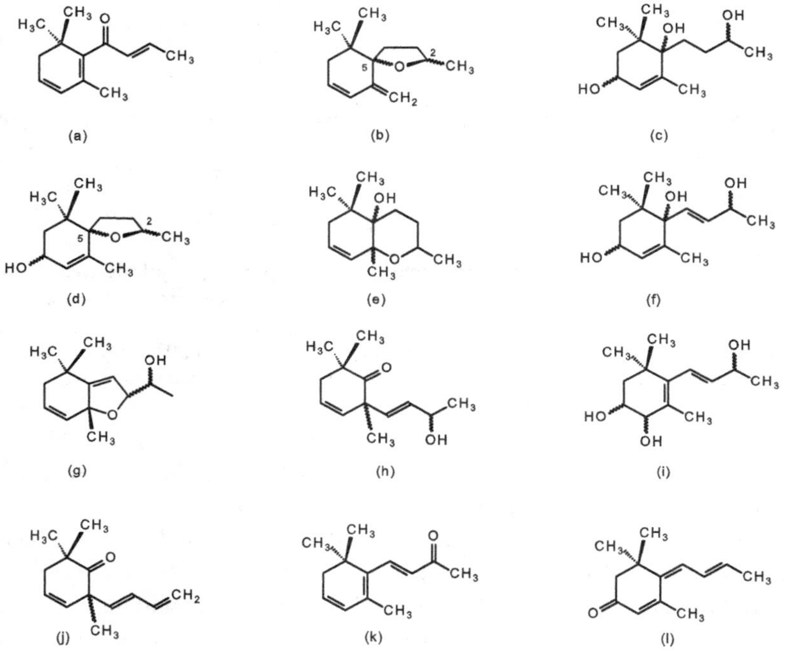

FIGURE 18.28 Representations of compounds isolated after treatment (but not before) of juice from the Nebbiolo grape (Chapter 14) and later from the wine itself. Concentrations were variable and isomeric mixtures evident. The compounds isolated were: (a) β-damsacenone; (b) vitispiranes; (c) megastigma-4-ene-3,6,9-triols; (d) theaspiranes; (e) hydroxydihydroedulans; (f) megastigma-4,7-diene-3,6,9-triols; (g) actinidols; (h) 2-(3-hydroxy-1-buten-1-yl)-2,6,6-trimethyl-3-cyclohexen-1-ones; (i) 4-(3-hydroxy-1-buten-1-yl)-3,5,5-trimethyl-3-cyclohexen-1,2-diols; (j) 2-(1,3-butadien-1-yl)-2,6,6-trimethyl-3-cyclohexen-1-ones; (k) 3-dehydro-β-ionone; and (l) megastigmatrienone D.

Although it might have been the case that some of the polyene-derived compounds isolated or identified were initially present as the aglycone portions of glycosides even in the grapes and were thus not detected until subsequently hydrolyzed, it is also possible that oxidation and rearrangement during fermentation produced them.

As shown (Figure 18.28), all of the isomers [(2R,5R)-*cis*, (2S,5S)-*cis*, (2S,5R)-*trans*, and (2R, 5S)-*trans*)] of vitispirane, which are presumably derived by rearrangement from the set of megastigma-4-ene-3,6,9-triols (i.e., isomers of 1-(3-hydroxybutyl)-2,6,6-trimethyl-2-cyclohexene-1,4-diol), appear to be present in some wines (but not all in the same

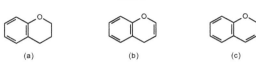

FIGURE 18.29 Representations of the basic fused-ring aryl-pyran systems: (a) chroman (or chromane); (b) 4H-chromene; and (c) 2H-chromene.

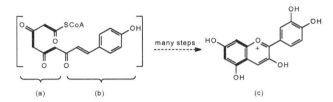

SCHEME 18.13 A representation of the use of three conjoined acetate units (a) derived from either malonyl-CoA or acetyl-CoA bound to a (b) cinnamyl unit (from phenylalanine *via* cinnamyl-CoA) leading, after many well-known steps (see Appendix 4 and http://www.enzyme-database.org/reaction/phenol/chalcone.html), to an anthocyanidin pigment (e.g.(c) cyanidin).

concentrations in different wines and even in different years of the "same" wines). Indeed, the same triols presumably yield the heterocyclic theaspirane and hydroxydihydroedulan isomers.

In addition, there is a host of compounds obtained from the more highly oxidized megastigmadienes (i.e., the megastigma-4,7-diene-3,6,9-triols [isomers of 1-(3-hydroxy-1-buten-1-yl)-2,6,6-trimethyl-2-cyclohexene-1,4-diol]). These include the heterocyclic actinidols as well as rearranged homocyclic ketones shown in Figure 18.28).

Again, as pointed out by Wamhoff and Gribble,[62] there is a wide variety of heterocyclic compounds in wine (most of which have already been discussed earlier as they occur in grapes) which have phenolic aryl groups attached to five- or six-membered heterocyclic systems. As shown in Appendix 4 in detail and more briefly in Figure 18.29, these heterocycles arise from a cyclization of an acetate-derived fragment attached to an amino acid–derived fragment (Scheme 18.13) and thus lead to a variety of benzodihydropyran (chroman) and benzopyran (chromene) derivatives.

Aryl rings simply fused to a pyran immediately adjacent to the oxygen [benzodihydropyran (chroman)] and unadorned with hydroxyl (–OH) groups are colorless. However, if they bear a carbonyl group conjugated (Appendix 1) with the aromatic ring, as in a chromanone, or a double bond conjugated with the oxygen of the pyran and the carbonyl group, as in a chromone, a pale yellow color is apparent. Addition of a phenyl group (again, lacking –OH substituents) adjacent to the oxygen of the pyran to produce 2-phenylchromone (flavone) or 2-phenylchromanone (flavanone), in the latter case, deepens the color and shifts it to slightly longer wavelength (Figure 18.30).

As noted already (Chapter 12) anthocyanins (those having a carbohydrate linkage attached to the hydroxyl at C3) and anthocyanidins, their corresponding aglycones, enjoy a wide display of colors as a consequence of extended conjugation through appropriately placed hydroxy (–OH) and methoxy (–OCH$_3$) groups (Figure 18.31).

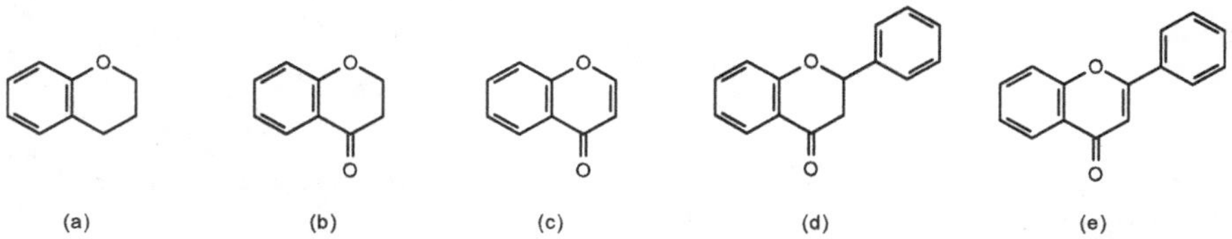

FIGURE 18.30 Representations of fused arylpyran systems. Those shown are: (a) benzodihydropyran (chroman); (b) 4-chromanone; (c) chromone; (d) flavanone; and (e) flavone.

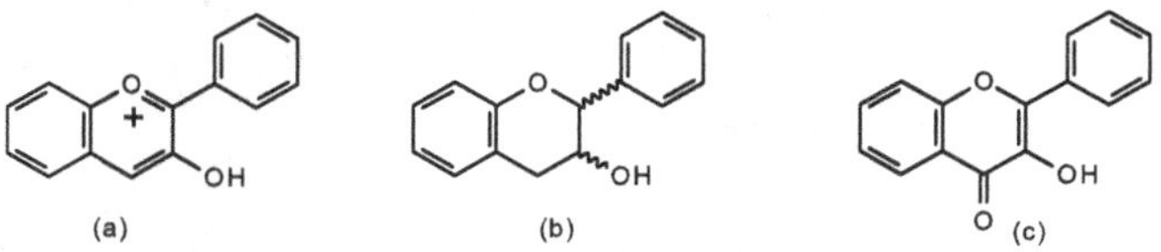

FIGURE 18.31 Representations of the unsubstituted parents of the various benzopyran systems present in flower, grape, and wine. Shown here are: (a) the pyrillium ion in the anthocyanidin parent; (b) the flavan-3-ol parent where, as usual, the wavy lines indicate undefined stereochemistry; and (c) the flavon-3-ol parent.

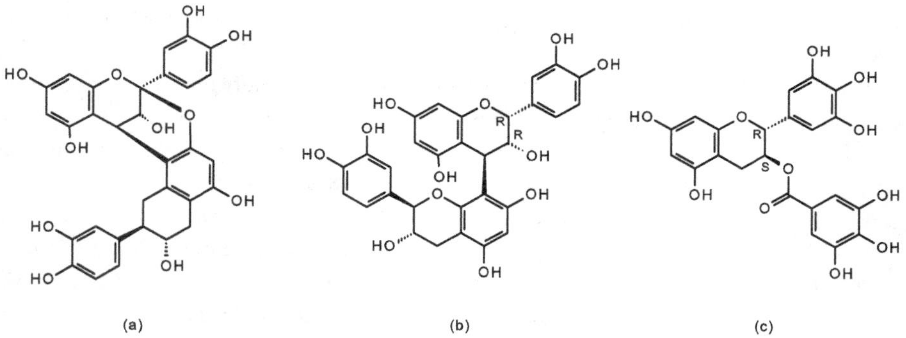

FIGURE 18.32 Representations of some tannins. Shown here are representations for: (a) procyanidin A1; (b) procyanidin B1: and (c) (+)-gallocatechin gallate.

As noted earlier, these phenolic materials are often present as glycosides which, on exposure to yeast enzymes, may undergo hydrolysis to produce the parent poly-hydroxylated heterocycle. The parent phenols are then free to undergo reaction with other components present in the fermenting medium, both in the presence of yeast and subsequently on standing. When these compounds react *via* phenolic coupling they reversibly give rise to a series of compounds called procyanidins. Further, the hydroxyl group at C3 in the flavan-3-ols is easily capable of esterification (it is neither phenolic as in the anthocyanidins nor enolic as in flavon-3-ols), and so it is not surprising to find that carboxylic acids such as gallic acid react and are found esterified there. These procyanidins, as well as the derivatives of gallic acid and additional phenolic dimers, trimers, and higher polymeric materials, are collectively known as *tanins* (Figure 18.32).

The phenolic content in wine, which contributes to both color and tannin formation, continues to be explored. As previously noted, locations within the grape (skin, seeds, pulp) appear to have different phenols both as glycosides and as free phenols. Keeping track of the most up-to-date list of the several hundred (at this writing) phenols and polyphenols, and the grapes from which they are derived, is daunting. However, a list (May 2015) is available (https://en.wikipedia.org/wiki/Phenolic_content_in_wine), and it appears to be modified and updated as warranted.

Despite the long lists of compounds of known structure, the complexity of the wine is clearly related to the complexity of the grape, and thus the number of identifiable components in grape juice and in wine continues to increase as the tools for their detection become ever more refined.[67] Thus, as indicated in Figures 18.33 through 18.35, the number of phenolic components that can be detected and identified with certainty in grape juice by the tools of high pressure liquid chromatography (HPLC) coupled with mass spectrometry (MS) is truly impressive.[68]

Relatively low dimeric, trimeric, and even some tetra- and pentameric tannins, with large numbers of phenolic hydroxyl (–OH) groups for their overall size, are often found to lend astringency to wine. White wines, and even some red wines that are meant to be drunk

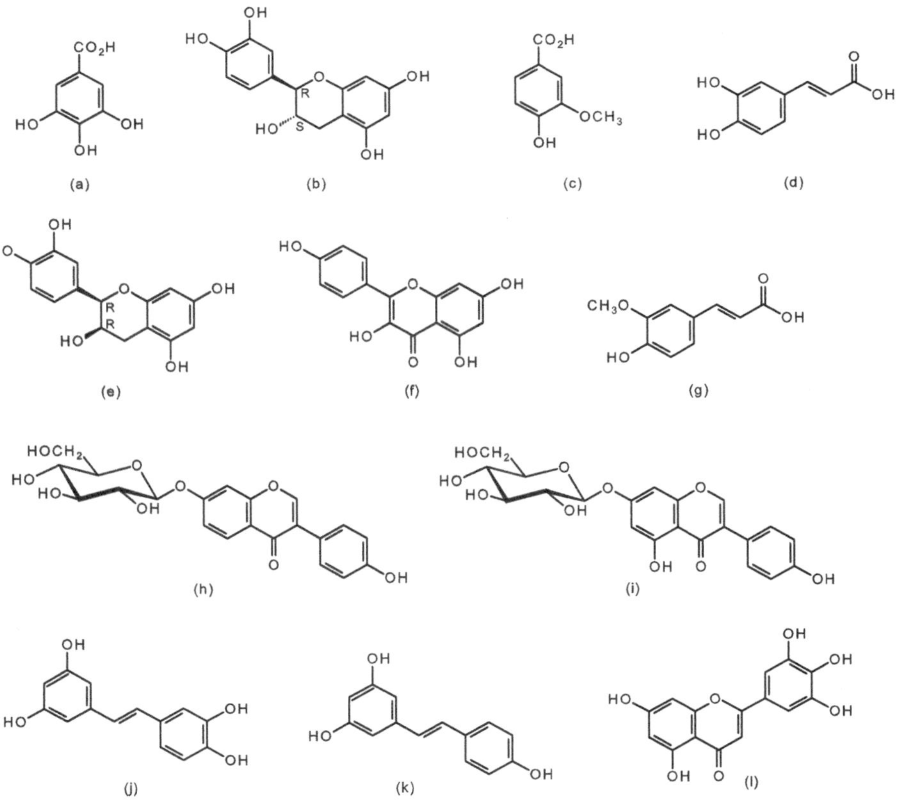

FIGURE 18.33 Representations of phenols found in grape juice which are also commonly found in wine. Shown are: (a) gallic acid; (b) (+)-catechin; (c) vanillic acid; (d) caffeic acid; (e) (–)-epicatechin; (f) kaempherol; (g) ferulic acid; (h) diadzein-7-O-glucoside (i) genistein-7-O-glucoside; (j) piceatannol; (k) *trans*-resveratrol; and (l) myricetin.

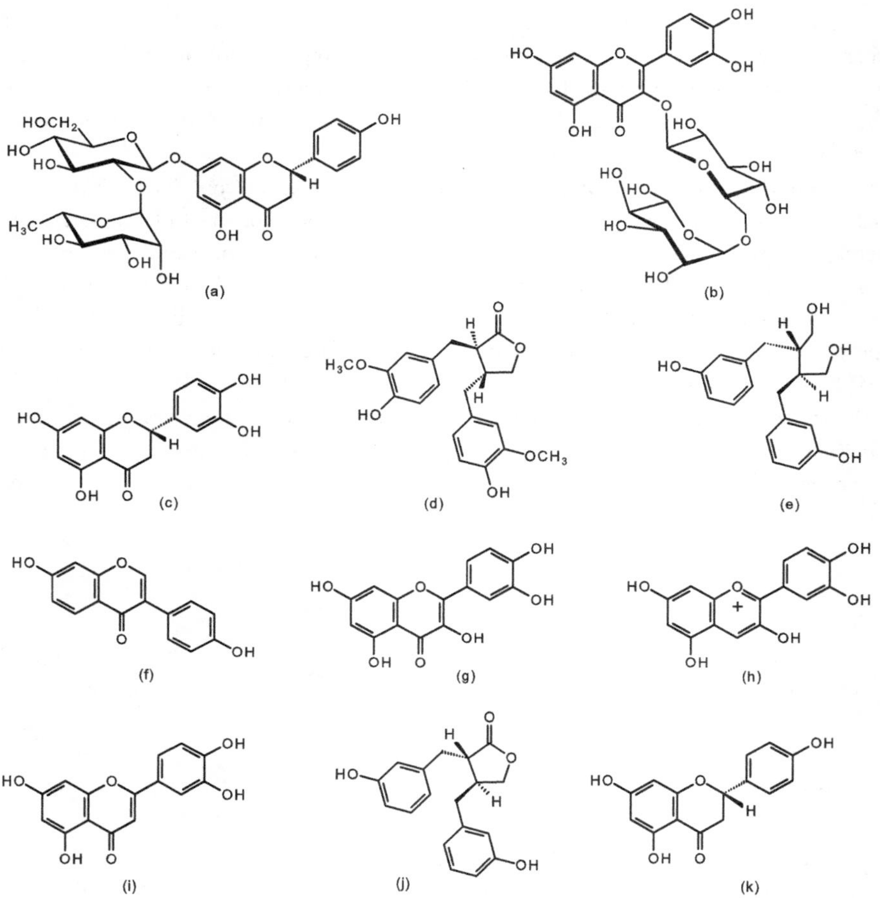

FIGURE 18.34 Representations of phenols found in grape juice which are also commonly found in wine. Shown are: (a) naringin; (b) rutin; (c) eriodictyol; (d) (−)-matairesinol; (e) (+)-enterodiol; (f) diadzein; (g) quercetin; (h) cyanidin; (i) luteolin; (j) enterolactone; and (k) naringenin.

"young," have such tannins present. As the wine ages at the appropriate temperature in wood (traditionally oak), the individual tannins grow larger by combination and rearrangement. That is, the smaller dimeric, trimeric, etc. tannins combine and: (a) polymerize into larger molecules (some of which are no longer soluble in the wine and coat the barrel); (b) rearrange to produce tannins with fewer phenolic hydroxyl groups for their larger size. A more mellow beverage results.

Additionally, again as pointed out by Wamhoff and Gribble,[62] there are other oxygen-containing heterocycles which arise from phenols. Phenols, particularly those with multiple hydroxyl groups on the aromatic ring are relatively easy to oxidize and, thus, are capable of scavenging radical species (while creating new radical species of lower energy). For example, the trisphenol (three −OH groups on the aromatic rings) resveratrol has been shown to yield products thought to be result of oxidation to a free radical followed by coupling reactions with at least one equivalent of resveratrol itself.

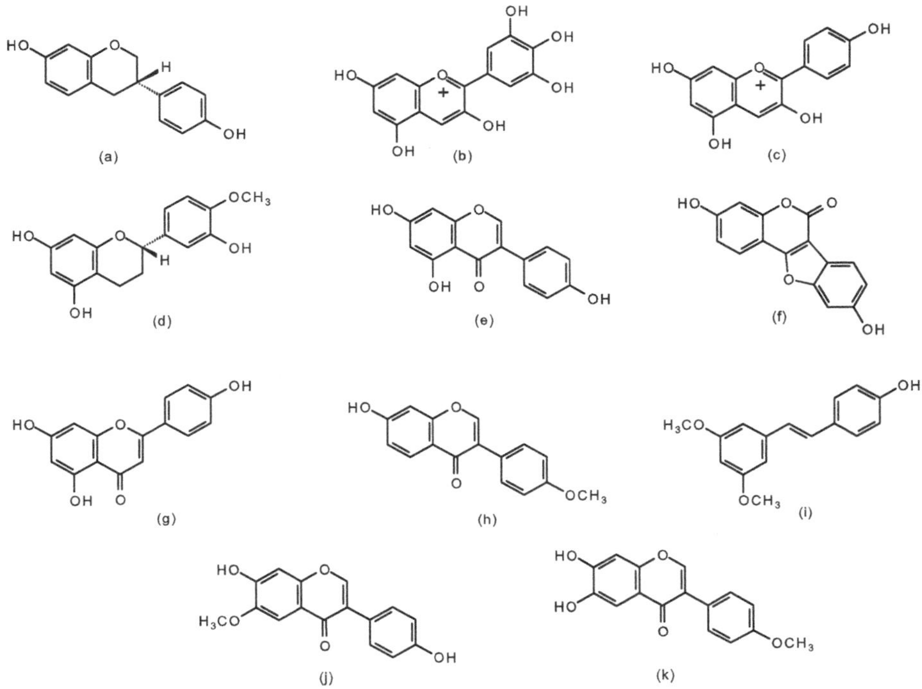

FIGURE 18.35 Representations of phenols found in grape juice which are also commonly found in wine. Shown are: (a) equol; (b) delphinidin; (c) pelargonidin; (d) hesperetin; (e) genistein; (f) coumesterol; (g) apigenin; (h) formonetin; (i) *trans*-petrosilbene; (j) glycitein; and (k) biochanin A.

This results in the destruction of the oxidized species with formation of ε-viniferin (Scheme 18.14).[69,70]

Other, more complicated materials, (e.g., the trimer α-viniferin) can also be formed. Thus, in the event the barrier to attack at the aromatic ring to form the dimer cannot be overcome (perhaps the temperature is too low) and/or the concentration of resveratrol is high enough, rather than attacking the aromatic ring, it appears that addition to another double bond of a third resveratrol can be effected. The result is the formation of a nine-membered ring—using only attack at double bonds and not at aromatic rings. Indeed, it is the apparent ability of resveratrol to use up reactive oxygen species in this and related ways that has caused it to be considered by some to be valuable.

Given the (obvious) nature of the living plants and yeast, the kinds of compounds necessarily present as a consequence of life are expected to be found. Thus the oxygen and nitrogen heterocycles, as well as purine and pyrimidine cores found in deoxyribonucleic and ribonucleic acids (DNA and RNA, respectively), are anticipated to be found in wine. Additionally, sulfur-containing compounds, based on sulfur amino acid metabolism (and sulfur dioxide, sulfite, and sulfate metabolism too), such as the cofactor biotin as well as the unique Chenin blanc (Chapter 14) tetrathiocane and trithiolane derivatives already reported, are also found along with the parent amino acids. A few of these well-known heterocycles are shown (Figure 18.37) in the structures of

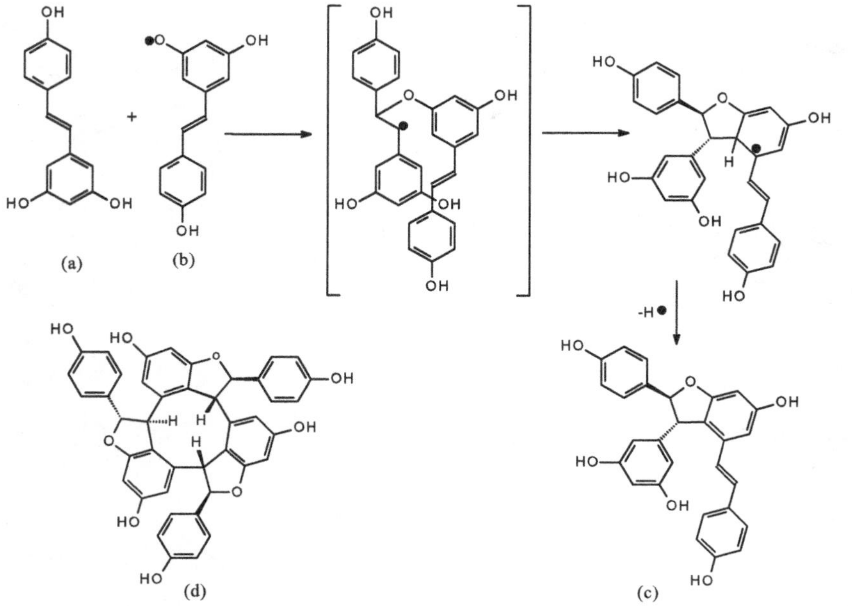

SCHEME 18.14 A representation of potential products derived from a presumed radical formed by oxidation of (a) *trans*-resveratrol. When reaction of one *trans*-resveratrol (a) with the radical from *trans*-resveratrol (b) occurs with addition to the double bond, subsequent cyclization can lead to a dimer (c) ε-viniferin or even a trimer, (d) (+)-α-viniferin.

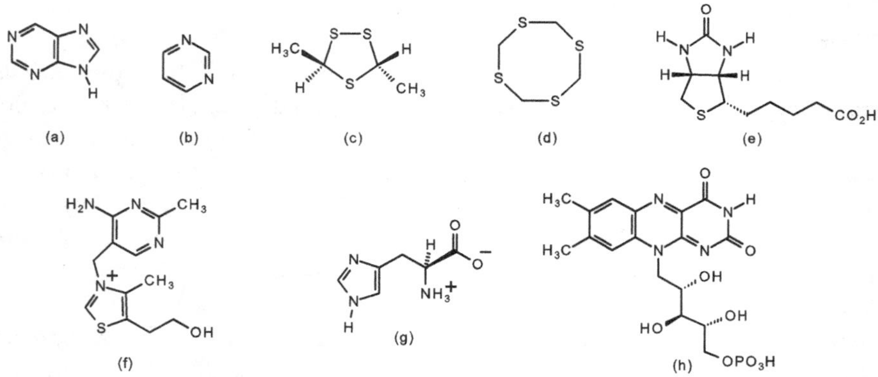

FIGURE 18.37 Representations of a few nitrogen and sulfur containing heterocycles present by isolation or as examples of systems expected as a result of dealing with living species. Shown here are: (a) the purine nucleus present in adenine (A) and guanine (G); (b) the pyrimidine nucleus present in cytosine (C), thymine (T) and uridine (U); (c) *trans*-3,5-dimethyl-1,2,4-trithiolane; (d) 1,3,5,7-tetrathiocane; (e) biotin (coenzyme R); (f) thiamine (vitamin B_1) (g) histidine (His, H); and (h) flavin mononucleotide (FMN).

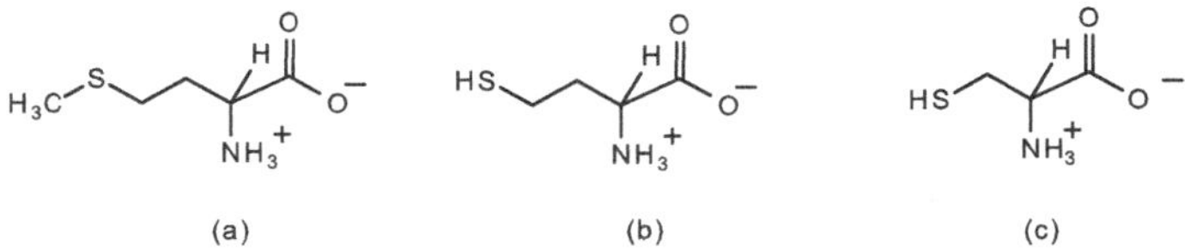

FIGURE 18.36 Representations of common sulfur containing amino acids: (a) methionine (Met, M); (b) homocysteine; and (c) cysteine (Cys, C).

the sulfur-containing amino acids methionine (Met, M), homocysteine and cysteine (Cys, C) (Figure 18.36).

Much of the story outlined above concerning the compounds produced in the wine both as a consequence of modification of those originally present in the grape must by the added yeast as well as those present as a consequence of the yeast getting on with its own life in terms of protein production, etc., need to be understood in terms of the latter. That is, it is quite clear that as ethanol (CH_3CH_2OH) is produced, the yeast changes to accommodate that poison. Thus, it comes as little surprise that in a study of the response of the genome of *S. cerevisiae* to stress as a function of the concentration of ethanol throughout a 15-day wine fermentation, it was found that about 40% of the yeast genome significantly changed expression levels.[71] About 2/3 of the fermentation response genes (FRS) had not previously been implicated in global stress response, and about 25% of that group had no functional annotation as to protein production. Although glucose concentration in the must remained high (a source of osmotic stress, i.e., stress because of the different pressures on opposite membrane sides), and although the yeast responded to ethanol formation genetically, eventually the yeast could not manage, and it died.

NOTES AND REFERENCES

1. It is not the purpose of this work to provide an exposition on fungi in general or yeast in particular. The much more detailed account provided in specialist works such as that edited by Feldman, H. *Yeast, Molecular and Cell Biology*, 2nd Ed.; Wiley-Blackwell: Weinheim, Germany, 2012 is recommended to the interested reader.

2. A strain refers generally to offspring of a genetically modified—intentionally or otherwise—parent. In some families, the parent and the accompanying offspring are used together. Some interesting selections can be (at this writing, 2016) found at http://winemaking.jackkeller.net/strains.asp (accessed Apr 7, 2017).

3. Recall from Chapter 13 that "the amount of sugar is traditionally measured in "brix" (one degree "brix" = 1 gram of sucrose in 100 grams of solution)." The amount of sugar will correlate with the quantity of alcohol that can be obtained from the grape—given the specific yeast to be employed (and other factors).

4. Feldman, H. *Yeast, Molecular and Cell Biology*, 2nd Ed.; Wiley-Blackwell: Weinheim, Germany, 2012; p 1.

5. Goffeau, A.; Barrell, B. G.; Bussey, H.; Davis, R. W., Dujon, B.; Feldmann, H.; Galibert, F.; Hoheisel, J. D.; Jacq, C.; Johnston, M.; Louis, E. J.; Mewes, H. W.; Murakami, Y.; Phillippsen, P.; Tettelin, H.; Oliver, S. G. *Science* **1996**, *274*, 546.

6. Westhead, E. W.; McLain, G. *J. Biol. Chem.* **1964**, *239*, 2464.

7. Reed, G. H.; Poyner, R. R.; Larsen, T. M.; Wedekind, J. E.; Rayment, I. *Cur. Opin. Struct. Biol.* **1996**, *6*, 736. DOI:10.1016/S0959-440X(96)80002-9; and Larsen, T. M.; Wedekind, J. E.; Rayment, I.; Reed, G. H. *Biochem.* **1996**, *35*, 4349. DOI:10.1021/bi952859c.

8. Just as pyruvate ($CH_3COCO_2^-$) can be utilized to produce, among others, carbon dioxide (CO_2) and acetyl CoA (CH_3CO-CoA) and/or acetaldehyde (CH_3CHO), both glycerone phosphate (dihydroxyacetone monophosphate) and glyceraldehyde-3-phosphate can be reduced to glycerol-3-phosphate [glycerol-3-phosphate dehydrogenase (NAD^+) EC 1.1.1.8] rather than going on to pyruvate. The phosphate can be hydrolyzed to glycerol ($HOCH_2CH(OH)CH_2OH$). The pathway for formation of glycerol, involving transfer of a hydride from NADH to the carbon of the carbonyl, is similar to that shown in Scheme 18.8 for the reduction of acetaldehyde to ethanol. Glycerol is a common constituent of wine. See, e.g., Albertyn, J.; van Tonder, A.; Prior, B. A. *FEBS Lett.* **1992**, *308*, 130.

9. At this writing a number of yeasts used in the wine industry can be found through the collection at the University of California, Davis. The contact information may change but it is currently http://wineserver.ucdavis.edu/ (accessed Apr 7, 2017). A larger database of fungi is maintained by the US Department of Agriculture and can currently be found at http://nt.ars-grin.gov/fungaldatabases/ (accessed Apr 7, 2017). Data bases are current as of May 2016.

10. Feldman, H. *Yeast, Molecular and Cell Biology*, 2nd Ed.; Wiley-Blackwell: Weinheim, Germany, 2012; Ch. 2 and 3, pp 5–58.

11. Kurtzman, C. P.; Fell, J. W.; Boekhout, T., Eds.; *The Yeasts: A Taxonomic Study*, 5th Ed.; Elsevier: Oxford, UK, 2010.

12. Pretorius, I. S. *Yeast* **2000**, *16*, 675 (Tailoring Wine Yeast for the New Millennium).

13. Quideau, S.; Snyder, S. A., Eds. *Tetrahedron* **2015**, *71*, 2955 *et seq*. (Chemistry in the Vine and Wine Sciences).

14. Ribéreau-Gayon, P.; Dubourdieu, D.; Donèche, B.; Lonvaud, A. *Handbook of Enology*, Vols. 1&2; John Wiley & Sons: Chichester, England, 2006.

15. A "strain" is rank in taxonomy lying beneath "species" as a genetic variant. The minor differences in strains are distinguishable and often reproducible.

16. The number is vague since until 2013 asexually reproducing fungi (anamorphs) and telemorphs (fruiting body—sexually reproducing states) were given different names. The older literature does not, and even some current workers do not, accept the change. See Hawksworth, D. L. *MycoKeys* **2011**, *1*, 7.

17. Pretorius, I. S. *Yeast* **2000**, *16*, 699.

18. Gérard, E *J. Pharm. Chim.* **1895**, *1*, 601.

19. Barton, D. H. R.; Kempe, U. M.; Widdowson, D. A. *J. Chem. Soc. Perkin* **1972**, *1*, 513.

20. Honeywell, E. M.; Bills, C. E. *J. Biol. Chem.* **1932**, *99*, 71 and references therein.

21. Schwenk, E.; Alexander, G. J. *Arch. Biochem. Biophys.* **1958**, *76*, 65.

22. Barton, D. H. R.; Moss, G. P. *Chem. Comm.* **1966**, 261. DOI: 10.1039/C19660000261.

23. Lees, N. D.; Bard, M.; Kirsch, D. R. *Crit. Rev. Biochem. Mol. Biol.* **1999**, *34*, 33.

24. Carrau, F. M.; Medina, K.; Boido, E.; Farina, L.; Gaggero, C.; Dellacassa, E.; Versini, G.; Henschke, P. A. *FEMS Microbiol. Lett.* **2005**, *243*, 107.

25. Hock, R.; Benda, I.; Schreier, P. *Zeitschrift für Lebensmitteluntersuchung und Forschung* (now *Eur. Food Res. Tech.*) **1984**, *179*, 450.

26. Misawa, N.; Shimada, H. *J. Biotechnol.*, **1998**, *59*, 169.

27. Jackson, B. E.; Hart-Wells, E. A.; Matsuda, S. P. T. *Org. Lett.* **2003**, *5*, 1629.

28. Takahashi, S.; Yeo, Y.; Greenhagen, B. T.; McMullin, T.; Song, L.; Maurina-Brunker, J.; Rosson, R.; Noel, J. P.; Chappell, J. *Biotechnol. Bioeng.* **2007**, *97*, 170.

29. Herrero, O.; Ramón, D.; Orgjas, M. *Metabol. Eng.* **2008**, *10*, 78.

30. Huang, B.; Guo, J.; Sun, L.; Chen, W. *Integr. Biol.* **2013**, *5*, 1282.

31. Liu, L.; Redden, H.; Alper, H. S. *Curr. Opin. Biotechnol.* **2013**, *24*, 1023.

32. Wriessnegger, T.; Pichler, H. *Prog. Lipid Res.* **2013**, *52*, 277.

33. Fossati, E.; Narcross, L.; Ekins, A.; Falgueyret, J.-P.; Martin, V. J. J. *PLOS One* **2015**, DOI:10.1371/journal.pone.0124459.

34. DeLoache, W. C.; Russ, Z. N.; Narcross, L.; Gonzales, A. M. Martin, V. J. J.; Dueber, J. E. *Nat. Chem. Biol.* **2015**. DOI:10.1038/nchembio.1816.

35. Meadows, A. L.; Hawkins, K. M.; Tsegaye, Y.; Antipov, E.; Kim, Y.; Raetz, L.; Dahl, R. H.; Tai, A.; Mahatdejkul-Meadows, T.; Xu, L.; Zhao, L.; Dasika, M. S.; Murarka, A.; Lenihan, J.; Eng, D.; Leng, J. S.; Liu, C.-L.; Wenger, J. W.; Jiang, H.; Chao, L.; Westfall, P.; Lai, J.; Ganesan, S.; Jackson, P.; Mans, R.; Platt, D.; Reeves, C. D.; Saija, P. R.; Wichmann, G.; Holmes, V. F.; Benjamin, K.; Hill, P. W.; Gardner, T. S.; Tsong, A. E. *Nature* **2016**, *537*, 694.

36. Fleet, G. H. *Internat. J. Food Microbiol.* **2003**, *86*, 11.

37. Romano, P.; Fiore, C.; Paraggio, M.; Caruso, M.; Capece. A. *Int. J. Food Microbiol.* **2003**, *86*, 169.

38. Ames, J. M.; Mac Leod, G. *J. Food Sci.* **1985**, *50*, 125.

39. Lin, M.; Liu, X.; Song, H.; Li, P.; Yao, J. *J. Sci. Food Agric.* **2014**, *94*, 882.

40. Broadly, a heterocycle is any ring-containing compound where more than one element is found in the ring. Generally, the cycle contains carbon and one or more other elements. Please see Appendix 1.

41. Giardina, B. J.; Stanley, B. A.; Chiang, H.-L. *Proteome Sci.* **2014**, *12*, 9. http://www.proteomesci.com/content/12/1/9 (accessed Apr 7, 2017)

42. Terrell, E.; Cliff, M. A.; VanVuuren, H. J. J. *Molecules* **2015**, *20*. 5112

43. Sarry, J.-E.; Günata, Z. *Food Chem.* **2004**, *87*, 509.

44. Loscos, N.; Hernandez-Orte, P.; Cacho, J.; Ferreira, V. *J. Agric. Food Chem.* **2007**, *55*, 6674.

45. Yonekura-Sakakibara, K.; Fukushima, A.; Nakabayashi, R.; Hanada, K.; Matsuda, F.; Sugawara, S.; Inoue, E.; Kuromori, R.; Ito, T.; Shinozaki, K.; Wangwattana, B.; Yamazaki, M.; Saito, K. *Plant J.* **2012**, *69*, 154.

46. Brouillard, R.; Dubois, J.-E. *J. Am. Chem. Soc.* **1977**, *99*, 1359.

47. Pina, F.; Oliveira, J.; de Freitas, V. *Tetrahedron*, **2015**, *71*, 3107.

48. Colagrande, O.; Silva, A.; Fumi, M. D. *Biotechnol. Prog.* **1994**, *10*, 2.

49. Schisler, D. A.; Janislewicz, W. J.; Boekhout, T.; Kurtzman, C. P. In Kurtzman, C. P.; Fell, J. W.; Boekhout, T., Eds.; *The Yeasts: a Taxonomic Study*, 5th Ed.; Elsevier: Oxford, UK, 2010; Ch. 4, p 45 ff.

50. Tipper, D. J.; Bostian, K. A. *Microbiol. Rev.* **1984,** *48,* 125.

51. Robinson, J. *The Oxford Companion to Wine;* Oxford University Press: New York, 2006; p. 485.

52. Robinson, J., *The Oxford Companion to Wine;* Oxford University Press: New York, 2006; p 423.

53. Allen, S. H. G.; Patil, J. R. *J. Biol. Chem.* **1972,** *247,* 909.

54. Creighton, D. J.; Rose, I. A. *J. Biol. Chem.* **1976,** *251,* 61; and Creighton, D. J.; Rose, I. A. *J. Biol. Chem.* **1976,** *251,* 69.

55. Shimazu, Y.; Uehara, M.; Watanabe, M. *Agric. Biol. Chem.* **1985,** *49,* 2147.

56. Bell, J. K.; Yennawar, H. P.; Wright, S. K.; Thompson. J. R.; Viola, R. E.; Banaszak, L. J. *J. Biol. Chem.* **2001,** *276,* 31156.

57. Garcia-Quintáns, N.; Blancato, V.; Repizo, G.; Magni, C.; López, P. in *Molecular Aspects of Lactic Acid Bacteria for Traditional and New Applications;* Mayo, B.; López, P.; Péres-Martínez, G., Eds.; Research Signpost: Kerala, India, 2008; Chapter 3, p 65.

58. Garcia-Quintáns, N.; Repizo, G.; Martin, M. Magni, C.; López, P. *Appl. Environ. Microbiol.* **2008,** *74,* 1988.

59. Lambrechts, M. G.; Pretorius, I. S. *S. Afr. J. Enol. Vitic.* **2000,** *21,* 97.

60. Molina, A. M.; Swiegers, J. H.; Varela, C.; Pretorius, I. S.; Agosin, E. *Appl. Microbiol. Biotechnol.* **2007.** DOI:10.1007/s00253-007-1194-3.

61. Styger, G.; Prior, B.; Bauer, F. F. *J. Ind. Microbiol. Biotechnol.* **2011,** *38,* 1145.

62. Wamhoff, H.; Gribble, G. W. In *Advances in Heterocyclic Chemistry,* Vol. 106; Katritzky, A. R., Ed.; Elsevier: New York, 2012; Ch. 3, p 185 ff.

63. Ilc, T.; Halter, D.; Miesch, L.; Lauvoisard, F.; Kriegshauser, L.; Ilg, A.; Baltenweck, R.; Hugueney, P.; Werck-Reichhart, D.; Duchéne, E.; and Nicolas, N. *New Phyologist* **2016.** DOI: 10.1111/nph.14139.

64. Mendes-Pinto, M. M. *Arch. Biochem. Biophys.* **2009,** *483,* 236.

65. Strauss, C. R.; Dimitriadis, E.; Wilson, B.; Williams, P. J. *J. Agric. Food Chem.* **1986,** *34,* 145.

66. Eggers, N. J.; Bohna, K.; Dooley, B. *Am. J. Enol. Vitic.* **2006,** *57,* 226.

67. Cristini, C.; Lange, H.; Bianchetti, G. *J. Nat. Prod.* **2016.** DOI:10.1021/acs.jnatprod.6b00380.

68. Sapozhnikova, Y. *Food Chem.* **2014,** *150,* 87. This particular study was "to develop and evaluate a simple, fast, high throughput multiclass analytical method for the simultaneous determination of a wide range of polyphenolic compounds of different classes in liquid samples (juice, tea and coffee)." Standard polyphenolic compounds were obtained commercially, and none of the grape juice, green tea, or coffees were identified beyond the statement ... "Grape juice, green tea, and ground coffee (premium dark and French roast) were purchased from local supermarkets. Juice samples were used as received."

69. Szewczuk, L. M.; Lee, S. H.; Blair, I. A.; Penning, T. M. *J. Nat. Prod.* **2005,** *68,* 36.

70. Keylor, M. H.; Matsuura, B. S.; Stephenson, C. R. J. *Chem. Rev.* **2015,** *115,* 8976 for a thorough study of the chemistry and biology of resveratrol-derived natural products.

71. Marks, V. D.; Ho Sui, S. J.; Erasmus, D.; van der Merwe, G. K.; Brumm, J.; Wasserman, W. W.; Bryan, J.; van Vuuren, H. J. J. *FEMS Yeast Res.* **2008,** *8,* 35.

19

Finishing the Wine

THE END OF fermentation, signaled by density measurements, the alcohol-driven death of the *Saccharomyces cerevisiae* strain that was used, the cessation of evolution of carbon dioxide, and the generally accepted passage of the several weeks over which time the fermentation has been permitted to extend, is followed by the previously discussed (Chapter 16) process of racking.

The racking, as noted earlier, will separate most of the precipitated solids that are present or have developed during the fermentation process (e.g., accumulated seed and twig pieces not previously removed, insoluble carboxylic acid salts, dead yeast cells, and other solids [the lees]) from the fermented juice. But the wine may not yet be clear. Indeed, the wine may need racking once or twice more for clarification before a final filtration to produce the appropriate bright and clear beverage-quality wine. The last, or even a penultimate racking, might be done into an oaken vessel and should be done into oak if a red wine is being finished (European or American oaks are commonly used, but with different results, *vide infra*). However, it is important that regardless of the color of the wine each racking operation be done as carefully as possible to exclude transfer of solids and oxygen. At this stage of finishing, the oxygen will probably not be utilized in biochemical processes, barring the presence of microbial life, and normal oxidation of phenols and alcohols in the wine will have been inhibited by the presence of carbon dioxide (which replaced the oxygen in the solution during fermentation). Thus, if oxygen is introduced, it is likely that unwanted oxidation products might form.

The final racking for white wines (excluding Champagne, other "sparkling" wines, and some specialty beverages to be considered later) is generally carried out so that the beverage can rest for a few months (often with cooling to inhibit deleterious processes occurring as a result of aging) before filtering and bottling. If, on standing, the wine has not become clear enough (e.g., a haze remains), a fining agent such as bentonite, a silicate clay with additional interspersed cations (commonly monovalent potassium, K^+, and/or sodium, Na^+, divalent calcium, Ca^{2+}, and/or magnesium, Mg^{2+}, and trivalent aluminum, Al^{3+}, and/or iron, Fe^{3+}) or kaolin, a simple aluminum silicate clay, or even egg white can be added. Clay fining agents effect clarification by providing a large surface area to which the constituents of the haze can be attracted and cling; with egg white, a web of proteinaceous material to which the particles can cling and aggregate is formed.

With regard to the fining steps, it is important to recognize: (a) that additional chemistry continues to occur during the period of standing; (b) fining agents, used to carry away unwanted materials that have a deleterious effect on the beverage may also bring contaminants to the beverage that will change its taste; and (c) oxygen, heat, and light need to be avoided as they will change the wine. And, of course, the tank (glass, plastic, stainless steel, oak, etc.) in which the wine is allowed to stand before bottling will play a role and may change the wine too.

As to standing and cooling the white wine during the aging process, it will be remembered that, as the temperature increases, the rates of most reactions also increase (Appendix 1). With white wines, it is generally found that the crisp and fruity tastes deteriorate on warming, and so the wine is stored and served cold. The deterioration is likely due to ester hydrolysis (which occurs slowly under acid catalysis) and oxidation of aldehydes and sulfur-containing flavorants as well as polymerization of cyanidins beyond the dimers. As noted in the description of the grapes earlier in this work, some of the unique characteristics of those flavorants, commonly found in the white wines, are more fragile than those of the reds. However, some of the cyanidins which might provide a more harsh taste before they dimerize or even trimerize do require time to undergo those further reactions. As a consequence, a short period (a few weeks) of standing to allow those reactions and to allow more complete settling of lees is considered valuable.

Fining by the addition of the bentonite clay (Figure 19.1)[1], the composition of which was defined above, can also introduce unwanted materials. Thus, the crude clay, even though it may be purified before use, has a structure (silicate sheets separated by cations) that allows for introduction of impurities accompanying the cations. A typical silicate sheet, composed of overlapping tetrahedra that have the silicon (Si) nuclei at the center of tetrahedra and the oxygen (O) nuclei at the apices, is shown in Figure 19.2.

Although frequently hydrated with water hydrogen bonded to the oxygen nuclei of the silicate network, alcohols as well as thiols and phenols can also hydrogen bond and can be removed from the wine treated with this clay. Remains of yeast-derived carbohydrates and chitins (from cell walls) remaining after most of the lees have been removed can also cling to the surface and/or be retained by the filter pad for additional purification of the wine. However, the cations and other impurities that might also lie between the clay sheets can be transported into the wine. Careful choice of the source of the clay is important.

The whites of chicken eggs contain globular (sphere or globe-like) proteins called albumin and globulin. Albumin proteins are largely water-soluble, relatively small (80 or so amino acids), generally negatively charged phosphate-bearing polymers. Globulin proteins are larger water-insoluble proteins. Both albumin and globulin proteins have large numbers of active sulfur, oxygen, and nitrogen sites for hydrogen bonding and so readily form many hydrogen bonds to phenolic and positively charged oxygen (e.g., in cyanidin) sites. Those bonds, while individually weak, are in aggregate strong enough to encourage agglomerates to form and settle, thus clarifying the wine. Because there are relatively few tannins in white wines that need removal in this way, egg whites are used less often for white than for red

FIGURE 19.1 A photograph of a sample of bentonite clay courtesy of Mr. James Ladd, Department of Geology, Temple University.

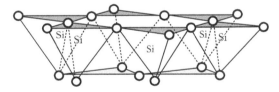

FIGURE 19.2 A representation of the silicate sheet structure of bentonite clay. Si refers to the element silicon, while the open circles represent oxygen atoms.

wines. In both cases, a change in the color of the wine occurs because the materials that have been removed are commonly highly conjugated and may have phenolic protons.

The *uncontrolled* presence of oxygen, heat, and light are generally to be avoided regardless of the nature of the final beverage to be produced. Although it can be argued that, for example, a deliberately heated, oxidized, "practically indestructible" wine such as Madeira (Chapter 21) is an exception, the value of the words "uncontrolled" and "deliberately" should not be taken lightly. Indeed, although reactions in wine generally occur slowly relative to those that are used in the laboratory, the entire production of wine requires, as with chemical processes in general, "deliberate control."

Nonetheless, it is worthwhile reiterating that the uncontrolled presence of oxygen is deleterious because of, among other things, the potential oxidation of phenols[2] to produce insoluble tannin-like products. The process serves, at a minimum, to remove many of the more flavorful components. Ethanol (CH_3CH_2OH) itself can be oxidized to acetaldehyde (CH_3CHO) and then to acetic acid (CH_3CO_2H), and unless wine vinegar is desired, the outcome may be found to be unpalatable.

Similarly, uncontrolled application of heat will serve to degrade the beverage, either through accelerating reactions that are unwanted or by encouraging evaporation of desirable high-vapor-pressure odoriferous and flavorful components.

Finally, in this vein, the highly conjugated double bond systems of aromatic phenols, terpenoid, and nonterpenoid polyenes and their oxygenated derivatives absorb light in the visible region of the spectrum (giving rise to color), as well as in the ultraviolet (UV) and infrared (IR) regions of the spectrum. While it might be argued that, in the short term, visible (and infrared) light may not be particularly deleterious to the wine, it is certainly true that the higher energy UV radiation (e.g., in direct sunlight) will encourage reactions that may be unwanted. The glass bottles will not absorb much of the UV radiation, but nonetheless, a green bottle is green because it is transmitting green visible light and absorbing (at least) red and blue visible light, which is thus not getting to the contents. It is argued that removing light at the red (closer to the infrared) and blue (closer to the ultraviolet) wavelengths is better than otherwise and that some antioxidants which may be inhibiting unwanted changes (such as vitamin C) are destroyed by sunlight. While there may be substance to the argument that even visible light just might be detrimental to some flavorants (and wine should be stored in cool dark places), it is not known what those flavorants are. Additionally, there is usually no (below the limits of detection) vitamin C in wine[3] (and vitamin C in water absorbs in the UV [λmax ≈ 259 nm]).[4] However, although aqueous solutions of both *cis*- and *trans*-resveratrol have maxima in the UV at 206 and 304 nm (the latter absorbing more strongly than the former) the isomerization from *trans* to *cis* is readily effected by sunlight.[5] Thus, again, sunlight does need to be excluded, and dark rooms (cellars) are good storage places. Unwanted reactions are less likely to occur in the dark and at low temperature (Figures 19.3 and 19.4).

It has also been argued that green glass hides sediment well, and wines that age may have accumulated more sediment. Young wines can be stored in colorless bottles since they are meant to be drunk before the deleterious effects of light on their normally diminished unsaturated components will be noted.

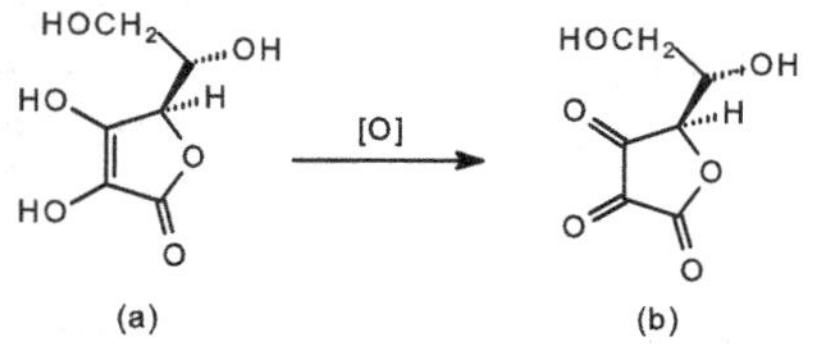

FIGURE 19.3 A representation of the oxidation of ascorbic acid (vitamin C) (a) to the corresponding dehydroascorbic acid (b) by a nonspecific oxidizing agent.

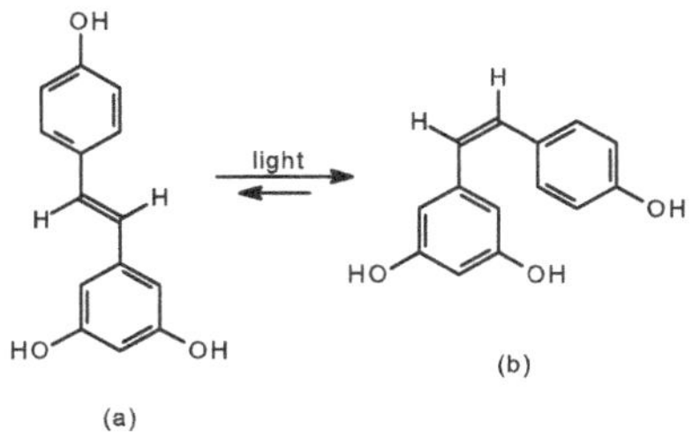

FIGURE 19.4 A representation of the reversible ultraviolet (UV) light (hν) promoted interconversion between *trans*-resveratrol (a) and its corresponding *cis*-resveratrol (b) isomer.

It is not uncommon for white wines (and even some rosé wines) to be aged in stainless steel and/or glass-lined steel vessels before bottling. This is based on understanding that the surfaces of stainless steel and glass are easily maintained and will neither contribute to, nor detract from, the flavor of the beverage. As long as the surfaces of those vessels are maintained, the assumption is a good one. Plastic-lined metal or glass and heavy plastic vessels are less widely used because it is held that some of the plasticizer used in the formulation of the plastic might remain in the solid and subsequently be extracted from the plastic into the wine. Of course this would depend upon the type of plastic utilized and the skill with which the vessel was formed, but in principle, there is no reason that Teflon® or polyethylene-coated vessels could not be used.

For some white wines, and for all but the least expensive red wines (the fining of which might nonetheless have involved one or more oak barrel rackings or racking with oak chips present), aging in oak produces enhanced flavors and is considered a requirement. And not just any oak is to be used. That is, one of the "traditional" three hard oaks is thought to be best. The North American white oak (*Quercus alba*), the European (English and French) oak (*Q. robur*) also called the pedunculate (i.e., short-stalk leaf) oak, and the sessile (i.e., no stalk leaf) oak (*Q. petraea*) are preferred. Those oak varieties and others are all largely composed (40–50%) of cellulose [thousands of glucose units attached to each other *via* (β1→4) linkages] (Figure 19.5), (25–30%) hemicellulose, and the remainder (lignins, tannins, etc.) that serve to distinguish them.

To some extent there are also small differences in the hemicellulose—which is similar to cellulose but has smaller linear chains of glucose monomers as well as branched chains of glucose and other hexoses (e.g., galactose, mannose). Generally hemicelluloses are more easily hydrolyzed than cellulose itself under the acidic conditions provided by the wine.

The major differences in the oaks that are used appears to lie in the (20–25%) lignins, the (2–10%) tannins, and some (<5%) smaller molecules. The lignins are largely polyphenolic [C_6] and phenylpropanoic [C_6–C_3] or cinnamyl [C_6–C_3] networked structures which occasionally include carbohydrates. Some related lignans which are dimers and trimers of the C_6–C_3 alkenols are also present. The tannins consist of variable amounts of condensed proanthocyanidin tannins, similar to tannins present in the wine, and hydrolyzable tannins, less common in wine, which are carbohydrate esters of gallic acid and include compounds such as tannic acid itself and castalagin (*vide infra*). The smaller molecules include a variety

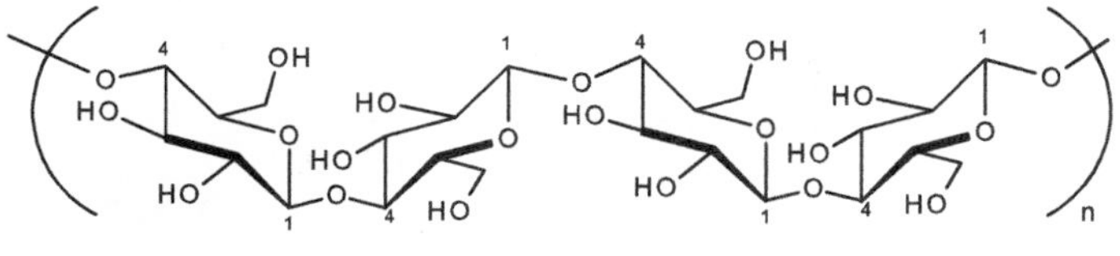

FIGURE 19.5 A representation of cellulose, shown here as a repeating fragment of a structural unit made of glucose molecular representations each linked β (1→4) to its adjacent glucose representation (n ≈ 10^5).

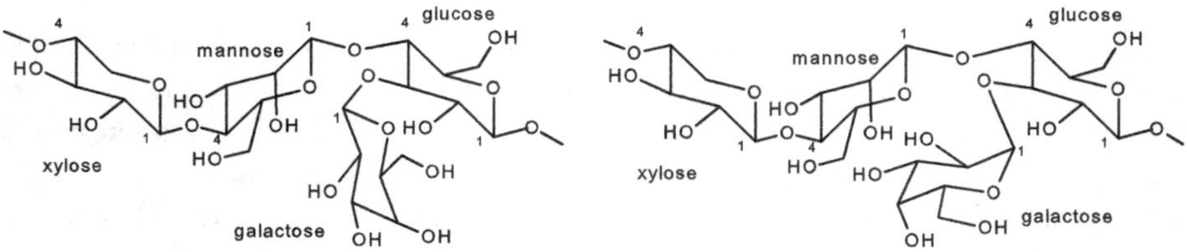

FIGURE 19.6 On the left, a representation of a repeating fragment of structural carbohydrate units from –xylose-β(1→4)-mannose-β(1→4)-glucose-β(1→3)-galactose– and on the right, a similar representation with the same repeating fragments but with an α(1→3) linkage between galactose and glucose at the right-hand terminus.

of polyphenols, simple flavanol and amino acid–derived compounds that might have resulted from the Maillard reaction (*vide infra*) between those nitrogenous materials and carbohydrates and that can be extracted on soaking and liberated on heating (toasting, roasting, or charring).

Other oaks and indeed other woods that might be used appear to produce different flavors from the same fermented juices and, interestingly, there is some evidence that individual preferences, expressed by those choices and others may become important market contributors.[6,7]

Cellulose is difficult to hydrolyze, and significant work is underway to effect its hydrolysis, since plant waste could, in principle, serve as a source of glucose which might be fermented for ethanol or other products suitable for fuel.

The hundreds of individual carbohydrates in the oak hemicellulose (see Figure 19.6) are more easily available than glucose in cellulose. Apparently, enzymes in the wine or treatment of the oak prior to addition of the wine (such as toasting) aid in releasing pieces of the carbohydrates into the wine. Xylans are also found in the hemicellulose and are of some interest, because some of them are made up of mixed carbohydrate structures to which a lignin or lignan type C_6–C_3 alcohol or carboxylic acid is attached. The xylan (Figure 19.7) shown has a coniferic acid [*E*-ferulic acid, (2*E*)-3-(4-hydroxy-3-methoxyphenyl)acrylic acid] bonded to the carbohydrate backbone.

While it is generally true that glycoside hydrolases and glycosyltransferases can be found in the genome of many organisms, the enzymes associated with these functions are quite specific. That specificity often generates different overall results as the enzymes in different wines interact with the same samples of oak. However, in the same vein, it is not surprising to

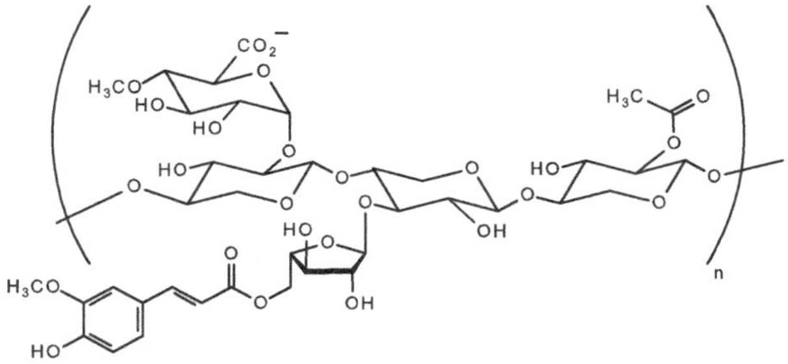

FIGURE 19.7 A representation of a xylan made up of mixed carbohydrate and coniferic acid units. The value of "n" may be large.

find that some of the compounds produced from cellulose and hemicellulose in a given oak cask are the same with different wines, but their relative concentrations are different.

When samples of European oak (*Quercus robur*) were ground to "sawdust-like particles" and extracted with a 1:1 ethanol:water solution, products could be isolated which, on silylation (for analysis by gas chromatography/mass spectrometry [GC-MS]) proved to be rich in "hydroxyl derivatives of cyclohexane (including isomers of inositol, 5-deoxy-*myo*-inositol, and tetrahydroxycyclohexane)." Further, saccharides and various acids including "malic, vanillic, syringic . . ." were found.[8] All of the compounds identified and reported are shown in Figure 19.8.

The complex set of compounds found in the amino acid–derived lignin portion of the body of the oak are similar to the C_6–C_3 fragments of xylans. Most are simply highly cross-linked polymers of the C_6–C_3 tyrosine (Tyr, Y) and phenylalanine (Phe, F) derived alcohols (called monolignols, usually consisting of *p*-coumaryl alcohol, coniferyl alcohol, and sinapyl alcohol) in a network (Figures 19.9 and 19.10). Carbohydrates may also be present (Figure 19.11). This is the case with all of the oaks used, although French and European oaks are said to possess more lignin than American oak.

Currently, efforts continue to find ways to depolymerize lignins common to hardwoods (as well as lignins in general) into "well-defined" aromatic chemicals. The conditions used are very different from the less acidic environment found in wine, and under the more vigorous treatment, some of the aromatic derivatives produced are the same regardless of the source. The pathway to the C_6–C_3 aromatics in lignin and their smaller and less complicated lignan relatives is generically the same (Figure 19.12).[9,10]

The alcohols (lignanols) incorporated, for the most part, into both lignin (Figure 19.13) and lignan are derived, as noted earlier, from phenylalanine (Phe, F) and tyrosine (Tyr, Y) which, after conversion to cinnamic and coumaric acids, respectively, are further oxidized and methylated (on the aromatic hydroxyl groups) and then reduced, *via* aldehydes, to the corresponding alcohols. Dimerization and oligomerization of lignanols, presumably by free radical processes, yields the amorphous polyphenolic low-molecular-weight lignans. Polymerization and extensive crosslinking of lignanols with or without carbohydrates present yields lignins. It is likely that the enzymes that control the final lignin formation are similar throughout various oaks and other plants, but the details of their activity as well as the specific lignanols are a function of the specific plant.

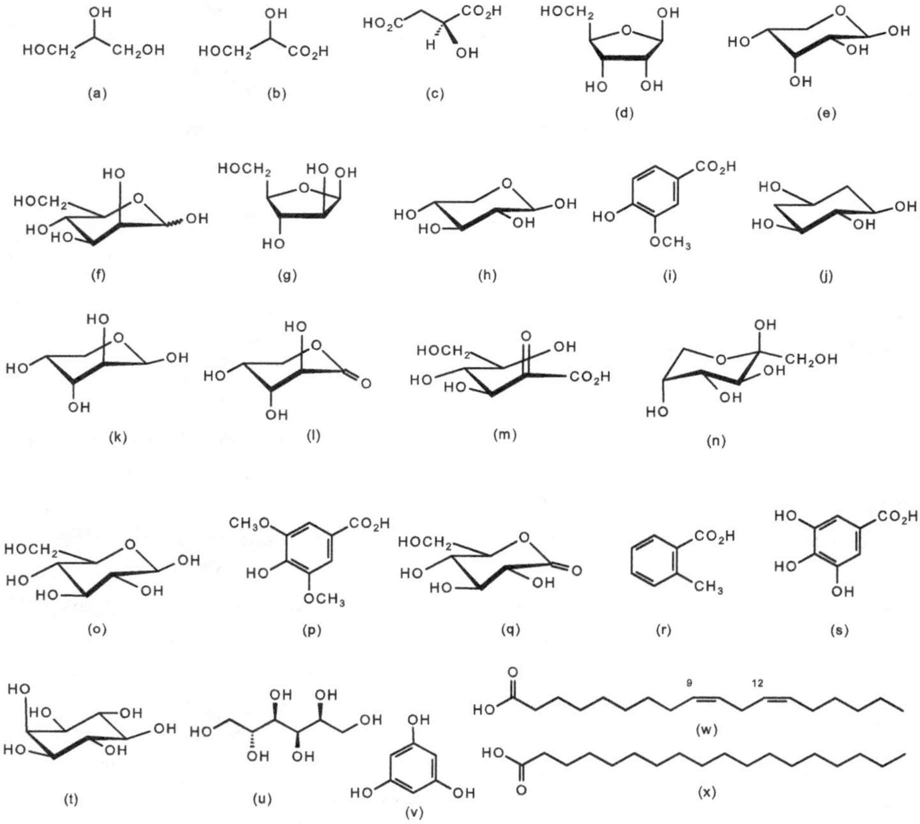

FIGURE 19.8 Representations of compounds isolated from samples of European oak (*Quercus robur*) on ethanol-water extraction. The isolates were identified as: (a) glycerol; (b) glyceric acid; (c) L-(−)-malic acid; (d) β-D-(−)-ribofuranose; (e) β-D-(−)-ribopyranose; (f) D-(+)-mannopyranose; (g) β-D-(−)-arabinofuranose; (h) β-D-(+)-xylopyranose; (i) vanillic acid; (j) 1,2,3,5-tetrahydroxycyclohexane; (k) β-D-(−)-arabinopyranose; (l) arabonic acid lactone; (m) fructosonic acid; (n) D-fructopyranose; (o) D-glucopyranose; (p) syringic acid; (q) D-gluconic acid lactone; (r) 2-methylbenzoic acid; (s) gallic acid; (t) inositol; (u) sorbitol; (v) phloroglucinol; (w) linoleic acid; and (x) octadecanoic acid.

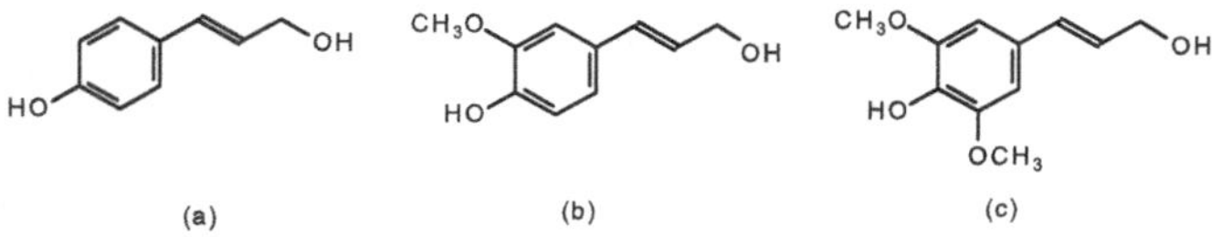

FIGURE 19.9 Representations of the C_6–C_3 tyrosine (Tyr, Y) and phenylalanine (Phe, F) derived alcohols: (a) *para*-coumaryl alcohol; (b) coniferyl alcohol; and (c) sinapyl alcohol found in xylans.

A significant amount of work has been undertaken regarding the lignin oligomers (an oligomer has only a few monomeric or single units in contrast to a polymer, which is composed of many single monomeric units), since it appears that such compounds occur in wine aged in toasted oak casks and are extracted into the wine from the oak. As recently pointed

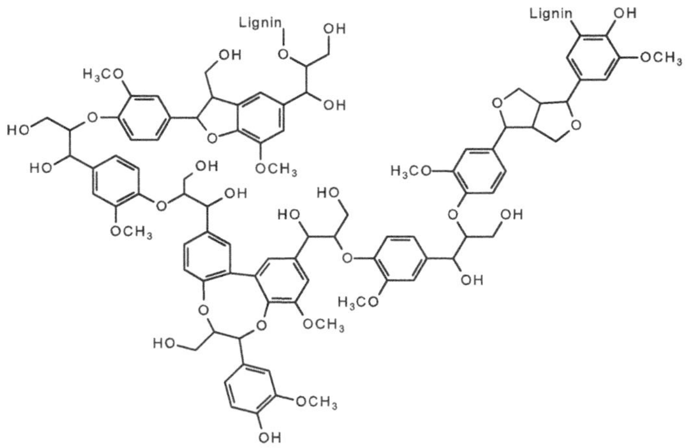

FIGURE 19.10 A representation of a lignin fragment with aryl–aryl bonding but lacking carbohydrate bonding.

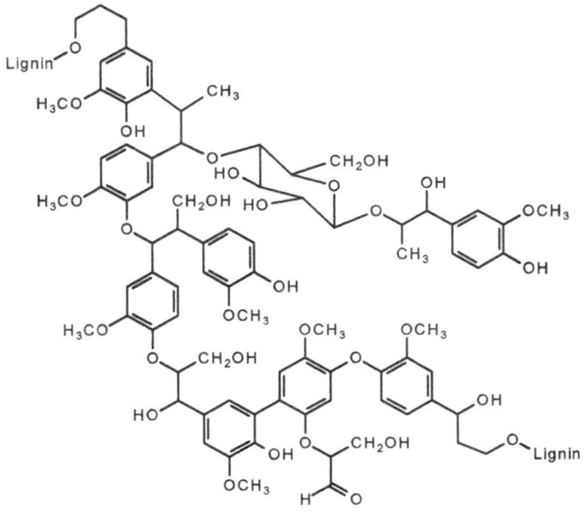

FIGURE 19.11 A representation of a lignin fragment with aryl–aryl as well as carbohydrate bonding.

out,[11–14] monolignols that give rise to lignans and neolignans (the difference being in the linkage of one monolignol unit to the next)[15,16] are often found in wine beverages.[17] Some examples are shown in Figure 19.14.

Procyanidin tannins (proanthocyanidins) are reported to be found in the heartwood extracts of oak (*Quercus petraea* and *Q. robur*) and apparently undergo oxidative depolymerization to procyanidins and prodelphinidins.[18] Subsequent oxidation to cyanidins and delphinidins is likely (Figure 19.15).

The hydrolyzable tannins, some of which are shown in Figure 19.16, on the other hand are generally gallic acid esters of carbohydrates. A pentagalloyl glucose, on complete hydrolysis, would yield gallic acid and glucose, as well as a glucose attached to gallic acid monomers. The monomers, which themselves, have bonded to each other (i.e., in castalagin and vescalagin, two of a class known as ellagitannins), are also shown.[19]

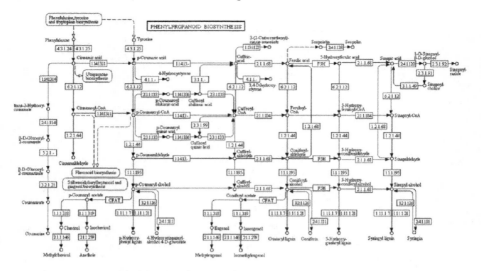

FIGURE 19.12 A portion of the Phenylpropanoid biosynthesis–Reference pathway from map00940m leading to lignin and lignan and to be found at http://www.genome.jp/kegg-bin/show_pathway?map00940. Used with permission of the Kyoto Encyclopedia of Genes and Genomes (http://www.kegg.jp). The numbers in rectangles along the arrows refer to Enzyme Commission (EC) numbers (see http://www.enzyme-database.org/) and link to that database for more information.

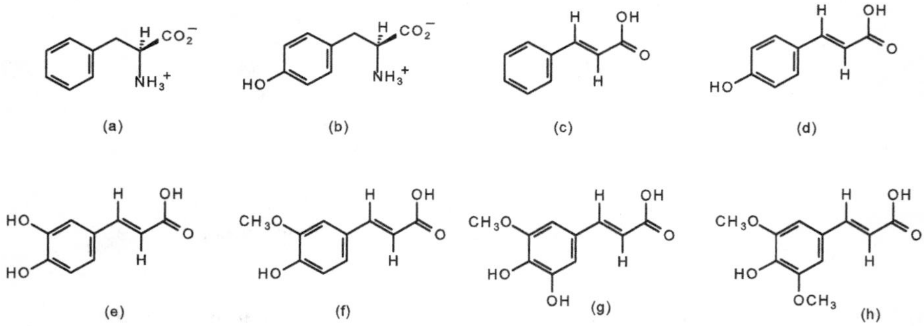

FIGURE 19.13 A representation of the amino acids (a) phenylalanine (Phe, F) and (b) tyrosine (Tyr, Y) along with the acids derived from them which, on reduction, lead to the alcohols found in lignin. The acids are: (c) cinnamic acid; (d) coumaric acid; (e) caffeic acid; (f) ferulic acid; (g) 5-hydroxyferulic acid; and (h) sinapinic acid.

Because hydrolyzable tannins such as the ellagitannins have a large number of hydroxyl groups (−OH), they are soluble in aqueous alcohol and are extracted from oak (barrel staves or chips). And since they are esters they can undergo acid-catalyzed hydrolysis at the pH (~3.3) of most wines. This leads to the presence of gallic acid and some less complex derivatives of partial hydrolysis as well as to the alcohols (i.e., sugars, from which they are derived). A family (called Roburins) of dimeric ellagitannins has also been foud.[20–22] Roburin A is reported to have the structure shown in Figure 19.17 (i.e., a dimer composed of two vescalagins), and a significant amount of effort was expended in their isolation and structure

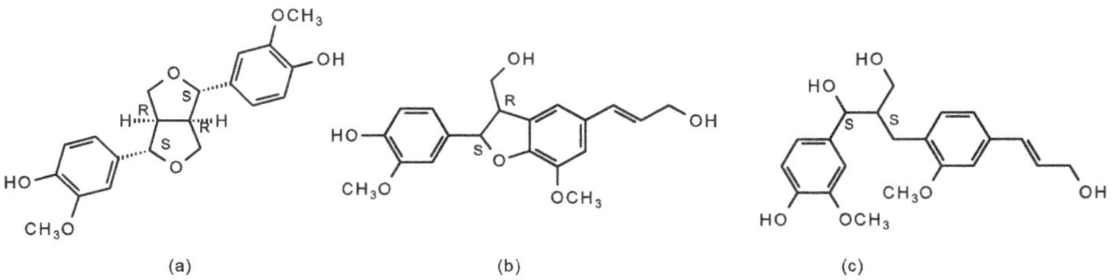

FIGURE 19.14 Representations of lignan components found in some wines that were aged in oak. The compounds shown are: (a) (+)-pinoresinol, (a lignane); (b) (+)-dehydrodiconiferyl alcohol (a neolignane); and (c) (1S,2S)-rel-1-(4-hydroxy-3-methoxyphenyl)-2-[[4-[(1E)-3-hydroxy-1-propen-1-yl]- 2-methoxyphenyl]methyl]-1,3-propanediol (an oxyneolignane).

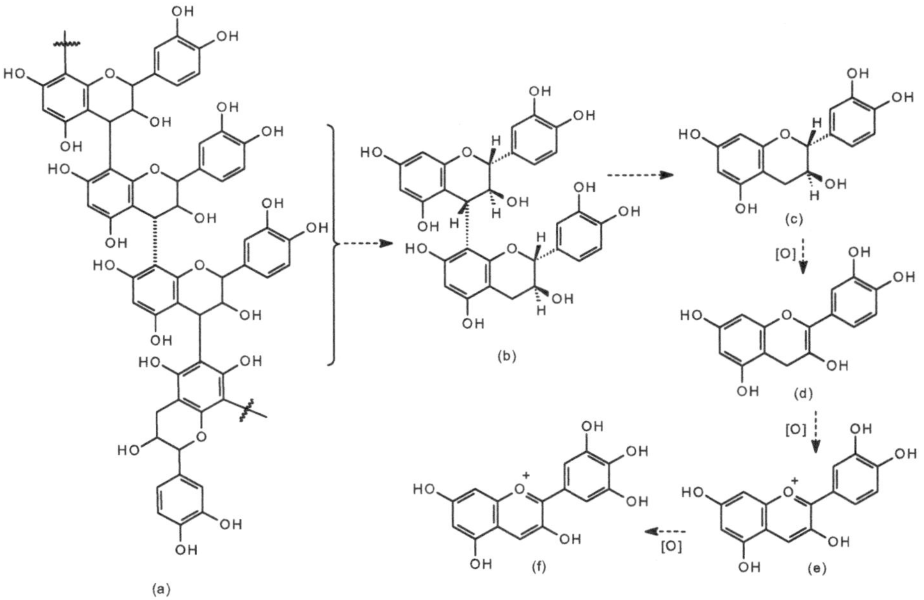

FIGURE 19.15 A representation of a polymeric procyanidin tannin (a proanthocyanidin) (a) undergoing partial depolymerization to the proanthocyanidin dimer B_3 (b). Further depolymerization leads to (c) (+)-catechin which, on oxidation leads to (d) quercetin. Continuing oxidation produced (e) cyanidin and (f) delphinidin.

proof.[23] These esters and similar compounds such as gallocatechol (gallocatechin) itself constitute a major portion of oak extracts that move into the wine. The hydrolysis to produce gallic acid (Equation 19.1) and the partially hydrolyzed carbohydrate ester backbone that remains does not occur at all of the esterified positions simultaneously, but rather sequentially. Under the acidic conditions and in the presence of enzymatic hydrolases and glycosyl transferases and as function of the stereochemistry and the conformation of the carbohydrate,[23] the different esters will undergo hydrolysis differently. In this way a vast variety of partially hydrolyzed carbohydrate derivatives, each potentially contributing to the flavor of the product and changing with time, results.

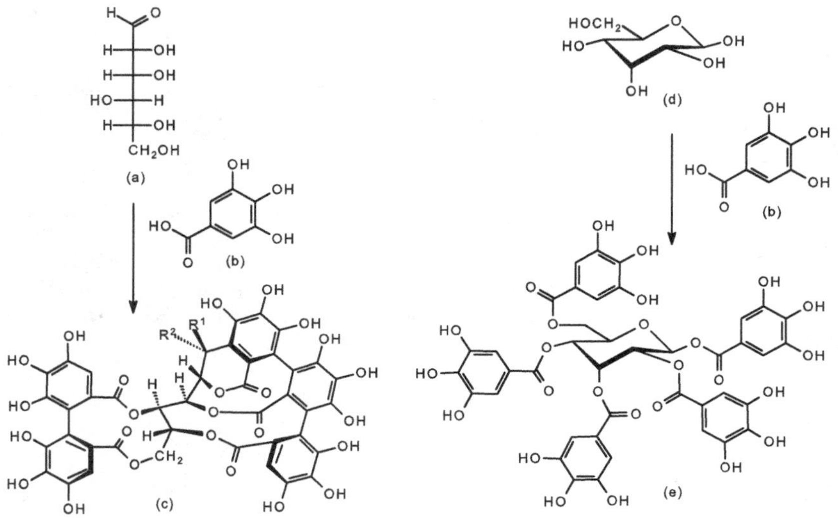

FIGURE 19.16 Representations of gallic acid–associated tannins. On the left, a representation of the Fischer projection of D-(+)-glucose (a) and gallic acid (b) and their polygallic acid ester product with fused gallic acids. The compounds shown are the isomers (c) vescalagin, R^1 = OH, R^2 = H) and castalagin (R^1 = H, R^2 = OH). On the right is a representation of β-D-glucopyranose (d), gallic acid (b) and a simpler pentagalloyl glucose hydrolyzable tannin product (e).

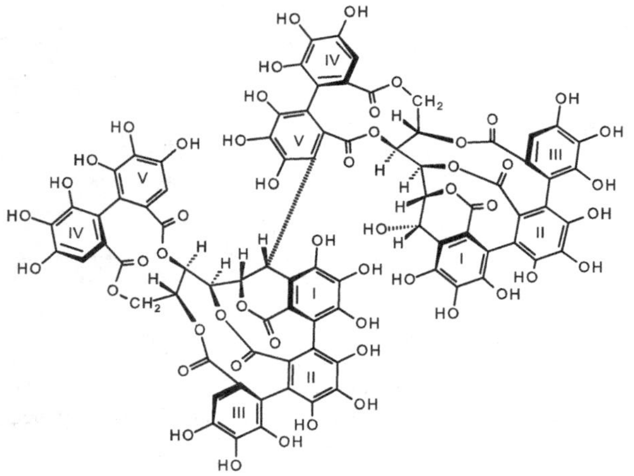

FIGURE 19.17 A representation of the presumed structure of Roburin A (see text).

Further, the large variety of hydroxyl-bearing substituents on phenols, carbohydrates and other hydroxyl (–OH) bearing molecules that are in the wine resulting from partial hydrolysis and reduction processes serves to help remove unwanted aldehydic and ketonic materials as well as to add to the rich variety of compounds that define the wine. Thus, acetaldehyde (CH_3CHO) generated by oxidation of ethanol (CH_3CH_2OH) resulting from adventitious oxygen leaking through barrel walls or entrained during fining can undergo reaction to form acetals or other side products generated from reaction between anthocyanins (that

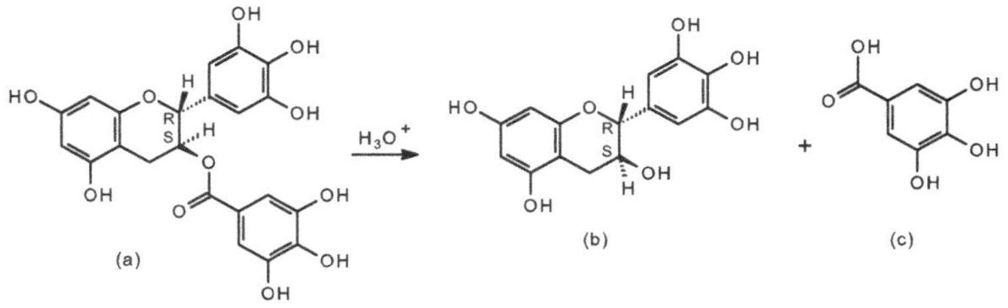

EQUATION 19.1 A representation of the catalyzed hydrolysis (aqueous acid or enzyme esterase) of (+)-gallocatechin gallate (a) to gallocatechin (gallocatechol) (b) and gallic acid (c).

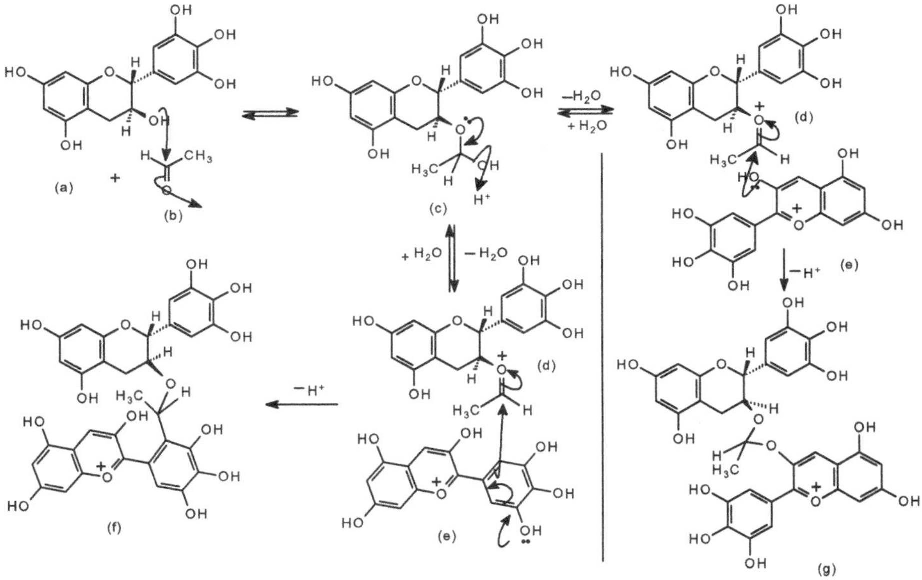

SCHEME 19.1 A representation of the reaction between gallocatechin (gallocatechol) (a) and acetaldehyde (ethanal) (b), which leads first to a hemiacetal (c). Loss of water from (c) leads to an oxonium cation (d). On the left, following one path, the cation (d) if attacked by a carbon (an aryl carbon from the anthocyanidin delphinidin (e) is shown) results in formation of an ether (f). On the right, the cation (d) if attacked by an oxygen (a phenolic oxygen from the anthocyanidin delphinidin (e) is shown) results in formation of an acetal (g).

have hydrolyzed to carbohydrate and anthocyanidin) and flavanols. As shown in Scheme 19.1, the anthocyanidin delphinidin and the antioxidant flavanol gallocatechol can form an acetal as well as an ether on reaction with acetaldehyde.

The formation of such species (acetals and ethers) not only serves to remove the aldehyde from the solution (as the more complex large molecules may precipitate), but it also sequesters phenols into tannin-like large compounds which, again, may be separable from the bulk of the wine.

While there is less information about extraction of compounds from oak casks by aqueous ethanol solutions before toasting the oak, it is nonetheless interesting to note that in the few examples that have been reported there seems to be general agreement that (a) not only do the specific structures and concentrations of the ellagitannins vary from oak variety to oak variety, but even the specific trees and their age makes a difference.[24] As might be guessed, based on those observations, it has been found that in some cases, examination of the compounds in the wine from a specific cask, can be utilized to determine where the tree, used in the cooperage, was grown.[25]

Of course it might be imagined that wood chips, on the one hand, and staves used by the cooper to make the cask, on the other would provide the same compounds. So it is interesting to note that although French oak casks and oak chips used in stainless steel vats give similar results, more of the ellagitannins (but the same ones) are extracted into wine stored in the casks compared to the chips over the same time period.[22]

In the same vein, it was concluded on examination of eight volatiles, [i.e., 4-allyl-2-methoxyphenol (eugenol), *cis*-5-butyl-4-methyldihydro-2(3*H*)-furanone (*cis*-oak lactone), *trans*-5-butyl-4-methyldihydro-2(3*H*)-furanone (*trans*-oak lactone), 4-ethylphenol, 4-ethyl-2-methoxyphenol (4-ethylguaiacol), 5-methyl-2-furaldehyde (5-methylfurfural), 2-furaldehyde (furfural), and 2-methoxyphenol (guaiacol) (Figure 19.18)] found in five hundred and ten (510) wines aged for (at least) 6, 12, and 18 months in oak barrels, that only relatively small differences in concentrations of the volatiles were present. But, while some of the differences that were found appeared to be a function of both geographic origin of the oak and type of oak, the length of time the wine spent in the barrel was most important.[25,26]

In addition to these flavoring components, and based upon "antioxidant capacity of different woods used in cooperage," it was concluded that "the major contributors to the antioxidant capacity were... phenolic acids (Figure 19.19), including 3,4,5-trihydroxybenzoic acid (gallic acid), 3,4-dihydroxybenzoic acid (protocatechuic acid), (2*E*)-3-(2-hydroxyphenyl)acrylic acid (*para*-coumaric acid), and 2,3,7,8-tetrahydroxychromeno[5,4,3-cde]chromene-5,10-dione (ellagic acid) and all the ellagitannins..."[26] (The structures for some ellagitannins are shown in Figures 19.16 and 19.17.)

Many more compounds common to all oaks examined (albeit in different amounts) have been isolated from "toasted" or charred wood by extraction with aqueous ethanol and other solvents, such as Freon 11 (trichlorofluoromethane).[27] So using additional extraction

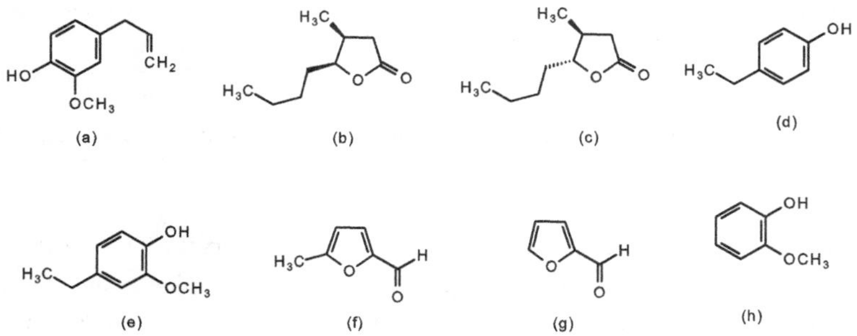

FIGURE 19.18 Representations of volatiles in wines aged in oak barrels. Large numbers of wines aged in oak have detectable amounts of all or most of these volatiles, albeit at varying concentrations. The compounds represented are: (a) eugenol; (b) *cis*-oak lactone; (c) *trans*-oak lactone; (d) 4-ethylphenol; (e) 4-ethylguaiacol; (f) 5-methylfurfural; (g) furfural; and (h) guaiacol.

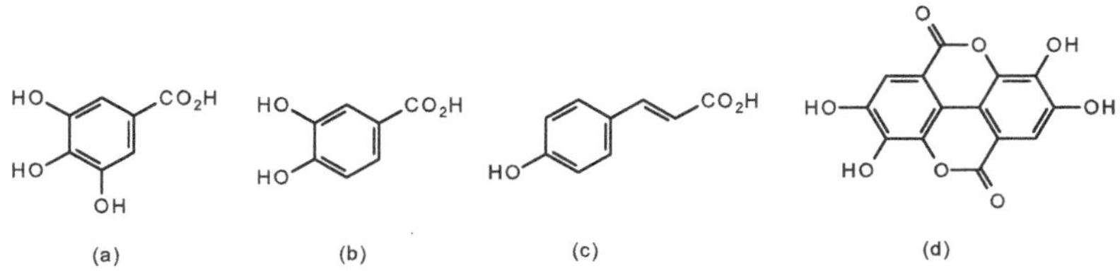

FIGURE 19.19 Representations of the phenolic acids which might be oxidized in place of other flavoring constituents and are thus responsible for inhibition of oxidation in wine. Those shown are: (a) gallic acid; (b) protocatechuic acid; (c) *para*-coumaric acid; and (d) ellagic acid.

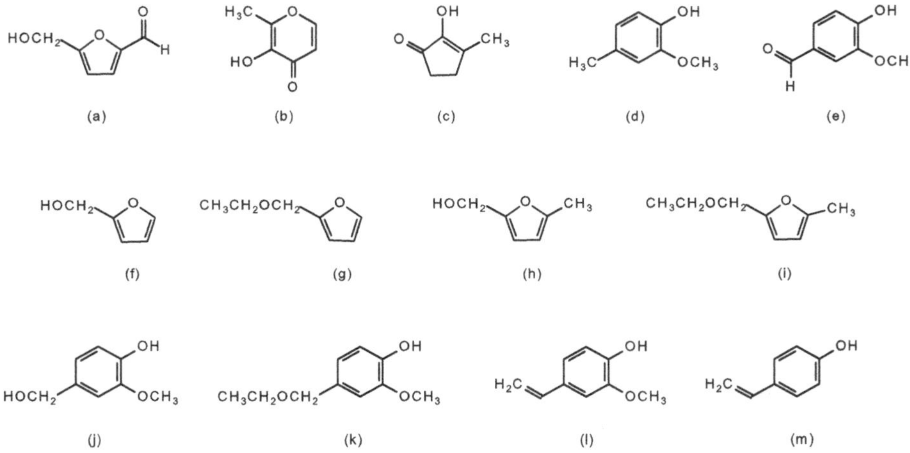

FIGURE 19.20 Representations of the additional compounds isolated from oak by different extraction methods. Shown are: (a) hydroxymethylfurfural; (b) maltol; (c) cyclotene; (d) 4-methylguaiacol; (e) vanillin; (f) furfuryl alcohol; (g) furfuryl ethyl ether; (h) 5-methylfurfuryl alcohol; (i) 5-methylfurfuryl ethyl ether; (j) vanillyl alcohol; (k) ethyl vanillyl ether; (l) 4-vinylguaiacol; and (m) 4-vinylphenol.

techniques, it appears that heating the hemicellulose produces a wealth of products in addition to the eight volatile compounds already noted (Figure 19.19) from water-ethanol extraction. Specifically, as shown in Figure 19.20, both phenolic and heterocyclic (pyranoid and furanoid) compounds are also present.

Interestingly, however, as analytical techniques have again improved, it has been possible to isolate small quantities of bitter principles generated from lignans and found in wine stored in oak casks. Thus, as shown in Figure 19.21, Quercoresinosides A (1) and B (2) along with the known glucopranoside (3) are found at very low concentration in aqueous-ethanol extract of toasted *Quercus petraea* wood.[28] These bitter compounds were found in a "taste-guided" approach.[28]

The authors concluded that "... as a likely result of coopering and microbiological variability, few consistent origin or seasoning effects of the compounds resulting from coopering or microbiological action during oak maturation were observed."[27] Thus, the flavor components were common throughout, but the amount present could not be correlated with the source of the oak wood.

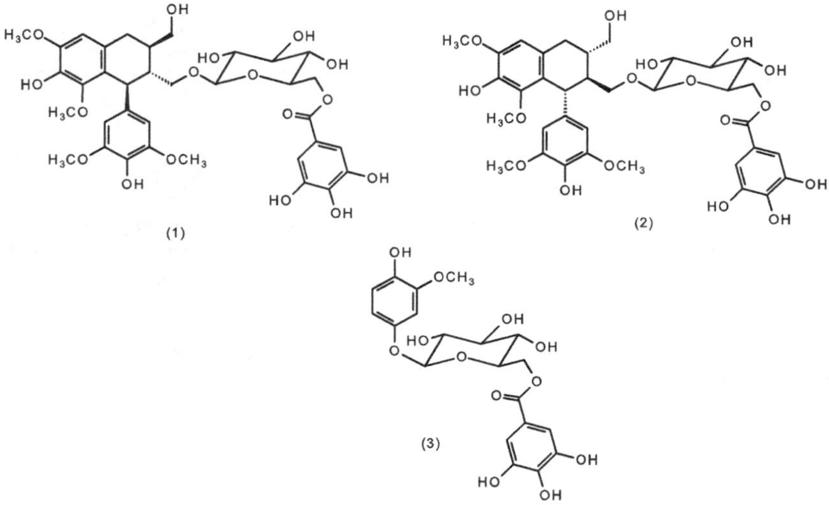

FIGURE 19.21 Representations of the Quercoresinosides A and B [shown as **1**, (+)-(8*R*,8'*R*,7'*S*)-lyoniresinol-9'-O-(6"-galloyl)-β-D-glucopyranose and **2**, (−)-(8*S*,8'*S*,7'*R*)-lyoniresinol-9'-O-(6"-galloyl)-β-D-glucopyranose] along with **3**, 3-methoxy-4-hydroxyphenol 1-O-β-D-(6'-O-galloyl)-glucopyranoside as reported isolated from toasted *Quercus petraea* heartwood. See Sindt, L.; Gammacurta, M.; Waffo-Teguo, P.; Dubourdieu, D.; Marchal, A. *J. Nat. Prod.* **2016**. DOI: 10.1021/acs.jnatprod.6b00142.

Finally, in this vein, it is worth noting that many of the smaller fragments (e.g., furan derivatives, esters) described as found in wine that has been stored in toasted casks are common, general, thermal degradation products of wood. Indeed, thermal decomposition of wood at 450 °C has yielded (GC-MS) one hundred and four (104) tabulated products, many of which are found in wine.[29] Interestingly, more gentle treatment with careful monitoring of toasting of different oaks produced a total of thirteen (13) oak-related volatiles. Of them, a recent report monitored differences in 2-methoxyphenol (guaiacol), 2-methoxy-4-methylphenol (4-methyl guaiacol), furan-2-carboxyaldehyde (furfural), 5-methyl-2-furancarboxaldehyde (5-methyl furfural), 4-allyl-2-methoxyphenol (eugenol), 2-methoxy-4-[(1*E*)-propen-1-yl]phenol (isoeugenol), and 4-hydroxy-3-methoxybenzaldehdye (vanillin) which had consistently different values. The others, isomers of 5-butyl-4-methyodihydrofuran-2(3*H*)-one (*trans*- and *cis*-oak lactone), 4-ethyl-2-methoxyphenol (4-ethylguaiacol), 2-methyl-4-vinylphenol (4-vinylguaiacol), 4-vinylphenol, 2-methoxy-4-vinylphenol (syringol), "were either not monitored in all of the barrel extracts or were found inconsistently at low levels or not at all in these barrels."[30]

Representations of the compounds mentioned above are provided in Figure 19.22.

As part of aging in toasted casks, and in addition to all of the changes that have occurred in the grape juice and what has been extracted from the oak, it is held that changes that might have occurred to some of the components in the wine result from the reaction of amino acids with carbohydrates. The reaction between amino acids and carbohydrates, initially between the amino acid glycine (Gly, G) and glucose, but subsequently expanded to include a few other amino acids and other carbohydrates, was first reported by Maillard in 1912.[31] He noted that decarboxylation occurred (CO_2 evolved), the solution yellowed and, with increasing speed, turned dark brown.[31] He postulated that this reaction might be

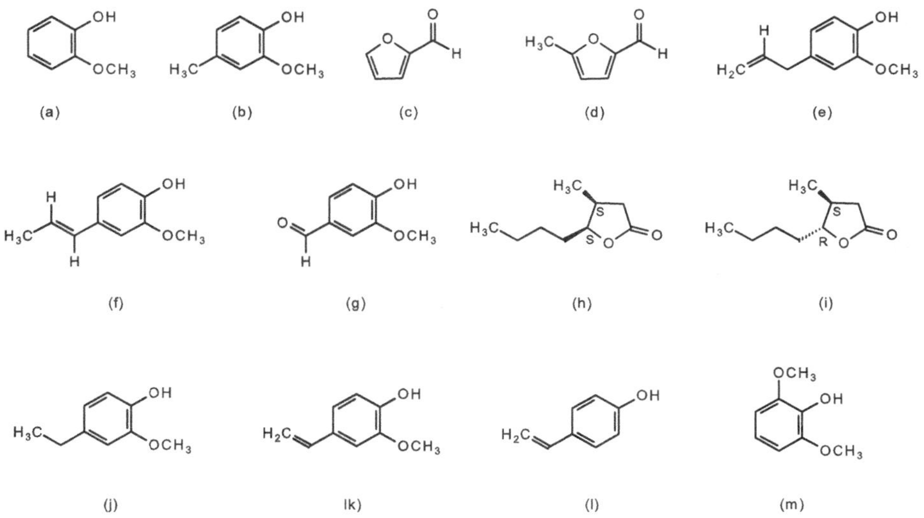

FIGURE 19.22 Representations of oak volatiles monitored as a result of toasting oak. As noted in the text, some of the volatiles were consistently present, albeit at varying concentrations, whilst others were less consistent. Those that were consistent were: (a) guaiacol; (b) creosol; (c) furfural; (d) 5-methylfurfural; (e) eugenol; (f) isoeugenol; and (g) vanillin. Oak volatiles, either not monitored or inconsistently present were reported to be: (h) *cis*-oak lactone; (i) *trans*-oak lactone; (j) 4-ethylguaiacol; (k) 4-vinylguaiacol; (l) 4-vinylphenol; and (m) syringol.

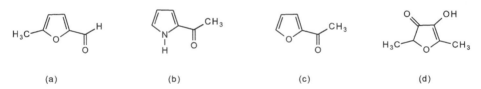

FIGURE 19.23 Representations of some of the products of the Maillard reaction between carbon-labeled alanine (Ala, A) and carbon-labeled glucose also found in wines. The products are: (a) 5-methylfurfural; (b) 2-acetyl-1*H*-pyrrole; (c) 2-acetylfuran; and (d) 4-hydroxy-2,5-dimethyl-3(2*H*)-furanone.

physiologically important. It was subsequently found that products derived from glucose (and other carbohydrates bearing carbonyl groups), on reaction with amino acids would lead to products similar (and in some cases identical) to those found in wine exposed to toasted casks. However, it is also generally clear that, while the Maillard reaction might be occurring, the products could be formed in another way or in other ways.[32–34] Indeed, in a series of studies, Yaylayan and coworkers[35] studied the reaction between isotopically labeled [^{13}C]-glucoses and [^{13}C]-alanines to yield a complex family of products. They suggested pathways to sugar degradation products and specifically accounted for 5-methylfurfural, 2-acetyl-1*H*-pyrrole, 2-acetylfuran, and 4-hydroxy-2,5-dimethyl-3(2*H*)-furanone (Figure 19.23). Not all of the products from potential Maillard reactions are found in wines that are aged in toasted casks. Indeed, even when some of them are present they may be formed in quantities below the current level of detection.

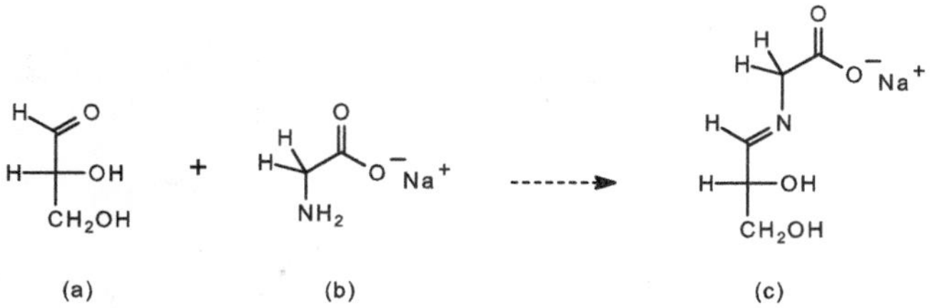

EQUATION 19.2 A representation of the attempted reaction between glyceraldehyde (a) and glycine (Gly, G) sodium salt (b) which might be expected to produce the imine (c). The reaction is reported to give a mixture of "amorphous brown products (see text).

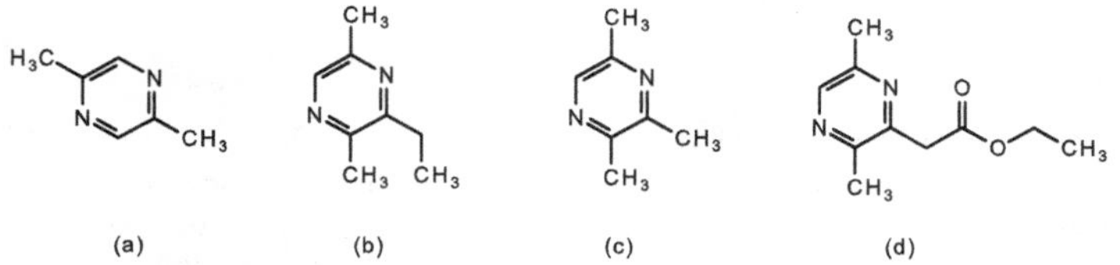

FIGURE 19.24 Representations of pyrazines found in oak-treated wine and thought to arise by reaction of amino acids or their derivatives with suitable small aldehydic precursors. The pyrazines shown are: (a) 2,5-dimethylpyrazine; (b) 2,5-dimethyl-3-ethylpyrazine; (c) 2,3,5-trimethylpyrazine; and (d) ethyl 2,5-dimethyl-3-pyrazinylacetate.

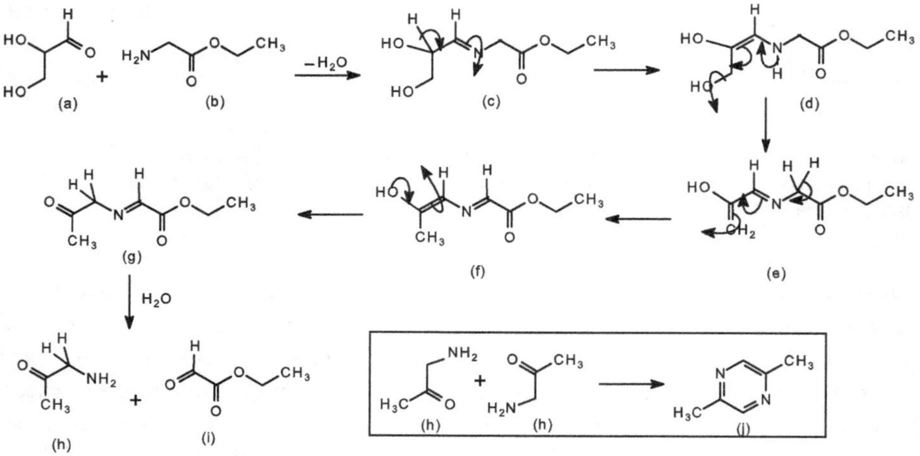

SCHEME 19.2 A representation of a reaction between glyceraldehyde (a) and ethyl glycine (b) where an imine (c) is the initial product which rearranges to a hydroxyimine (d). The hydroxyimine, on loss of water to an alkene (e) is suggested to tautomerize to produce an enol (f) which, on ketonization produces an imine (g) potentially derived from aminoacetone (h) and the ethyl ester of glyoxylic acid (i). Dimerization of aminoacetone (h) produces 2,5-dimethylpyrazine (j).

Early attempts to define pathways and products that consistently result from the Maillard reaction have not proven successful. Simplifying the reactants to glyceraldehyde and the sodium salt of glycine (Gly, G) (Equation 19.2) as a model has also been attempted. The results led to the conclusion that "... the formation of oligomeric compounds ... cannot be excluded ..." as only "... amorphous brown products ..." resulted.[36] The same pattern of many products, presumably arising in a variety of ways along enzymatic and nonenzymatic pathways, continues to prevail.[37]

In addition to the furan derivatives obtained (Figure 19.23) pyrazine derivatives have also been found (Figure 19.24). Scheme 19.2 provides a path to 2,5-dimethylpyrazine from the reaction of the ethyl ester of glycine with glyceraldehyde. In addition, proposals for paths to the formation of 2,5-dimethylpyrazine, 2,5-dimethyl-3-ethylpyrazine, 2,3,5-trimethylpyrazine, and ethyl 2,5-dimethyl-3-pyrazinylacetate (ionic and radical) have been provided.[38-40]

Lastly, the beverages derived from the aging in toasted casks, often with a final filtration or fining, are bottled and sealed, either ready for consumption (generally in colorless or green clear glass bottles) or for in-bottle-aging (generally in green bottles). The latter will allow some of the more reactive phenols to continue to undergo rearrangements and some tannins to polymerize and settle; a final mellowing before consumption!

NOTES AND REFERENCES

1. For more on bentonite, please see Prayongphan, S.; Ichikawa, Y.; Kawamura, K.; Suzuki, S.; Chae, B.-G. *Comput. Mech.* **2006**, *37*, 369. DOI 10.1007/s00466-005-0676-3.

2. Musso, H. *Angew. Chem. Internat. Edit.* **1963**, *2*, 723.

3. Naidu, K. A. *Nutr. J.* **2003**, *2*, 7.

4. Dabbagh, H. A.; Azami, F.; Farrokhpour, H.; Chermahini, A. N. *J. Chil. Chem. Soc.* **2014**, *59*, 2588.

5. Camont, L; Cottart, C.-H.; Rhayem, Y.; Nevet-Antoine, V.; Djelidi, R.; Collin, F.; Beaudeux, J.-L.; Bonnefont-Rousselot, D. *Anal. Chim. Acta* **2009**, *634*, 121.

6. Schoenfeld, B. "The Wrath of Grapes." *New York Times*, May 28, 2015. http://www.nytimes.com/2015/05/31/magazine/the-wrath-of-grapes.html?_r=2 (accessed Apr 8, 2017).

7. Heyman, S., "A Popular App Charts Changing Tastes in Wine." *New York Times*, July 2, 2015. http://nyti.ms/1LHE7ii (accessed Apr 8, 2017).

8. Pisarnitskii, A. F.; Rubeniya, T. Y.; Rutitskii, A. O. *Appl. Biochem. Microbiol.* **2006**, *42*, 514.

9. Deuss, P. J.; Scott, M.; Tran, F.; Westwood, N. J.; de Vries, J. G.; Barta, K. *J. Am. Chem. Soc.* **2015**, *137*, 7456. DOI: 10.1021/jacs.5b03693.

10. Jouanin, L; Lapieere, C., Eds.; Lignins Biosynthesis, Biodegradation and Bioengineering. In *Advances in Botanical Research, 61;* Elsevier: New York, 2012.

11. Dima, O.; Morreel, K.; Vanholm, B.; Kim, H.; Ralph, J.; Boerjan, W. *The Plant Cell* **2015**, *27*, 695.

12. Davin, L. B.; Lewis, N. G. *Plant Physiol.* **2000**, *123*, 453.

13. Boerjan, W.; Ralph, J.; Baucher, M. *Ann. Rev. Plant Biol.* **2003**, *53*, 519.

14. Davin, L. B.; Lewis, N. G. *Curr. Opinion Biotech.* **2005**, *16*, 398.

15. Pilkington, L. I.; Barker, D. *Nat. Prod. Rep.* **2015,** *32,* 1369. DOI:10.1039/c5np00048c; and Tepanno, R. B.; Kusari, S.; Spiteller, M. *Nat. Prod. Rep.* **2016,** *33,* 1044. DOI:10.1039/c6np00021e.

16. The nomenclature of lignans, neolignans, and related compounds has been carefully defined. https://www.degruyter.com/view/IUPAC/iupac.72.0041. by the International Union of Pure and Applied Chemistry (IUPAC) and appears in Moss, G. P *Pure Appl. Chem.* **2000,** *72,* 1493 ff. DOI:10.1351/pac200072081493 and see also http://www.chem.qmul.ac.uk/iupac/.

17. Nurmi, T.; Heinonen, S.; Mazur, W.; Deyama, T.; Nishibe, S.; Adlercreutz, H. *Food Chem.* **2003,** *83,* 303 and Marchal, A.; Cretin, B. N.; Sindt, L.; Waffo-Téguo, P.; Dubourdieu, D. *Tetrahedron* **2015,** *71,* 3148.

18. Vivas, N.; Nonier, M.-F.; Pianet, I.; de Gaulejac, N. V.; Fouquet, E. *Compt. Rend. Chemie* **2006,** *9,* 1221.

19. The stereochemical assignments of these ellagitannins have been examined and revised. See Matsui, Y.; Wakamatsu, H.; Omar, M.; Tanaka, T. *Org. Lett.* **2015,** *17,* 46. DOI:10.1021/ol530212v.

20. Hervé de Penhoat, C. L. M.; Michon, V. M. F.; Peng, S.; Viriot, C.; Scalbert, A.; Gage, D. *J. Chem. Soc. Perkin Trans.* **1991,** 1653.

21. Garcia-Estévez, I.; Escribano-Bailón, M. T.; Rivas-Gonzalo, J. C.; Alcalde-Eon, C. *J. Agric. Food Chem.* **2012,** *60,* 1373.

22. Jourdes, M.; Michel, J.; Saucier, C.; Quideau, S.; Tiessedre, P.-L. *Anal. Bioanal. Chem.* **2011,** *401,* 1531.

23. Ardèvol, A.; Rovira, C. *J. Am. Chem. Soc.* **2015,** *137,* 7528. DOI: 101021/jacs.5b01156.

24. Mosedale, J. R.; Charrier, B.; Crouch, N.; Janin, G.; Savill, P. S. *Ann. Sci. For.* **1966,** *53,* 1005.

25. Gougeon, R. D.; Lucio, M.; Frommberger, M.; Peyron, D.; Chassagne, D.; Alexandre, H.; Feuillat, F.; Voilley, A.; Cayot, P.; Gebefügi, I.; Hertkorn, N.; Schmitt-Kopplin, P. *PNAS* **2009,** *106,* 9174. DOI: 10.1073/pnas.0901100106.

26. Garde-Cerdán, T.; Lorenzo, C.; Carot, J. M.; Esteve, M. D.; Climent, M. D.; Salinas, M. R. *Food Chem.* **2010,** *122,* 1076.

27. Alañón, M. E.; Castro-Vázquez, L.; Díaz-Maroto, M. C.; Hermosín-Gutiérrez, I.; Gordon, M. H.; Pérez-Coello, M. S. *Food Chem.* **2011,** *129,* 1584.

28. Spillman, P. J.; Sefton, M. A.; Gawel, R. *Aust. J. Grape and Wine Res.* **2004,** *10,* 216.

29. Sindt, L.; Gammacurta, M.; Waffo-Teguo, P.; Dubourdieu, D.; Marchal, A. *J. Nat. Prod.* **2016.** DOI: 10.1021/acs.jnatprod.6b00142.

30. Faix, O.; Fortmann, I.; Bremer, J.; Meier D. *Holz als Roh- und Werkstoff* **1991,** *49,* 213.

31. Collins, T. S.; Miles, J. L.; Boulton, R. B.; Ebeler, E. S. *Tetrahedron* **2015,** *71,* 2971.

32. Maillard, L.-C. *Compt. Rend. Hebdomadaires des Seances de l'Acad. Sci.* **1912,** *154,* 66.

33. Cutzach, K.; Chatonnet, P.; Henry, R.; Dubourdieu, D. *J. Agric. Food Chem.* **1999,** *47,* 1663.

34. Pripis-Nicolau, L.; de Revel, G.; Bertrand, A.; Meujean, A. *J. Agric. Food Chem.* **2000,** *48,* 3761.

35. Oliveira, C. M.; Ferreira, A. C. S.; De Freitas, V.; Silva, A. M. S. *Food Res. Int.* **2011,** *44,* 1115.

36. Yaylayan, V. A.; Keyhani, A. *J. Agric. Food Chem.* **2000,** *48,* 2415 and earlier papers.

37. Angrick, M. *Z. Naturforsch.* **1983,** *38b,* 530.

38. Nursten, H. E. *The Maillard Reaction:Chemistry, Biochemistry and Implications;* The Royal Society of Chemistry, 2005. ISBN 978-0-85404-964-6. DOI 10.1039/9781847552570.

39. Yuan, C. W.; Marty, C.; Richard, H. *Sci. des Aliments* **1989,** *9,* 125.

40. Totlani, V. M.; Peterson, D. G. *J. Agric. Food Chem.* **2005,** *53,* 4130.

41. Guerra, P. V.; Yaylayan, V. A. *J. Agric. Food Chem.* **2010,** *58,* 12523. DOI:10.1021/jf103194k.

20

Sealing the Bottles

IT HAS BEEN suggested that containers made of clay (e.g., the amphora of the Bronze Age) were adopted for use during the thousands of years of winemaking that preceded the ability to produce suitable glass vessels. Sealing the amphora, as reported by archeologists and historians, was accomplished with clay or leaves covered with clay, rags, wax, pine resin (producing retsina),[1] and even today's popular choice, cork. With the exception of the latter, where only a small amount of air can leak in, it appears that too much air would enter and the flavor of the wine would change.

In part, the effort to seal the amphora was futile, as the clay amphora would leak too. But waxes and resins helped seal out air and, in the process, often changed the flavor of the beverage.[2,3] Again, historically, it appears from analysis of the contents remaining in the old vessels that various flavoring agents, such as berries, fruits, leaves, flowers, and even metals such as lead were intentionally added to wines to suit the tastes of the consumer. Nonetheless, oxidation and bacteria (e.g., *Acetobacter aceti*, known to convert ethanol [CH_3CH_2OH] to acetaldehyde [CH_3CHO] and thence to acetic acid [CH_3CO_2H]) would often make the beverage unpalatable (by today's tastes). So, tastes were adjusted to fit the beverage available!

It was also found that wines that had additional ethanol present were resistant to bacterial action, so tastes (even into the twentieth century) were developed for "fortified" wines (*vide infra*, Chapter 21) such as Port, Sherry, and Madeira that were to be shipped in casks. More recently shipment of the latter in glass bottles (since late in the nineteenth century) along with cork stoppers have become common. Most recently, synthetic (i.e., polymer) stoppers and aluminum screw caps have been used for all of these beverages because most wine is produced to be consumed within a few years of its bottling. This fairly recent change has arisen as an accommodation to large-scale production, long-distance shipping, and storage in commercial sales facilities, none of which encourage saving wine for aging.

So, currently, the beverage having been prepared can be found with natural cork, synthetic "cork," or aluminum formed screw cap (with plastic liner between cap and bottle, usually a polyvinyl chloride [PVC] or other polymer film so that a tight seal is made). Of these three related closure methods, it appears that the last is becoming common for young wines that are to be drunk and not ever considered held for aging. The potential for extraction of monomer from the cap lining polymer is considered negligible. The same may be true for those closed with synthetic "cork." Indeed, as with the aluminum screw cap, there is no evidence for intended longevity.

The synthetic "cork-like" material is usually a polyethylene or mixed polyethylene–polypropylene material which may have an additional polymer coating so as to make it somewhat less dense at the surface where it mates to the glass bottle. This aids removal and, if necessary, reinsertion. While it is possible that some air might leak into the bottle, a secure closure would preclude that from happening, and regardless, bottles with polymer closures are not generally expected to be held for long-term maturation. Further, although it is unlikely, some of the hydrocarbons in the polymer mixture could be leached into the wine.

As to the cork, this natural material is the outer bark of the oak *Quercus Suber*. The tree, common to Spain and Portugal, is carefully stripped of its bark (it is stated that the trees live for more than 100 years and can be stripped about every 10 years),[4] which is then spread out to air dry. After air-drying, the bark is treated by boiling in large vats to remove contaminants and after redrying, is cut into strips and punched into cylinders for bottle closures. Interestingly, it is claimed[4] (without presenting any details as to how to accomplish that feat) that gas chromatographic analysis can detect 2,4,6-trichloroanisole (*vide infra*), the presence of which results in wine referred to as "corked" and considered non-potable.[5] Further cleansing and processing is undertaken before the corks are ready for commercial use.

The composition of cork has been determined.[6] Although there was some variation, the overall mean chemical composition of the samples examined showed that cork consisted of 16.2% extractives (with dichloromethane, ethanol, and water used sequentially and in that order), suberin (42.8%), and lignin (22.0%), with the remainder, apparently, polysaccharides composed, in decreasing abundance, of glucose, xylose, arabinose, mannose, galactose and rhamnose (Figure 20.1).

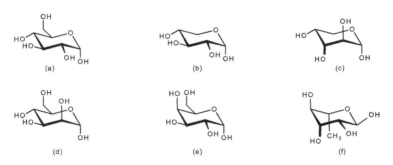

FIGURE 20.1 Representations of some of the carbohydrates (sugars) isolated as components of polysaccharides found in cork. As shown, they are: (a) α-D-glucopyranose; (b) α-D-xylopyranose; (c) α-D-arabinopyranose; (d) α-D-mannopyranose; (e) α-D-galactopyranose; and (f) α-L-rhamnopyranose.

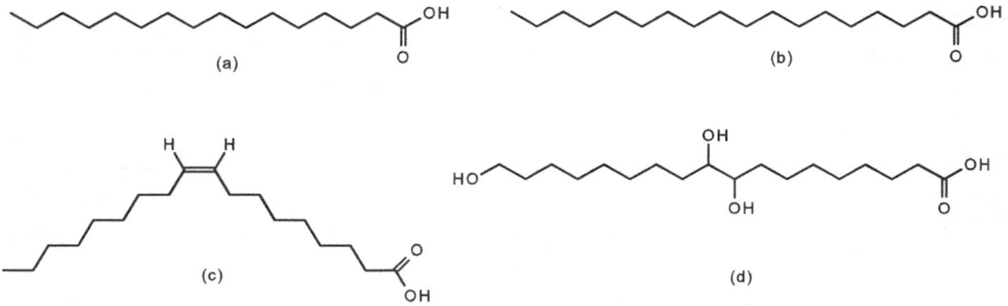

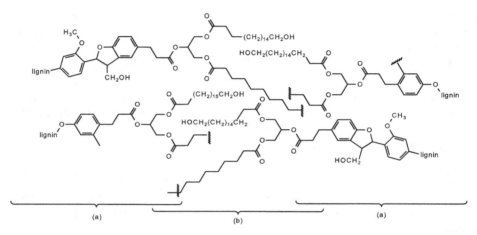

FIGURE 20.2 Representations of some of the fatty acids whose esters make up a major portion of the suberin lamellae. The acids shown are: (a) palmitic acid; (b) stearic acid; (c) oleic acid; and (d) 9,10,18-trihydroxystearic acid.

FIGURE 20.3 A representation of the structure of the suberin lamellae (b) lying between lignin domains (a).

The extractives consisted of long-chain lipids (fatty acids and alcohols, Figure 20.2) which might be derived from suberin (Figure 20.3), triterpenes (i.e., cerin, friedelin [Figure 20.4]) and phenolic compounds. Except for the triterpenes, the other compounds (i.e., the acids, alcohols and phenols) could be derived from lignin and suberin fragmentation.

Suberin lamellae, lying alongside the primary cell wall lignin is believed to consist largely of fatty acid already shown in Figure 20.2 and phenylpropanoic acid (as in lignin) esters of glycerol (1,2,3-propanetriol [$HOCH_2CH(OH)CH_2OH$]) similar to the structure of the waxy polymers (cutin) found in the cuticle of the leaves (Figure 20.3).[7,8]

The composition of lignin with regard to aryl amino acid composition has been discussed (Chapter 19) and, again, briefly can be abbreviated as shown in Figure 20.5.

It is reasonable to assume that at least some of the sugars shown in Figure 20.1 that were extracted by aqueous alcohol from the cork can be equally well extracted by the aqueous alcoholic solution called wine. Indeed, the fatty acids shown in Figure 20.2 might also be extracted. Further since the porous nature of the cork might well allow slow leakage of air into the bottled beverage, oxidation of the sugars as well as the phenolic lignin components

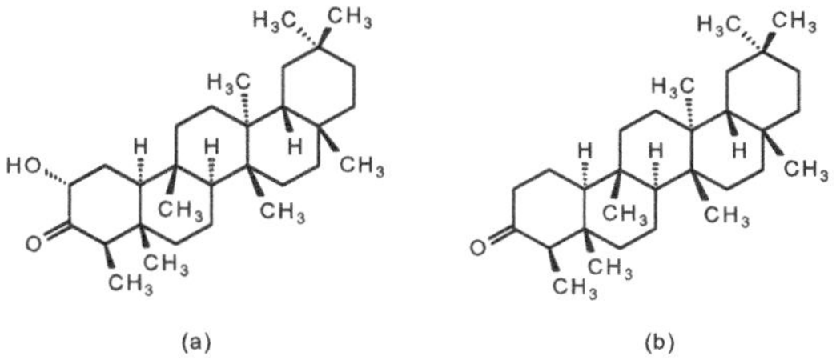

FIGURE 20.4 Representations of two triterpenes isolated on extraction of cork. As shown these are (a) cerin and (b) friedelin.

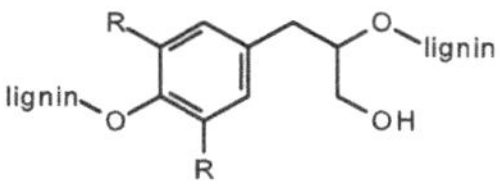

FIGURE 20.5 A broadly generic representation of a portion of a lignin. Throughout, R may be either hydroxy (–OH) or methoxy (CH_3O–).

might also occur on standing. This contributes to the flavor of the beverage. It does not appear that any clean way currently exists to test how much of the final product comes from the cork.

However, it does appear the oak *Quercus Suber* concentrates or sufficiently absorbs environmental chlorophenol or precursors from the environment so that (apparently) indigenous fungi convert it to 2,4,6-trichlorophenol which is methylated to the corresponding O-methyl ether (anisole) (Figure 20.6). This leads to wine reputed to have deleterious odors ("corked") that are undesirable.

Interestingly, recent long-term experiments that have come to fruition show that oak trees (*Quercus sessiliflora* was used among others) concentrate terrestrial chlorine.[9] Further, it had been known for some time that chlorophenols were common in the environment and that filamentous fungi could be isolated from cork that effected the methylation of 2,4,6,-trichlorophenol.[10] Additionally, it has been shown that 2,4,6-trichlorophenol and similar compounds can be produced from isomers of hexachlorocyclohexane. One of the isomers of hexachlorocyclohexane (with three adjacent chlorines axial and three equatorial) is commonly found in halogenated pesticides (including the commercially available Lindane®); along with other isomers isolated from the chlorination of cyclohexane and related materials.[11] Interestingly, chlorination of the poorly defined earthy material called humic acid and indigenous phenols often found there with fungal (e.g., *Caldariomyces fumago*) chloroperoxidase (EC 1.11.1.10)[12] also produces trichlorophenol.

However produced, chloroanisole and related compounds result in expensive waste for the wine industry. Indeed, it appears that despite efforts to detect contaminated corks the large-scale bottling and corking operations preclude industrial organizations undertaking that work and the waste is simply absorbed as part of the cost of doing business.

Of greater interest, however, is the very recent report[13] that trichloroanisole (TCA) has been detected in a wide variety of foods and beverages *surveyed for odor loss* (emphasis added) suggesting that it may always be present, but it is detected only when there is a problem with the beverage or food itself.

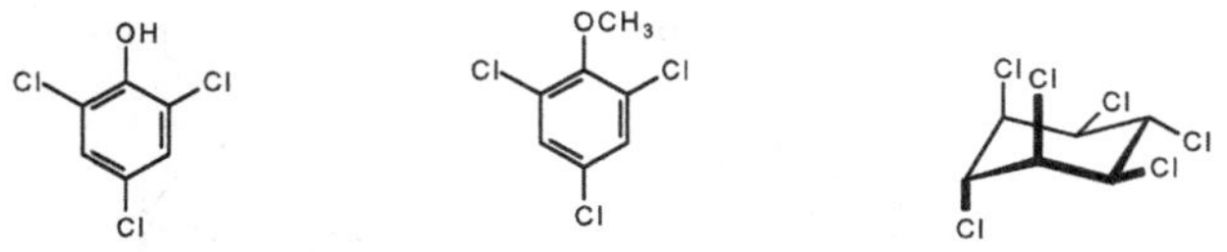

FIGURE 20.6 Representations of chlorinated materials common to the environment. The first 2,4,6-trichlorophenol (a) has been shown to be present in the environment and may be naturally occurring as both plants and animals can pick up chlorine from the soil. A fungus (see text) has been shown to convert (a) to 2,4,6-trichloroanisole (TCA) (b) by methylation. Such methylation of phenols is common. One of the hexachlorocyclohexane isomers (c) apparently continues to contaminate the environment.

NOTES AND REFERENCES

1. Robinson, J. *The Oxford Companion to Wine;* Oxford University Press: New York, 2006; p 568.

2. McGovern, P. E.; Zhang, J.; Tang, J.; Zhang, Z.; Hall, G. R.; Foreau, R. A.; Nuñez, A.; Butrym E. D.; Richards, M. P. Wang, C.; Cheng; G.; Zhao, Z.; Wang, C. *Proc. Nat. Acad. Sci. U.S.* **2004,** *101,* 17593.

3. McGovern, P. E. Glusker, D. L.; Exner, L. J. Voigt, M. M. *Nature* **1996,** *381,* 481.

4. Cork Institute of America. http://www.corkinstitute.com/home.html (accessed Apr 8, 2017) and Corticeira Amorim. http://www.amorimcork.com/en/natural-cork/raw-material-and-production-process/ (accessed Apr 8, 2017).

5. Although, it has been suggested that remediation of the cork taint is possible, the details of the treatment of the wine that is corked to consummate that desired end is apparently neither simple nor certain.

6. Pereira, H. *BioResources* **2013,** *8,* 2246.

7. Bernards, M. A. *Can J. Bot.* **2002,** *80,* 227).

8. Gandini, A.; Neto, C. P.; Silvestre, A. J. D. *Prog. Polym. Sci.* **2006,** *31,* 878.

9. Montelius, M.; Thiry, Y.; Marang, L.; Ranger, J.; Cornelis, J.-T.; Svensson, T.; Bastviken, D. *Environ. Sci. Technol.* **2015,** *49,* 4921.

10. Álvarez-Rodríguez, M. L.; López-Ocaña, L.; López-Coronado, J. M.; Rodríguez, J.; Martínez, M. J.; Larriba, G.; Coque, J.-J. R. *Appl. Environ. Microbiol.* **2002,** *68,* 5860.

11. Keiji, T.; Norio. K; Minoru, N. *Pesticide Biochem. Physiol.* **1979,** *10,* 96.

12. Czaplicka, M. *Sci. Total Environ.* **2004,** *322,* 21.

13. Takeuchi, H.; Kato, H.; Kurahashi, T. *Proc. Nat. Acad. Sci. U.S.* **2013,** *110,* 16235.

SECTION VI
Special Wines

21

Specialized Wines

PART A. PORT, SHERRY, AND MADEIRA

Specialized wines will have often have passed through similar sequences of grape maturation, harvest, and fermentation (which may or may not be carried to completion) typical of more normal wines but are, nonetheless, treated somewhat differently. Wines developed for shipment, such as Port, Madeira, and Sherry, discussed here, and wines developed from grapes infected with the fungal ascomycete known as *Botrytis cinerea* (*aka* the Noble Rot), wines produced from frozen grapes (Ice wine), and wines produced from grapes similar to those grown in the Champagne region of France and destined to become "sparkling wines," to be discussed subsequently, are all slightly different from the general types already described. And yet, because the compounds initially found in the grapes enjoy the same precursors and are doubtlessly very similar save for their terroir and their individual genomic and epigenomic differences, all but the final treatments they undergo are somewhat similar. Indeed, in that vein, Port and Madeira wines apparently originated in Portugal and Sherry in Spain, and the preferred grapes from which the preferred beverages are made continue to grow in and or near the initial locations. Other growing regions do try, with greater or lesser success, to create similar beverages.

Port, Sherry, and Madeira wines are all called "fortified" beverages. They are generally higher in alcohol content and other flavorings than those produced by typical fermentation processes and have one or two additional steps that are associated with their processing.

The first step involves producing a "distilled beverage." In this process a portion of the wine produced in the usual way by fermentation is subjected to the process of distillation in the presence of air. In that process the wine is heated above the temperature at which it vaporizes (different components vaporizing at different temperatures) and then the vapors produced are removed, condensed, and separated from the residue. Low-boiling materials such as water (H_2O, the major constituent of wine), methyl alcohol (methanol [CH_3OH] of which there are traces), acetaldehyde (ethanal [CH_3CHO] produced by oxidation of ethanol), and perhaps surprisingly, ethyl alcohol (ethanol [CH_3CH_2OH]), as well as other low-boiling components (e.g., some esters) are removed. The residue is rich in high-boiling flavorful components, some more desirable than others. Many of the more stable materials produced in the fermentation remain, and some new compounds are formed from the particular vintage as a consequence of reactions induced between components by heating.

This "pot residue" is now added back to some of the original wine from which it came, but since ethyl alcohol (a minor component compared to water in the initial beverage) was removed, a second step, the re-addition of ethyl alcohol (and if desired, other flavoring components) to the extent desired by the vintner and allowed by law, is now effected.

The wine is now fortified against further change, although oxygen and some reactive species might remain. Regardless, it is generally held that the most reactive components have been removed either through the distillation or by reaction between the retained components. While some chemistry might continue between the untreated wine to which the distillation residue was added, it is minor relative to what has already been done.

But what has been done?

With the widespread availability of analytical techniques such as high pressure liquid chromatography, mass spectrometry, and newer forms of spectroscopy such as nuclear magnetic resonance (NMR) and Fourier transform infrared (FT-IR) (see Appendix 1), it has become apparent that some comments about what happens can be made. Thus, various anthocyanins, whose specific structure (i.e., substitution pattern of hydroxyl groups and sugars) and concentration, were a function of grape, terroir, and subsequent handling, have changed. In addition to hydrolysis to anthocyanidins and as anticipated, these species continue to react to form products resulting from condensation reactions between anthocyanins and anthocyanidins and flavanols as well as amongst themselves and between those heterocycles and other wine components (e.g., acetaldehyde; propionaldehyde and other aldehydes; pyruvic acid and other acids such as *para*-coumaric acid, caffeic acid, and ferulic acid; and other phenols). Further, these materials oligomerize, resulting in color changes as well as formation of tannin-like polymeric residues.[1–3] Many of the normal flavoring constituents and low-boiling aroma constituents of the wine remain since only a portion was distilled.

(a) Port

For Port wine, the original (and genuine) beverage is a fortified red wine produced in the Douro Valley in the north of Portugal. Although there are a number of "Port grapes," it is argued that the "best" red grape is the "Touriga Nacional," which is similar to Cabernet Sauvignon.[4] For this wine in particular, and less common for Sherry, the addition of a distillation residue and additional ethanol is effected part way through the fermentation process of the "next" batch. This interrupts the fermentation by killing the yeast, but of course, not all of the glucose has been fermented. Thus Port wine is generally sweet.

Indeed, as early as 1980[5] it became clear that many components of various Port wines could be isolated and identified by the chromatographic and mass spectrometric tools available. Thus, it was reported[6] that 141 components were identified in Ruby and Tawny Ports, both in the headspace above the liquid and in the beverage itself. In addition to the usual alcohols, acids aldehydes, ketones, etc. expected on the basis of the grape and the action of *Saccharomyces cerevisiae* already discussed, notable isolates which the authors acknowledged do not account for the odor included those shown in Figure 21.1

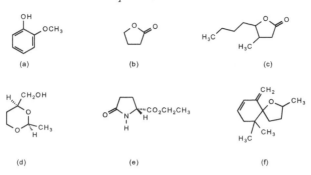

FIGURE 21.1 Notable isolates included (a) 2-methoxyphenol (guaiacol); (b) dihydro-2(3H)-furanone (γ-butyrolactone); (c) β-methyl-γ-octalactone (whisky lactone); (d) cis-4-hydroxymethyl-2-methyl-1,3-dioxolane; (e) (R)-ethyl 5-oxopyrrolidene-2-carboxylate (ethyl pyroglutamate); and (f) 2,10,10-trimethyl-6-methylene-1-oxaspiro[4.5]dec-7-ene (vitispirane).

More recent work[7] has produced evidence of reasonable concentrations of sulfur-containing compounds. A "young ruby Port" was found to contain small amounts of dimethyl sulfide, ethyl (methylthio)acetate, 2-mercaptoethanol, 2-(methylthio)ethanol, 2-methyltetrahydrothiophen-3-one, ethyl 3-(methylthio)propionate, acetic acid 3-(methylthio)-propyl ester, 3-(methylthio)-1-propanol, cis-2-methyltetrahydrothiophen-3-ol, 3-(ethylthio)-1-propanol, trans-2-methylhydrothiophen-3-ol, 4-(methylthio)-1-butanol, 3-(methylthio)-1-hexanol, dimethyl sulfone, benzothiazole, 3-(methylthio)-1-propionic acid, N-3-(methylthiopropyl)acetamide, and bis-(2-hydroxydiethyl) disulfide (Figure 21.2).

In contrast, concentrations of most of the same compounds in an "old tawny Port" were below the limits of detection, with only 2-(methylthio)ethanol, 3-(methylthio)-1-propanol, dimethyl sulfone, and 3-(methylthio)-1-propanoic acid being found.

The same group[7] noted that the following set (Figure 21.3) of flavorful norterpenoid compounds were present in the Port wines: (2E)-1-(2,6,6-trimethyl-1,3-cyclohexadien-1-yl)-2-buten-1-one (β-damascenone), (3E)-4-(2,6,6-trimethyl-1-cyclohexen-1-yl)-3-buten-2-one (β-ionone), 2,2,6-trimethylcyclohexanone (TCH), 1,1,6-trimethyl-1,2-dihydronaphthalene (TDN), and 2,10,10-trimethyl-6-methylene-1-oxaspiro[4.5]dec-7-ene (vitispirane), and the tetra-terpenoid β-carotene, as well as its oxidized relatives lutein, neoxanthin, and violaxanthin. The terpenoids found in Port and from which the norterpenoids are presumably generated are shown in Figure 21.4.

Additionally, the anthocyanins and flavanols characteristic of red wine and, as shown in Scheme 21.1, procyanidins, generated by the polymerization of flavan 3-ols (such as catechin and/or epicatechin) might subsequently react on standing and/or heating, with anthocyanins (although the latter continued to bear their carbohydrate links). This would almost certainly result in color changes and formation of insoluble residues in the mature Port wine.

Given the observation that 4-vinylphenol is frequently present as a consequence of the action of yeast during fermentation, it is not surprising that some of the 4-vinylphenol could react with anthocyanins when the two are allowed to remain together and heat is provided to induce reaction (Scheme 21.2). Thus, despite the fact that the reaction is not an electrophilic addition, given the circumstances the adduct shown is not unreasonable.[8]

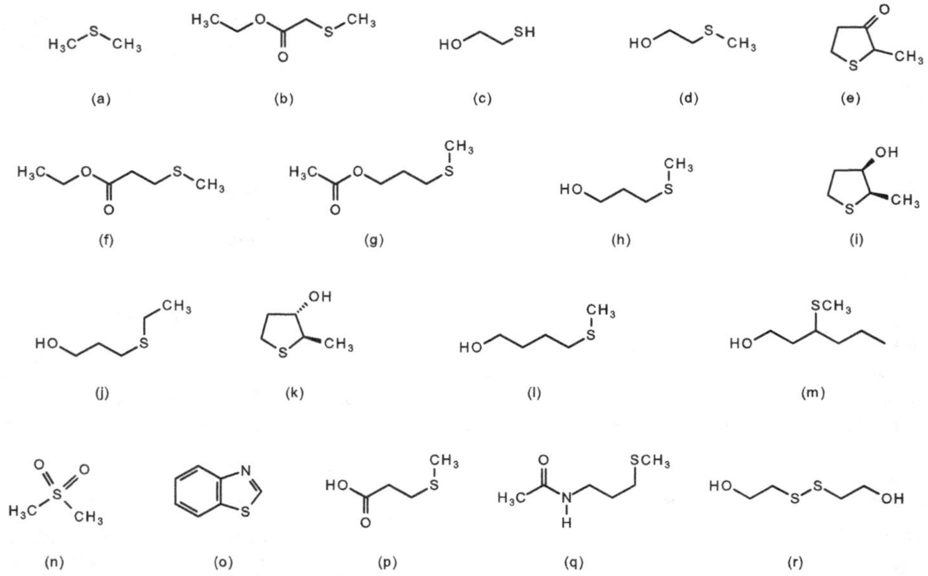

FIGURE 21.2 Representations of sulfur-containing compounds found in Port wine. Shown are: (a) dimethyl sulfide; (b) ethyl (methylthio)acetate; (c) 2-mercaptoethanol; (d) 2-(methylthio)ethanol; (e) 2-methyltetrahydrothiophen-3-one; (f) ethyl 3-(methylthio)propionate; (g) 3-(methylthio)-1-propanol acetate; (h) 3-(methylthio)-1-propanol; (i) *cis*-2-methyltetrahydrothiophen-3-ol; (j) 3-(ethylthio)-1-propanol; (k) *trans*-2-methyltetrahydrothiophen-3-ol; (l) 4-(methylthio)-1-butanol; (m) 3-(methylthio)-1-hexanol; (n) dimethyl sulfone; (o) benzothiazole; (p) 3-(methylthio)-1-propanoic acid; (q) *N*-3-(methylthiopropyl)acetamide; and (r) *bis*-(2-hydroxydiethyl) disulfide.

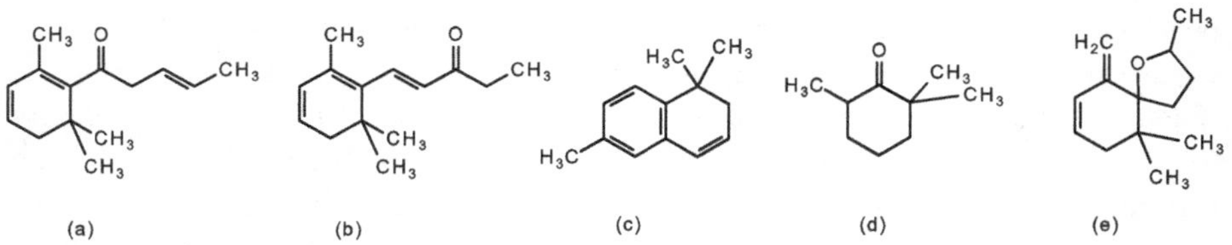

FIGURE 21.3 Representations of norterpenoids reported to be found in Port wine. Shown are: (a) β-damsacenone; (b) β-ionone); (c) 2,2,6-trimethylcyclohexanone (TCH); (d) 1,1,6-trimethyl-1,2-dihydronaphthalene (TDN); and (e) vitispirane.

Subsequently (in 1997) several reports of additional malvidin-3-glycoside (oenin) adducts with simple two- and three-carbon compounds were reported. Of course, given the plethora of simple fragments that might be utilized to produce these derivatives, it was not clear and, indeed, does not yet appear clear, as to their provenance. These vitisins A and B (Figure 21.5) were also believed found in young wines as well as in those allowed to mature.[9,10]

Following on the heels of those reports, there soon appeared another suggesting that, on standing, pyruvic acid (CH_3COCO_2H) could also react with anthocyanins, and again using malvidin-3-glucoside as a typical anthocyanin, the product was shown as a carboxylic acid (Figure 21.6).[11]

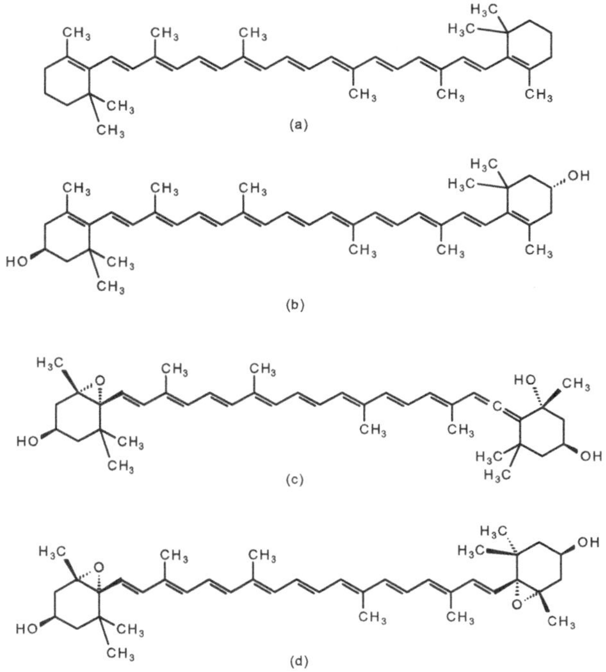

FIGURE 21.4 Representations of terpenoids reported to be found in Port wine. Shown are: (a) β-carotene; (b) lutein; (c) neoxanthin; and (d) violaxanthin.

Then, just after the turn of the century, Port wine, made from the grapes of the Douro Valley in northern Portugal, and aged for about a year was analyzed. Again, anthocyanin derivatives, originally dark red but turning yellow, appeared to be present. The change was attributed to aging. Interestingly, it was found that pyruvic acid adducts of malvidin 3-glucoside, but with further substitution on the C-6 glucose hydroxyl by acetyl and coumaroyl functionality, were present. Indeed, not only were the same kinds of derivatives attached to malvidin 3-glucoside found; delphinidin 3-glucoside could also be inferred as being present (Figure 21.7).[12]

Subsequently, dimers of anthocyanins with flavones, and also bearing benzylthiol units (Figure 21.8) as well as benzylthiol attachments to what appear to be acetaldehyde-derived two-carbon units and also attached to anthocyanins, were also found in red wines that had been stored and exposed to oxygen.[13,14] Since many of the compounds thought to be flavanol-anthocyanin adducts are incapable of isolation from the fortified wine,[15] reasonable structural assumptions are made and the likely compounds synthesized. The synthetic materials are compared mass spectrometrically (MS) with the MS of the likely separated component peak.[16]

Similarly, the earlier suggestion that a vinylphenol might be involved in condensation reaction with malvidin (*vide supra*) received substance in that Pinotin A was isolated from aged Pinotage wine (Figure 21.9).[17]

Given the difficulties in isolation, and the idea that fragments might be put together that would have mass spectra identical to some of the fragments, there has been an efflorescence

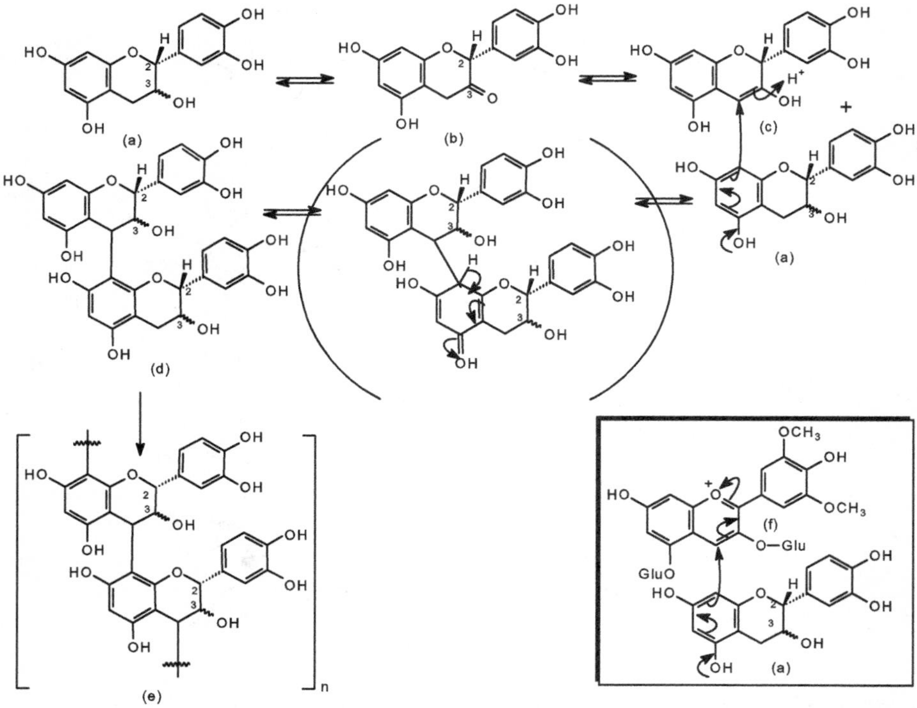

SCHEME 21.1 A representation accounting for the formation of procyanidins from reaction of (a) catechin (2R,3S)- or epicatechin (2R,3R)-2-(3,4-dihydroxyphenyl)-3,4-dihydro-2H-1-benzopyran-3,5,7-triol, on oxidation to the corresponding 3-ketone (b) and enolization to (c) with itself to yield a dimer (d). The process can produce polymers (e). The boxed example suggests a process involving malvidin-3,5-diglucoside (oenin) (f) reacting with catechin or epicatechin (a) to provide similar product.

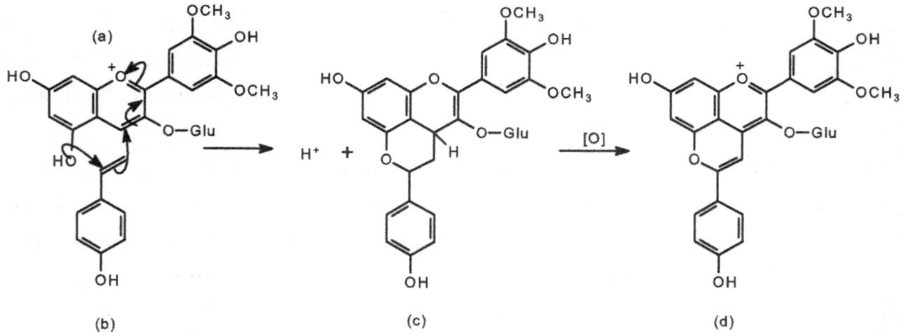

SCHEME 21.2 A representation of a potential (or actual) reaction between malvidin-3-glucoside (a) and 4-vinylphenol (b) to produce an adduct (c) which, on oxidation, yields the ion (d). The latter (a presumed vitisin) was identified as present through mass spectrometry.

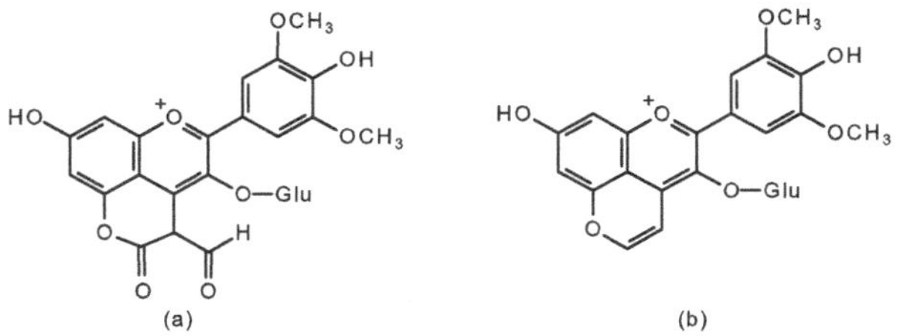

FIGURE 21.5 Representations of the structures proposed for the vitisins, compounds presumably formed from malvidin-3-glucoside and not-yet-defined small carbon fragments. The reported structure for vitisin A (a) is on the left and that for vitisin B (b) on the right.

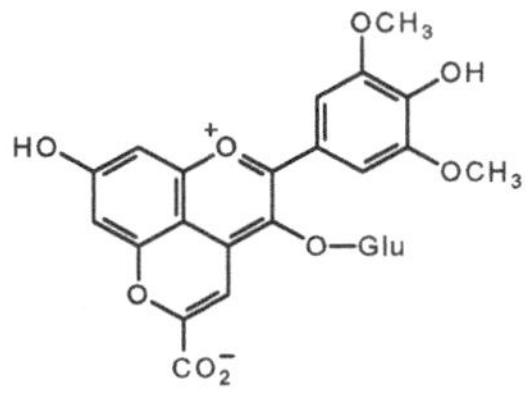

FIGURE 21.6 A zwitterionic representation of another derivative of malvidin 3-glucoside (oenin) which is presumed to arise from the reaction with pyruvic acid. This material is also named vitisin A.

of discovery of new materials in Port wines. These include isolates where aldehydes, such as propanal, lie between anthocyanin and flavanol (Figure 21.10),[18] where hydroxycinnamic acids (e.g., *para*-coumaric acid, caffeic acid, and ferulic acid) have condensed with the initial products of the reaction between anthocyanin and a two-carbon fragment (e.g., acetaldehyde) or a three-carbon fragment (e.g., pyruvic acid) to yield a vitisin A (Figure 21.6) like fragment which on further reaction with one of the cinnamic acid derivatives (Figure 21.11) would give rise to a member of the family of "portisins" such as that shown in Figure 21.12.[19,20] The family of such materials is large as a consequence of the multitude of anthocyanins differing both in sugar attached at C3 (and sometimes at C5) and hydroxy and methoxy substitution patterns.[20–23]

All of the expressions for these compounds have been invoked not only because they are present in fortified Port wine, but also because the color and flavor changes occurring as a consequence of the reaction processes, including polymerization and oxidation while aging, are now clearer.

(b) Sherry

For Sherry, the original (and genuine) beverage is a fortified wine produced in Andalusia, Spain, which can be light or dark, depending upon the specific grape, as well as both the time and the conditions under which the beverage is permitted to age in the barrel.

In contrast to Port wine, Sherry, properly produced only from the Palomino Fino grape variety, is found in light, medium, and dark styles from those white-to-light yellow grapes. All of the Fino, Oloroso, and Amontillado types have usually been fermented to completion and then treated with a distillate (as was the case with Port wine) to fortify them. The dynamic

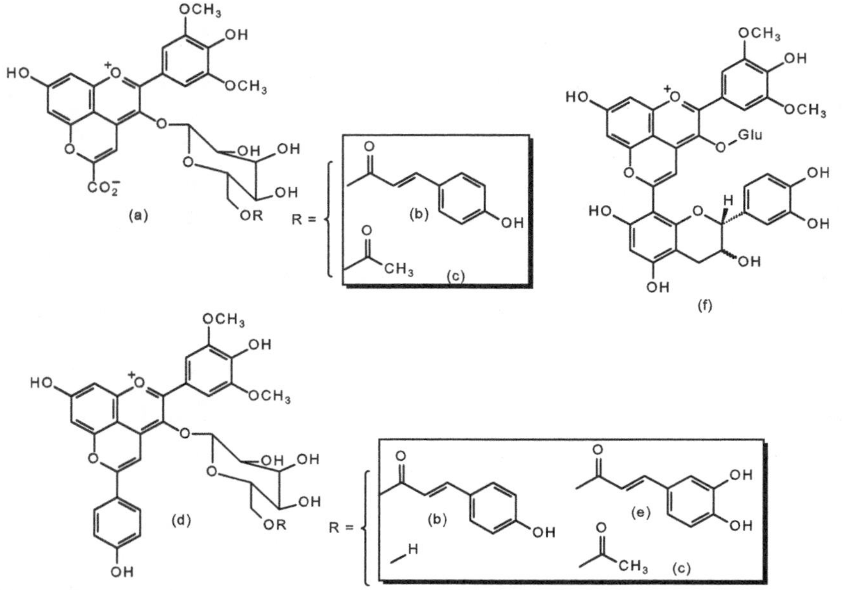

FIGURE 21.7 Representations of derivatives of malvidin 3-glucoside (oenin) having reacted with pyruvate (a) bearing a coumaroyl (b) and acetyl (c) substituents at C-6 of the glucose and (d) the product between malvidin 3-glucoside and *para*-vinylphenol (see Scheme 21.2) where, again the glucose is substituted at C-6 with a coumaroyl (b); an acetyl (c); a caffeoyl (e), or a proton (H^+). Finally, there is shown (f) a representation of the product of what is presumed to be derived from malvidin 3-glucoside and pyruvate having undergone further reaction with catechin or its C-3 epimer, epicatechin.

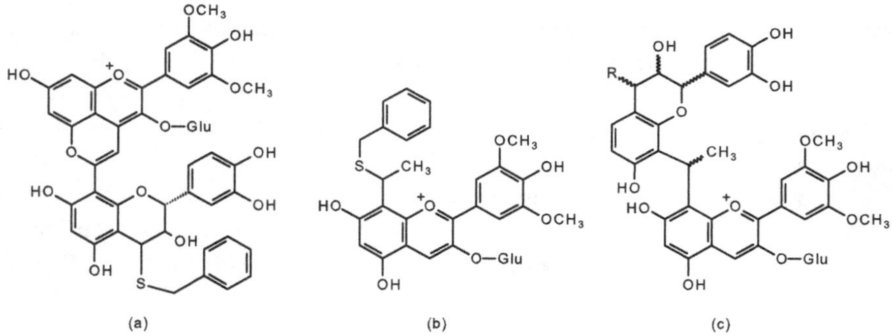

FIGURE 21.8 Representations of several compounds bearing benzylthiol units (a) as attached to an anthocyanin-flavonoid compound and (b) to the aromatic A-ring of malvidin 3-glucoside (oenin). Also shown (c) is an example of an anthocyanin-flavanol dimer (where R = H or additional flavanol units) bonded through a two-carbon unit (rather than directly attached).

racking process distinguishes these types. The somewhat more porous North American Oak barrels used for aging these beverages encourages the development of a thin growth of yeast (called a *flor*) on the surface of the liquid in the barrel that limits the amount of oxygen reaching the liquid below it. The direct consequence of the presence of the flor with regard to

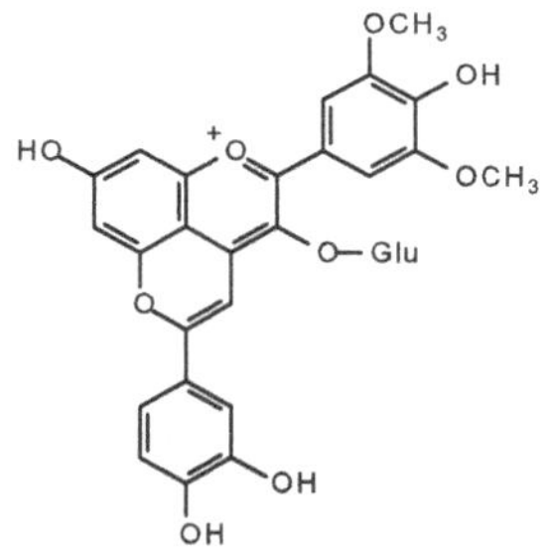

FIGURE 21.9 A representation of the presumed reaction product (Pinotin A) between malvidin 3-glucoside (oenin) and 4-vinylcatechol reportedly isolated from aged wine produced from the Pinotage grape.

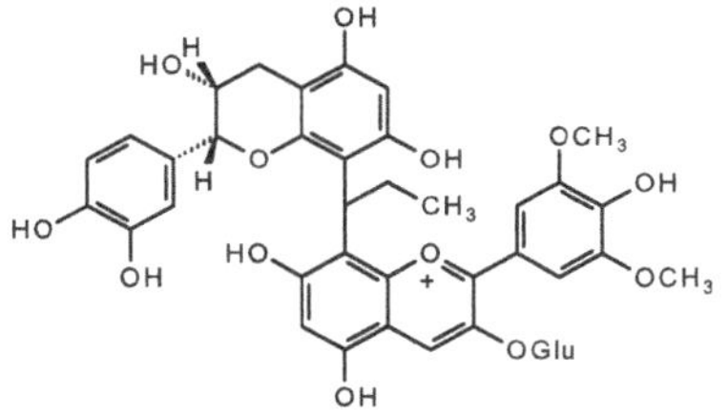

FIGURE 21.10 Representation of 8-[1-[(2R,3R)-2-(3,4-dihydroxyphenyl)-3,4-dihydro-3,5,7-trihydroxy-2H-1-benzopyran-8-yl]propyl]-3-(b-D-glucopyranosyloxy)-5,7-dihydroxy-2-(4-hydroxy-3,5-dimethoxyphenyl)-1-benzopyrylium ion as reported isolated from a Port wine. The three-carbon fragment lying between the lower malvidin 3-glycoside (oenin) portion and the upper flavanol is presumed to be derived from propanal (CH_3CH_2CHO).

flavor is that aldehydes produced during the normal fermentation are not oxidized further, and so the wine takes on flavors reflecting their presence.

The flor will remain as long as the alcohol concentration is kept low enough, but should more alcohol be added, the flor is destroyed, oxidation will begin, and the flavor and color change.

The Sherry matured under the flor is generally pale and dry (all of the sugar having been permitted to ferment) and is usually referred to as "Fino." If aged under the flor and then exposed to air, the Sherry is darker and is the variety called "Amontillado." "Oloroso" Sherry is the darkest and most highly oxidized of the Sherry varieties. It also has the highest alcohol content and occasionally is sweetened.

Figure 21.13 has representations of compounds in all three Sherry varieties.

As shown in Figure 21.14, an early chromatographic investigation of the three varieties arising, as noted, by manipulative techniques from the same grape provides evidence that the differences lie in changes in the concentration of constituents rather than in kind.

Interestingly, while more than three hundred (300) volatiles have been reported present in Sherry, only a small subset appears to be involved as measured by human olfactory receptors. As discussed in Chapter 12, there are serious questions about how many human olfactory receptors are capable of discriminating olfactory stimuli![24] In that vein, a recent analysis of Amontillado Sherry from the Palomino Fino grape has been completed, and

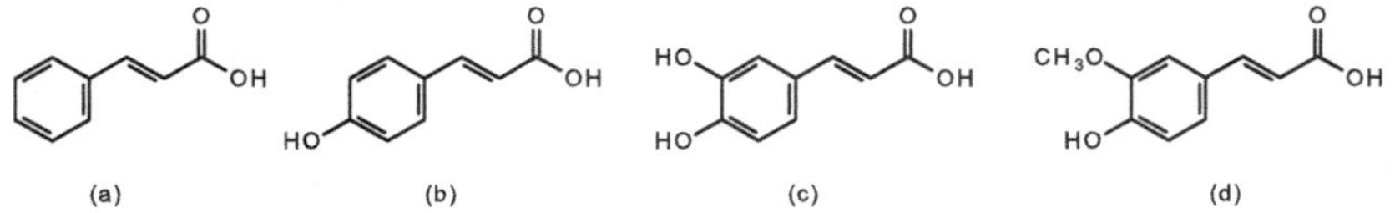

FIGURE 21.11 Representations of cinnamic acid (a) as well as hydroxylated cinnamic acids (b) *para*-coumaric acid, (c) caffeic acid, and (d) ferulic acid. These acids are reported as having apparently condensed with anthocyanins to produce a set of compounds collectively referred to as "portisins" (see He, J.; Santos-Buelga, C.; Silva, A.M.S.; Mateus, N.; de Frietas, V. *J. Agric. Food Chem.* **2006**, *54*, 9598).

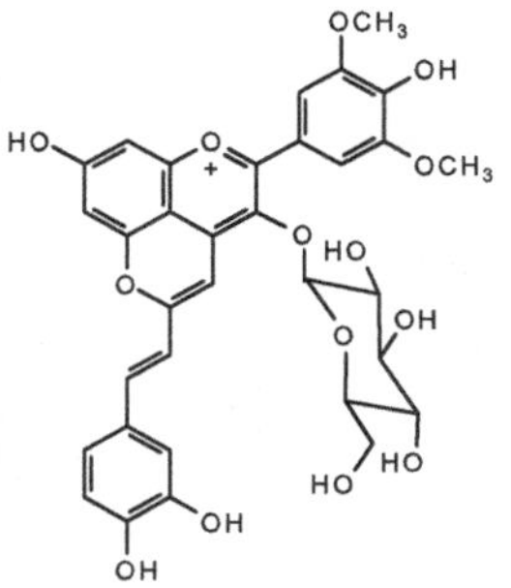

FIGURE 21.12 Representation of a portisin, 5-[2-(3,4-dihydroxyphenyl)ethenyl]-3-(β-D-gluco-pyranosyloxy)-8-hydroxy-2-(4-hydroxy-3,5-dimethoxyphenyl)-pyrano[4,3,2-de]-1-benzopyrylium ion (trivially called vinylpyranomalvidin 3-*O*-glycoside-catechol). This is presumed to arise from reaction between, e.g., pyruvic acid (as in vitisin A, Figure 21.6) followed by decarboxylation, oxidation, and subsequent reaction with a caffeic acid (Figure 21.11) equivalent.

thirty-seven (37) "odor-active" compounds isolated and identified.[25] The most odor-active volatile constituents as measured there are shown in Figures 21.15 and 21.16.

When the most odor-active compounds found in the Sherry as shown above are compared to the compounds found in Palomino fino grape must, from which the Sherry was produced (after the use of glycosidases to liberate the free alcohols),[26] it is clear that either major changes occurred during production or the analytical techniques could not be usefully compared. The compounds reported to be found in the Palomino fino grape must are shown Figures 21.17 and 21.18, and it is worthwhile noting that, despite the wealth of terpenes shown, "the (must had) a nonfloral aroma profile, benzylic alcohol being the main component present in the bound fraction." Further, "while the (glucosidases) were able to increase the concentrations of seven terpenes in wine, the individual compounds remained below their olfactory detection thresholds."[26]

Interestingly, the lower-molecular-weight alcohols, aldehydes, etc. are clearly present throughout. Again, however, as it cannot be emphasized enough, minor changes in technique can and often will produce variation in the final product.

(c) Madeira

For Madeira, the original (and genuine) beverage is a fortified white (tawny) wine produced in the Madeira Islands off the coast of Portugal. The originally preferred white grape is called Malvaisa.[27] But there is more to it than that, and there is more than one Madeira wine.[28]

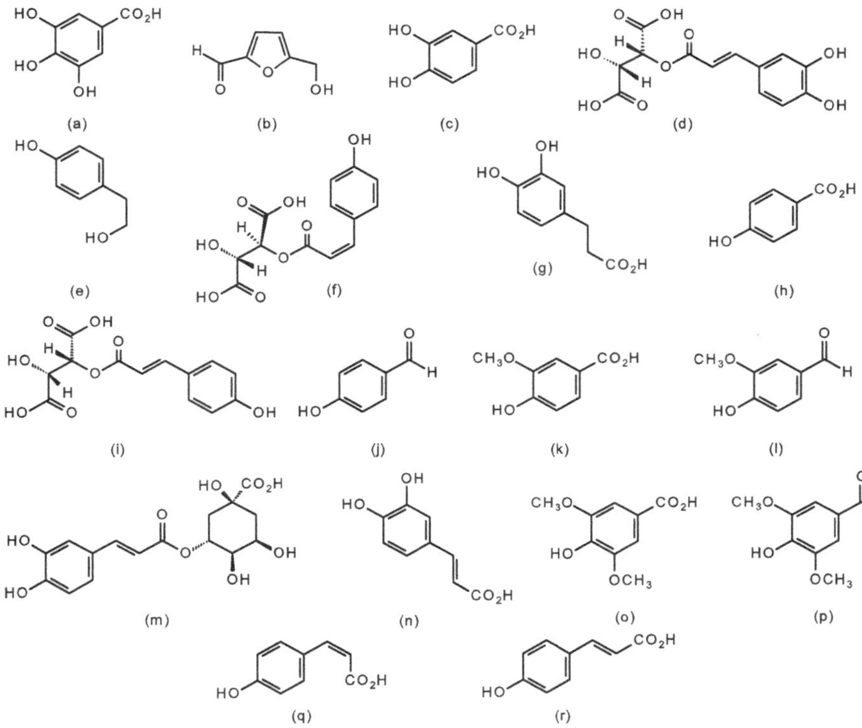

FIGURE 21.13 Representations of compounds reported found in all three Sherry varieties (Fino, Oloroso, and Amontillado), but in different concentrations. Those shown include: (a) gallic acid; (b) 3-hydroxymethyl-2-furaldehyde (HMF); (c) protocatechuic acid; (d) caftaric acid; (e) tyrosol; (f) *cis-para*-coutaric acid; (g) hydrocaffeic acid; (h) *para*-hydroxybenzoic acid; (i) *trans-para*-coutaric acid; (j) *para*-hydroxybenzaldehyde; (k) vanillic acid; (l) vanillin; (m) chlorogenic acid; (n) *trans*-caffeic acid; (o) syringic acid; (p) syringaldehyde; (q) *cis-para*-coumaric acid; and (r) *trans-para*-coumaric acid.

There are currently five (5) varieties of *V. vinifera* grape used to produce Madeira fortified wine on the island of Madeira in the archipelago. Each leads to a different outcome, as they are treated differently. There are four (4) white grape varieties (Sercial, Boal, Verdelho, and Malvasia) and one red grape variety (Tinta Negra) that are now used.

By choosing the right grape, grown in the right location and allowing the fermentation to proceed just far enough, each produces a different beverage which, when subsequently enriched appropriately, provides a unique flavor. In general, the Sercial variety is used to produce a dry wine with the lowest residual sugar content and the highest acidity of the Madeiras. The other Madeiras are produced with increasing residual sugar so that Boal yields a medium sweet beverage, Verdelho one that is medium dry, and Malvasia is the sweetest of them. Tinta Negra, perhaps the most widely grown grape, produces a sweet pale-red wine.

Clearly, both the amount of sugar present in the grape variety and the extent to which fermentation is allowed to proceed are the initial major differences.

The distinguishing characteristic of all of the Madeira wines is the aging process which, it is claimed, was originally found desirable when casks of wine from the island, treated with

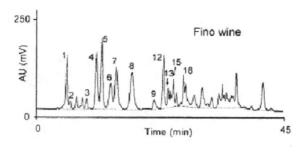
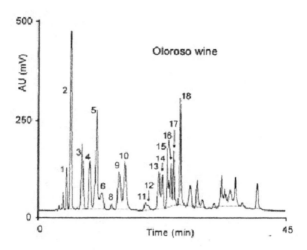
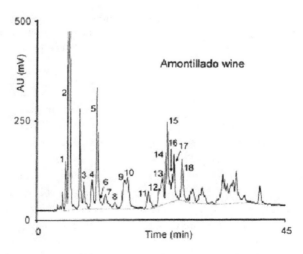

FIGURE 21.14 Chromatograms of Sherry wines. Peaks: 1 = gallic acid; 2 = hydroxymethylfuran; 3 = protocatechuic acid; 4 = caftaric acid; 5 = tyrosol; 6 = *cis-para*-coutaric acid; 7 = hydrocaffeic acid; 8 = *para*-hydroxybenzoic acid; 9 = *trans-para*-coutaric acid; 10 = *para*-hydroxybenzaldehde; 11 = vanillic acid; 12 = chlorogenic acid; 13 = caffeic acid; 14 = vanillin; 15 = syringic acid; 16 = *cis-para*-coumaric acid; 17 = syringaldehyde; 18 = *trans-para*-coumaric acid. (---) basis line. (A UV detector was used with a C_{18} reverse phase column using methanol–acetic acid–water as the moving phase.) Reprinted with permission, from Moreno, M. V. G.; Barroso, C. G. *J. Agric. Food Chem.* **2002**, *50*, 7556. Copyright 2002, American Chemical Society.

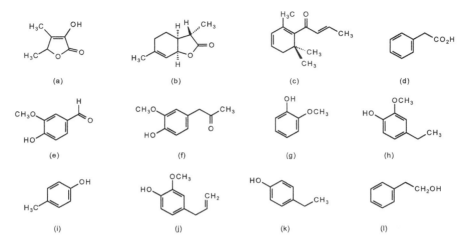

FIGURE 21.15 Representations of some of the olfactory stimuli reported to be present in Amontillado Sherry from the Palomino Fino grape. The compounds shown are: (a) sotolon; (b) wine lactone; (c) β-damascenone; (d) phenylacetic acid; (e) vanillin; (f) acetovanillone; (g) guaiacol; (h) 4-ethylguaiacol; (i) *para*-cresol; (j) eugenol; (k) 4-ethylphenol; and (l) 2-phenylethanol.

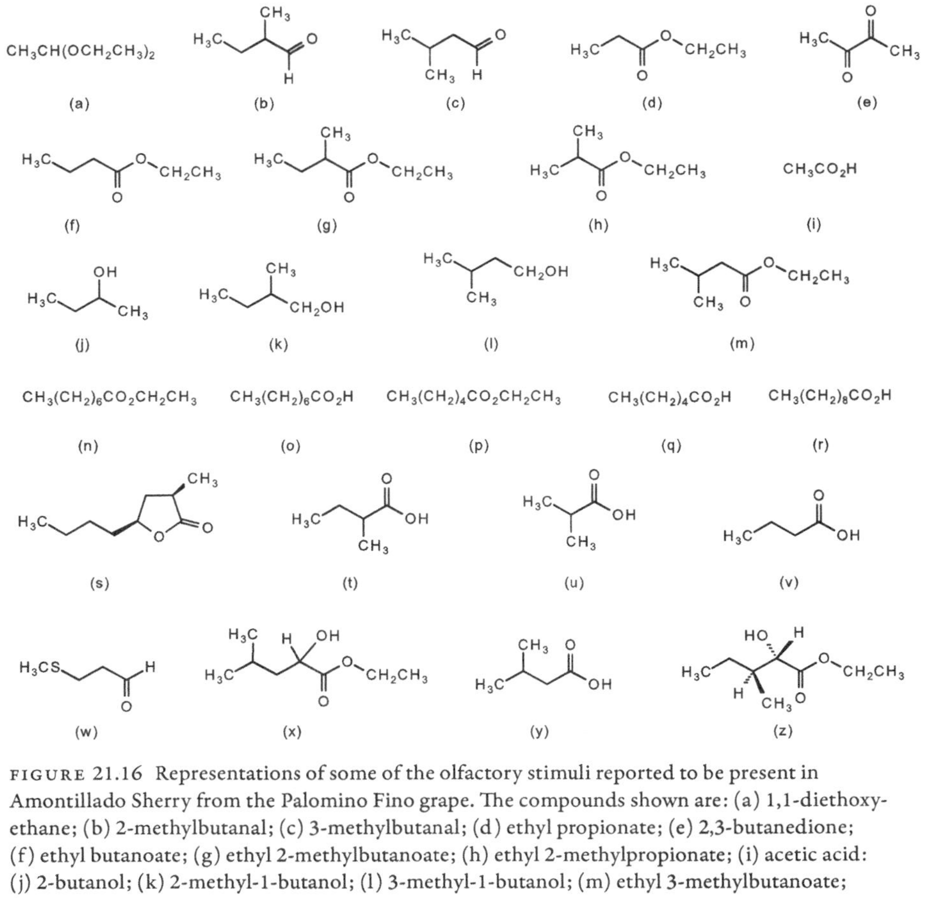

FIGURE 21.16 Representations of some of the olfactory stimuli reported to be present in Amontillado Sherry from the Palomino Fino grape. The compounds shown are: (a) 1,1-diethoxyethane; (b) 2-methylbutanal; (c) 3-methylbutanal; (d) ethyl propionate; (e) 2,3-butanedione; (f) ethyl butanoate; (g) ethyl 2-methylbutanoate; (h) ethyl 2-methylpropionate; (i) acetic acid: (j) 2-butanol; (k) 2-methyl-1-butanol; (l) 3-methyl-1-butanol; (m) ethyl 3-methylbutanoate; (n) ethyl octanoate; (o) octanoic acid; (p) ethyl hexanoate; (q) hexanoic acid; (r) decanoic acid; (s) *cis*-whisky lactone; (t) 2-methylbutanoic acid; (u) methylpropanoic acid; (v) butanoic acid; (w) 3-(methylthio)propanal; (x) ethyl 2-hydroxy-4-methylpentanoate; (y) 3-methylbutanoic acid; and (z) ethyl (2S,3S)-2-hydroxy-3-methylpentanoate.

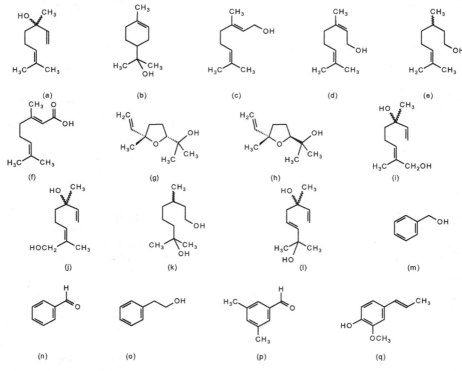

FIGURE 21.17 Representations of terpenes and some aromatic ring–containing compounds found to be present in the Palomino fino grape must from which Sherry, apparently lacking many of them, was produced. The compounds shown are: (a) linalool; (b) α-terpineol; (c) geraniol; (d) nerol; (e) citronellol; (f) geranic acid; (g) *cis*-linalool oxide; (h) *trans*-linalool oxide; (i) (*E*)-2,6-dimethyl-2,7-octadien-1,6-diol; (j) (*Z*)-2,6-dimethyl-2,7-octadien-1,6-diol; (k) 3,7-dimethyl-1,7-octanediol; (l) (*E*)-3,7-dimethyl-1,5-octadien-3,7-diol; (m) benzyl alcohol; (n) benzaldehyde; (o) 2-phenylethanol; (p) 3,5-dimethylbenzaldehyde; and (q) *trans*-isoeugenol.

distillate—thus making a fortified wine—were ship-bound for extended periods under conditions of heat and ship's motion. Current shipping practices do not allow replication of the previous work, and so two different paths are used to produce a similar beverage.

The wines are fortified by adding distillate or "mature beverage" and additional alcohol, and after this fortification, one of two quite different processes (i.e., estufagem and canteiros) is applied.

The estufagem process is the faster and is said to simulate the effects of a lengthy tropical sea voyage in aging barrels. The wine is placed in large vats or tanks and heated by stirring and immersing coils through which hot water (*ca.* 50 °C [120 °F]) is passed for at least three (3) months. The resulting solution is moved into oak casks where, with occasional movement, it rests and undergoes oxidation for a minimum of three (3) months. If the wine is bottled after this relatively short treatment, regulations have prohibited sale before the end of October of the second year following grape harvest.

The slower process involves wooden beam-supported oaken casks and is referred to as canteiros. Here, the casks are placed on the beams at the top of the winery, near the roof,

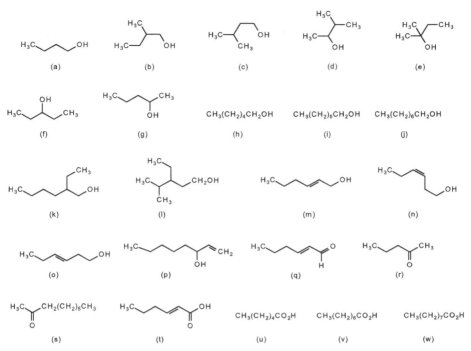

FIGURE 21.18 Representations of low-molecular-weight alcohols, aldehydes, ketones, and carboxylic acids found to be present in the Palomino fino grape must from which Sherry, apparently lacking many of them, was produced. The compounds shown are: (a) 1-butanol; (b) 2-methyl-1-butanol; (c) 3-methyl-1-butanol; (d) 3-methyl-2-butanol; (e) 2-methyl-2-butanol; (f) 3-pentanol; (g) 2-pentanol; (h) 1-hexanol; (i) 1-heptanol; (j) 1-octanol; (k) 2-ethyl-1-hexanol; (l) 3-ethyl-4-methyl-1-pentanol; (m) (E)-2-hexen-1-ol; (n) (Z)-3-hexen-1-ol; (o) (E)-3-hexen-1-ol; (p) 1-octen-3-ol; (q) (E)-2-hexenal; (r) 2-pentanone; (s) 2-nonanone; (t) (E)-2-hexenoic acid; (u) hexanoic acid; (v) octanoic acid; and (w) nonanoic acid.

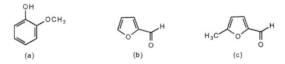

FIGURE 21.19 Representations of: (a) guaiacol; (b) furfural: and (c) 5-methyl-2-furfural reported present in all Madeira wines aged in previously used barrels.

where temperatures during the summer might reach 35 °C (95 °F) and the humidity is high. They are allowed to stand for at least two years so that both oxidation of the contents and its enrichment by oak extractables leached from the solid wood can take place. Guaiacol, furfural, and 5-methyl-2-furfural (Figure 21.19) are present in all wines aged in previously used barrels.

Interestingly, albeit at different concentrations, a measured set of norisoprenoids and isoprenoids (mostly terpenols) were all present in Boal, Malvasia, Sercial, and Verdelho musts. It is presumed that the compounds, perhaps as glycosides, were originally present in the grapes from which the musts were generated. The compounds were reported as linalool

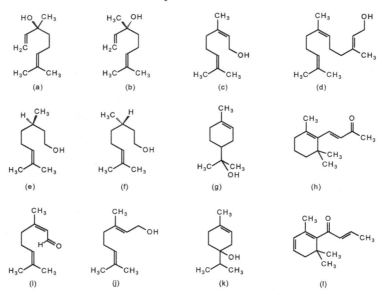

FIGURE 21.20 Representations of terpene and terpene-derived odorants found in all Madeira wine musts from Boal, Malvasia, Sercial, and Verdelho grape varieties. The compounds shown are: (a) (S)-(+)-linalool; (b) (R)-(−)-linalool; (c) nerol; (d) farnesol; (e) (R)-(+)-citronellol; (f) (S)-(−)-citronellol; (g) α-terpineol; (h) (E)-β-ionone; (i) neral; (j) geraniol; (k) 4-terpineol; and (l) (E)-β-damsacenone.

(isomer not specified), α-terpineol, citronellol (isomer not specified), β-damascenone, nerol, geraniol, 4-terpineol, neral, β-ionone, and farnesol.[29] Structural representations are shown in Figure 21.20.

Furthermore, of this set of compounds, seven (7) of them (viz. linalool, α-terpineol, citronellol, geraniol, neral, β-damascenone, and β-ionone) turned out to be sufficient to account for more than 90% of the variation between the four grapes and to thus "adequately describe the samples according to variety."[29]

More recent work has shown subsequently that it is very clear that differences can be observed "within and between grape varieties (white and red) as well as grape fractions (pulp and skin)." And, "a strategy of instrumental and chemometric tools is suitable to discriminate grapes based on variety... independently of grape fractions, and using grape skin fraction(s) (it) is possible to discriminate grapes from different environments or terroirs." Further... "insights into varietal volatile profile were observed based on... different environmental conditions."[30]

Thus, it was found that a variety of terpenes, in addition to those already shown in Figure 21.20, could also be isolated from the pulp and skin of both white and red varieties, but in different concentrations. Indeed, in some cases compounds expected to be present on the basis of previous analyses were below the detection limits (or absent). Structural representations of the terpenes (in addition to those already known and shown above) and related compounds (e.g., dilakylated cinnamaldehyde derivative, lilial, and the fatty acid derived methyl dihydrojasmonate) also found are provided in Figure 21.21, and as usual, many of these compounds are also common to other grape varieties.

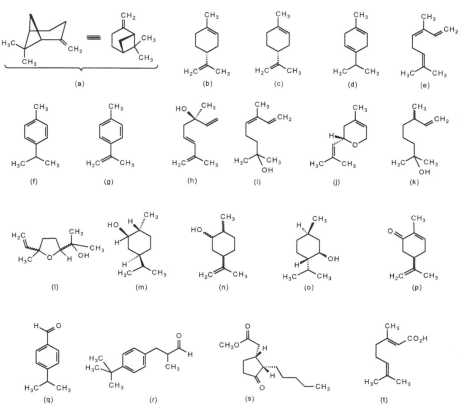

FIGURE 21.21 Representations of compounds found in the grapes (including skin) of all of the varieties of grape from which Madeira wines are prepared. The compounds shown are: (a) (+)-β-pinene; (b) (S)-(−)-limonene; (c) (R)-(+)-limonene; (d) γ-terpinene; (e) β-ocimene; (f) *para*-cymene; (g) *para*-cymenene; (h) hotrienol; (i) ocimenol; (j) nerol oxide; (k) dihydromyrcenol; (l) linalool oxide; (m) 4-carvomenthol; (n) 2-methyl-5-(1-methylethenyl)cyclohexanol; (o) L-(−)-menthol; (p) DL-carvone; (q) cumin aldehyde; (r) lilial; (s) methyl dihydrojasmonate; and (t) *trans*-geranic acid.

Interestingly, a number of sesquiterpenes were also found in the skin of the Madeira varieties examined, but the one sesquiterpene already listed in Figure 21.20 [i.e., (2E,6E)-3,7,11-trimethyldodeca-2,6,10-trien-1-ol (farnesol)] was not reported. The sesquiterpenes were not found in the pulp! Structural representations of the compounds reported, α-ylangene, α-cubebene. β-bourbonene, β-cubebene, β-caryophyllene, epi-bicyclosesqiphellandrene, isoledene, aromadendrene, β-curjunene, α-caryophyllene, alloaromadendrene, α-humulene, γ-muurolene, γ-selinene, α-amorphene, epizonarene, germacrene D, α-muurolene, α-bisabolene, bicyclogermacrene, δ-cadinene, calamenene, (Z)-nerolidol and guaiazulene, are shown (Figures 21.22a and 21.22b).

In the same vein, C-13 norterpenoids were also found in the skin and in the pulp of some Madeira varieties, and again similarly, concentration differences were observed. Representations of the reported compounds (i.e., vitispirane I, vitispirane II, α-ionene (1,1,6-trimethyl-1,2-dihydronaphthalene [TDN]), (E)-β-damascenone, β-cyclocitral and β-ionone) are shown in Figure 21.23.

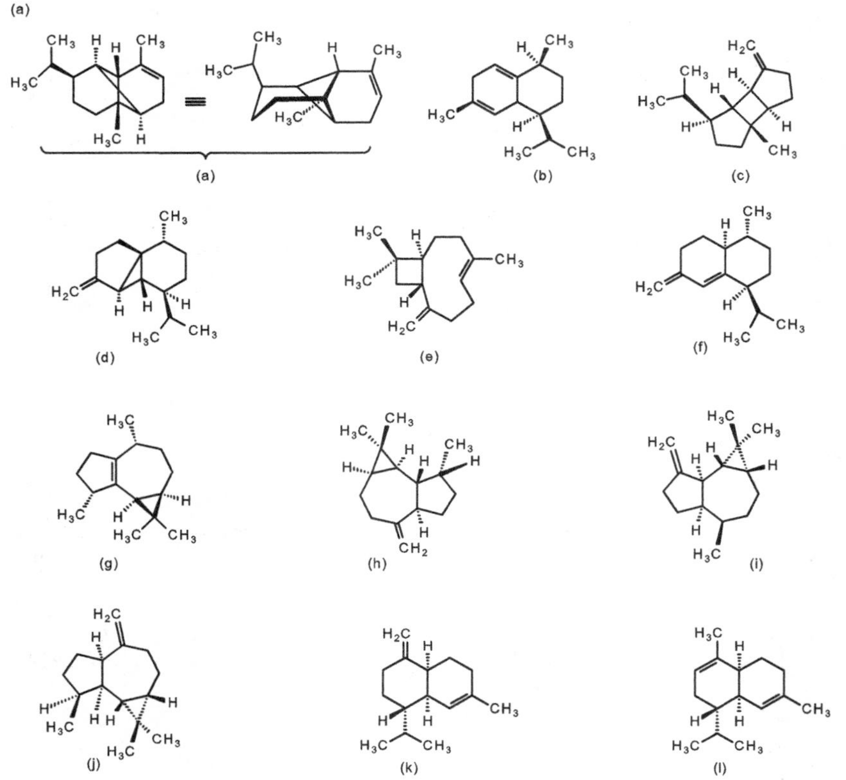

FIGURE 21.22a Representations of some of the sesquiterpenes found in the skin but not found in the pulp of the grapes of all of the varieties from which Madeira wines are prepared. The compounds shown here are: (a) α-ylangene; (b) α-cubebene; (c) β-bourbonene; (d) β-cubebene; (e) β-caryophyllene; (f) epi-bicyclosesquiphellandrene; (g) isoledene; (h) aromadendrene; (i) β-gurjunene; (j) alloaromadendrene; (k) γ-muurolene; and (l) α-muurolene.

Most recently, and of some general interest, amino acids, and amines formed from them as a consequence of their degradation found in fortified wines and allowed to evolve under the *estufagem* protocol, have been examined.[31] It was found that while arginine (Arg, R) was the most abundant of the 18 identified, all of amino acids (except for asparagine [Asn, N]) decreased during the standard heating process in the protocol. Interestingly, biogenic amines such as 2-phenylethylamine, 3-methylbutanamine (isopentylamine), and 1,5-pentanediamine (cadaverine), normally present at low concentrations and expected on amino acid degradation did not appear to be formed, but rather new (and as yet unidentified) products resulted (Figure 21.24).

PART B. *BOTRYTIS CINEREA* (NOBLE ROT)

When Sauvignon blanc grapes (usually from the Graves region of Bordeaux) or the Tokaji Aszu from Hungary or some of the late harvest German Rieslings are intentionally infected

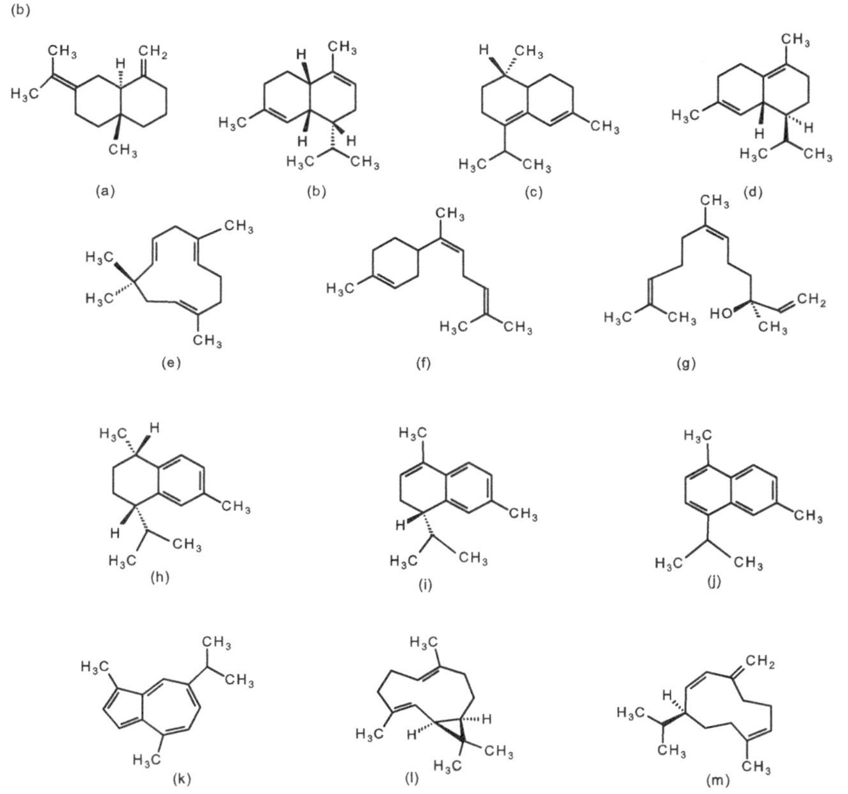

FIGURE 21.22b Representations of some of the sesquiterpenes found in the skin but not found in the pulp of the grapes of all of the varieties from which Madeira wines are prepared. Structural representations of the compounds shown here are for: (a) γ-selinene; (b) α-amorphene; (c) epizonarene; (d) δ-cadinene; (e) α-humulene (aka α-caryophyllene); (f) α-bisabolene; (g) (Z)-nerolidol; (h) calamenene; (i) α-calacorene; (j) cadalene; (k) guaiazulene; (l) bicyclogermacrene; and (m) germacrene D.

with the fungus *Botrytis cinerea,* the grapes, while on the vine, become partially raisin-like. It has been found possible to produce sweet, distinctly flavored wines from these grapes.

There are conflicting stories associated with the initial use of the fungus, but somehow, it appears advantageous to relate it to royal European vineyards. For that and other possible advertising purposes, the infecting fungus *Botrytis cinerea* is also referred to as the "Noble Rot." Interestingly, it appears that the infection causes grapes to dehydrate while maintaining glucose and other sugar (e.g., arabinose, xylose) levels. At the appropriate time the harvest is carefully made, choosing only the grapes experience has dictated will produce the most flavorful product. More wine grapes are needed to make the same amount of juice because of dehydration.

Dessert wines made from Noble Rot grapes are more viscous and sweet. Some even have higher alcohol content than expected as a consequence of emendation.

Sommeliers, as (presumed) connoisseurs, are said to often use words such as "honey" and "ginger" to describe the flavors that *Botrytis cinerea* adds to wine. It is clear that constituents present as a consequence of the presence of *Botrytis cinerea* prior to fermentation induce

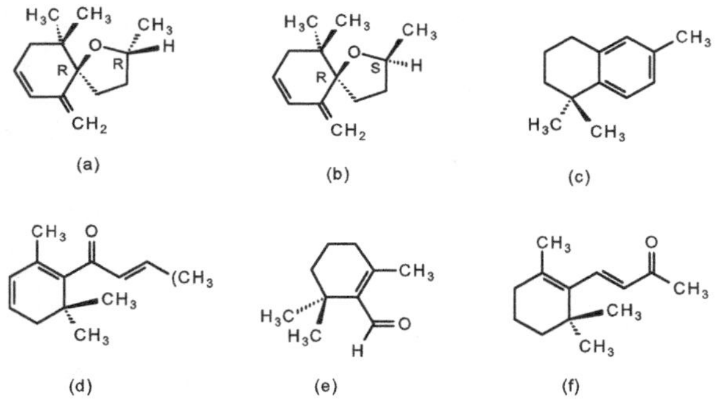

FIGURE 21.23 Representations of C-13 norterpenoids found in the grapes of all of the varieties from which Madeira wines are prepared. The compounds shown are: (a) vitispirane I; (b) vitispirane II; (c) α-ionene or 1,1,6-trimethyl-1,2-dihydronaphthalene (TDN); (d) (E)-β-damascenone; (e) β-cyclocitral; and (f) β-ionone.

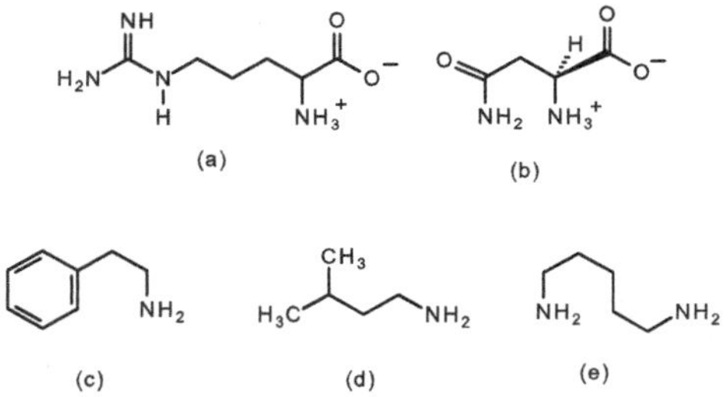

FIGURE 21.24 Representations of nitrogenous compounds expected and either found or not found in an analysis of Madeira formed under *estufagem* protocol. All of the expected amino acids (not shown) were present. Arginine (Arg, R) (a) was the most abundant, and asparagine (Asn, N) (b) did not decrease during the protocol as others did. The typical "biogenic amines," 2-phenethylamine (c), 3-methylbutanamine (d), and cadaverine (e) were expected and not found. As yet unidentified nitrogenous materials were present.

formation of new compounds which may provide new odors and flavors. Indeed, as pointed out in another connection by Proust, particular odors are particularly evocative.[32]

Somewhat more current, botrytized wine volatiles were compared to dry white wine volatiles made from the same grape (Semillon and Sauvignon blanc grapes were used). The comparison was made by gas chromatographic separation of volatiles and detection of the odors of the compounds so separated. Although a number of unidentified compounds were present, the list of those known is impressive.[33]

While there were a few differences in the compounds detected among the different volatile components, the major finding was that the differences in aroma could be accounted for by a few potent compounds. Representations of the identified compounds, *viz*. 3-metcapto-1-hexanol, 2-ethyl-4-hydroxy-5-methyl-3(2H)-furanone (homofuraneol), 4-hydroxy-2,5-dimethyl-3(2H)-furanone (furaneol), 4-hydroxy-5-methyl-3(2H)-furanone (norfuraneol), 3-hydroxy-4,5-dimethylfuran-2(5H)-one (sotolon), 3-(methylthio)-propanal (methional), and phenylacetaldehyde are shown in Figure 21.25.

Additional compounds also found, but not noted as "truly odoriferous" in the same distinguishing sense, are 2-methyl-3-furanthiol, 4-mercapto-4-methyl-2-pentanone, ethyl

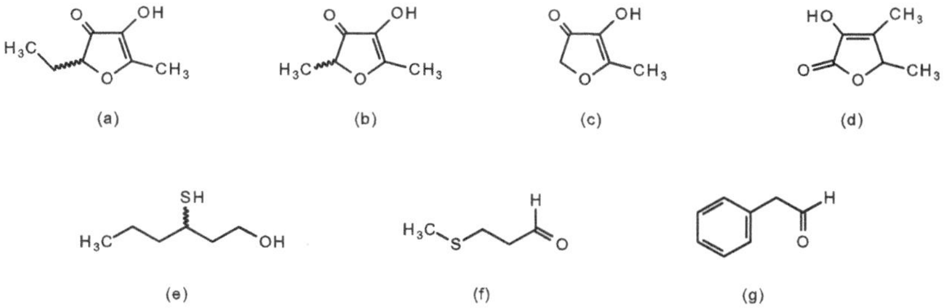

FIGURE 21.25 Representations of compounds whose differences in aroma at various concentration changes were sufficient to account for differences in various examples of wines from grapes attacked by *Botrytis cinerea* fungus. The compounds are: (a) homofuraneol; (b) furaneol; (c) norfuraneol; (d) sotolon; (e) 3-mercapto-1-hexanol; (f) methional; and (g) phenylacetaldehyde.

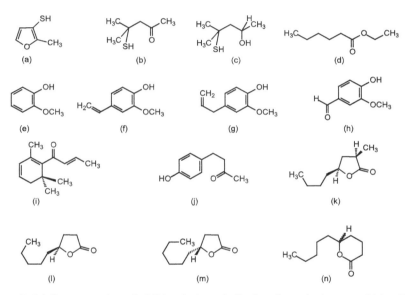

FIGURE 21.26 Representations of additional wine volatiles from botrytized grapes which, while somewhat important in their respective odors, are also reputed to be less dramatically odoriferous at their respective concentrations than those whose representations are shown in Figure 21.24. The representations shown here are: (a) 2-methyl-3-furanthiol; (b) 4-mercapto-4-methyl-2-pentanone; (c) 4-mercapto-4-methyl-2-pentanol; (d) ethyl hexanoate; (e) guaiacol; (f) 4-vinylguaiacol; (g) eugenol; (h) vanillin; (i) β-damsacenone; (j) raspberry ketone; (k) whisky lactone; (l) γ-nonalactone; (m) γ-decalactone; and (n) δ-decalactone.

hexanoate, 4-mercapto-4-methyl-2-pentanol, 2-methoxyphenol (guaiacol), 2-methoxy-4-vinylphenol (4-vinylguaiacol), (3R,5S)-5-butyldihydro-3-methyl-2(3H)-furanone (whisky lactone), 2-methoxy-4-allylphenol (eugenol), (5S)-5-pentyldihydro-2(3H)-furanone (γ-nonalactone), (2E)-1-(2,6,6-trimethyl-1,3-cyclohexadien-1-yl)-2-buten-1-one (β-damascenone), 4-hydroxy-2-methoxybenzaldehyde (vanillin), (5R)-5-hexyldihydro-2(3H)-furanone (γ-decalactone), tetrahydro-6-pentyl-2H-pyran-2-one (δ-decalactone), and 4-(4-hydroxyphenyl)-2-butanone (raspberry ketone). Structural representations of these are shown in Figure 21.26.

Among the more dramatic revelations concerning Noble Rot is the recent publication of a genomic analysis.[34] The analysis of this necrotrophic (i.e., deriving nutrients from killed host cells) fungus, discussed more fully below, revealed an "expansion in number and diversity of *B. cinerea*-specific secondary metabolites..." some of which are truly interesting and apparently unique (to this fungus) compounds.[34]

Necrotrophs infect many plants under a wide variety of conditions and succeed, in part, by secreting toxins, as well as enzymes that degrade walls. In that vein, it was found that *B. cinerea* "produce(s) dedicated infection structures in order to invade plants." The fungus demonstrates plasticity in the complexity of these structures that are necessary to breach the host cuticle.[34] Interestingly, *B. cinerea* appears to succeed so well because it has genes that encode putative enzymes for the degradation of complex plant carbohydrates including what appears to be preference for pectin (i.e., cell wall heteropolysaccharides).

Groups of phytotoxic metabolites produced by *B. cinerea* have been identified as including sesquiterpenes related to (1*S*,3a*R*,4*S*,6*R*,7*S*,7a*S*)-4-(acetyloxy)octahydro-7a-hydroxy-1,3,3,6-tetramethyl)-1*H*-indene-1,7-dicarboxaldehyde (botrydial)[35] and (2*R*,3*S*)-3-hydroxy-3-[(2*S*,3*S*,4*S*,5*R*,6*S*)-3-hydroxy-5-{[(2*E*,4*S*)-4-hydroxy-2-octenoyl]oxy}-2,4,6-trimethyltetrahydro-2*H*-pyran-2-yl]-2-methylpropanoic acid (botcinic acid) and its derivatives including a class of lactones, the boctinolides such as bactinolide itself [i.e., (2*S*,3*S*,4*S*,5*S*,6*R*,7*R*,8*S*)-5,6,7-trihydroxy-2,4,6,8-tetramethyl-9-oxo-3-oxonanyl (2*E*)-4-hydroxy-2-octenoate] derived from the acid and with varying substitution patterns.[36] These metabolites have been shown to be phytotoxic, producing cell collapse, fungal penetration, and subsequent colonization of plant tissue.[37] Representation of the structures of the sesquiterpenes are shown in (Figure 21.27).

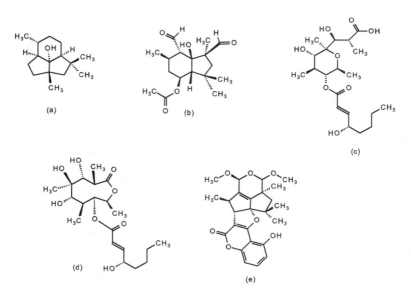

FIGURE 21.27 Representations of phytotoxic metabolites isolated from grapes infected with *Botrytis cinerea*. The structures shown have been assigned to (a) presilphiperfolanol; (b) botrydial; (c) botcinic acid; (d) botcinolide; and (e) hypocrolide A.

As a result of continuing examination of this limited family of sesquiterpenoids, new and expanded metabolites, such as (1S,3S,4R,5aS,12aS,14aS)-4,5,5a,13,14,14a-hexahydro-11-hydroxy-1,3-dimethoxy-4,13,13,14a-tetramethyl-1H,3H,6H-pyrano[5″,4″,3″:1′,7′]indeno-[4′,3a′:4,5]furo[3,2-c][1]benzopyran-6-one (hypocrolide A) continue to be isolated from related fungi.[38]

With regard to the enzymes that degrade walls, and given that in general the defense response of plants to infection is the production of reactive oxygen species (frequently abbreviated as ROS), it was expected that induction of ROS during infection by *Botrytis cinerea* would be found. Thus it was of some interest to examine the genome for genes involved in ROS tolerance. Remarkably, it was found that *B. cinerea* simply did not sense oxidative stress so that the normal plant defense was frustrated. It remains unknown how that stress is avoided in *B. cinerea*. In a similar vein, *B. cinerea* is "renowned" for the ability to acidify its environment through the secretion of organic acids ... in particular ... oxalic acid (HO_2CCO_2H).[34] Oxaloacetate acetyl hydrolase (EC 3.7.1.1) activity is commonly associated with the accumulation of oxalic acid, since the hydrolase converts oxaloacetate back to oxylate and acetate. The detection of the hydrolase provides a path for the production of the acid from the Krebs cycle intermediate (which normally goes on to citrate).

Further, the "*B. cinerea* genome (is) especially suited for pectin decomposition" as it is reported to have a large set of enzymes that are likely involved in 1,3-glucan degradation.[34] In that vein it was reported that the genome was searched for genes encoding key enzymes needed for the biosynthesis of peptides, polyketides, terpenes, and alkaloids, and some likely candidates were found. It was subsequently noted that there were major metabolic changes when grapes were infected with *B. cinerea*. There appear to be simultaneous influences of the plant defense system and fungal growth. Major differences were detected in amino acid concentrations between uninfected and infected grape berries.[39] Most recently, the genetic and molecular basis of botrydial biosynthesis from farnesyl diphosphate has been commented upon.[40]

It has recently been reported[41] that small RNA (sRNA) molecules derived from *B. cinerea* can act as effectors[42] to suppress host immunity and hijack the plant's own gene-silencing mechanism. Although this work was carried out using the model plant *Arabidopsis thaliana*, it is likely that the same or similar processes occur on various *Vitis vinifera* as well.

Interestingly, in the classic compendium *Handbook of Enology*[43] it is carefully pointed out that "Noble rot requires specific environmental conditions. The many studies undertaken have not yet been able to define these conditions." Indeed although it appears that the variety of the grape infected is related to the speed with which the progress of the infection progresses, the penetration of the cuticle and the loss of defense mechanisms have not successfully been related to that progress.[43]

As to further observational details, the same source notes that "... (t)he infection process, from healthy grape to *pourri rôti* (i.e., matured withered), lasts from 5 to 15 days" and that many years of observation in Sauternes vineyards (France) have shown that the first symptoms of attack appear 15–20 days before maturity. Interestingly, it has been found that the ratio of the concentration of glycerol to gluconic acid apparently represents the length of internal and external development phases of the parasite. This led to the idea that the ratio may serve to constitute a Noble Rot quality index. In vintages with favorable

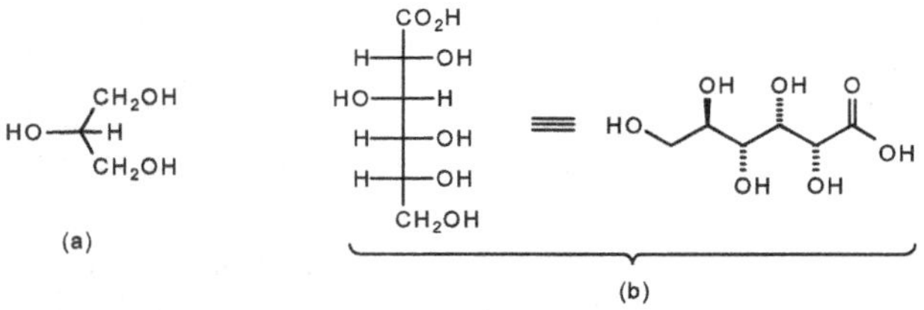

FIGURE 21.28 Representations of glycerol (a) and gluconic acid (b), grape levels of which are reported to be indicative of the extent of *Botrytis cinerea* parasite development.

climatic conditions for the fungus, rapid grape desiccation from the *pourri plein* (i.e., matured but not yet withered) stage onward leads to an elevated glycerol to gluconic acid ratio.[43] Gluconic acid (Figure 21.28) is an oxidation product of glucose, and glycerol is a reduction product of glyceraldehyde, a product of glycolysis, but details of the process(es) establishing the paths to the two products in the presence of the fungus and why the ratio is important are lacking.

Since terroir is a major (if not the major) determinant in the quality of the wine a grape can produce, it is also worthwhile noting that terroir will play a critical role in the health of the fungus *Botrytis cinerea*. Thus, it has been pointed out that . . . "The influence of noble rot . . . is greatly variable, depending on the occurrence of favorable seasonal condition for the mould (*sic*) infection."[44] The same authors noted that ". . . Fruity esters, carbonyl compounds, phenols, lactones and acetamides greatly changed among Botrytized wines." It is clear that the same compounds normally expected as a function of grape (the wine referred to in the above mentioned study is referenced as a "Recioto di Soave" and is presumably from the Soave region in northeastern Italy—but the grape itself might be a Chardonnay or Garganega, a Pinot grigio variety). Representations of the compounds used for the purposes of comparison are listed and shown below. In addition to those shown, the relative amounts of residual sugars (of the Botrytized and control) were compared, and the Botrytized vinification was significantly higher. Further, there were differences in simple alcohols and esters, including ethyl acetate ($CH_3CH_2OCOCH_3$), 1-butanol ($CH_3CH_2CH_2CH_2OH$), 1-propanol ($CH_3CH_2CH_2OH$), 2-methyl-1-propanol [$(CH_3)_2CHCH_2OH$], 2-methyl-1-butanol [$CH_3CH_2CH(CH_3)CH_2OH$], 3-methyl-1-butanol [$(CH_3)_2CHCH_2CH_2OH$], 2-phenylethanol ($C_6H_5CH_2CH_2OH$), 3-methylthiopropanol (methionol, $CH_3SCH_2CH_2CH_2OH$), benzyl alcohol ($C_6H_5CH_2OH$), 1-hexanol [$CH_3(CH_2)_4CH_2OH$], isopentyl acetate [3-methylbut-1-yl ethanoate, $CH_3CO_2CH_2CH_2CH(CH_3)_2$], 2-phenylethyl acetate ($CH_3CO_2CH_2CH_2C_6H_5$), ethyl butanoate ($CH_3CH_2CH_2CCO_2CH_2CH_3$), ethyl hexanoate [$CH_3(CH_2)_4CO_2CH_2CH_3$], ethyl octanoate [$CH_3(CH_2)_6CO_2CH_2CH_3$], ethyl decanoate [$CH_3(CH_2)_8CO_2CH_2CH_3$], butanoic acid ($CH_3CH_2CH_2CO_2H$), hexanoic acid [$CH_3(CH_2)_4CO_2H$], octanoic acid [$CH_3(CH_2)_6CO_2H$], decanoic acid [$CH_3(CH_2)_8CO_2H$], phenylacetaldehyde ($C_6H_5CH_2CHO$), benzaldehyde ($C_6H_5CHO$), and as shown with the representations below, in Figure 21.29, 1-octene-3-ol, ethyl cinnamate, furan-2-carboxyaldehyde (furfural), 4-hydroxy-3,5-dimethoxybenzaldehyde (syringaldehyde), 3,7-dimethylocta-1,6-dien-3-ol (linalool), 1-methylethyl-4-methyl-3-cyclohexen-1-ol (4-terpinenol). (2E)-3,7-dimethyl-2,6-octadien-1-ol (geraniol), (2E)-1-(2,6,6-trimethyl-1,3-cyclohexadien-1-yl)-2-buten-1-one (β-damascenone), 2-methoxy-4-vinylphenol (4-vinylguaiacol), 4-vinylphenol, and 2-methoxy-4-(2-propen-1-yl)-phenol (eugenol).

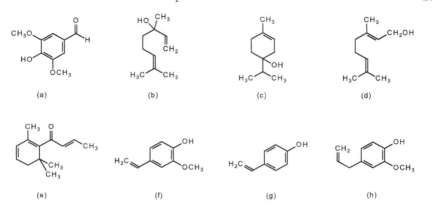

FIGURE 21.29 Representations of compounds both expected and normally found in the grape (e.g., a Chardonnay or Garganega) without infection by the fungus *Botrytis cinerea* and with infection by the fungus. Comparison as to relative quantities appear unchanged. The compounds shown are: (a) syringaldehyde; (b) linalool; (c) 4-terpinenol; (d) geraniol; (e) β-damsacenone; (f) 4-vinylguaiacol; (g) 4-vinylphenol; and (h) eugenol.

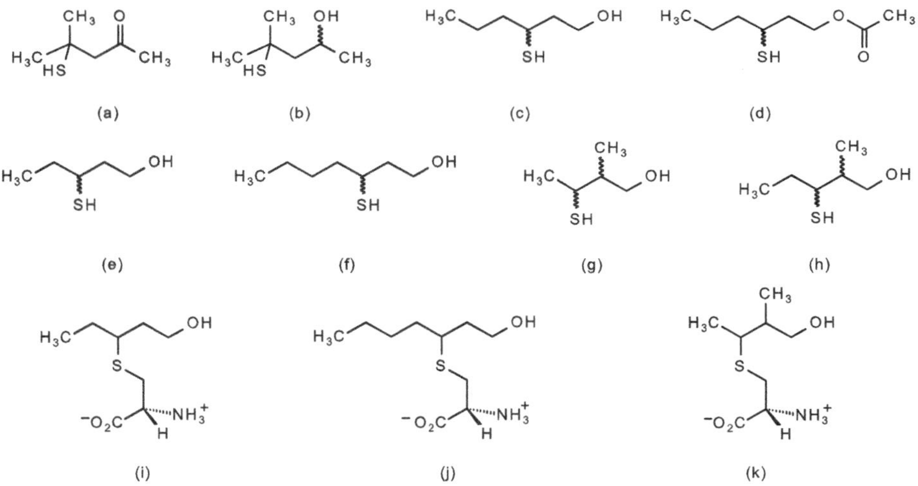

FIGURE 21.30 Representations of thiol and L-cysteine (Cys, C) related compounds normally found in the Sauvignon blanc uninfected grapes used for the Noble Rot and in grapes infected with *B. cinerea*. The relative amounts (when both were detected) of these compounds were different in the infected and uninfected samples. The compounds shown are (a) 4-mercapto-4-methyl-2-pentanone; (b) 4-mercapto-4-methyl-2-pentanol; (c) 3-mercapto-1-hexanol; (d) 3-mercapto-1-hexanol acetate; (e) 3-mercapto-1-pentanol; (f) 3-mercapto-1-heptanol; (g) 3-mercapto-2-methyl-1-butanol; (h) 3-mercapto-2-methyl-1-pentanol; (i) (S)-3-(1-pentanol)-L-cysteine; (j) (S)-3-(1-heptanol)-L-cysteine; and (k) (S)-3-(2-methyl-1-butanol)-L-cysteine.

Again, it is not surprising that the same compounds (some of which were seen to be found in the grapes themselves and in other varieties of *V. vinifera*) are found in both the grapes that are infected with the fungus *B. cinerea* and those that are not. It appears that it is the differences in relative concentrations that matter most.

There are several additional major differences between *B. cinerea*–infected grapes and (apparently) uninfected grapes taken from the same bunch. For example the concentrations of thiols and the L-cysteine derivatives (or progenitors) of those thiols common to Sauvignon blanc, discussed there (Chapter 14) and which, taken together, are apparently major contributors to the flavor of this grape are found in quite different concentrations when the grape has enjoyed infection by *B. cinerea*.[45] Representations of those thiols are shown in Figure 21.30.

Representations of additional volatiles which, for the most part, are present in both Botrytized and uninfected wines are shown in Figures 21.31a, 21.31b, 21.31c, and 21.32,. In general it appears that different strains of *Botrytis cinerea* as well as the extent of infection of

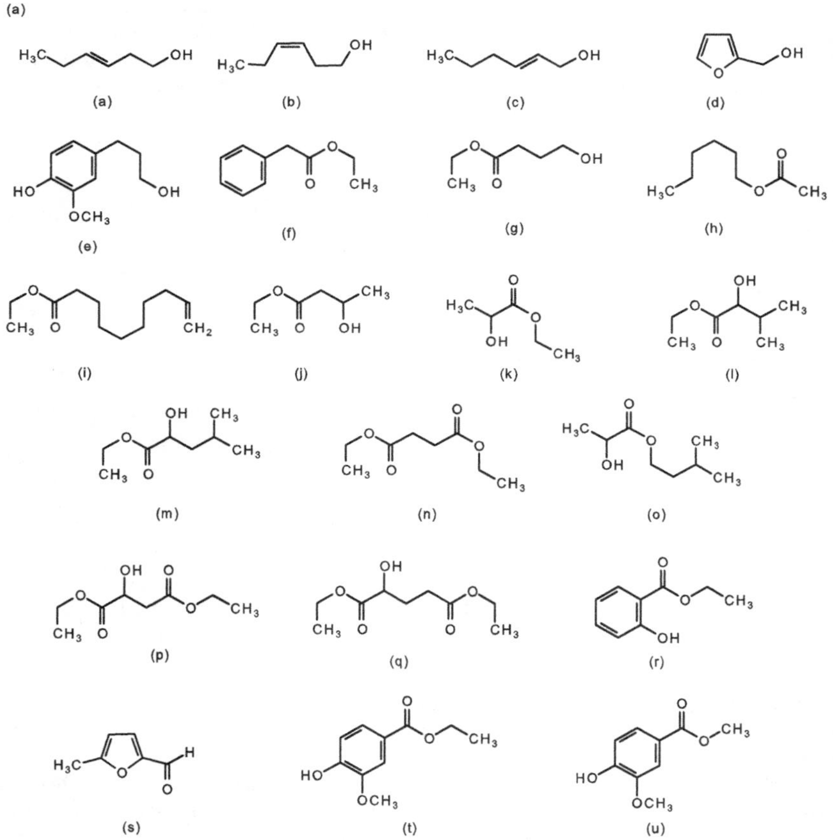

FIGURE 21.31a Representations compounds normally found in the grapes used for the Noble Rot. The relative amounts of these compounds were different in the infected and uninfected samples. Those shown are: (a) *E*-3-hexene-1-ol; (b) *Z*-3-hexene-1-ol; (c) *E*-2-hexene-1-ol; (d) furfuryl alcohol (furfuranol); (e) homovanillic alcohol; (f) ethyl 2-phenylacetate; (g) ethyl 4-hydroxybutanoate; (h) hexyl acetate; (i) ethyl 9-decenoate; (j) ethyl 3-hydroxybutanoate; (k) ethyl lactate; (l) ethyl 2-hydroxy-3-methylbutanoate; (m) ethyl 2-hydroxy-4-methylpentanoate; (n) diethyl succinate; (o) 3-methylbutyl lactate; (p) diethyl malate; (q) diethyl 2-hydroxyglutarate; (r) ethyl salicylate; (s) 5-methylfurfural; (t) ethyl vanillate; and (u) methyl vanillate.

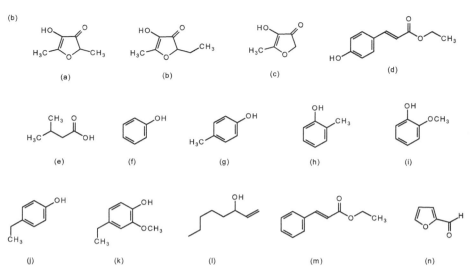

FIGURE 21.31b Representations compounds normally found in the grapes used for the Noble Rot. The relative amounts of these compounds were different in the infected and uninfected samples. Those shown are: (a) furaneol; (b) homofuraneol; (c) norfuraneol; (d) ethyl coumarate; (e) 3-methylbutanoic acid; (f) phenol; (g) *para*-cresol; (h) *ortho*-cresol; (i) guaiacol; (j) 4-ethylphenol; (k) 4-ethylguaiacol; (l) 1-octene-3-ol; (m) ethyl cinnamate; and (n) furfural.

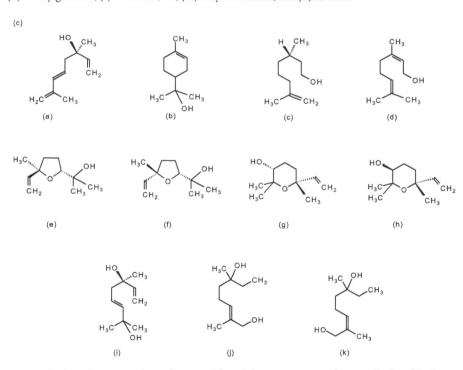

FIGURE 21.31c Representations of isoprenoids and C-13 norisoprenoids normally found in the grapes used for the Noble Rot. The relative amounts of these compounds were different in the infected and uninfected samples. Those shown are: (a) hotrienol; (b) α-terpineol; (c) citronellol; (d) nerol; (e) *trans*-linalool oxide—furan form; (f) *cis*-linalool oxide—furan form; (g) *cis*-linalool oxide—pyran form; (h) *trans*-linalool oxide—pyran form; (i) (R)-(E)-2,6-dimethyl-3,7-octadien-2, 6-diol; (j) *trans*-8-hydroxydihydrolinalool; and (k) *cis*-8-hydroxydihydrolinalool.

any given strain has a marked effect on the final product and, "in particular... the interaction between the levels of *B. cinerea* infection and degree of grape withering is very important to modulate the aroma . . ." (*vide supra*). However, individual taste and odor detection varies. Thus, the analytical chemistry done to identify many of the more prevalent components (i.e., those of concentrations high enough to be detected) is important, as are the observations showing that the same compounds in different ratios produce odors whose combination is distinctive.

Thus, again, while there are compounds that are reported present in both infected and uninfected grapes, a few are different (particularly, as noted above, toxins produced by the fungus). The particular flavor(s) reported to be associated with Noble Rot appear to result from differences in concentrations of the same (relatively) few compounds.

In the same vein, Chardonnay grapes infected by *Botrytis cinerea* were compared to the same grapes that were uninfected. The particular grape is used to produce Champagne (i.e., Chardonnay grapes specifically from the Champagne region of France), and the grape was examined because, in the presence of *Botrytis cinerea,* it was held that proteins involved in the foam stabilization of sparkling wines were missing (i.e., degraded or their formation repressed). Indeed, in the event, the proteins isolated were largely of grape origin (e.g., invertase and pathogenesis-related (PR) proteins (*pathogenesis-related proteins* are plant proteins that are formed when a plant is attacked by a pathogen). However, numerous grape proteins had disappeared, and pectinolytic enzymes (i.e., enzymes that facilitate hydrolysis of cell wall polysaccharides [pectins]) were found in the wine from botrytized grapes. It was shown that the low-molecular-weight "foam-active" proteins wanted in the champagne beverage disappeared on infection and that the foaming was thus inhibited.[46]

PART C. CHAMPAGNE

It is generally appreciated that although a wine might be a "sparkling wine," having been (perhaps) prepared in a manner at least similar to that of a sparkling wine actually labelled "Champagne," it is, nonetheless, not Champagne. First, in order to be "Champagne," the wine must be produced from grapes grown in the Champagne region of France. Second, to be labeled "Champagne," the grapes (grown in specific areas of Champagne) are said to be treated in specific traditional ways or by specific well-defined methods. These methods apparently include specific vineyard practices, specific pressing regimes, and the requirement that the final fermentation take place in the bottle.

Nonetheless, the term "Champagne" is sometimes loosely used as a synonym for sparkling wine; it may be unlawful to do so.

The black grape cultivars, Pinot noir and Pinot Muenier, whose skins are separated from the juice early in the extraction process, along with the white grape Chardonnay, are those from which Champagne is made. The compounds found in Pinot noir and Chardonnay have been discussed (Chapter 14), and the Pinot Muenier, apparently a somewhat more acidic grape than Pinot noir,[47] has not yet been investigated as thoroughly as the others. It has been reported to have a high ergosta-5,7,22-trien-3β-ol (ergosterol, Figure 21.33) content (used as an indicator of fungal contamination).[48] The initial fermentation is apparently the same as with other grape cultivars.

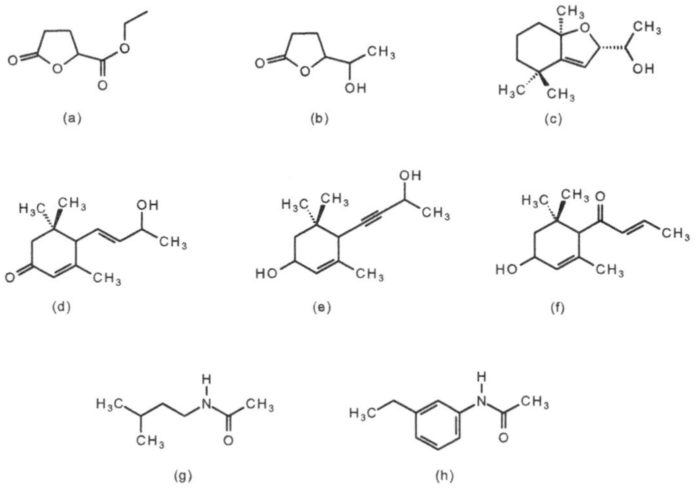

FIGURE 21.32 Representations of novel compounds present in the grapes infected with Noble Rot. These compounds are modified from those normally present by additional oxidation and or other modification. Those shown are: (a) ethyl tetrahydro-5-oxo-2-furancarboxylate; (b) sherry lactone; (c) actinidol; (d) 3-oxo-α-ionol; (e) 3-hydroxy-7,8-dihydro-β-ionol; (f) 3-hydroxy-β-damascone; (g) N-(3-methylbutyl)acetamide; and (h) N-(3-ethylphenyl)acetamide.

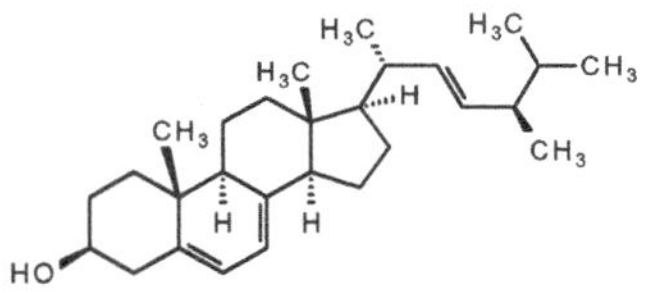

FIGURE 21.33 A representation of the structure of ergosterol.

To make the sparkling wine, a second fermentation with fresh yeast is carried out. The second fermentation occurs after the wine has been bottled! Thus, the bottled beverage (usually with a crimp-type bottle cap) is opened, a bolus of sugar (usually sucrose) and additional *Saccharomyces cerevisiae* added and the bottle resealed. It is generally understood that the second batch of *S. cerevisiae* might or might not be the same strain of the yeast originally used in the first fermentation and that each vineyard producing Champagne (and possibly other sparkling wines) has its own recipe. To be compliant with the *appellation d'origine contrôlée* as defined by the **Institut national de l'origine et de la qualité** which regulates what can be labelled as Champagne, the bottle is aged for a minimum of 1.5 years on the *lees*.

When the aging is considered sufficient, the vintage is riddled. In the riddling (*remuage*) the bottles are inclined at about a 45-degree angle on a riddling rack. The rack consists (at a minimum) of holes, bored in a board and able to hold the neck of a champagne bottle. It can, of course be more elaborate but it must simply hold the angled bottles with their necks pointed downward. Then, the Riddler (i.e., (s)he who riddles) on a daily cycle, rotates every bottle a few degrees from its original position in the same direction and with a twist that raises the bottle's bottom and lowers the bottle's neck slightly. As the weeks pass, the bottles, beginning at a 45-degree angle, are now at about a 60-degree angle.

During the process, the solids (e.g., the lees and insoluble metabolites) resulting from the secondary fermentation gradually move into the neck of the bottle where they form a plug. When visual inspection by the Riddler confirms that the secondary fermentation is complete (the plug no longer grows) and the wine is clear, the bottle is carefully removed from the rack and the neck and shoulders of the bottle are plunged into an ice–salt (or colder) sub-zero (e.g., about −15 °C or 5 °F) bath for several minutes. Depending upon the concentration of alcohol and the other constituents (e.g., sugar, tanins, phenols) in the solution (colder temperatures are needed for higher alcohol concentrations) a water-ice plug holding the spent yeast and sediment in the neck results, so that when the crown cap of the bottle is removed, the carbon dioxide pressure within the bottle forces the frozen sediment and ice out of the bottle.[49]

Now, some wine from a previous vintage is added to fill the bottle, and the bottle is quickly corked to maintain the pressure. As desired some additional sugar (presumably sucrose dissolved in the liquor from the prior vintage) may be added. The amount of sugar, *le dosage*, (the dosage, grams/liter, gL^{-1}) added determines the sweetness rating. For example, "Brut Nature" may not have more than 3 gL^{-1} of sugar, "Extra Brut" may have up to 6 gL^{-1}, and "Brut" up to 12 gL^{-1}. "Extra dry," "Dry," etc. have increasing amounts of sugar.

There are several interesting features in this practice that need consideration.

First, it might be presumed that the alcohol concentration in the wine to which fresh sugar and yeast is added had not reached a concentration that was toxic to the yeast, since if the yeast died as soon as it was added, the second fermentation would not occur. However, since the second fermentation does occur it is likely that the changes taking place in the yeast (for example in the cell walls to allow carbohydrates in and ethanol and other products out) have not yet occurred in the newly added yeast. Thus, the freshly added yeast does not initially suffer the same fate as the older yeast, the cell walls being robust, until it too has metabolized some glucose and produced alcohol.

Another alternative explanation is that the strain of the freshly added *S. cerevisiae* is one that is more tolerant to ethanol, or a different yeast (which may or may not be a strain of *S. cerevisiae*) is used only for the secondary fermentation. Indeed, as noted earlier in the discussion of *Botrytis cinerea* (Noble Rot) where some specific proteinaceous secondary metabolites were formed, it is likely that formation of specific metabolites occurs in the second fermentation too.[50,51] Thus, it is not surprising to find the suggestion that, with regard to the second fermentation, "... most of these compounds contribute positively to the aroma, taste and foaming properties of sparkling wines."[52]

Second, the release of carbon dioxide (CO_2) results in the effervescence occurring when the chilled bottle is opened and has been the subject of considerable investigation.

On the one hand, gases dissolved in liquids (ideal gases in ideal liquids, ideally) can be shown to follow (approximately for real gases and liquids) Henry's Law. That is, the amount of a dissolved gas in a liquid is proportional to the partial pressure of that gas above the liquid. The constant converting the proportionality into an equality is Henry's Gas Law Constant.[53] Thus, if a liquid is bottled under a pressure of a gas, the gas will dissolve in the liquid and will subsequently be released if the partial pressure of the gas, on opening the bottle, is less than what it was when the bottle was closed. So, if seltzer (a solution of carbon dioxide [CO_2] in water) is bottled under a pressure of carbon dioxide (CO_2) of several atmospheres (atm) pressure and it is subsequently opened in a room where the carbon dioxide pressure is less than it was when bottled (currently [November 2015] 400.16 ppm at Mauna Loa Observatory),[54,55] then the carbon dioxide will be released and bubbles will form. Henry's constant for a given gas is only a constant at a given temperature. A new Henry's constant is needed for each different temperature. To a first approximation and considering components in addition to carbon dioxide, bottles of soda that have just over one atmosphere total pressure at about 4 °C (39 °F) will have about three atmospheres total pressure at about 20 °C (70 °F).

Additionally, since carbon dioxide is being released, the equilibria between carbon dioxide, carbonic acid, bicarbonate, and carbonate will shift to produce more carbon dioxide (in accord with Le Châtelier's Principle)[56] which, as proton (H^+) transfer is required, will be influenced by the pH of the solution.[57]

The bubbles formed in champagne, derived from carbon dioxide, have been studied in some detail using high-speed photography and the change in size and shape as the bubbles move from nucleation to the surface noted.[58] Subsequently, work in the same laboratory undertook to attempt to measure the composition of the aerosol formed by droplets carried to the surface as bubbles rose through the champagne flute. The release of carbon dioxide (it is noted that about 5 L of CO_2 will eventually escape from the liquid in a 0.75-L champagne bottle) carries champagne components (some of which are surfactants [i.e., surface tension–lowering materials] which, in this case, are probably proteins and lipids formed during fermentation) into the vapor above the bulk liquid. And bubbles breaking at the surface force some of the surfactants into droplets, adding to the fragrance. The aerosol and the champagne bulk liquid were compared using mass spectrometry.

In the event, hundreds of components in the mass-to-charge (m/z) range of about 150–1,000 amu (atomic mass units), assuming Z = 1, were found preferentially partitioning in the aerosol, and some structural suggestions—fatty acids (i.e., even-carbon-numbered carboxylic acids C_{10} to C_{24} range) and their esters, norisoprenoids (i.e., C_{13} derivatives), and some lower-molecular-weight alcohols and ethers—were made. Definitive assignments were not made.[59]

A recent effort, necessitating many approximations, has been made in an attempt to determine the global number of bubbles likely to nucleate in a flute into which a freshly opened bottle of champagne has been poured. The following are among the interesting observations reported:[60] (a) The temperature-dependent number of bubbles can be determined... in 100 mL of champagne ... "one million bubbles seems to be a reasonable approximation ... if you resist drinking." (b) As a function of temperature ... "several tens of thousands of bubbles may ... be saved if champagne is served in a tilted flute." (c) ... "The global number of bubbles likely to nucleate in a flute increases with the champagne temperature." (d) "If champagne is ... served on the wall of a tilted flute, dissolved CO_2 is better preserved and ... several tens of thousands of bubbles should additionally form."

PART D. ICE WINE

As pointed out earlier (Chapter 14) the Gewürztraminer grape is among the few that are often used to prepare ice wine. Grapes that are going to be used for ice wine are allowed to mature on the vine well past the time when mature grapes might ordinarily be harvested until the grapes freeze in winter weather. It is sometimes difficult to permit this prolonged period of over-ripening on the vine without the grapes rotting before they freeze.

When the water in the grape freezes, cell walls rupture, cellular water freezes out, and the grape is dehydrated. The components of the fragmented cells are excluded from the aqueous phase by its solidification into ice crystals, as the crystal lattice cannot form if they are included. The components excluded remain available to be acted upon by yeast or to be found unchanged in the resulting wine. Thus, while frozen, the grapes (generally harvested by hand) are pressed and the ice separated from the viscous juice. The yield of concentrated juice is significantly less than would be were the grapes not frozen. Since this concentrate lacks sufficient water to effect easy penetration of yeast cell walls, it is often difficult to get fermentation to begin, and finding the right strain of *Saccharomyces cerevisiae* requires care.[61]

Interestingly, in the Gewürztraminer ice wine the concentrations of glycerol, gluconic acid, glucose and, dramatically, fructose, (Figure 21.34) are much higher than in the normally harvested and produced beverage.[61]

In addition, there are odorants which are dramatically enhanced in the ice wine relative to those found in wine produced from the same grapes but not allowed to freeze on the vine. These include tetrahydro-4-methyl-2-(2-methylpropenyl)-2*H*-pyran (*cis* rose oxide), 3,7-dimethyl-6-octen-1-ol (citronellol), (*E*)-3,7-dimethyl-2,6-octadien-1-ol (geraniol), ethyl acetate, and 1,1-diethoxyethane (the diethyl acetal of acetaldehyde) as shown in Figure 21.35.

And there are odorants which are dramatically reduced. These include the sweet-smelling esters, ethyl butanoate, ethyl hexanoate, and ethyl octanoate, along with 2-methoxy-4-ethenyl-phenol (4-vinylguaiacol) as seen in Figure 21.36.

As to the final beverage, it is claimed "... sweet and fruity (aromas) were dominant in standard and late harvest wines, while in ice wine they diminished, and with increased terpenic, floral, chemical, pungent and ripe fruit series established a more complex profile."[61]

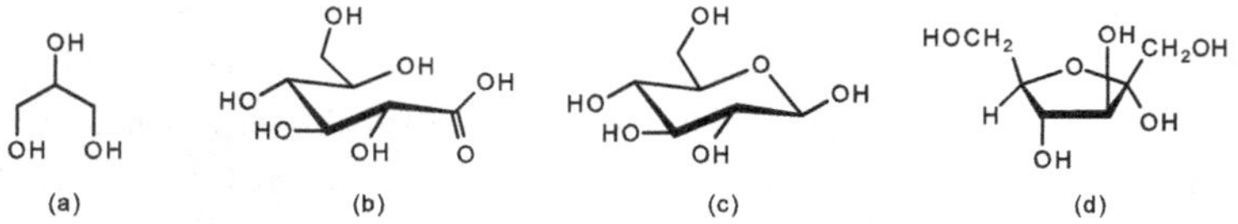

FIGURE 21.34 Representations of species whose concentration in Gewürztraminer ice wine is enhanced over that normally found. The compounds are: (a) glycerol; (b) D-gluconic acid; (c) β-D-glucopyranose; and (d) α-D-fructofuranose.

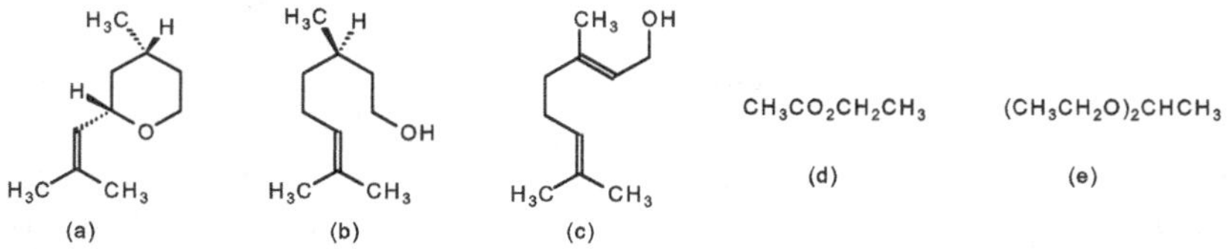

FIGURE 21.35 Representations of odorants whose concentration in Gewürztraminer ice wine is enhanced over that normally found. The compounds are: (a) *cis*-rose oxide; (b) (*R*)-(+)-citronellol; (c) geraniol; (d) ethyl acetate; and (e) 1,1-diethoxyethane.

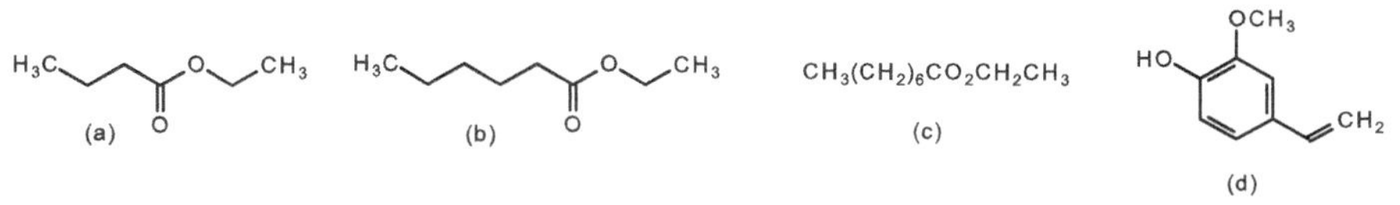

FIGURE 21.36 Representations of odorants whose concentration in Gewürztraminer ice wine is reduced over that normally found. The compounds are: (a) ethyl butanoate; (b) ethyl hexanoate; (c) ethyl octanoate; and (d) 4-vinylguaiacol.

NOTES AND REFERENCES

1. Somers, T. C.; *Phytochem.* **1971,** *10,* 2175, *et. seq.*

2. Sommers, T. C.; Vérette, E. In *Modern Methods of Plant Analysis, New Series,* Vol 6; Linskins, H. F.; Jackson, J. F., Eds.; Springer: Berlin, 1988; p 219 ff.

3. Liao, H.; Cai, Y.; Haslam, E. *J. Sci. Food Agric.* **1992,** *59,* 299.

4. Robinson, J. *Vines, Grapes and Wines;* Alfred A. Knopf: New York, 1986, p 216.

5. Simpson, R. G. *J. Sci. Food Agric.* **1980,** *31,* 214.

6. Williams, A. A.; Lewis, M. J.; May, H. V. *J. Sci. Food Agric.* **1983,** *34,* 311.

7. Moreira, N.; Guedes de Pinho, P. In *Advances in Food and Nutrition Research,* Vol. 63; Jackson, R. S., Ed.; Academic Press: Waltham, MA, 2011, Ch. 5, pp 119 ff.

8. Fulcrand, H.; Cameira do Santos, P.-J.; Sarni-Manchado, P.; Cheynier, V.; Favre-Bonvin, J. *J. Chem. Soc. Perkin Tran.* **1996,** *1,* 735.

9. Bakker, J.; Bridle, P.; Honda, T.; Kuwano, H.; Saito, N.; Terahara, N.; Timberlake, C. F. *Phytochem.* **1997,** *44,* 1375.

10. Bakker, J.; Timberlake, C. F. *J. Agric. Food Chem.* **1997,** *45,* 35.

11. Fulcrand, H.; Benabdeljalil, C.; Rigaud, J.; Cheyner, V.; Moutounet, M. *Phytochem.* **1998,** *47,* 1401. Interestingly, some databases show this compound as Vitisin A.

12. Mateus, N.; de Pascual-Teresa, S.; Rivas-Gonzalo, J. C.; Santos-Belga, C.; de Freitas, V. *Food Chem.* **2002,** *76,* 335.

13. Atanasova, V.; Fulcrand H.; Cheynier, V.; Moutounet, M. *Anal. Chim. Acta* **2002,** *458,* 15.

14. Pissarra, J.; Lourenço, S.; González-Paramás, A. M.; Mateus, N.; Buelga, C. S.; Silva, A. M. S.; De Freitas, V. *Food Chem.* **2005,** *90,* 81 for a discussion of the acetaldehyde component.

15. Our skills at isolating a few compounds of complex structure amongst the thousand or so present for complete analysis are not yet sufficient. Hence, small samples are injected into chromatographic systems where they are separated into their multitude of components, and analysis of the separated materials by mass spectroscopy (MS) is effected. The mass spectrum of each separated component (see Appendix 1) is unique. So, with a belief that identification can be confirmed, components are synthesized and their mass spectra compared. If the spectra are identical, then the structure of the isolate is considered identical to that synthesized.

16. Salas, E.; Atanasova, V.; Poncet-Legrand, C.; Meudec, E.; Mazauric, J. P.; Cheynier, V. *Anal. Chim. Acta* **2004,** *513,* 325.

17. Schwarz, M.; Jerz, G.; Winterhalter, P. *Vitis* **2003,** *42,* 105.

18. Mateus, N.; Oliverira, J.; Santos-Buelga, C.; Silva, A. M. S.; de Freitas, V. *Tetrahedron Lett.* **2004,** *45,* 3455.

19. Olivera, J.; De Freitas, V.; Silva, A. M. S.; Mateus, N. *J. Agric. Food Chem.* **2007,** *55,* 6349.

20. He, J.; Santos-Buelga, C.; Silva, A. M. S.; Mateus, N.; de Frietas, V. *J. Agric. Food Chem.* **2006,** *54,* 9598.

21. Figueiredo-González, M.; Cancho-Grande, B.; Simal-Gándara, J.; Teixeira, N.; Mateus, N.; de Frietas, V. *Food Chem.* **2014,** *152,* 522.

22. Mateus, N.; Silva, A. M. S.; Rivas-Gonzalo, J. C.; Santos-Buelga, C.; de Freitas, V. *J. Agric. Food Chem.* **2003,** *51,* 1919.

23. Figueiredo-González, M.; Reegueiro, J.; Cancho-Grande, B.; Simal-Gándara, J. *Food Chem.* **2014,** *143,* 282.

24. Bushdid, C.; Magnasco, M. O.; Vosshall, L B.; Keller, A. *Science* **2014,** *343,* 1370.

25. Marq, P.; Schieberle, P. *J. Agric. Food Chem.* **2015,** *63,* 4761.

26. Genoves, S.; Gil, J. V.; Valles, S.; Sasas, J. A.; Manzanares, P. *Am. J. Enol. Vitic.* **2005,** *56,* 188.

27. Robinson, J. *The Oxford Companion to Wine;* Oxford University Press: New York, 2006; p 423.

28. Perestrelo, R.; Albuquerque, F.; Rocha, S. M.; Câmara, J. S. In *Advances in Food and Nutrition Research,* Vol. 63; Jackson, R. S., Ed.; Academic Press: Waltham, MA, 2011, Ch. 7, pp 207 ff.

29. Câmara, J. S.; Herbert, P.; Marques, J. C.; Alves, M. A. *Anal. Chimica Acta* **2004,** *513,* 203.

30. Perestrelo, R.; Barros, A. S.; Rocha, S. M.; Câmara, J. S. *Microchem. J.* **2014,** *116,* 107. DOI:10.1016/j.microc.2014.04.010.

31. Pereira, V.; Pereira, A. C.,: Trujillo, J. P. P.; Cacho, J.; Marques, J. C. *J. Chem.* **2015.** ID 494285. DOI:10.1155/2015/494285.

32. "But, when nothing subsists of an old past ... frailer but more enduring, more immaterial, more persistent, more faithful, smell and taste still remain for a long time ... on their almost impalpable droplet, the immense edifice of memory." Proust, M. *Remembrance of Things Past,* Vol. 1, *Swann's Way;* Vintage: New York, 1913.

33. Sarrazin, E.; Dubourdieu, D.; Darriet, P. *Food Chem.* **2007,** *103,* 536.

34. Anselem, J. et al. *(66 additional authors) PLoS Genet.* **2011,** *7,* e1002230.

35. Which, it has been suggested, is derived from the sesquiterpenol (2aS,4aS,5R,7aS,7bR)-decahydro-1,1,2a,5-tetramethyl-7bH-cyclopent[cd]inden-7b-ol (presilphiperfolanol). See Wang, C.-M.; Hopson, R. Lin, X.; Cane, D. E. *J. Am. Chem. Soc.* **2009,** *131,* 8360; and Wedler, H. B.; Permberton, R. P.; Tantillo, D. J. *Molecules* **2015,** *20,* 10781.

36. Tani, H.; Koshino, H.; Sakuno, E.; Cutler, H. .; Nakajima, H. *J. Nat. Prod.* **2006,** *69,* 722.

37. Colmenares, A. J.; Aleu, J.; Durán-Patrón, R.; Collado, I. G.; Hernández-Galán, R. *J. Chem. Ecol.* **2002,** *28,* 997; and Morage, J.; Dalmais, R.; Izquierdo-Bueno, I.; Aleu, J.; Hanson, J. R.; Hernández-Galán, R.; Viaud, M.; Collado, I. G. *ACS Chem. Biol.* **2016.** DOI:10.1021/acschembio.6b00581.

38. Yuan, Y.; Feng, Y.; Ren, F.; Niu, S.; Liu, X.; Che, Y. *Org. Lett.* **2013,** *15,* 6050.

39. Hong, Y. S.; Martinez, A.; Liger-Belair, G.; Jeandet, P.; Nuzillard, J.-M.; Cilindre, C. *J. Exp. Bot.* **2012,** *63,* 5773.

40. Moraga, J.; Dalmais, B.; Izquierdo-Bueno, I.; Aleu, J.; Hanson, J. R.; Hernández-Galán, R.; Viaud, M.; Collado, G. *ACS Chem. Biol.* **2016.** DOI:10.1021/acschembio.6b00581; and see also Pinedo, C.; Wang, C.-M.; Pradier, J.-M.; Dalmais, B.; Choquer, M.; Le Pêcheur, P.; Morgant, G.; Collado, I. G.; Cane, D. E.; Viaud, M. *ACS Chem. Biol.* **2008,** *3,* 791.

41. Weiberg, A.; Wang, M.; Lin, F.-M.; Zhao, H.; Zhang, Z.; Kaloshian, I.; Huang, H.-D.; Jin, H. *Science* **2013,** *342,* 118.

42. An effector is (usually) a small molecule that selectively binds to a protein and regulates its activity.

43. Ribéreau-Gayon, P.; Dubourdieu, D.; Donèche, B.; Lonvaud, A. *The Handbook of Enology*, Vols. 1&2, 2nd Ed.; Wiley: Chichester, UK, 2006; Volume 1, Chapter 10, p 283 ff.

44. Tosi, E.; Azzolini, M.; Lorenzini, M.; Torriani, S.; Fedrizzi, B.; Finato, F.; Cipriani, M.; Zapparoli, G. *Eur. Food Res. Technol.* **2013,** *236,* 853. DOI:10.1007/s00217-013-1943-8.

45. Thibona, C.; Shinkarukb, S.; Jourdesa, M.; Bennetauc, B.; Dubourdieua, D.; Tominaga, T. *Anal. Chim. Acta* **2010,** *660,* 190.

46. Cilindre, C.; Jégou, S.; Hovasse, A.; Schaeffer, C.; Castro, A. J.; Clément, C.; Van Dorsselaer, A.; Jeandet, P.; Marchal, R. *J. Proteome Res.* **2008,** *7,* 1199.

47. Regner, F.; Stadlbauer, A.; Eisenheld, C.; Kaserer, H. *Am. J. Enol. Vitic.* **2000,** *51,* 7.

48. Porep, J. U.; Mrugala, S.; Pour Nikfardjam, M. S.; Carle, R. *Food Bioproc. Technol.* **2015,** *8,* 1455.

49. The freezing temperature of a solution which is about 15% ethanol in water is of the order of −15 °C (Smithsonian Physical Tables, 9th Ed.; Forsythe, W. E., Ed., Knovel, 2003. ISBN 978-1-59123-539-8, Table 126, p 132 and The Engineering Toolbox. http://www.engineeringtoolbox.com/ethanol-water-d_989.html (accessed Apr 9, 2017). However, the freezing temperature of the water in the beverage will almost certainly be somewhat lower as carbon dioxide, bicarbonate, and other "impurities" are also present.

50. Troton, D.; Charpentier, M.; Robillard, B.; Calvyrac, R.; Duteurtre, B. *Am. J. Enol. Vitic.* **1989,** *3,* 175.

51. Martínez-Rodriguez, A. J.; Carrascosa, A. V.; Polo, M. C. *Int. J. Food Microbiol.* **2001,** *68,* 155.

52. Penach, V.; Valero, E.; Gonzalez, R. *Int. J. Food Microbiol.* **2012,** *153,* 176.

53. Henry, W. *Phil. Trans. R. Soc. London* **1803,** *93,* 29. As originally stated "... water takes up, of gas condensed by one, two, or more additional atmospheres, a quantity which, ordinarily compressed, would be equal to twice, thrice, &c. the volume absorbed under the common pressure of the atmosphere." Subsequently more precisely restated to reflect the observation that the absorbed gas was studied at constant temperature and that the gas to be absorbed (whatever its partial pressure) was in equilibrium with the liquid at that temperature.

54. http://co2now.org/current-co2/co2-now/ (accessed Apr 9, 2017).

55. The abbreviation ppm refers to parts per million (or one part in 10^6 or 1 in a million = 1,000,000). To a first approximation, 400 ppm means that there are 400 parts per million atmospheres (atm) partial pressure of carbon dioxide in one atmosphere total pressure (which is mostly nitrogen and oxygen). This number does not include the fact that carbon dioxide, dissolved in water, is in equilibrium with both carbonic acid and its ionization products, the bicarbonate and carbonate anions which serve to allow more carbon dioxide to dissolve than would be the case were it unreactive.

56. Broadly, Le Châtelier's Principle argues that if a system is at equilibrium and the equilibrium is perturbed by some change, the system will adjust to reestablish the equilibrium.

57. As described in Appendix 1, the pH is a measure of the acidity of the solution. The more acidic the solution the lower the pH. Where the data has been made available, it appears that most Champagne has a pH close to 3.0.

58. Liger-Belair, G.; Jeandet, P. *Langmuir* **2003,** *19,* 5771.

59. Liger-Belair, G.; Cilindre, C.; Gougeon, R. D.; Lucio, M.; Gebefügi, I.; Jeandet, P.; Schmitt-Kopplin, P. *PNAS* **2009,** *106,* 16545.

60. Liger-Belair, G. *J. Phys. Chem. B* **2014,** *118,* 3156. DOI:10.1021/jp500295e.

61. Lukić, I.; Radeka, S.; Grozaj, N.; Staver, M.; Peršurić *Food Chem.* **2016,** *196,* 1048 report choosing "Excellence-FTH" (Lamothe Abjet, Bordeaux, France) as well as rehydration and incubation of a hydrated sample of the concentrate with a yeast activator (Vitactif, Lamothe Abjet) which is then added to the batch for fermentation (at 17 °C for 45 days!).

SECTION VII
Drinking the Wine

22

Drinking the Wine

THE BOTTLED BEVERAGE before you is to be opened.

This work has already described the bottle (colorless or not), the closure (screw cap, synthetic cork and cork), and the contents (the wine).

If the wine is not a table wine (*vin ordinaire* or *vin de pays*) which is simply enjoyed in a family or informal surrounding where the details of the container into which it is poured are less important, then it is generally found that: (a) clear colorless glass or crystal is used so that the visual appeal of the beverage can be enjoyed; (b) the bowls of wine glasses (except for sparkling wines and dessert wines) will be tapered upward from the stem into a bulbous shape which diminishes again at the top; and (c) the rim of the glass will be thin enough to allow it to be unnoticed when the wine is sipped. It is held that these are important, and in particular, the shape of the glass helps retain the more volatile constituents for the consumer's enjoyment.

Bowls used in glasses for red wines are more rounded so that when half full, the surface area is large. For white wines, this is considered less important, and of course, for Champagne and other sparkling wines, where conical flutes are used, a small surface area is avoided to enhance the flow of bubbles.

As the wine briefly stands, perhaps having been swirled, it is often found that "legs" or "tears" of wine are seen to form on the wall above the surface. Their appearance is, in part, a function of temperature as well as the alcohol content[1] of the wine and the resulting surface tension of the liquid.[2] Then, using capillary action,[2] the liquid climbs the side of the glass.

Both alcohol and water evaporate, but the alcohol evaporates faster, so more liquid is drawn up from the bulk. The wine thus moves up the side of the glass and forms droplets that run back down the glass.

The surface area of the liquid poured into the glass is important because it is where air comes into contact with the wine, and so, generally, it is best to fill the glass about halfway. In the balloon-shaped glass this will maximize the surface area. Now, the volatile components (low-molecular-weight esters, terpenes, diterpenes, norisoprenoids, and other components with vapor pressures high enough to escape the surface of the liquid) must rise to meet the nasal receptors. For some wines, potentially less pleasant constituents (e.g., some thiols) will have the opportunity to undergo oxidation. Other components may rearrange, undergoing structural change, and subsequently changing their pre-rearrangement odor. Still others may be lost to the vapor. See Appendix 5 for some observations about structure and odor. This process is referred to as "allowing the wine to breathe."

During this period, not only do the more volatile components rise out of the liquid, but oxidation, enhanced by swirling the wine in the glass, may also occur. Over time it has become clear that the flavor of the beverage does change as it is exposed to oxygen [air is normally about 78% nitrogen (N_2) and 21% oxygen (O_2) with traces of other gases such as water vapor (H_2O), argon (Ar), carbon dioxide (CO_2), etc. making up the last 1% or so.[3] The change in the taste of the beverage on oxidation is preferred by some and not others. However, it is generally agreed that over-oxidation is to be avoided. As may be recalled (Chapter 12) with regard to olfaction detection, multiple different gene-encoded olfactory receptors, each recognizing different odorants singly and in combination, are involved in defining the odor of the beverage. Each of the gene-encoded nasal receptors is defined by the DNA of the animal detecting the odor. Of course, this means that each individual detects the odor differently, but since there are similarities in the overall structure of the detectors, it is likely that agreement on specific odors can be reached. If a gene is missing or damaged the odor may fail to be detected or detected improperly. When an odor of a fruit or flower is detected, for example, what is being smelled corresponds to members of the terpene, sesquiterpene, and other hydrocarbon species (Appendix 2), as well as esters (Appendix 1 for synthesis from carboxylic acids and alcohols and Appendix 3 for a list of esters) evaporating from the fruit or flower.[4]

The initial experience of the entry of wine into the mouth is called "mouthfeel." Mouthfeel refers to the texture of the beverage in the mouth. It can be argued that the lipids and fatty acid esters (particularly those of glycerol [1,2,3-propanetriol]) found in some wines are more suitable to the mouth surfaces than, for example, phenols and tannins. Small quantities may make a difference. The same or a similar kind of receptors may also allow for temperature detection.

Salty and sour tastes require ion, and usually cation [e.g., sodium (Na^+), calcium (Ca^{2+}), proton (H^+)], transport through the appropriate ion channels found throughout the oral cavity. The G-protein coupled receptors (Chapter 12) that allow the other tastes (sweet, bitter, or savory) of the wine to be transmitted across membranes are found on the tongue and the lining of the cheek as well as the esophagus and the soft palate. The soft palate is connected to the nasal cavity *via* the nasopharynx so that taste and odor are inextricably intertwined.

It is generally agreed that almost everyone suffers to greater or lesser degrees from loss of taste (and odor) perception with age. It is this, as well as other normal biological differences, that accounts, at least in part, for the preferences in wines. Nonetheless, wine flavor and aroma, however subjective, have served as foundation for countless discussions and debates, as well as numerous articles listing the merits of various vintages. It is therefore reasonable to ask again to what are the differences attributed.

So, once again, hopefully it will be remembered (Chapter 13 and Appendix 1) that "…wine, even the very best, is composed of 85–90% water and 8–14% alcohol by volume…"

The water found in the wine is the same as water everywhere (although there will be differences in mineral content etc., the water itself is simply H_2O). The other major component

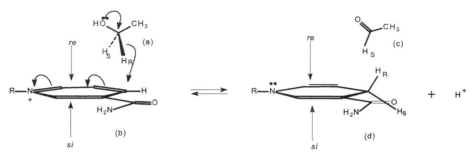

FIGURE 22.1 On the left, a cartoon representation of the transfer of a proton (H_R) from (a) ethanol to the (as shown) top or "re" face of the pyridinium ring (b) of the nicotinamide adenine dinucleotide (NAD^+) cofactor used by the NAD^+/NADH family of oxidoreductase enzymes (EC 1.1.1. -). On the right, a cartoon representation of the products acetaldehyde (c), a proton (H^+), and the reduced pyridinium ring of NADH (d). The proton, lost from the oxygen of the aldehyde, (c) if regained, would allow the process to reverse and NADH (d) to undergo oxidation to NAD^+ (b) and the acetaldehyde (c) to be reduced to ethanol (a). The abbreviations "re" and "si" are linked to the abbreviations "R" and "S" as discussed in Appendix 1.

is ethanol (ethyl alcohol, CH_3CH_2OH), and the third component (normally less than 4 % unless additional sugars have been added too) consists of the rich plethora of a thousand or so chemical compounds that provide the additional sensory experience.

It is generally well appreciated that ethanol (ethyl alcohol, CH_3CH_2OH) is a central nervous system (CNS, i.e., the brain and spinal cord parts of the human nervous system) depressant (i.e., it lowers the speed at which information is transmitted through nerves, reducing stimulation) which can have serious health effects.[5] Indeed, it has been argued[5] that even on a single occasion serious health issues can result from excess ethanol consumption, and well-documented effects on heart, liver, pancreas, and even the immune system are available. So, in addition to the societal warnings, it is important to know that the metabolism of ethanol has been carefully studied.[6]

Ethanol (CH_3CH_2OH) is metabolized in every cell as well as by intestinal microbiota. There is more than one pathway. Alcohol dehydrogenases (EC 1.1.1.1 and EC 1.1.1.2), also called alcohol:NAD^+ oxidoreductase, along with other members of the class (EC 1.1.1 -) which might use $NAD(P)^+$ as the cofactor rather than NAD^+, are widely distributed. All of them work in essentially the same way. However, there are some stereochemical differences which are a function of the way the enzyme holds the substrate (i.e., concerning the face of the nitrogen-containing ring to which—and later, from which—the hydrogen on the hydroxyl bearing carbon is transferred) (Figure 22.1).[7] Ethanol (CH_3CH_2OH) is converted to ethanal (acetaldehyde, CH_3CHO) while NAD^+ is converted to NADH (Figure 22.2). A proton (H^+) is also liberated.

The cytochrome P450 hemoproteins which use the central iron (Fe) atom of the heme to participate in oxidation reactions are also available, and in particular, cytochrome P450 2E1 (EC 1.14.13.n7), an enzyme which is known to involve the generation of (often) cellularly deleterious reactive oxygen species, is involved.[8,9]

Further, the enzyme catalase (EC 1.11.1.6) must be involved too, since the reactive oxygen species require decomposition, and this enzyme is widely available to that end.

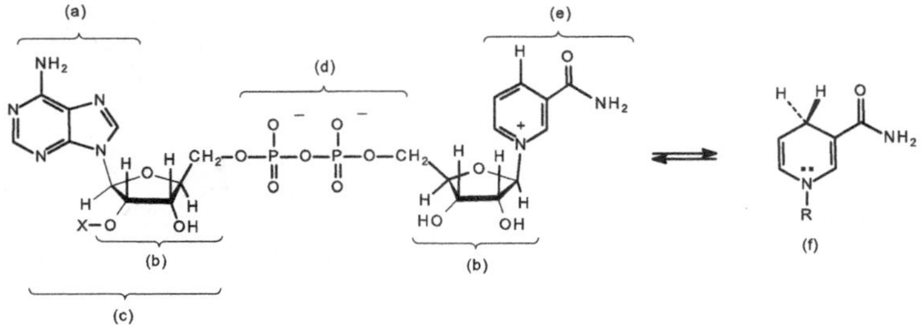

FIGURE 22.2 A cartoon representation of NAD$^+$ and NAD(P)$^+$ with (left to right) the adenine (a), ribose (b) bearing either a hydrogen (X = H) or phosphate (X = PO$_3^{2-}$) on the oxygen of the adenosine unit (c). This is followed by a "bridging" diphosphate (d) attached, on the right, to the ribose (b) that bears the pyridinium ring of the nicotinamide (e). On the far right is shown only the reduced ring of the nicotinamide (f).

Thus, it should be clear that ethanol, which proved toxic to the yeasts producing it, is also probably somewhat toxic to other organisms too. In part, the toxicity results from the formation of acetaldehyde (CH$_3$CHO) shown in Figure 22.1. Acetaldehyde is a Group 1 carcinogen[10] which reacts readily with many important biological entities; it is destroyed by further oxidation to acetate, written here as acetic acid (CH$_3$CO$_2$H). The oxidation is effected by acetaldehyde dehydrogenase (EC 1.2.1.10), as well as by more generic aldehyde dehydrogenases (EC 1.2.1.3–EC 1.2.3.5), and the product is usually acetyl coenzyme A (CH$_3$COSCoA), a more useful form of acetate, but one whose generation is not the result of a simple oxidation alone. However, the aldehyde dehydrogenases do need to react quickly, as acetaldehyde reacts with the mitochondria (cellular organelles involved in energy production and respiration), proteins, amino acids, and the amino and hydroxyl groups of the nucleic acids. Some of the reactions are not readily reversed, and so severe liver and brain damage can result (unless the acetaldehyde is removed). Indeed, it has recently been pointed out[11] "... Removal of acetaldehyde is essential to the metabolism of ethanol and critical to the survival of the animal. Ethanol could not be metabolized unless acetaldehyde was removed since the equilibrium constant is far in the direction of ethanol. Survival of the organism following ethanol intake depends on the removal of highly toxic acetaldehyde, either by reduction back to ethanol, oxidation to acetate, or removal via respiration, kidney excretion or protein binding." Fortunately the acetaldehyde can be removed.

The trace materials that afford wine its color, odor, and flavor have been described, and it should be clear that while there are many variations on the themes of esters, terpenes, acids, ethers, alcohols, norterpenes, thiols, amines, etc. which have been discussed in some detail, it is largely the interaction of these materials with the sensory equipment brought to them that is important. Their metabolism is largely irrelevant compared to that of ethanol.

For example, the phenol resveratrol (Figure 22.3) is among the minor constituents of wine receiving considerable attention. There is relatively little resveratrol in wine (generally agreed to be somewhere between 1 and 10 mg L^{-1}), but apparently more (assayed as stilbenes) in grape skins.[12] Nonetheless, the idea of something in the wine that accounts for the "French Paradox"[13] and the easy isolation of resveratrol has led to extensive testing of that phenol and related compounds in the hope that it is related to longevity.

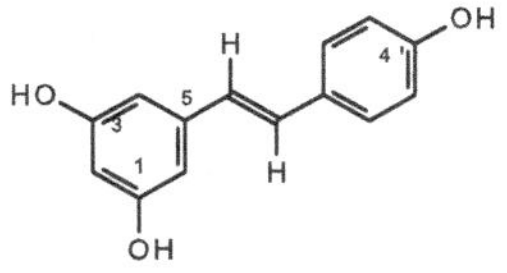

FIGURE 22.3 A representation of (*E*)-5-(4-hydroxystyryl)-benzene-1,3-diol (*trans*-resveratrol) which, in popular culture, is now occasionally referred to as "RES" or "Res-Q."

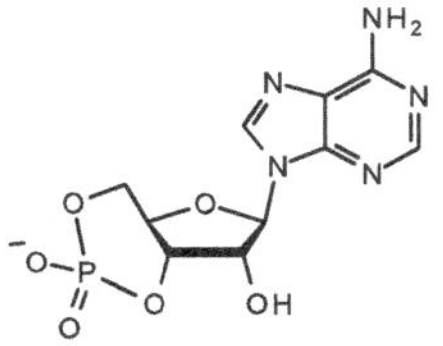

FIGURE 22.4 A representation of cyclic adenosine monophosphate (cAMP).

As is typical of such research, one of the more pressing questions remains, *viz.*, what is the target—or what are the targets—affected by consuming resveratrol? In general that target or those targets are elusive save for the suggestion that resveratrol mimics triggers for calorie restriction—an intervention recognized as protecting against age-associated metabolic diseases.[14]

A widely reported recent attempt to find the target (in mice) was recently made available.[15] In mice and culture media, a case was made that resveratrol ameliorates aging-related metabolic phenotypes by inhibiting cyclic adenosine monophosphate (cAMP) phosphodiesterases. Cyclic adenosine monophosphate (cAMP) (Figure 22.4) belongs to a class of compounds called second messengers. Second messengers transfer some signals received on the outside of cells from agents that cannot past through the cell wall, to the receptors on the inside of cells. So, by avoiding destruction by phosphodiesterases, metabolic function is said to be increased along with physical stamina and general well-being.

The quantities of resveratrol utilized (from 20 to 400 mg/kg body weight for the mice or, to a first approximation, about 1–20 grams/day for a 70-kg human) are far in excess of what is found in red wine. Of course there are at least two caveats. First, the activation may be small and significant, but large quantities were needed to clearly observe the process, and second, mice and men do have their differences.

Indeed, as more recently pointed out "... There are many mechanisms and pathways supposedly regulated by RES (*sic*) but the evidence in humans is very poor. In this regard, marketing and media coverage have moved on much faster than research."[16]

Thus, it is of some interest to note that a small (but much larger than many others) study of 119 human subjects in a year-long, randomized, double-blind, placebo-controlled trial of resveratrol for Alzheimer disease has been reported.[17] A variety of biomarkers were measured, and resveratrol safety and tolerability at large doses (beginning with 500 mg every 13 weeks and ending with 1000 mg (1 g) twice daily, [*NB.* normally the concentration of resveratrol in red wine lies between 1 and 10 mg/liter, *vide supra*]). Resveratrol and several metabolites (listed as 3G-RES, 4G-RES, and S-RES without further identification)[18] were monitored.

The results of this study in humans who, it was claimed, tolerated the resveratrol about as well as the placebo, included the following: (a) Both resveratrol and its major metabolites (*vide supra*) were measurable in plasma and cerebral spinal fluid (i.e., the fluid surrounding the brain and spinal cord). (b) Several biomarkers watched for progression of Alzheimer disease declined more in the placebo group than in the resveratrol-treated group (i.e., there is a suggestion that the disease progressed further in the placebo group). (c) Brain volume loss was increased by resveratrol treatment. As the authors note, however, the "biomarker trajectories must be interpreted with caution . . . (since) . . . they do not indicate benefit."

Similar problems have afflicted all of the investigative measures concerning the multitude of flavoring, coloring, and potential beneficial (or harmful) compounds at the concentrations found in wines. Measuring the effect of the truly miniscule quantities found in wine is a challenging problem, and the results may neither clearly affirm nor refute hypotheses initially formulated.

The belief, unsubstantiated, continues,

NOTES AND REFERENCES

1. It is common to measure the quantity of alcohol in bottled wine by using volume rather than weight. Thus, most wine is about 12% to 15% alcohol by volume. However, desert wines, such as ice wine, may have as much as 25% alcohol by volume. Fortified wines are similar to desert wines. Finally, the word "proof" as used in the United States is a value given as twice the alcohol content, thus a beverage that is 50% ethyl alcohol by volume is said to be 100 proof.

2. In all liquids, regardless of their composition, the bulk liquid acts on the surface in a way that pulls the surface into the bulk. This process minimizes the surface area and is one manifestation of surface tension. Ethyl alcohol in water has a lower surface tension than water. A different manifestation of surface tension is seen at the glass–liquid interface, where the portion of the liquid coming in contact with the solid is attracted to the surface and adheres, pulling itself along the surface. This process is called capillary action.

3. Oxygen, O_2, (really "dioxygen" in the atmosphere since atomic "mono-oxygen, O" is not found there) as normally observed in the atmosphere exists in a "state" called the "ground state." The ground state of oxygen is one in which there are two (2) unpaired (and thus not in a bond) electrons. The ground state of oxygen with two unpaired electrons is called the "triplet state" (the name of the state refers to the number of unpaired electrons +1). In general, states with unpaired electrons are more reactive than those with paired electrons. A consequence of the reactivity of triplet oxygen is life as we know it on the planet Earth. There are well-known carefully controlled biological (and other) processes that occur in the absence of oxygen. Processes occurring in the absence of oxygen are referred to as "anaerobic" and have been occurring during wine fermentation. Interestingly, it has been speculated that these anaerobic processes were present on the planet before the atmosphere became as rich in oxygen as we currently find it. Nonetheless, it is the reactivity of triplet oxygen that accounts for the oxidation of wine (and many other processes). Hopefully, it will be remembered that the presence of oxygen is to be avoided during fermentation and subsequent treatment of wine. Lengthy exposure to oxygen will change the flavor. So, prior to opening the bottle, efforts are made to avoid oxygen. Once

the bottle is opened, a little oxidation may remove some of the unwanted flavorants by increasing their molecular weight, introducing new bonds that make the constituents less volatile, and producing new compounds by rearrangement or other pathways. Should the wine (particularly those wines rich in phenols) stand in the presence of atmospheric oxygen for too long the taste, by current standards, deteriorates.

4. The ester that gives a banana its odor is called 3-methyl-1-butanol acetate (isoamyl acetate). The primary odor of an orange comes from octyl acetate. A table of structures assigned to ester odorants is found in Appendix 3. It is important to recognize that there is a true difference between the main odorant of a fruit and the odor of the fruit itself. So, while 3-methyl-1-butanol acetate is, for example, easily recognized as "banana smell" and is occasionally used in confections to produce the odor of banana, the many overlapping and more subtle odors of a ripe banana (an unripe banana also has a "different smell") are readily recognized because receptors in addition to those for 3-methyl-1-butanol have been activated.

5. National Institutes of Health (a) National Institute on Alcohol Abuse and Alcoholism (http://www.niaa.nih.gov) 5635 Fishers Lane, MSC 9304, Bethesda, MD 20892-9304, USA, and (b) National Institute on Drug Abuse (http://www.drugabuse.gov) 6001 Executive Boulevard, Room 5213, Bethesda, MD 20892-9561, USA.

6. An excellent review article has appeared. See Seitz, H. K.; Mueller, S. In *Metabolism of Drugs and Other Xenobiotics;* Anzenbacher, P.; Zanger, U. M., Eds.; Wiley-VCH Verlag GmbH&Co KGaA: Weinheim, Germany, 2012; Chapter 18.

7. Nambiar, K. P.; Stauffer, D. M.; Kolodziej, P. A.; Benner, S. A. *J. Am. Chem. Soc.* **1983,** *105,* 5886. Transfer from either the *re* or the *si* face has been observed in different cases.

8. Hayashi, S.; Watanabe, J.; Kawajiri K. *J. Biochem*. **1991,** *110,* 559.

9. An interesting report has appeared showing that some cytochrome P450 enzymes (including CYP2E1) are inhibited by some nonvolatile red wine components, including resveratrol. See Piver, B.; Berthou, F.; Dreano, Y.; Lucas, D. *Toxicol. Lett*. **2001,** *125,* 83.

10. As determined by the International Agency for Research on Cancer (IARC). The IARC is an intergovernmental agency forming part of the World Health Organization (WHO) of the United Nations (UN). Group 1 compounds are known to be carcinogenic to humans. That is, it has been decided that there is enough evidence to conclude a Group 1 compound can cause cancer in humans.

11. Deitrich, R. A.; Petersen, D.; Vasiliou, V. In *Acetaldehyde Related Pathology: Bridging the Transdisciplinary Divide: Novartis Foundation Symposium 285;* Chadwick, D. J.; Goode, J., Eds.; Novartis Foundation, Inc. 2007 (ISBN: 978-0-470-05766-7).

12. LeBlanc, M. R. Cultivar, Juice Extraction, Ultra Violet Irradiation and Storage Influence the Stilbene Content of Muscadine Grapes (Vitis Rotundifolia Michx). Ph.D. Thesis, Louisiana State University, 2006.

13. A phrase that summarizes the report of the apparently paradoxical observation that, statistically, the French enjoy a life relatively free of coronary heart disease although their diet is rich in saturated fat (a diet statistically linked elsewhere to coronary problems).

14. Tennen, R. I.; Michishita-Kioi, E.; Chua, K. F. *Cell* **2012,** *148,* 387.

15. Park, S. J.; Ahmad, F.; Philp, A.: Baar, K.; Williams, T.; Luo, H.; Ke, H.; Rehmann, H.; Taussig, R.; Brown, A. L.; Kim, M. K.; Beaven, M. A.; Burgin, A .B.; Manganiello, V.; Chung, J. H. *Cell* **2012,** *148,* 421.

16. Tomé-Carneiro, J.; Larrosa, M.; González-Sarrías, A.; Tomás-Barberán, F. A.; García-Conesa, M. T.; Espín, J. C. *Curr. Pharma. Design* **2013**, *19,* 6064.

17. Turner, R. S.; Thomas, R. G.; Craft, S.; van Dyck, C. H.; Mintzer, J.; Reynolds, B. A.; Brewer, J. B.; Rissman, R. A.; Raman, R.; Aisen, P. S. *Neurology* **2015,** *85,* 1383.

18. 3G-RES and 4G-RES refer to glucuronic acid derivatives of resveratrol at the 3- and 4-hydroxyl groups (see Figure 22.3), while the S-RES refers to the 3-sulfate derivative, all three of which are said to be common metabolites in human, rat, and mouse resveratrol metabolism. See Yu, C.; Shin, Y. G.; Chow, A.; Li, Y.; Kosmeder, J. W.; Lee, W. S.; Hirschelman, W. H.; Pezzuto, J. M.; Mehta, R. G. von Breemen, R. B. *Pharm. Res.* **2002,** *19,* 1907.

Epilogue

"A GOOD GLASS of wine is a multisensory experience. There's the aroma of the wine entering your mouth, a burst of volatiles meeting olfactory and taste receptors; then there's the tart acidity of the tannins paired with the sweetness of natural sugars, packaged in a luxurious drink whose complex chemical palette results from a union of grapes, yeast, and oak." (Perkel, J. M. *Biotechniques* **2015**, *58* (1), 8).

Again, this is not the book I set out to write. However, now that it is written, I have learned that in addition to other pleasures that might be derived from the sensory experience, the drinking of wine is a social phenomenon and should be enjoyed in that way. I have come to believe that if you, the individual drinking the wine, find a wine you like, then despite the expert testimony of truly well-qualified and knowledgeable sommeliers whose tastes almost certainly differ from yours, the wine you have found you like is the wine for you.

Salut! "לחיים"

Appendix 1

A CHEMISTRY PRIMER

Wine, even the very best, is composed of 80–90% water and 8–14% alcohol by volume. This means that the measurement is comparing volumes, not weights. This is important because a given volume of alcohol weighs less than the same volume of water. The alcohol is ethyl alcohol—also called ethanol.

The numbers corresponding to percent composition of the solution must add up to 100%. Since there is more to wine than simply ethanol and water, it is reasonable to ask what else in this mixture is present in the solution not accounted for by sum of the percentages of ethanol and water. It is this small amount of complex organic molecules that distinguishes a fine or great wine from something less.

Careful study of the wines produced in some of the great (and not so great) vineyards around the world has produced evidence that there are frequently more than a thousand different chemical constituents (in addition to ethanol and water) in the symphony comprising some excellent wines. Some of these chemical constituents are common to most wines and include substances about which this primer is concerned. It is hoped that this primer will enable you to learn something about their structures, chemistry, and what we know about how they interact with us. Some of the more common compounds discussed here include: an aldehyde, acetaldehyde; alcohols, such as glycerol, sorbitol, and mannitol in addition to ethanol; acids, such as acetic, tartaric, malic, citric, and succinic, as well as amino acids—largely found in the form of peptides; esters of the alcohols with the acids and the amino acids; minerals (including compounds containing potassium, nitrogen, phosphorus, sulfur, magnesium, and calcium); phenols; and, generally, some sugars (as sucrose, although a few other sugars are also there).

And what, exactly do all those words mean? And do I want to know? And is it important?

If you continue reading, the meaning of the words will become clear. And, since you are reading this, it is supposed you want to know.

However, you do not need to know any of this to enjoy a glass of wine.

As to importance, if you are producing wine from a vineyard and selling it for human consumption, you will almost certainly want to know what and how much of these and many other of the constituents in the wine you are selling are present. If you are attempting to produce consistently good-to-excellent wines, that consistency requires quality control, and such control only comes with knowledge of the variables and continued monitoring of materials.

And now ... for the chemistry ...

Aside from dark energy and dark matter about which we know very little, the elements we do know something about can be organized into the form of a table designed to emphasize periodicity of properties. A standard, more-or-less current, depiction of the Periodic Table appears below as Figure A1.1.

These eighteen (18) columns and seven (7) rows (not counting the two "extra" rows at the bottom [Lanthanides and Actinides] about which almost nothing more will be said here) constitute the building blocks of everything. That's all there is. Everywhere!

The elements themselves are built up by the addition of one proton (+), a positively charged entity, and one electron (−), which bears a negative charge (to keep the system electrically neutral). One at a time, beginning with hydrogen (H) and simply adding one of each, (giving an *atomic number*) and some number of neutrons, to make the next element, the periodic table can be built. Protons are positively charged particles (1.67×10^{-19} coulomb) and have a rest mass of about 1.67×10^{-27} kg; 1 kg = 1 kilogram ≈ 2.2 pounds = 2.2 lb.). Neutrons are particles which have no charge and a mass slightly greater than that of the proton. Electrons, which are negatively charged to match the positive charge of the proton, have a rest mass of about 1800 times less than that of the proton and have properties of both particles and waves. These atomic particles are all brought together to form the elements. Current accepted theory argues that protons and neutrons are bound together

H																	He
Li	Be											B	C	N	O	F	Ne
Na	Mg											Al	Si	P	S	Cl	Ar
K	Ca	Sc	Ti	V	Cr	Mn	Fe	Co	Ni	Cu	Zn	Ga	Ge	As	Se	Br	Kr
Rb	Sr	Y	Zr	Nb	Mo	Tc	Ru	Rh	Pd	Ag	Cd	In	Sn	Sb	Te	I	Xe
Cs	Ba	Lu	Hf	Ta	W	Re	Os	Ir	Pt	Au	Hg	Tl	Pb	Bi	Po	At	Rn
Fr	Ra	Lr	Rf	Db	Sg	Bh	Hs	Mt	Ds	Rg							

La	Ce	Pr	Nd	Pm	Sm	Eu	Gd	Tb	Dy	Ho	Er	Tm	Yb
Ac	Th	Pa	U	Np	Pu	Am	Cm	Bk	Cf	Es	Fm	Md	No

FIGURE A1.1 A representation (*ca* 2012) of the Periodic Table of the Elements. This representation is presented with the consent of Professor Eric Scerri and appeared in *The Periodic Table,* Oxford University Press, 2012.

A Chemistry Primer

by something called the "strong force," which is required for the stability of nuclei. Although there are different ways a table can be arranged, in this particular arrangement, atomic numbers (corresponding to the number of protons) increase across the rows from left to right across the columns until the so-called Noble Gases, when the row ends. The next row begins again on the left in a new column. Elements in columns have similar chemical properties.

In the upper left-hand corner of the periodic table is the symbol for hydrogen (H). The hydrogen atom, represented by this symbol, has been shown to have one proton (+) and one electron (−) for a net charge of zero. The element is sometimes abbreviated as (^{1}H) and is commonly called protium (Greek, *first*). Protium has no neutrons! If a neutron is added, the mass of the particle is nearly doubled; the species can be written as ^{2}H. It is now called deuterium (sometimes the symbol "D" is used, and its name is derived from the Greek, *deuteros*, second). Since the neutron has no charge, and since one electron has remained, the deuterium atom is neutral. If two neutrons are added, the system, while still neutral, becomes less stable. The symbol for this unstable material (which nonetheless has been isolated) is ^{3}H, and its name is tritium (Greek, *third*). Tritium is radioactive, which means that it spontaneously decomposes, and about every 12 years, half of the original has disappeared (one atom at a time, to form helium, with the simultaneous emission of energy). Can more neutrons be added to go beyond tritium? Of course they can! However, *isotopes* (same atomic number but different atomic mass) of hydrogen thus formed are not very long lived (only about 10^{-22} seconds) before decomposition. You might wish to visit nucleonica (http://www.nucleonica.net/unc.aspx), a site where the description of the work for adding more and more neutrons is found.

This leads to a very important issue. With the advent of the ability to search for information in a variety of places, it is very important to attempt to find whose word you are taking for anything. If it is a reputable source of information, it will be clear that you can access the details of the investigation, read about the experiment(s) done, question the author(s), and then decide for yourself if the conclusions are warranted. In the last analysis, it might even be possible to replicate the experiment described that led to the conclusions presented. So, when I have attempted to describe something, I may have gotten it wrong, and so don't take my word for it. Nonetheless, I am attempting to be honest, and it should be clear that in going from hydrogen, (1 proton and 1 electron) to Roentgenium (Rg, number 111 with 111 protons and 111 electrons) all the other elements are encountered.[1]

Now, with regard to all of the elements lying between hydrogen (H) and the latest (not yet shown on the Table), characterized livermorium (Lv), with 116 protons and electrons, and the addition of protons and electrons to go from one element to the next, the role of the neutron has been ignored. As might be gathered from earlier comments about adding neutrons to a single proton, the story is pretty much the same for all of the protons and neutrons as the number of protons increases. Too few neutrons for the number of protons present can be added but such species are unstable. Neutrons can be added beyond the number of protons but there is a limit, again because of stability. Interestingly, it has been found experimentally, that there are so called "islands of stability" where some optimum number of neutrons and protons are more stable than any other number of neutrons with the given number of protons. Some numbers of protons (i.e., some elements) have multiple long-term numbers of stable isotopes. Many do not. It is currently held that lead-204 is the heaviest long-term stable element and since the atomic number (the number of protons) of lead is 82, ^{204}Pb must have 122 neutrons (204 − 82 = 122) and, of course, 82 electrons.

While it is true that the elements in the Periodic Table are all we seem to have, it would be a dull world if all of the pure elements lacked the ability to participate in reactions with other pure elements. As it is, many of the elements participate in compound formation with other elements. This participation takes place through the presence of a force lying between the elements participating that holds the nuclei together and, in that way, allows them to form compounds.

The force that holds two (or more) elements together in a compound is called a "bond." The force that holds the nucleus (protons and neutrons) together in the nucleus of atoms is, as noted earlier, called the "strong force."

Based on experimental observations, it is generally argued that there are two extreme types of bonds. One type is called "covalent," and the other is called "ionic." The first type results from the "sharing" of electrons between elements. The second type results from one element losing one or more electrons (becoming positively charged) whilst another element picks up one or more of the electrons lost and becomes negatively charged. The oppositely charged species are attracted to each other. With few exceptions, the model[2] of such extremes is considered to be "ideal." A current model[3] for bonds in real compounds and between elements that is used is that bonds are formed by the overlap of "orbitals," initially associated with atoms ("atomic orbitals' or AOs), and once the overlap occurs, molecules (with "molecular orbitals" or MOs) result. What are these "orbitals"?

There are fixed levels (just like shelves in a library) that can be occupied by electrons associated with each element. The shelves, which represent energy levels, are called *orbitals* and are at different heights (energies) depending upon the element.[3,4] In all of these models, it is argued that the shelf can hold no more than two electrons, and since the electrons appear to have a property called "spin," the two electrons must have opposite spins if they are in the same orbital (i.e., on the same "shelf"). In the working model, orbitals cannot be created, nor can they be destroyed. This means that if one atomic orbital (an AO) of one element interacts with one atomic orbital (an AO) of a second element (since 1 + 1 = 2) two molecular orbitals (MOs) must have been created!

In this quantum mechanical model, each electron is considered as a wave, and when the wave from one electron on hydrogen, for example, interacts with the wave of the other electron from a second hydrogen, the two waves can interact in a positive way (+) or in a negative (-) way. If they interact in a positive way, there is overlap, and the bond is made. If they interact in a negative way, then the nuclei remain in the proper position but no bond can form, and there is a "node" (a place where there is change in sign of the wave) between the nuclei. The former is constructive or "bonding," and the latter is destructive or "antibonding."

The energies of the AOs and MOs are determined experimentally. Energy is absorbed (taken up) when the element is heated or otherwise supplied with energy (e.g., an electric current is passed through it), and higher orbitals are occupied by the electrons. Then the energy is emitted (given off) as the electrons fall back to their lower orbitals. Sometimes, some of the energy that is being emitted is seen in the visible (VIS) part of the electromagnetic spectrum, and white light (a mixture of all of the colors in the visible spectrum) is seen or a specific color is produced. Thus, for example, sodium vapor lights are yellow.

It seems that any orbital (AO or MO) can contain a maximum of two electrons. So when two hydrogen atoms, each with one electron in an atomic orbital (AO), approach each other and reach a given distance (74 pm = 0.74×10^{-10} m = 0.74×10^{-8} cm = 0.074 nm)[5], energy is given off and a bond forms if, and only if, the wave functions have the same sign (positive overlap)! The bond is very strong and has been measured at about 4.5 electron volts (eV) (about 104 kcal/mol or 436 kJ/mol).[5] Of course there must also be an antibonding orbital, because orbitals cannot be created nor destroyed (1 AO + 1 AO = 2 MO) which results from the signs of the wave functions being different. The picture normally shown involves two hydrogen atoms approaching each other from a great distance apart. As they draw together energy decreases for the bonding orbital and a bond is formed. If the nuclei get too close, the energy rises dramatically. This gives an average distance of the two hydrogen nuclei of 74 pm[5] but clearly they can still move a little around this value—it's an "average." A cartoon representing this is shown as Figure A1.2.

Thus, as two hydrogen atoms approach each other, if the wave functions of the AOs have (by convention) the same sign, a bond will form. Two electrons, with opposite spins, can occupy the new bonding MO.

Another cartoon representation showing the bonding and antibonding orbitals and their relationship to the energy of the system is shown in Figure A1.3.

Now, we can consider some of the material necessary for more complete comprehension.

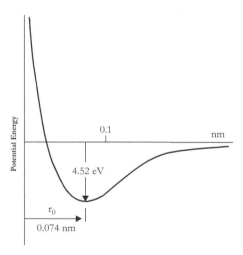

FIGURE A1.2 A representation of the hydrogen molecule (H_2) as a combination of two hydrogen atoms (H•) as they approach (right-to-left). There is an energy minimum reached at a distance of 74 pm (0.074 nm) where, we argue, a bond has formed.

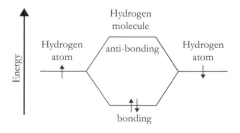

FIGURE A1.3 A cartoon representation of the interaction of two spin-paired (i.e., opposite spins) electrons of the same energy. Their combination into a bonding orbital, it is argued, lowers the total energy of the system. So, bonding is better than not bonding because it is more stable.

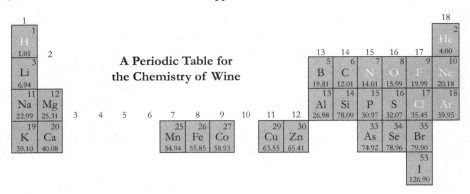

FIGURE A1.4 A representation of an abbreviated periodic table to emphasize elements important to the discussion of the chemistry of wine.

An abbreviated version of a periodic table needed from this point forward is shown in Figure A1.4.

As noted previously, wine is largely water, a compound, not an element. Compounds are made up by elements combining with each other. The resulting combination cannot be separated back into the elements making up the compound by physical means because the elements making up the compound are attached to each other by bonds. Just as elements are different than compounds, compounds are different than mixtures.

Mixtures can be physically separated into the compounds and/or elements that make up the mixture. Thus, the mixture of water and the ethanol in wine can be separated by physical methods (e.g., distillation, a process which involves converting liquids to their vapors and, based on differences in the temperatures at which they boil, condensing the separated vapors back into the liquid state). So, ethanol boils at 82 °C (180 °F), while water boils at 100 °C (212 °F), and the mixture can be separated. However the hydrogens and oxygen in water, a compound, cannot be separated from each other by such physical means, nor can the carbons, oxygen, and hydrogens in ethanol be separated from each other in this way. Indeed, if they were separated we would no longer have water and ethanol but rather carbon, hydrogen, and oxygen Many people are willing to accept now (although it was not always so) that the chemical formula for the molecule water is H_2O. There are two hydrogen atoms connected to one oxygen atom (although, a priori, it might be that one hydrogen is connected to *both* one other hydrogen and an oxygen, *viz.* "H–H–O" rather than "H–O–H"). But, based upon the experimental observation that *both hydrogens in water are identical and indistinguishable* from each other, there is only one possibility for water, and that is H–O–H. Is the water molecule, H_2O, linear or bent? That is, what is the angle subtended between the two hydrogens attached to the oxygen? Is it 180° (linear) or not? In that vein, it has been stated that "Although more investigation has been done on water than on any other liquid, there still remains uncertainty about the details of its structure ... many of the experimental results can be interpreted in different ways."[6] Direct measurement by X-ray crystallography on crystals of ice demonstrates that the structure is bent with an average H–O–H bond angle of about 105 degrees.[7] Also, water molecules are bonded to each other by another (weaker) set of bonds called "hydrogen bonds." Bonds within molecules are intramolecular; bonds between molecules are intermolecular. The covalent H–O intramolecular bonds use one electron from the hydrogen (H) and one electron from the oxygen (O) in the bonding orbital. When broken symmetrically, or homolytically, one electron leaves with the hydrogen and one leaves with the oxygen. When broken heterolytically (or nonsymmetrically), both of the electrons stay with the oxygen. In the latter, the proton (H+), a cation, results along with the hydroxide ion (HO−), an anion.

By contrast, the intermolecular H–O bonds are made using electrons from oxygen only! A standard H–O intramolecular bond in water has a bond strength of about 460 kJ/mol (about 110 kcal/mol), and the weaker intermolecular "hydrogen bonds" have bond strengths of only about 23 kJ/mol (5.5 kcal/mol). The intramolecular bonds are about 96 pm long on the average and the intermolecular bonds are longer (shorter in ice than in liquid water and shorter in liquid water than in water vapor) with less-well-defined lengths. A number of "cartoon" type representations of the picture this presents can and have been made. A simple one between only two water molecules is shown in Figure A1.5 where the solid lines represent intramolecular bonds from the covalent sharing of electrons between hydrogen and oxygen, and the dashed line represents an intermolecular bond.

Of course, water is not "flat" or two dimensional, it is a three-dimensional substance and so, in three dimensions, can be represented as shown in Figure A1.6.

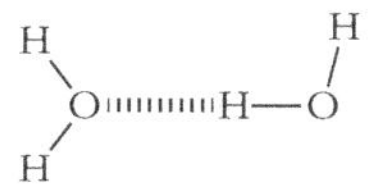

H
 \
  O⋯⋯⋯H—O
 /   \
H   H

FIGURE A1.5 A cartoon representation of two water (H₂O) molecules cojoined by a hydrogen bond from the oxygen on the left to the proton on the right.

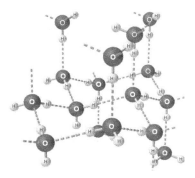

FIGURE A1.6 A cartoon representation of a three-dimensional lattice network of water molecules.

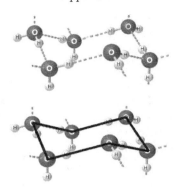

FIGURE A1.7 Cartoon representations of "chair" like figures representing water clusters.

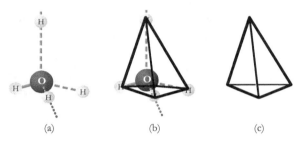

FIGURE A1.8 Three representations of distorted tetrahedra. In (a) the hydrogen nuclei (H) in the white spheres are of two types. One hydrogen on the left and a second hydrogen facing the viewer are drawn to represent two covalently bonded hydrogen atoms. The other two hydrogen nuclei (above oxygen and to the right) represent hydrogen-bonded hydrogens from adjacent water nuclei. In (b) the representation of (a) is repeated, but a tetrahedron is superscribed. In (c) only the tetrahedron from (b) is shown.

It should be clear that there is a variety of three dimensional "chair" like structures present (e.g., Figure A1.7).

Every oxygen is bonded to four (!) hydrogens; that is two to which it is directly bonded by covalent bonds to form a single water molecule and two to which oxygen is hydrogen bonded to adjacent water molecules using electrons from oxygen. This array of four items (H) around a fifth (O) leads to a distorted tetrahedron (Figure A1.8).

An undistorted tetrahedron (with four identical faces) is a "Platonic solid." Indeed, forming a tetrahedron is the lowest-energy way to put any four identical objects around and equidistant from a fifth and equidistant from each other. Because the objects are identical, the final configuration corresponds to the surface of a sphere (from the center looking outward). From a chemical perspective and allowing for distortions because of unlike species, the tetrahedral configuration is quite general, and putting four objects around a carbon, nitrogen, sulfur, or phosphorus (see the Periodic Table for the Chemistry of Wine) allows even generally similar

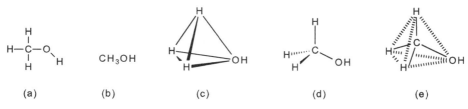

FIGURE A1.9 Representations of methane (CH₃OH): (a) a "flat" two dimensional representation showing, however, that carbon has four (4) bonds, oxygen two bonds, and each hydrogen only one bond; (b) a condensed representation with the representation of (a) understood; (c) a tetrahedral representation of methanol showing the four attachments to the central carbon but not showing the central carbon atom; (d) a tetrahedral arrangement showing an understood tetrahedron but the substituents at approximately the correct internal tetrahedral angles; and (e) the same representation as (c) overlapped with (d).

attachments to adhere to the common theme of a tetrahedron. However, just as for the single water molecule, H_2O, there are only two hydrogen atoms attached to form one strong covalent bond each to the oxygen, so we will find that, in neutral molecules, it is generally observed that: hydrogen (H) has only one bond; carbon (C) four bonds; nitrogen (N) and phosphorus (P) three bonds (although phosphorus is often found with four or five); oxygen (O) and sulfur (S) two bonds; and halogens [fluorine (F), chlorine (Cl), bromine (Br), and iodine (I)] one bond (but there are exceptions).

Consider wood alcohol (methyl alcohol, CH_3OH, also called methanol), a poisonous compound that has one carbon, four hydrogens, and one oxygen. Since each hydrogen must have one bond, oxygen must have two bonds, and carbon four, it is reasonable to attach one hydrogen to the oxygen and then attach the other bond from oxygen to carbon, leaving the three hydrogen atoms to form the other three bonds to carbon. This leads to representations shown in Figure A1.9.

In the left-most structure, it is clear that each hydrogen has one bond, three hydrogens are each bonded, once, to carbon, and one hydrogen is bonded, once, to oxygen. The oxygen has two bonds (as it did in water, H_2O), and the carbon has four bonds. However, the misimpression of a flat molecule is conveyed. To the right of that is a condensed formula representation following a general rule that the attachments to the carbon are followed by the OH functional group (a functional group, about which more will be said later, is a group that defines a class of compounds, in this case, with the –OH, the class is "alcohols"). The object in the center is a tetrahedral arrangement of only the attachments to the central carbon (which itself is not shown, but understood, as lying in the center of the tetrahedron). The final two structures reinforce the idea of the tetrahedron for a carbon atom with four substituents.

There is a significant amount of experimental evidence in concert with the notion that a carbon with four substituents is approximately tetrahedral. Thus, for example, treatment of the hydrocarbon (hydrocarbons are a class of compounds made up of molecules containing hydrogen and carbon only) methane, CH_4, with chlorine gas (Cl_2) in the presence of heat or light, results in the formation of molecules where the hydrogens of methane are, successively, replaced by chlorine (hydrogen chloride [HCl] is also formed) as shown in Figure A1.10. These compounds are members of a class of compounds called chloroalkanes.[8]

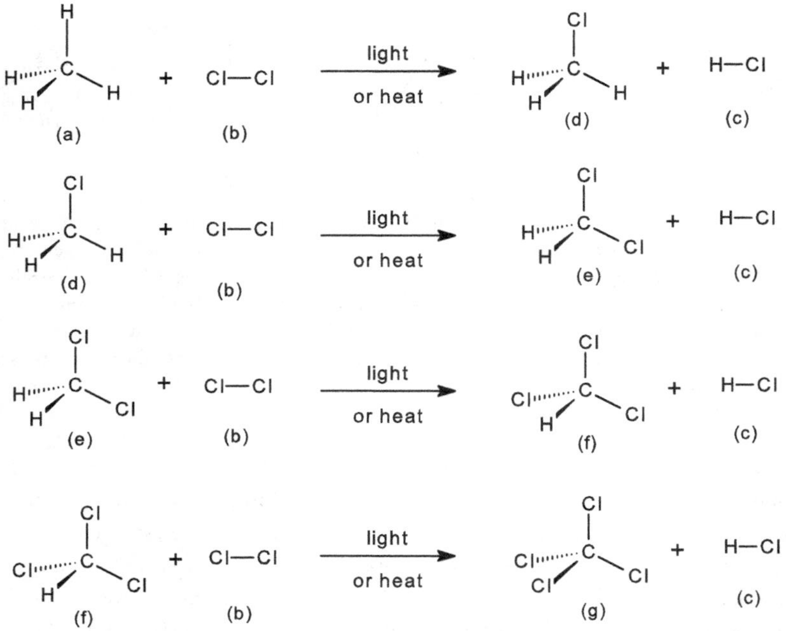

FIGURE A1.10 Representation of the chlorination of (a) the gas methane (CH_4) with (b) the gas chlorine (Cl_2) with ultraviolet light (hν) or heat (Δ). The products are the gas (c) hydrogen chloride (HCl) and the low-boiling liquid (d) chloromethane (CH_3Cl). Subsequently (d) reacts with (b) to generate (c) and (e), dichloromethane (CH_2Cl_2); (e) reacts with (b) to generate (c) and (f), trichloromethane (chloroform, $CHCl_3$); if enough (b) is present, (f) will react with (b) to generate (c) and (g), tetrachloromethane (carbon tetrachloride, CCl_4).

Consider the case of dichloromethane (methylene chloride, CH_2Cl_2), the product of the second reaction shown above. It had been pointed out early that if dichloromethane were flat, then there should be two molecules with the same empirical formula, CH_2Cl_2 (compounds with the same formula but different arrangement in space are called isomers). If flat, there would be one isomer with a chlorine–carbon–chlorine internal angle of 90° and one with a chlorine–carbon–chlorine internal angle of 180°. The experimentally found value is about 112° and there is only one such *isomer* (Figure A1.11).

Returning to the images of water (Figure A1.12), hopefully it is clear that there are two different ways (as noted above) that protons (^1H) are bonded to oxygen (^{16}O). That is there are those two strong (H to O covalent) bonds (energy is about 460 kJ/mol). And there are weaker bonds (one or two) between water molecules where a proton of one water is "hydrogen bonded" (nearly a factor of 20 times weaker, about 23 kJ/mol) to the oxygen of another water. Here, the proton involved in hydrogen bonding lacks an electron for covalent bonding (it's already in use) and donates electrons to attract the proton.[9]

The ability to hydrogen bond leads to an important consequence. The mass of one hydrogen (^1H) is 1 atomic mass unit (amu) [because there is only one proton one electron and no neutrons], and the mass of the most common isomer of one oxygen (^{16}O) is 16 amu (eight protons, eight electrons and eight neutrons). The mass of water (H_2O) is 18 amu (16 + 2). By definition, the boiling

A Chemistry Primer

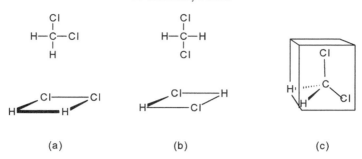

FIGURE A1.11 A demonstration that the structure assigned to dichloromethane is best met by a tetrahedral configuration. In (a) where a carbon (not shown) lies in or near the center of the plane of the drawing, the Cl–C–Cl angle would be about 90°. In (b) where a carbon (not shown) lies in or near the plane the Cl-C-Cl angle would be about 180°. The reported angles in (c), from the National Institute of Science and Technology are H–C–Cl ≈ 180°; Cl–C–Cl ≈ 112°; and H–C–H ≈ 112° [http://cccbdb.nist.gov/exp2.asp?casno=75092, August, 2013].

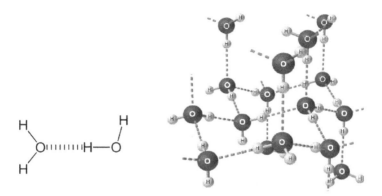

FIGURE A1.12 A repeat of the cartoon representation of two water (H_2O) molecules cojoined by a hydrogen bond from the oxygen on the left to the proton on the right and the cartoon representation of a three-dimensional lattice network of water molecules.

temperature of water is 100 °C (373.23 K). Now, consider the case of methane, CH_4, which has four protons (^1H) as shown in Figure A1.13 at 1 amu each and one carbon (^{12}C) [six protons, six electrons, and six neutrons] for a total of 12 amu. Thus, the molecular weight of the compound methane is 16 amu (12 + 4). The boiling temperature of methane is −164 °C (109 K). All of the electrons from carbon and from hydrogen are used in the covalent bonding. So, since methane has no ability to hydrogen bond as water does, we argue that, at least in part, the "elevated" boiling temperature of water is due to the hydrogen bonding network found in the liquid state having been destroyed in the vapor state. That such a relatively small energy per molecule should have such a relative large effect can be related, by analogy, to the great strength of a rope that is composed of a large number of individual strands each of which is relatively weak (Figure A1.14).

There are additional issues to which some attention must be directed before returning to a fuller discussion of what is happening in setting, ripening, harvesting and subsequent manipulation of the grape to make wine.

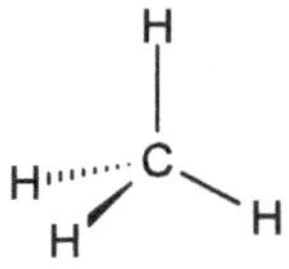

FIGURE A1.13 A representation of methane (CH₄).

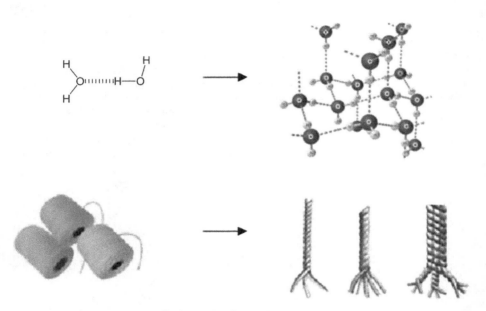

FIGURE A1.14 A comparison of hydrogen bonding with an aggregate of threads. Although each hydrogen bond or each thread is relatively weak, when they are combined the total is strong. With water, the boiling temperature is much higher than predicted on the basis of molecular mass; with threads, when intertwined, the total allows a heavy weight to be held or lifted.

BREAKING BONDS

One issue relates again to the covalent bond between the hydrogen and oxygen in water. As noted above, the breaking of the hydrogen–oxygen covalent bond "costs" 460 kJ/mol (about 110 kcal/mol), a measure of its bond strength. Since energy is conserved, that amount of energy is, of course, also released when the bond is "made" rather than "broken". The "Conservation of Energy" is one of the fundamental laws of Nature. As noted earlier, bonds can be broken homolytically or heterolytically. For water, heterolysis yields the "proton"(H^+)" a "cation" and the hydroxide "anion" (OH^-).

It is very difficult to separate species of unlike charge in the gas phase. However, in solution, where hydrogen bonding can occur, separation of unlike charges with the "solvent" (the material in which the "solute" dissolves) participating ameliorates the cost of separation and the loss (and gain) of protons is relatively easy. So, water can be thought of as existing not only as H_2O but also as a mixture of H_3O^+ (the "hydronium" ion) and the hydroxide anion OH^- (Equation A1.1).

$$H_2O + H_2O \rightleftharpoons H_3O^+ + {}^-OH$$

EQUATION A1.1 A representation of the transfer of a proton from water to water generating, in equilibrium, the hydronium cation and the hydroxide anion. The protonation of water by water.

$$H_2O + H_2O \underset{}{\overset{K_{eq}}{\rightleftharpoons}} H_3O^+ + {}^-OH$$

$$K_{equilibrium} = K_{eq} = \frac{[products]}{[reactants]} = \frac{[H_3O^+][{}^-OH]}{[H_2O][H_2O]} \cong [H_3O^+][{}^-OH]$$

FIGURE A1.15 The repeated formation of the hydroxide and hydronium ion by proton exchange in water and an expression for the temperature-dependent constant (the equilibrium constant, K_{eq}) relating the reactants and products at equilibrium. The values in brackets are concentration terms (e.g., moles/liter for solutes in solution). Since the concentration of water (in the denominator) is large and, to a first approximation constant, its value is subsumed into the equilibrium constant K_{eq}.

In this chemical equation the statement is being made that water (H_2O) reacts with water (H_2O) to produce the hydronium ion (H_3O^+) by protonation on oxygen, and at the same time and as a consequence of the proton transfer, a hydroxide anion (OH^-) is also generated. The two arrows of equal length connecting the reactants (usually written on the left) with the products (usually written on the right) define the process as being in "equilibrium" (a state defined as one with equal numbers of species reacting in each direction at the same time so that concentrations of the reactants and products have no net change over time).

Systems that are at equilibrium at a given temperature enjoy the property of being able to maintain that equilibrium by self-adjusting! This observation is known as Le Châtelier's principle, and it is used to predict the effect of a change in conditions on a chemical system at equilibrium.[10] Generally, if the system experiences a change in concentration, then the equilibrium shifts to counteract the imposed change and a new equilibrium is established. This means, of course, that there is a constant ($K_{equilibrium}$) connecting reactants and products. The process, while quite general, is shown above for the representation of water reacting with itself to generate the positively charged hydronium ion [H_3O^+] and the negatively charged hydroxide anion [OH^-]. In the expression of the equilibrium (Figure A1.15), it is understood that terms in brackets (i.e., [term]) specifically refer to concentrations, usually expressed in moles (e.g., moles per liter for solutes in solvents). Generally, the concentration of water [H_2O] in the denominator is considered to be large and invariant and thus treated as a constant which is included in K_{eq}.

Those species that either (a) behave as if they were electron deficient (such as the proton, H^+) or (b) undergo reactions where electron rich species are consumed are called "acids." Alternatively, those species that either (a) behave as if there were electron rich (such as the hydroxide anion OH^-) or (b) undergo reactions where electron poor species are consumed are called "bases." Although it is not necessary, the acidity or basicity of a species is frequently referred to water, which is considered "neutral."

In order to communicate the extent of acidity or basicity, with water considered as neutral, a scale or range needed to be defined. So, just as the 20th century dawned (actually in 1909), Søren Sørensen, a Danish chemist, working at the Carlsberg laboratory in Copenhagen, Denmark, determined that the amount of acidic material (expressed as the "concentration of the acid, [H^+], in moles/liter of solution," where the brackets defined the term moles/liter) could be related to the passage of an electric current through the solution.[11] This was defined as the "pH" of the solution, where the "p" preceding the "H" is taken (by some) to mean the "power" of the hydrogen ion concentration and, by others, as a more general definition (i.e., "the decimal co-logarithm of" in this case the hydrogen ion concentration). Since the pH scale is a logarithmic scale (base 10), each integer is one power of 10 more acidic (or basic) than the one next to it. For example, the pH = $-\log_{10}[H^+]$, so for pure water, at a neutral pH = 7 (exactly), the hydrogen ion concentration [H^+] = 10^{-7}. There are many items to be found in our environment that have pHs spanning a large range of concentration. Some typical values for common items are shown in Figure A1.16.

Given the general definition of the symbol "p" (i.e., "the decimal co-logarithm of"), it is possible to consider an equilibrium constant for acids (i.e., $K_{eq} = K_a$) such that a new scale might be developed along the same lines and called the pK_a scale. So, what does it mean that lemon juice has a pH of 2? As defined above it, must mean that [H^+] = 1×10^{-2} and that lemon juice is more

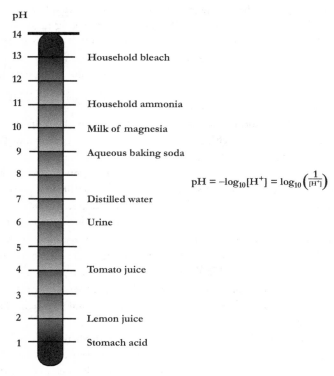

FIGURE A1.16 A representation of a pH scale with some common household acids and bases shown. The lower the pH number the more acidic the material, and the higher the pH number the more basic the material.

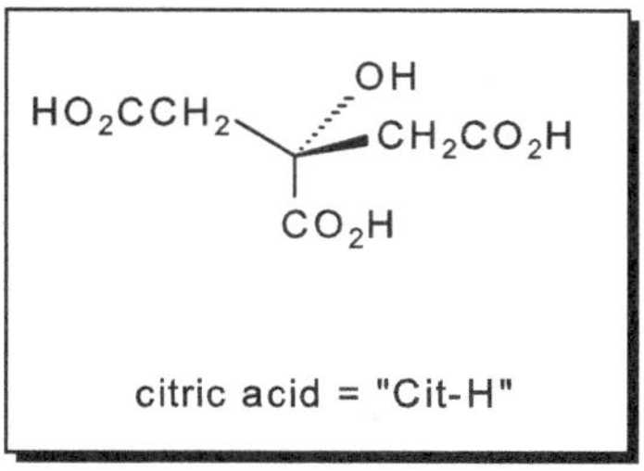

$$\text{Cit-H} + H_2O \xrightleftharpoons{K_a} \text{Cit}^- + H_3O^+$$

HO$_2$CCH$_2$—C(OH)(CO$_2$H)—CH$_2$CO$_2$H

citric acid = "Cit-H"

$$K_a = \frac{[\text{Cit}^-][H_3O^+]}{[\text{Cit-H}][H_2O]}$$

$$[H_3O^+] = K_a \frac{[\text{Cit-H}][H_2O]}{[\text{Cit}^-]}$$

FIGURE A1.17 A representation of the establishment of the ionization of citric acid (Cit-H) in water to produce [H_3O^+] and the citrate anion [Cit$^-$] and the representation of that ionization by an ionization constant K_a. Again, the concentration of water [H_2O] is generally considered large and invariant and is included in K_a.

acidic than neutral water (pH = 7.0 and that means [H^+] = 1×10^{-7}). Does this mean that there is acid in lemon juice? Indeed it does!

There is some acid in the citrus fruit lemon (call it "citric acid" "Cit-H") as shown in Figure A1.17. As shown, when citric acid dissolves in water, it ionizes to the cation [H_3O^+] and to the anion [Cit$^-$], but it ionizes only to the extent dictated by K_a (the ionization constant). This is an example of the general principle that most solutes that ionize do not ionize completely. Indeed, complete ionization is the exception rather than the rule. This is most likely because separation of unlike charges, even when the solvent is present to help, is costly. The acids that are present in wine are present in ionized and unionized equilibria. The subtle changes in those equilibria as a consequence of other constituents that may be present, temperature, age, the palate and condition of the consumer, etc. all play a role in its enjoyment.

THE FUNCTIONAL GROUPS AND CLASSES OF ORGANIC COMPOUNDS—OXIDATION AND REDUCTION

That feature (or those features) in each compound in addition to carbon and hydrogen (if only carbon and hydrogen are present, the compound is a hydrocarbon) is called (or are called if more than one is present) a functional group (or functional groups). The functional group exhibits some specific characteristic which is essentially independent of the molecule in which it is found. The value of the system of functional groups is one of classification.

Compounds of hydrogen and carbon only (hydrocarbons) fall into four broad categories. First, there are the alkanes. They can be in rings (cyclic) or not (acyclic), but regardless, each carbon is bonded once to either another carbon or to hydrogen. The geometry at carbon is approximately tetrahedral. While flat "two dimensional" drawings are common, it is incumbent upon the reader to understand that the molecules are three dimensional. Some examples are seen in Figures A1.18 and A1.19.

In the acyclic examples of in Figures A1.18 and A1.19 it is important to note the hydrocarbons can be looked at from right-to-left or left to right and up-to-down or down-to-up. It doesn't matter! It is also important to know that the molecules are three dimensional, and many drawings

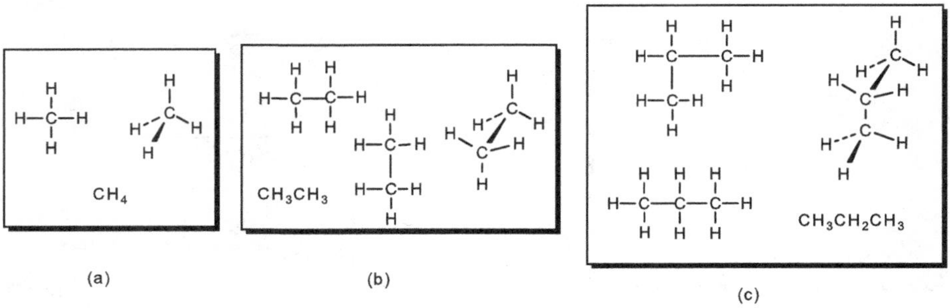

FIGURE A1.18 Representations of the hydrocarbons: (a) methane; (b) ethane; and (c) propane.

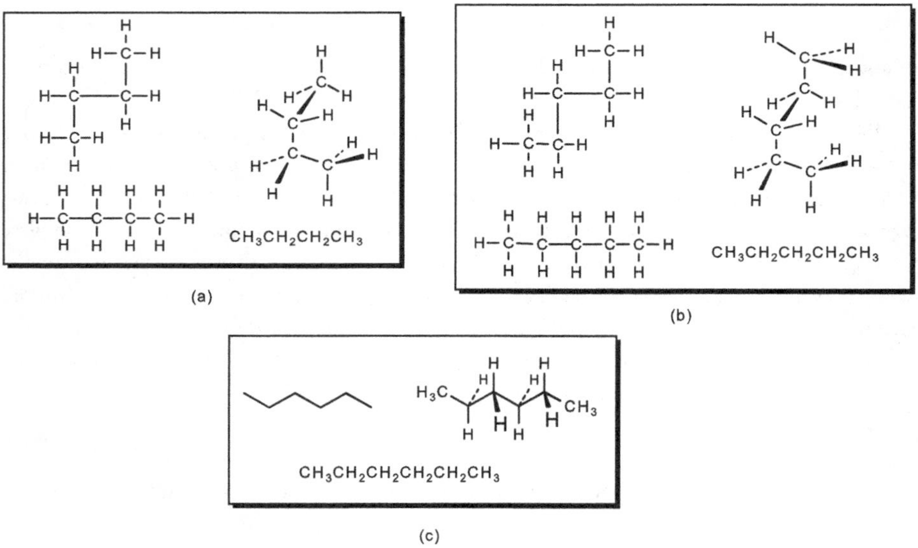

FIGURE A1.19 Representations of the hydrocarbons: (a) butane; (b) pentane; and (c) hexane. In (c) hydrogens attached to carbon bearing a heavy wedged bond are considered above the plane of the paper, whilst those with dashed wedged bonds ae below the plane of the paper. Regular lines are in the plane of the paper. In the upper left of the hexane cartoon (c) with a direction change at every carbon atom and a carbon at each terminus contains all of the unwritten but understood information of the more "filled-out" representations.

are attempted on the two-dimensional surface of paper to attempt to convey the three dimensionality of the subject. Some cyclic hydrocarbons, each clearly three dimensional, are shown in Figure A1.20. Just as was the case for looking at acyclic hydrocarbons from any direction, so it is for cyclic hydrocarbons. That is, one can count the carbons in either a clockwise or a counterclockwise direction.

As seen earlier for the chlorination of methane (Figure A1.10), if a hydrogen is replaced with a halogen, a haloalkane is obtained. Thus, as shown in Figure A1.21, if one hydrogen of butane is replaced with a chlorine, a chlorobutane is obtained. It should be clear that if a chlorine replaces

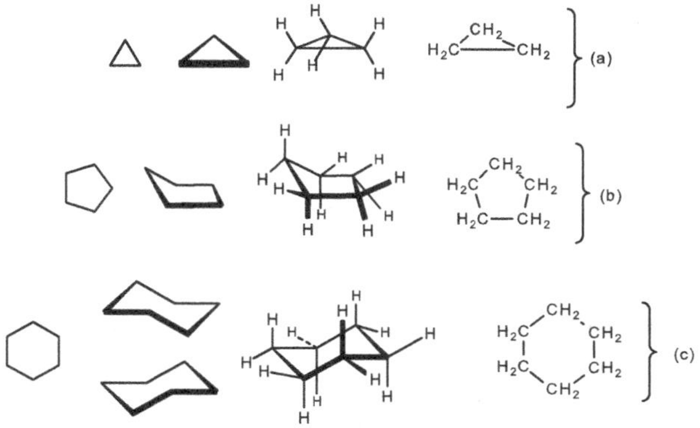

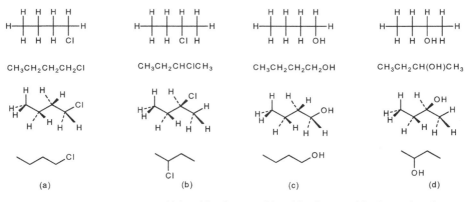

FIGURE A1.20 Representations of the carbocyclic hydrocarbons: (a) cyclopropane; (b) cyclopentane; and (c) cyclohexane. Each drawing contains the same information. In some of the drawings, the information is made more explicit (e.g., angles drawn so they are approximately tetrahedral), and an effort is made to convince the viewer of the three-dimensional nature of the structure. But whether drawn or not, the viewer needs to keep in mind the structure being viewed is three dimensional!

FIGURE A1.21 Representations of (a) 1-chlorobutane; (b) 2-chlorobutane; (c) 1-butanol, and (d) 2-butanol (also called isobutanol).

one of the hydrogens at either terminus, the result is exactly the same (since counting can be from right-to-left or from left-to-right) and all six of the hydrogens on either end of the chain, being at the corners of regular tetrahedra, are identical. Thus, 1-chlorobutane results. However, if a hydrogen on either of the other two carbons is replaced with a chlorine then a different isomer results. The empirical formula of the chlorobutane is the same (C_4H_9Cl) in either case, but the attachments in the second case are different than in the first case. Again, since the counting can be from either end, the new isomer is 2-chlorobutane. Clearly there can be no 3-chlorobutane, because one would simply count from the other end! In the event a ^-OH group is used in place of a halogen (F, Cl, Br, I), then the functional group which results is that of an alcohol. Just as replacement of a hydrogen in ethane leads to ethanol (ethyl alcohol) so replacement of a hydrogen (or halogen!) with ^-OH in butane (or chlorobutane) leads to butanol. Again there can be either 1-butanol or

2-butanol (also called "*iso*-butanol" because it is an *iso*mer of butanol). Both correspond to $C_4H_{10}O$ (Figure A1.21).

This leads us to an important interruption. Notice that the alcohols are richer in oxygen than the hydrocarbon. When there is an increase in the amount of oxygen or a decrease in the amount of hydrogen in a molecule, that will correspond to an oxidation. When there is a decrease in the amount of oxygen or an increase in the amount of hydrogen in a molecule, that will correspond to a reduction.

Now consider butane again. If hydrogen is lost from butane, an oxidation has occurred and keeping four bonds to carbon, there are a number of possible results. First, alcohols, corresponding to addition of oxygen (Figure A1.22) could occur.

Second, if two hydrogen atoms are lost a variety of products (loss of hydrogen being an oxidation) might be generated as shown in Figure A1.23. Thus, cyclobutane as well as several different alkenes (a family of compounds distinguished by possession of a carbon–carbon double bond) results. Indeed, introduction of a double bond between the first and second carbon atoms forms an alkene, 1-butene; introduction of a double bond between the second and third carbons forms two different alkenes, *cis*-2-butene also called *Z*-2-butene (with the –CH_3 groups on the same side of the double bond, where the "*Z*" is from the German "*zusammen*" or "together") and *trans*-2-butene also called *E*-2-butene (with the –CH_3 groups on opposite sides of the bond, where the "*E*" is from the German "*entgegen*" or "opposed").

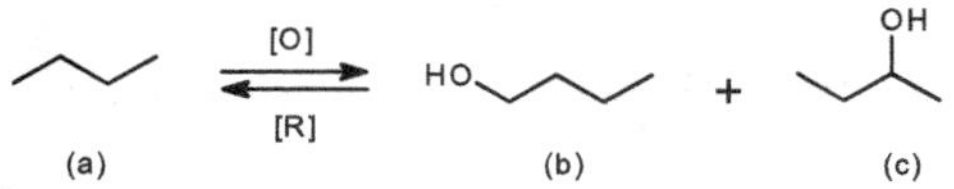

FIGURE A1.22 A cartoon representation indicating that oxidation of an alkane can produce an alcohol and reduction of an alcohol produce an alkane. The example shows butane (a) undergoing oxidation [O] to yield a primary alcohol (b) 1-butanol and a secondary alcohol (c) 2-butanol. The reverse reaction, a reduction [R], is also shown.

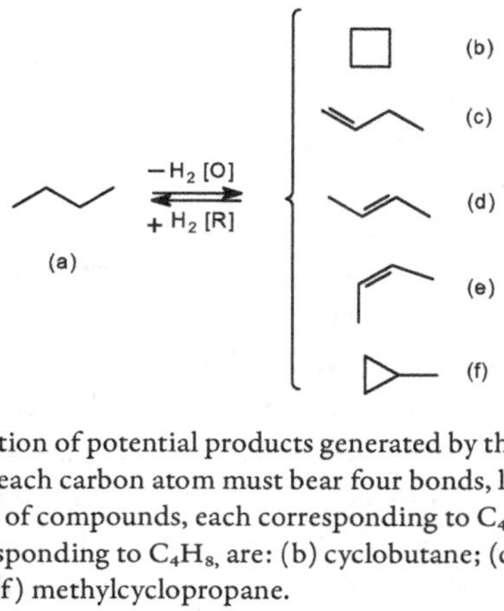

FIGURE A1.23 A representation of potential products generated by the loss of two (2) hydrogen atoms from (a) butane. Since each carbon atom must bear four bonds, loss of two hydrogen atoms from C_4H_{10} can yield a family of compounds, each corresponding to C_4H_8. The compounds from butane (a) are shown as corresponding to C_4H_8, are: (b) cyclobutane; (c) 1-butene; (d) *trans*-2-butene; (e) *cis*-2-butene; and (f) methylcyclopropane.

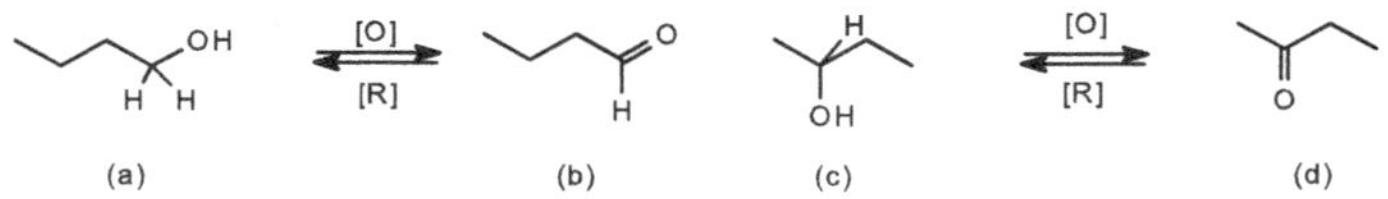

FIGURE A1.24 Representations of the oxidation [O] and reduction [R] of (a) 1-butanol to (b) butanal and the reverse as well as the oxidation [O] and reduction [R] of (c) 2-butanol to (d) 2-butanone.

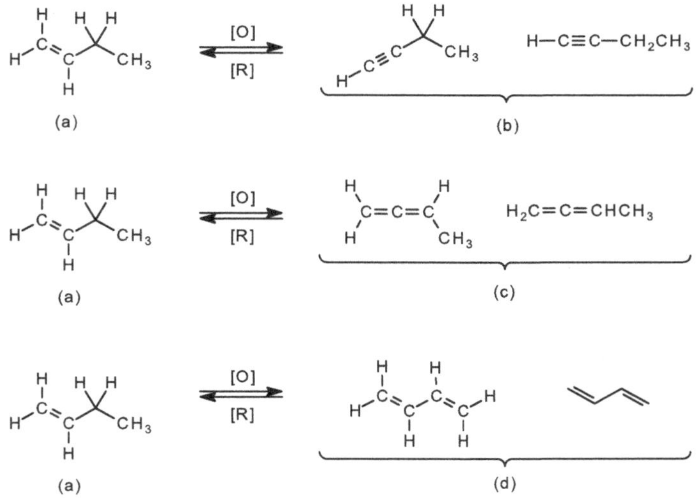

FIGURE A1.25 A representation of the various possible outcomes from oxidation [O] (hydrogen removal) from the alkene (a) 1-butene. In the first equation, 1-butene (a) could lose two hydrogen atoms from the adjacent carbons of the carbon–carbon double bond to yield the alkyne 1-butyne (b). The two other possibilities which accommodate loss of two hydrogen atoms from adjacent carbons lead to the diene (c) 1,2-butadiene or the diene (d) 1,3-butadiene.

Further oxidation of the 1-butanol (a primary alcohol because the –OH is on a carbon bonded to only one other carbon) yields butanal, an aldehyde (oxidation because there are fewer hydrogens than in the alcohol). It is not necessary to specify the position of the aldehyde (on the first carbon) because aldehydes must always be on the first carbon; there is always a hydrogen on the carbon of the carbonyl (C=O) group of an aldehyde. Oxidation of the 2-butanol (a secondary alcohol because the carbon bearing the –OH is bonded to two other carbons) yields a ketone (2-butanone). (Actually, it's not necessary to use the number in 2-butanone either but it is common to do so). These reactions are outlined in Figure A1.24.

Further oxidation (removal of hydrogen) of the alkene family member 1-butene yields a member of the alkyne family named 1-butyne (also called an acetylene) if the two hydrogens are removed from each of the carbons already bearing the double bond. Removal of other hydrogens might yield 1,2-butadiene or 1,3-butadiene, (members of the diene family) which are both isomers of the acetylene and all corresponding to C_4H_6! These reactions are outlined in Figure A1.25.

Similarly, as shown in Figure A1.26, removal of hydrogens from the carbons of the double bond of either of the alkenes *cis*-2-butene or *trans*-2-butene yields the alkyne 2-butyne (also known as dimethylacetylene). Removal of the proton from any of the methyl (–CH₃) groups of the alkenes and the proton on the adjacent double bond bearing that methyl (–CH₃) group also yields 1,2-butadiene.

Double bonds between adjoining carbon atoms, such as in 1,2-butadiene are called "cumulative" double bonds while those double bonds separated by a single bond as in 1,3-butadiene (above) are called "conjugated." Generally, compounds with cumulative double bonds are less stable that those with conjugated double bonds. The latter are commonly found in nature and associated with bright colors.

Interestingly, if there are three (3) double bonds inside a ring of six carbon atoms, and the double bonds are separated from each other by single bonds, the compounds that result have special stability. These compounds are called "aromatic." There is more than one such compound because the hydrogens that remain around the periphery can be replaced by other groups. Representations of the aromatic hydrocarbon benzene are shown in Figure A1.27. Notice that three structures are drawn and they are connected by double-headed arrows! The double-headed arrows do not mean equilibrium between the forms. They do mean that no single structure is sufficient, that is,

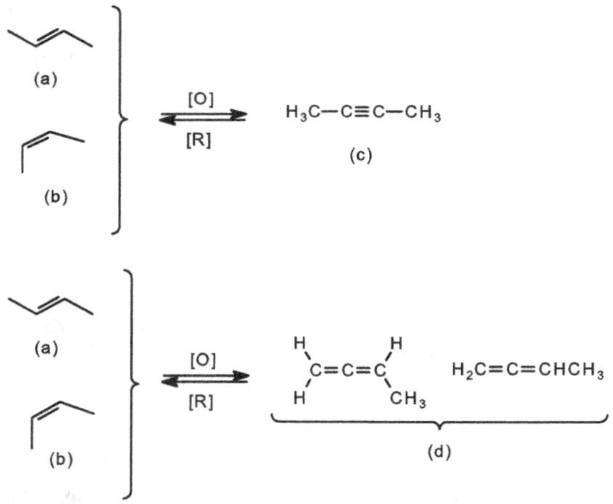

FIGURE A1.26 A representation of the various possible outcomes resulting from oxidation [O] (hydrogen removal) and its reverse [R] from the alkenes (a) *trans*-2-butene and (b) *cis*-2-butene. Removal of the two hydrogens on the carbon–carbon double bond from either (a) or (b) results, as shown in the first equation, in the formation of the alkyne 2-butyne (c). Alternatively, removal of one of the hydrogens from either end of the double bond and from the methyl group (CH₃) on the carbon atom at the adjacent terminus, leads to the "cumulative" diene, 1,2-butadiene (d).

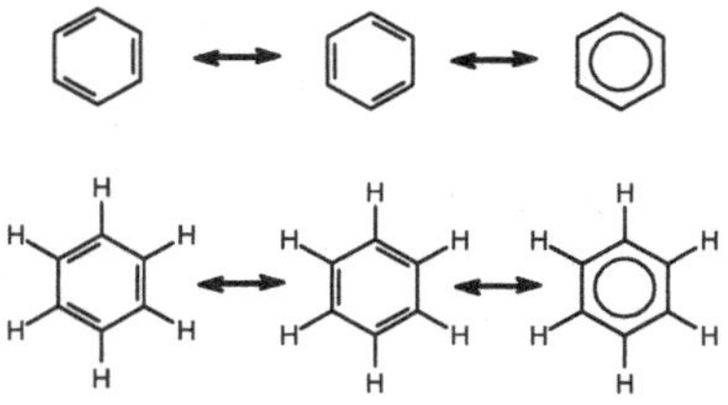

FIGURE A1.27 Representations of benzene.

A Chemistry Primer 327

the distance between the carbon atoms is the same (139 pm) whether a single bond (normally in, e.g., ethane, CH_3CH_3, 154 pm) or a double bond (normally in, e.g., ethene, $H_2C=CH_2$, 133 pm) is drawn! Such systems, where no single structure can be drawn to accurately represent the molecule in question and where the double-headed arrow is used exclusively, are said to be in *resonance* and the structures are "*resonance structures.*"

Aldehydes, but not ketones, are easily oxidized further. Oxidation of an aldehyde produces a carboxylic acid. As shown in Figure A1.28, oxidation of butanal leads to butanoic acid.

The general pattern relating methane to carbon dioxide, the former completely "reduced" and the latter, completely "oxidized" is shown in Figure A1.29. Oxidation of hydrocarbons leads to alcohols. Oxidation of alcohols leads to aldehydes or ketones. Oxidation of aldehydes leads to carboxylic acids. And, overall, complete oxidation of any hydrocarbon (such as methane to heat a house or butane to light a flame) leads to carbon dioxide, CO_2 (from the carbon that made up the hydrocarbon) and water, H_2O (from the hydrogen that, along with the carbon, made up the hydrocarbon).

Interestingly, as will be seen when the discussion of the chemistry of wine is resumed (or begun), the partial reversal of the process seen above accounts for not only the wine itself but also all that plants, absorbers of carbon dioxide, do for us. The issue of reactions and reactants has already been raised in other contexts, and among those that are important to the chemistry of wine are the formation of *esters*, the products of the reaction between alcohols and carboxylic acids.

As shown below in Figure A1.30 for the reaction between ethyl alcohol (ethanol) and acetic acid (ethanoic acid) to form ethyl acetate (ethyl ethanoate), not only is water lost, but by use of

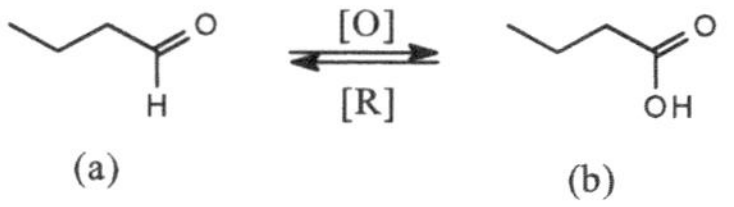

FIGURE A1.28 A representation of the oxidation [O] of butanal (a) to butanoic acid (b) and the reduction [R] of the acid (b) to the aldehyde (a).

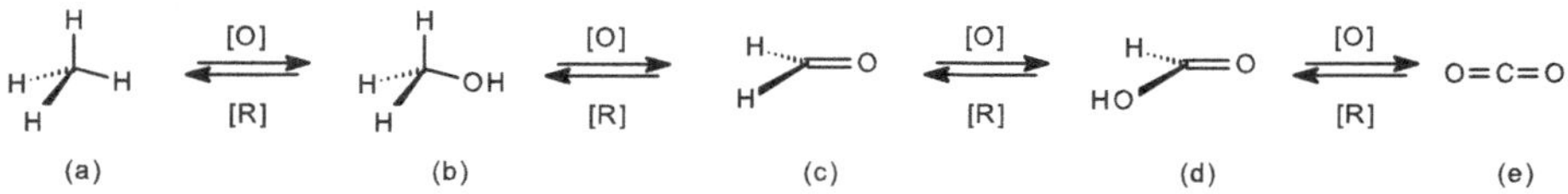

FIGURE A1.29 A representation of the stepwise conversion of (a) methane to (b) methanol and then to (c) methanal (formaldehyde), (d) methanoic acid (formic acid) and finally to (e) carbon dioxide. In the direction listed in the previous sentence, each step is an oxidation [O]. In the reverse direction, each step is a reduction [R].

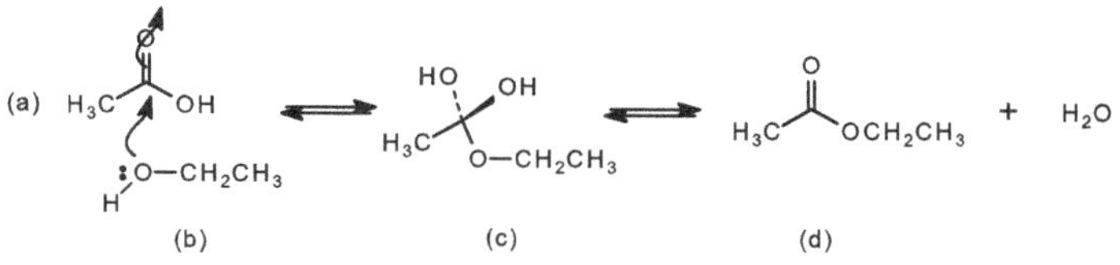

FIGURE A1.30 An abbreviated representation of a reaction between acetic acid (a) and ethanol (b) showing that the positive end of the carbonyl group (C=O) is being attacked by an electron pair on the oxygen of ethanol to form (c) the tetrahedral intermediate (a proton being lost from the oxygen of the ethanol and the same or another being added what was the oxygen of the carbonyl). Loss of water from the tetrahedral intermediate (c) produces the ethyl ester of acetic acid, ethyl acetate (d).

isotopes of oxygen (e.g., ^{18}O, *vide supra*), it is clear that it is one of the two oxygens of the carboxylic acid that is lost as water, whilst the oxygen originally present as part of the alcohol is retained. A cartoon pathway showing a portion of the way the reaction is known to occur (*via* a tetrahedral intermediate) is depicted.

In addition to carbon, hydrogen, and oxygen, nitrogen (and, later, phosphorus) needs to be considered as these elements are important in living systems. Nitrogen, lying between carbon (four bonds) and oxygen (two bonds) in the Periodic Table, should, reasonably, have three bonds to it, and in the neutral state, it does! Thus, for example, ammonia corresponds to NH_3, and gaseous nitrogen, N_2, is commonly written with three bonds connecting the two "Ns," (e.g., N≡N). Gaseous oxygen, O_2, is written with two bonds (O=O), and hydrogen gas, H_2, is written with one bond (H–H). Carbon, of course, is not a gas, and the network of four bonds from carbon-to-carbon-to-carbon results, among other *allotropes*, in diamond.

When nitrogen is found in organic compounds related to wine, it is usually present in the form of amines or compounds derived from amines (such as peptides) or in the heterocyclic bases (*vide infra*) found in the nucleic acids DNA and RNA (to be discussed later). Amines are basic (in contrast to carboxylic acids which are, of course, acidic). As a consequence, when an amine reacts, in an acid-base reaction, with a carboxylic acid, a salt is formed (since salts are defined as the product of an acid reacting with a base). So, as shown in Figure A1.31, when butanoic acid reacts with methylamine (methanamine) the compound methylammonium butanoate is formed. The transfer of the acidic proton of the carboxylic acid to the basic nitrogen of the amine has occurred.

However, under other conditions, often in the presence of a catalyst,[12] a different reaction occurs, water is lost, and an amide forms (Figure A1.32).

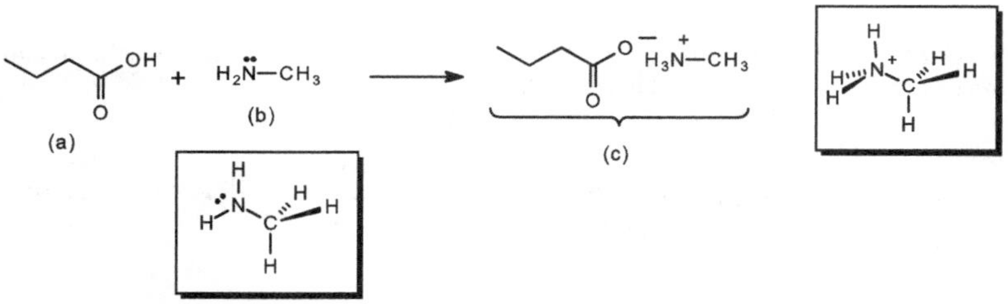

FIGURE A1.31 A representation of salt formation on reaction of butanoic acid (a) with methylamine (or methanamine) (b). The product (c) resulting from transfer of a proton from the acid to the base is the salt methylamonium butanoate.

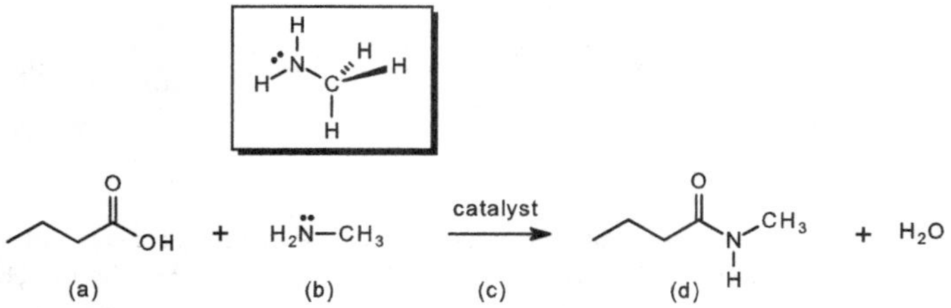

FIGURE A1.32 A representation of the reaction between (a) butanoic acid and methylamine (b) in the presence of a catalyst (c) to produce the amide N-methylbutanamide (d) and water.

A Chemistry Primer 329

Although it has not been specifically made clear earlier, there is no reason that different functional groups (or multiples of the same functional group) cannot be found in the same molecule. So, for example, ethanol (ethyl alcohol, CH_3CH_2OH) has, as already noted, one hydroxyl (–OH) group needed to place it in the class of alcohols. However, addition of a second –OH group on the carbon adjacent to the first (*viz.*, $HOCH_2CH_2OH$) produces a new compound, ethylene glycol (1,2-ethanediol), a poisonous component of antifreeze! Similarly, as shown in Figure A1.33, both 1-propanol and 2-propanol each has only one hydroxyl (–OH) group. However, the sweetening agent glycerol (1,2,3-propanetriol), has three such groups and is backbone to all lipids (triglycerides)!

Considering the case of the alcohol ethanol (ethyl alcohol, CH_3CH_2OH) again, there is no reason to restrict the attachment of another group to one identical to that already there. As shown in Figure A1.34, replacing a hydrogen on the methyl group (the carbon adjacent to the one bearing the hydroxyl) with a nitrogen that bears three methyl groups, produces a cation called "choline" (the *N,N,N*-trimethylethanol ammonium cation). As is the case with glycerol, choline, accompanied by a variety of anions (such as phosphate, *vide infra*) to preserve neutrality, is also naturally occurring and required for continuing good health. It is a B-complex vitamin as well as the precursor to the ubiquitous neurotransmitter acetylcholine.

Among those large number of compounds that have more than one functional group is a family containing both amine and carboxylic acid functionality and called therefore *amino acids*. Within that large subset of amino acids, those compounds that have an amino function on the same carbon as the carboxylic acid function are called alpha-amino acids. The basic amino group, receiving the proton from the acidic carboxylic acid group, results in a "double ion" (referred to as a "zwitterion"). There are twenty amino acids that make up the majority

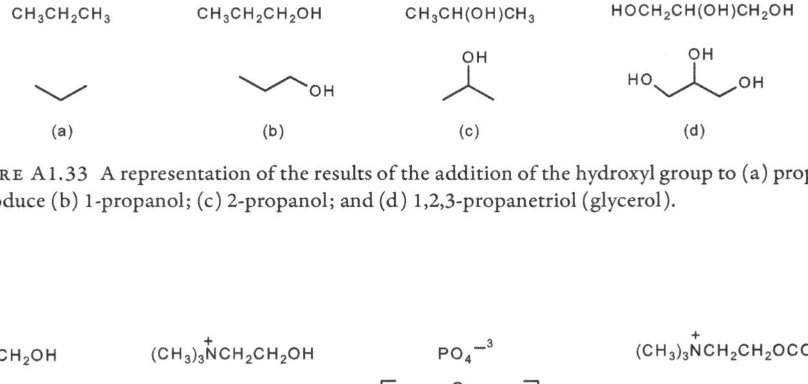

FIGURE A1.33 A representation of the results of the addition of the hydroxyl group to (a) propane to produce (b) 1-propanol; (c) 2-propanol; and (d) 1,2,3-propanetriol (glycerol).

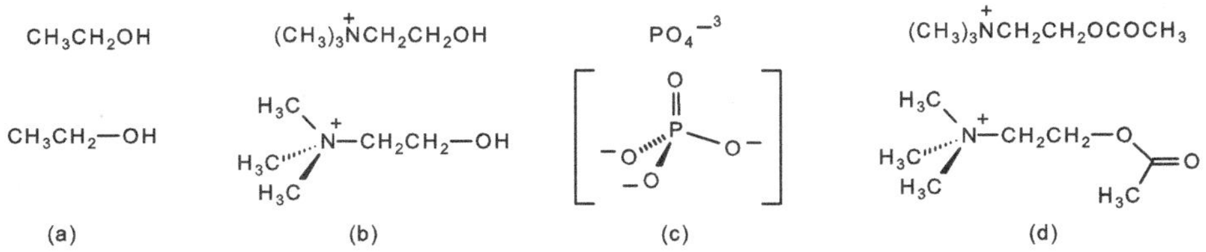

FIGURE A1.34 An idealized representation of the replacement of a hydrogen on the methyl (CH_3) group of ethanol (a) with an N,N,N-trimethylamine group to give (b) *N,N,N*-trimethylethanolammonium cation (*aka* choline). A representation of the phosphate anion (c) is also shown along with a representation of the neurotransmitter (d) acetyl choline.

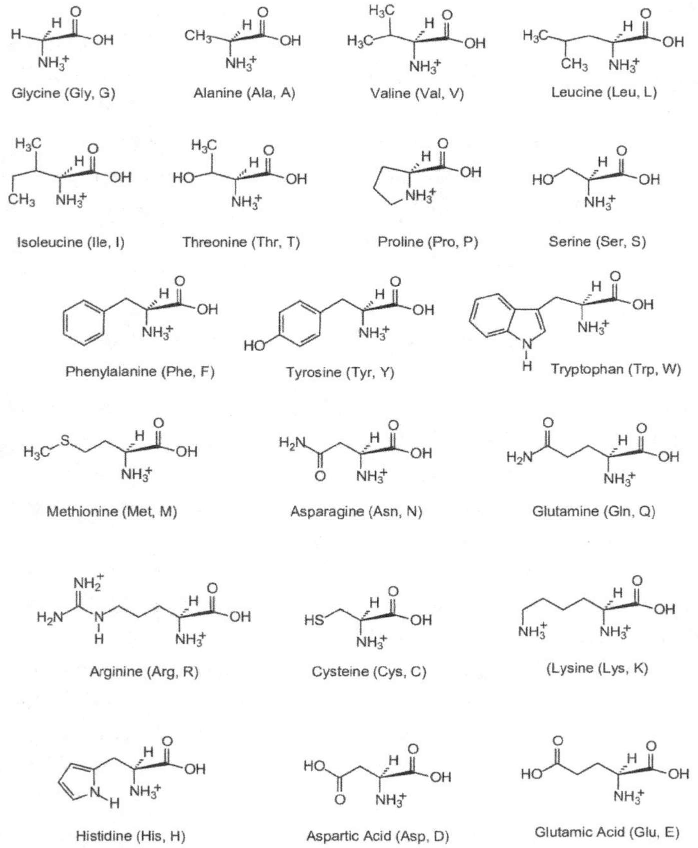

FIGURE A1.35 Representations of the twenty (20) most common amino acids, their names and their respective three- and single-letter abbreviations.

of amino acids found in all proteins in all living (plant, fungi, and animal) systems. They are shown in Figure A1.35.

Amino acids can be linked to each other *via* amide bonds which, in the case of amino acids, are referred to as "peptide bonds." Peptides are considered fragments (often up to about 50 connected amino acid fragments long) of proteins.

The proteins found in animal, fungus, and plant products are derived from (for the most part) these twenty amino acids, and since wine, while mostly water (H_2O) and ethanol (CH_3CH_2OH), contains plant and yeast (fungus) products, amino acids, proteins, and products derived from them are present in the wine.

Peptides are made up of linking these amino acids together through amide bonds. For example (Figure A1.36) consider two amino acids, glycine (Gly, G) and alanine (Ala, A).[13] When these two amino acids are joined together, different structures are obtained as a function of the order in which they are linked. By tradition, the amino end (positively charged) is on the

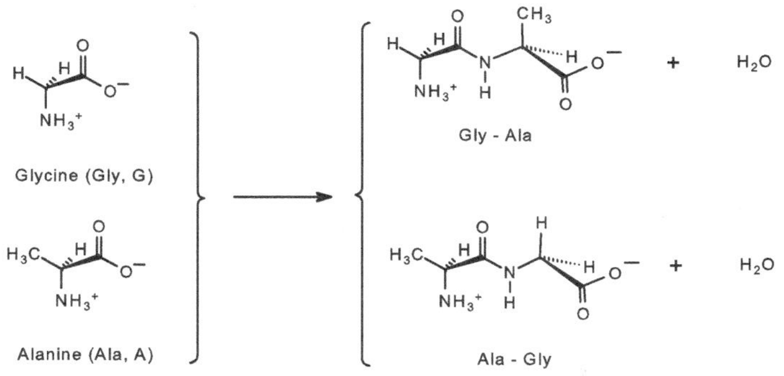

FIGURE A1.36 A representation of the combination of the amino acid glycine (Gly, G) and alanine (Ala, A) to yield dimers which are either Gly-Ala or Ala-Gly and thus are not identical. The DNA of the plant, fungus, or animal dictates in what order the assembly must occur to produce the desired fragment.

left and the carboxylic acid end (negatively charged) on the right. So, Gly-Ala and Ala-Gly are different!

The exact sequence of amino acids making up proteins (tens, hundreds, or thousands in any given protein) is specified by the genetic material (the deoxyribonucleic acid, DNA) of the plant, fungus, or animal. But the amino acids themselves are all the same regardless of the source, and the proteins made up of the amino acids are all the same whether found in nature or prepared synthetically.

Interestingly, at this writing, the "smallest" naturally found protein is recorded as being in the saliva of the Gila monster and is called TRP-Cage. That protein (1L2Y), as well as all that have been reported to date, from all sources along with the identifying information regarding the details of analysis, is found in the freely available Protein Data Bank (PDB) (http://www.ebi.ac.uk/thornton-srv/databases/pdbsum/). The representation from the PDB of TRP-Cage is shown on the left side of Figure A1.37, along with one (on the right) from the European Bioinformatics Institute (http://www.ebi.ac.uk/) which lists (in the same code as the PDB) the amino acids found in the TRP-Cage protein in the proper sequence. As can be gauged from the structure, and has become apparent from the geometry associated with the organic molecules, a simple "straight line" representation of structure is incorrect.

Just as we learned, as children, round pegs fit in round holes and square pegs fit in square holes, so it is that the shapes of organic molecules frequently determine our perceptions of the odor and taste of the material. Apparently this is because the shapes of olfactory receptors match (or do not match) the shapes of the molecules impinging upon them.

Part of the understanding of the organic chemistry of taste and smell that allows us to enjoy a good wine, or be put off by a poor one or perhaps one that has been oxidized or "corked"[14] is the shape of the organic molecules. For example, there is a group of volatile compounds called terpenoids (which have ten carbon atoms) which give odors, detectable by humans, to pine and other trees and shrubs. Among them are the two isomers of the ten-carbon terpene, carvone, shown in Figure A1.38.

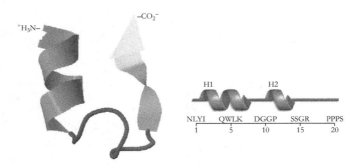

FIGURE A1.37 Two representations of the TRP-Cage protein. On the left is the representation from the Protein Data Bank (PDB); accession number is 1L2Y. The amino terminus is on the left and the carboxylate on the right. On the right is the same protein from the European Bioinformatics Institute. The letters of the amino acids are from the list of Figure A1.34 and thus show protonated asparagine (Asn, N) on the left to begin and the carboxylate anion of serine (Ser, S) on the right to end. The structure reported by Neidigh, J. W.: Fesinmeyer, R. M.; Andersen, N. H. *Struct. Biol.* **2002**, *9*, 425. Used under the terms of the charter of the PDB.

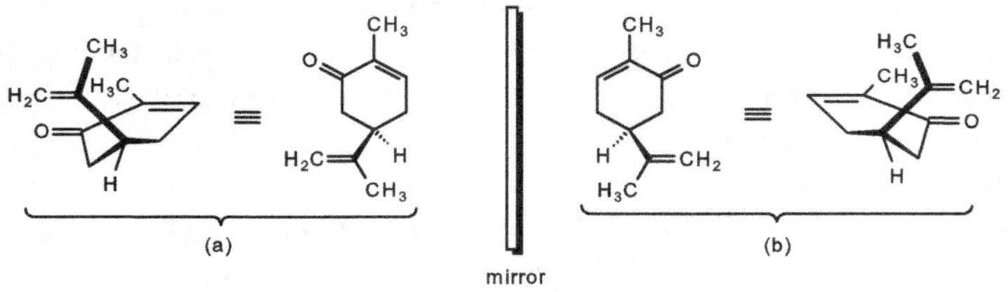

FIGURE A1.38 Representations of the terpenes (a) R-(−)-carvone, on the left, whose odor and taste are called "spearmint," obtained synthetically or from the aromatic herb *Mentha spicata*. On the right is the mirror image terpene (b) S-(+)-carvone, whose odor and taste are called "caraway." It too can be obtained synthetically or from the seeds of caraway (*Carum carvi*).

Molecules whose realized mirror images cannot be brought into superposition are called "*enantiomers.*" It is argued, in this case, that the reason these enantiomers are perceived as having different smells/tastes is suggestive of the necessity of olfactory receptors (see G-protein coupled receptors in Chapter 12) having a detector group or groups that allow them to respond more strongly to one enantiomer than to the other. Real objects (enantiomers) that cannot be brought into superposition with their real mirror images lack "mirror symmetry" and are thus "handed." The special name for this handedness in chemistry is "*chirality.*"

This kind of handedness is only one example of the general nature of the lack of mirror symmetry of any number of objects. Shoes, gloves, screws, nuts, and bolts are all common objects that lack a plane of symmetry. The consequence is that shoes and gloves only fit on one foot or hand or the other. A normal screw goes in when twisted clockwise and out in the reverse (mirror or counterclockwise) direction. Shoes and gloves cannot be exchanged, right-for-left (at least without discomfort), and screws, nuts and bolts all will fail if not tightened in the right direction.

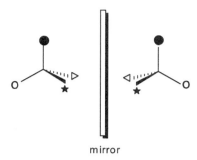

FIGURE A1.39 A representation of a real object in the form of an idealized tetrahedron and its real mirror image which cannot be brought into superposition since they lack mirror symmetry.

All of the amino acids shown above (except for glycine, Gly, G) also enjoy the lack of mirror symmetry. In chemistry there are many such objects; it is often possible to quickly identify a subset by noting that a carbon atom *with four different substituents* will not be capable of being brought into superposition with its realized mirror image. So, in the objects shown in Figure A1.39, the stars are above the plane of the paper (heavy wedge lines), the triangles below the plane of the paper (dashed lines), and the circles (dark and clear) are in the plane of the paper. The wedge/dash line notation is common. Any such real objects cannot be brought into superposition without breaking bonds.

It is, of course, reasonable to ask how we know which is which (and, by the way, what do the letters and signs preceding the name "carvone" have to do with it)? So, without reason (as it's not necessary here) the letters "*R*" and "*S*" refer to designations chemists have devised to indicate whether one should be called "right handed" or "*rectus*" (designated by "*R*") or "left handed" or "*sinister*" from the Latin meaning "left" (and designated by "*S*"). Some faces of molecules can also be considered the same way. Those faces have the designations "*re*" or "*si*" on the same basis. The signs (+) and (−) refer to a physical property that has nothing to do with the chemists or anyone else. It is a fact of nature and deals with the interaction of light with matter.

There has already been an introduction of the interaction of light with matter in the discussion of the "emission lines" observed when electrons in sodium vapor are excited by absorption of energy and its subsequent emission of energy seen as yellow-orange light in the visible region of the spectrum (page 310).

As shown in Figure A1.40, a light wave coming from a point source (out of the paper in the X direction toward the viewer) may be deconstructed by considering it as the sum of two mutually orthogonal vectors (i.e., the vector "A" in the YZ plane in the figure on the left and its deconstruction into Ay and Az as shown on the right). Both Ay and Az lie in the YZ plane. If either the Ay or the Az component could be blocked, then only one-half (½) the light would come through. This transmitted light is said to be "plane-polarized," and the wave corresponding to that light lies in one plane. The intensity of the light is also diminished (as in polarized sun glasses).

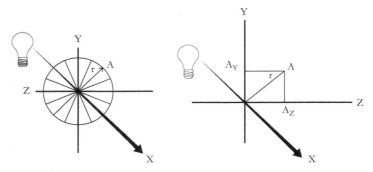

FIGURE A1.40 Two representations of the same light source with a beam in the X-direction (toward the viewer). On the left is a representation of a point along the beam of light indicating that the beam is composed of waves which have, at any given point along them, components in the x, y, and z planes. Thus, if, at a point "A" in the YZ plane of the beam moving in the X direction, the Y (or Z) component can be removed, what is left will be a beam traveling in the X direction and only in the Z (or Y) plane. The beam is plane polarized.

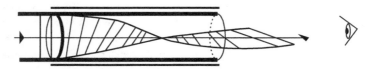

FIGURE A1.41 A representation of the twisting of the plane of plane-polarized light entering from the left as it passes through a container holding a solution of a chiral solute. There is a representation of an eye at the right end observing the change in rotation.

When a solution of chiral molecules interacts with the plane of plane-polarized light, the light is not absorbed by the solution equivalently in all directions. The result, as shown in Figure A1.41, is to cause the plane of the plane-polarized light to twist (or "rotate").

This property was first observed in the case of sugars by the French physicist Jean-Baptiste Biot in 1815 and exploited by the engineer William Nicol who, in 1828, noted that a crystal of Icelandic spar (calcite, calcium carbonate, $CaCO_3$) cut properly would produce polarized light. Thus, by convention, when a beam of polarized light passes through a solution made of molecules of one chirality, the isomer that rotates the plane of polarization counterclockwise from the point of view of the observer is called levorotatory and given a minus (−) sign; its mirror image, the dextrorotatory form of the same chiral molecule in solution in the same solvent, rotates the plane of polarization clockwise from the point of view of the observer and is given a plus (+) sign. Interestingly, the optical rotation is also a function of: (a) the wavelength of the light used for plane polarization (usually a sodium lamp is chosen because the intense yellow-orange sodium D-line at 589 nm is commonly available); (b) the temperature at which the measurement is taken; (c) the solvent in which the chiral material is dissolved; as well as (d) its concentration. Therefore these variables are reported when the measurement is taken. The commercial device for measuring this physical property of chiral molecules is called a polarimeter, and a cartoon sketch is shown in Figure A1.42.

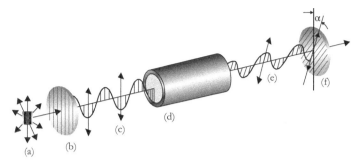

FIGURE A1.42 A cartoon representation of a polarimeter. Light from a known source (it is common to use a sodium lamp which has a bright yellow line at 589 nm [known as the D-line]) which is not polarized. This is shown as a filament (a) in the diagram. The unpolarized light goes through a polarizer (b), and the emerging light is now polarized. That incident light (c) passes through the polarimeter tube (d) which contains the solution of the optical isomer to be examined and emerges (e) with its plane of polarization having been "rotated." A second polarizer (f) which can be rotated is now used to move the beam back to its original plane. The direction of rotation, clockwise (+) or counterclockwise (−) as observed by the viewer through some angle "α" is then known and is adjusted for the concentration of the solute, the temperature at which the observation was made, and the solvent used before reporting.

With regard to wine, the use of polarized light has become very important, since two of the more prominent constituents of wine, both of which lend flavor and tartness to the grapes are (S)-(−)-malic acid [(S)-(−)-2-hydroxysuccinic acid] [[α]^{20}D −27 (c, 5.5 in pyridine)] and (2R,3R)-(+)-tartaric acid [[α]^{20}D +12 (c, 20.0 in H_2O)]. The latter is commonly available from the monopotassium salt deposited in fermenting grape juice.

The letter alpha (α) in the above expression (called *specific rotation*) refers to the extent to which the plane of plane polarized light is rotated. The sign (+) or (−) of the rotation refers to the direction (i.e., to the right or left of the viewer as she or he looks through the "analyzer"), and the superscript number (20 in both cases above) is the temperature (degrees Celsius) at which the experiment was undertaken. The subscript "D" refers, as already noted, to the emission spectrum of sodium (Na) metal vapor. It's the emission that gives sodium vapor street lamps their yellow tinge. Finally the small "c" preceding the solvent in which the sample is dissolved (pyridine and water, respectively) refers to the concentration in grams (gm) per milliliter (mL) of sample. Unless otherwise specified a tube of 1 decimeter (dm) length is used.

A relationship between electromagnetic radiation and matter has already been raised in regard to the excitation of sodium vapor to produce yellow light. Again, it was in the visible region of the spectrum to which reference was made. However, the visible region of the electromagnetic spectrum is only a small portion of the entire spectrum. As seen in Figure A1.43, our eyes (optical detectors) only work in the (approximately) 400–800 nm region of the spectrum. The rest of the usable (at the moment to us) spectrum is truly large and something about which you should know.

First, with regard to the visible (VIS) region of the spectrum, there are two different aspects of what we see. On the one hand, there are emitters of light, such as lamps, which generate light in the visible region of the spectrum by exciting electrons in gases or solids (e.g., hydrogen, neon,

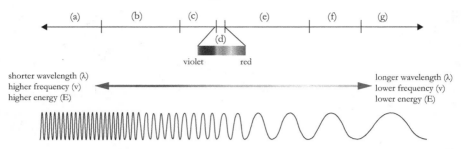

FIGURE A1.43 Representations of a portion of the electromagnetic spectrum. In the upper part of the Figure there are shown (left-to-right) the shorter to longer wavelength regions and the approximate size of the currently useful regions for: (a) gamma ray; (b) X-ray; (c) ultraviolet (UV); (d) visible (VIS); (e) infrared (IR); (f) microwave; and (g) radio wave spectroscopies. Beneath that representation, there is an arrow with shorter wavelength (λ = lambda), higher frequency (ν = nu), and therefore higher energy (E) at the left and longer wavelength (λ), lower frequency (ν), and therefore lower energy (E) on the right. At the bottom, there is a cartoon representation for the change in wavelength (λ).

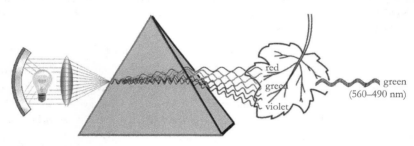

FIGURE A1.44 A representation of the colors of the visible spectrum observed by passing white light through a prism. The wavelength regions corresponding to red, green, and violet colors are shown striking a representation of a green leaf. Colors in the visible region of the spectrum that are not green are absorbed so that only green light is seen.

sodium vapor, electric filaments). Ordinary light bulbs emit white (visible) light (the mixture of all colors of the visible spectrum is "white") as well as heat (infrared radiation). There are now some light-emitting diodes (LEDs) that emit colored light directly.

Alternatively, portions of white light can be absorbed so that only colors specified by what was not absorbed are those that our eyes see. In this vein, white light can be passed through a prism or grating so that it can be deconstructed into the colors from which it is made. When one or more of those colors are absorbed by an object, what we see are only the colors that are left over. As shown in Figure A1.44, the colors falling on an object can be absorbed and only the colors not absorbed are seen. So, for example, the absorption of red and blue light by plant leaves is perceived as the color we call green.

Returning to the remainder of the electromagnetic spectrum, it is important to know that when absorption of energy occurs, particularly with molecules rather than atoms, the emission of that absorbed energy can occur in forms other than light and/or heat. That is, sometimes, work is done on or by the molecules that absorb the energy. This work can take the form of excitations of

A Chemistry Primer 337

electrons associated with the molecule that absorbed the light or with other molecules to which the energy is transferred. Alternatively, it may show up in vibrations of the bonds in the molecules. With enough energy added, molecular fragmentation can occur.

As discussed in the text (Chapter 7), the absorption of light is critical to growing the vines and, indeed, to all life on the planet. Examination of Figure A1.43 shows, the ultraviolet (UV) region of the spectrum (noted there as "c") lies beyond the violet end of the visible and into the higher-energy (shorter-wavelength) region. The energy associated with UV light reaching the upper atmosphere of our planet is high enough to harm most living organisms. Fortunately, most of the potentially damaging radiation is absorbed by the atmosphere.

The highest-energy part of the UV is called the "vacuum UV" (between 10–100 nm), and we are screened from it by nitrogen gas, N_2, (about 78% of the atmosphere) which has a nitrogen–nitrogen triple bond (N≡N). Significant amounts are also filtered around 200 nm wavelength by oxygen gas, O_2, (about 21% of the atmosphere) which has an oxygen–oxygen double bond (O=O) and by the ozone layer. Ozone, O_3, a very small component of the atmosphere but which strongly absorbs in the region of 200–310 nm also contains an oxygen–oxygen double bond (O=O$^+$O$^-$).

The principle of carbon compounds containing double and triple bonds (called *sites of unsaturation*) similar to the bonds in nitrogen (N_2), oxygen (O_2), and ozone (O_3) has already been referred to in the discussion of alkenes and alkynes. Thus, it should come as no surprise to find that such unsaturated compounds also absorb in the UV region of the spectrum.

However, carbon-containing unsaturated organic compounds absorb in different regions of the UV spectrum just as nitrogen, oxygen, and ozone do. Interestingly, in the case of carbon compounds, the more double and triple bonds, the longer the wavelength at which absorption occurs. So, as noted below in Figure A1.45, the compound ethylene absorbs UV radiation at about 163 nm; for 1,3-butadiene, with two conjugated (*vide supra*) double bonds, the absorption is at 217 nm; and for 1,3,5-hexatriene it's at 268 nm. The aromatic compound benzene absorbs ultraviolet radiation at 255 nm, while its two-ring relative naphthalene absorbs 315 nm. The naturally occurring compound β-carotene, found in carrots and tomatoes (and relatives derived from which are found in many wines), absorbs in the region of 470 nm, and its close relative, lycopene absorbs at 502 nm. For β-carotene and lycopene, those frequencies are in the blue-green region of the spectrum and so they are seen, respectively, in the yellow-orange region and red regions of the spectrum.

Representations of some brightly colored compounds are also provided in Figure A1.46.

The compounds shown in Figure A1.46 attracted the interest of naturalists and others over many years. However, it was not until M. S. Tswett, an assistant in the laboratories of Warsaw University, presented a lecture in 1903 on differential adsorption of pigments on various adsorbents using different solvents that the process now known as *chromatography* became established.

This process of chromatography as reported by Tswett allowed for the separation and purification of substances such as the pigments mentioned above because they were differentially adsorbed (held on surface) and they eluted (washed off a surface) at different times. The differential adsorption resulted (and continues to result) from the presence of different numbers and kinds of functional groups. The colors were easily detected by the naked eye. Tswett poured extracts of plants down columns of adsorbents such as alumina (aluminum oxide, Al_2O_3) and silica gel (SiO_2) as well as cellulose and many other materials to see which ones worked best, and he, and those following him, have continued similar experiments over the years. The pure materials

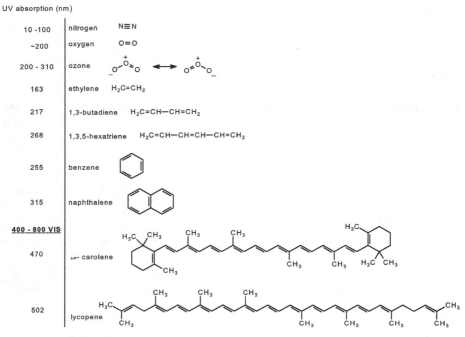

FIGURE A1.45 A representation of a set of examples of the decrease in energy (E) associated with the increase in the number of conjugated (i.e., single-bond-separated) double bonds in a molecule. As the energy decreases, the wavelength of light (λ) absorbed increases. For the highly conjugated polyenes β-carotene and lycopene the wavelength absorbed leaves orange-yellow and red, respectively as the colors transmitted and thus detected by our eyes.

isolated by subsequently washing the adsorbed compounds off the adsorbents (elution) were then subjected to analysis (by UV spectroscopy and elemental analysis). Other analytical techniques are now employed too.

However, the nature of chromatography has changed in several important ways. First, column chromatography can be replaced by other methods of separation. A small sample of them include: (a) thin-layer chromatography (TLC) for small samples. TLC utilizes thin layers of adsorbent on a plate of glass or plastic, and the solvent creeps up the plate, with material being differentially adsorbed; (b) gas-liquid partition chromatography, GLPC or simply GC, where vaporized samples pass over liquid adsorbed on some suitable substrate to be eluted back into the gas phase by a moving pressurized gas; (c) high pressure liquid chromatography, HPLC, where solvents containing the materials to be separated are pumped at high pressure over suitable columns of adsorbents; (d) ion exchange chromatography, where charged species move over charged surfaces, and the difference in charges allows for retention or passage; (e) gel electrophoresis, where proteins placed in gels are separated by charge and size differences; and (f) gel permeation chromatography (GPC), where the size of the particles makes a difference in the rate of passage through a column. An extensive variety of similar chromatographic techniques, where compounds are manipulated before separation is undertaken, are also now used.

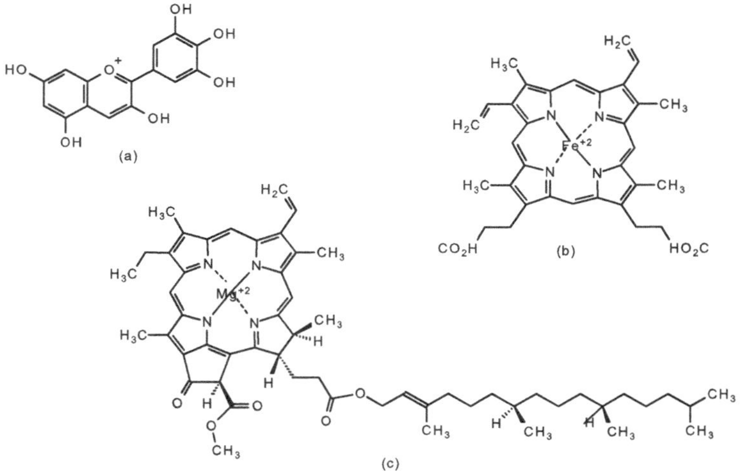

FIGURE A1.46 A representation of structures of some brightly colored compounds. The bright red anthocyanidin named malvidin (a) (absorbs at 538 nm—in the visible green region of the spectrum) is a coloring matter found coordinated with sugars in flowers and grapes and can also be isolated from some samples of wine. There is some nonspecified anion coordinated so that the positive charge is neutralized. Also shown is the structure of heme (b). The presence of iron (Fe^{2+}) bonded to the nitrogens should be noted. Heme, found in red blood cells, is itself red since there is broad absorption below 550 nm (yellow-green). Finally, the structure of (c) Chlorophyll a is also shown. Chlorophyll a is a component of green organisms. Note the presence of magnesium (Mg^{2+}) in the center of the same kind of ring system found in heme. Chlorophyll a absorbs visible light at about 665 and 430 nm (i.e., at the red and purple—opposite—ends of the visible spectrum); therefore it looks green.

The second major difference that has come about since the time of Tswett lies in the detectors that are used. Many interesting compounds have no color associated with them, and so different detectors have been developed.

For example, as shown in Figure A1.45, neither naphthalene nor benzene will absorb in the visible (VIS) region of the spectrum (and so they are colorless), but both absorb in the ultraviolet (UV) region, and so a UV detector might be used.

In gas chromatography, it is common to use a thermal conductivity (TC) detector. That detector is based on the principle that gasses that contain "impurities' conduct heat differently than gasses that do not. Therefore if a near simultaneous comparison is made between pure gas and gas carrying some sample as it comes off a column, it is clear when the sample is passing through the detector since the heat content of the impure gas differs from the pure one.

It is very common now to utilize a mass spectrometer (MS) (Figure A1.47) as a detector. With the MS detector, the effluent from the chromatography system is passed directly into the spectrometer. A tiny sample is vaporized and then, commonly, ionized by bombarding it with high-energy electrons so that it, the sample, is ionized. The ionized sample is focused into a beam by an electric field and passed into a mass analyzer. Commonly, the mass analyzer has a magnetic field so that the masses can be separated according to their mass-to-charge ratios (m/z) (m/z where m = mass and z = charge). Ions are finally detected by striking an analyzer and characterized

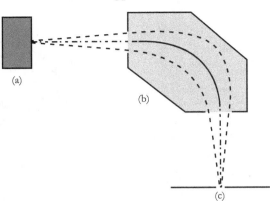

FIGURE A1.47 A cartoon representation of a mass spectrometer. All under vacuum, an ionized material leaves the source (a) and passes through a magnetic field (b). The ions are separated because charged materials interact with the magnetic field, and large particles move more slowly than do small particles. Small charged species (molecules or molecular fragments) reach the detector (c) before large charged species. Their masses are recorded at the detector.

according to their mass-to-charge ratios. Atomic, ionic, and molecular weights in MS are normally expressed in terms of atomic mass units (AMUs) based on $^{12}C = 12$ exactly.

It is common or becoming common today, and will be seen (or has been seen) in the text for wine samples to be chromatographically separated and then analyzed by MS. Generally, samples of material of known structure are used for comparison purposes and not only are they eluted from the gas or liquid chromatographic equipment with identical retention times (i.e., the amount of time spent on the column between injection and elution) but their masses and fragmentation patterns are seen to be identical. The analysis requires only a small amount of material (usually a few micrograms) which is destroyed in the process!

Consider, for example, the mass spectrum of ethanol (ethyl alcohol, CH_3CH_2OH), Figure A1.48. When ethanol (CH_3CH_2OH) is bombarded with high-energy electrons, either an electron is captured or an electron is knocked loose. In the first case a negatively charged particle is formed, and in the second case a positively charged particle is formed. With high-energy electrons, the latter is more likely than the former. The charged particle is usually written as ($CH_3CH_2OH^+$), and it is that particle which is considered as the parent ion. For ethanol, the m/z for the parent is 46 amu (i.e., 2 carbon atoms at 12 amu = 24 amu, six hydrogen atoms at 1 amu each = 6 amu, and one oxygen atom at 16 amu = 16 amu, so that 24 + 6 + 16 = 46 amu). However, some of the high-energy parent ions fragment or fall apart before reaching the detector. One possibility for the fragmentation is that it simply loses a hydrogen atom (•H) to give $C_2H_5O^+$ (m/z = 45 amu), while another possibility is that is loses a methyl group (•CH_3 = 15 amu) to give CH_2OH^+ (m/z = 31 amu) or something similar. Charged species to which these ions (representing these fragments) are assigned are seen in the spectrum (Figure A1.48).

Other compounds can be analyzed in the same way. So, a common ketone, acetone (CH_3COCH_3), has a mass spectrum shown as Figure A1.49. Again, initially, a positively charged parent species ($CH_3COCH_3^+$) is seen along with several fragments.[15]

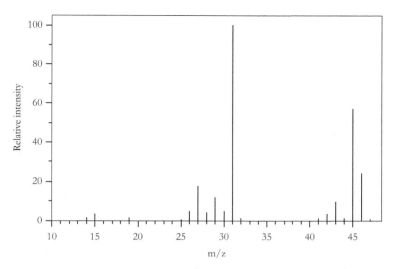

FIGURE A1.48 A representation of the low-resolution mass spectrum of ethanol (CH_3CH_2OH). The parent ion is assigned as the one with the highest mass (m/z = 46) but, clearly there is an ion corresponding to m/z 45 of higher intensity (generally a reflection of abundance) which could correspond to the loss of a hydrogen atom. The major fragment (most intense peak) at m/z 31 corresponds to the loss of 15 mass units from the parent, suggesting that a structure such as "CH_2OH^+" is possible. The most intense ion peak (major fragment) is known as the "base" peak. Here it is at m/z 31.

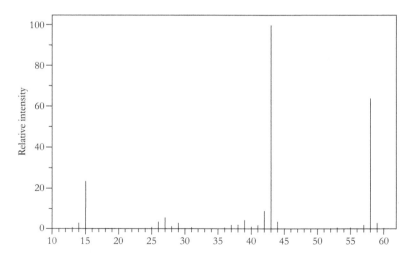

FIGURE A1.49 A representation of a low-resolution mass spectrum of acetone (2-propanone, CH_3COCH_3). The parent ion is seen at m/z 58 with a small satellite at m/z 59 which might be due to some ^{13}C isotope. The base peak at m/z 43 amu is common among compounds that contain the CH_3CO fragment and is often attributed to ($CH_3C\equiv O^+$) or something similar.

Again, as seen for ethanol, the first fragment lost appears to correspond to a methyl group (•CH_3) to give a fragment 15 amu less than the original 58 amu (58−15 = 43 amu) which presumably corresponds to ($CH_3C≡O^+$) or something similar.

With pure compounds (from chromatography) of known masses (from MS of pure compounds compared to what we have obtained) we might still ask how we know the details of the structure. The history of Organic Chemistry is replete with chemical means of structure proof (including synthesis) which can be applied to many compounds that have been known for some years. More recently, in addition to crystallographic methods which, require crystals (which may not be available), spectroscopic tools using portions of the electromagnetic spectrum outside the visible have become available to help solve the problem of structure. One such method involves the use of the infrared (IR) portion of the spectrum and another the radio region of the spectrum. The latter is called Nuclear Magnetic Resonance (NMR) and is similar to Magnetic Resonance Imaging (MRI) where individual nuclei that possess a property called "spin" can be detected. The IR is considered first.

The emission of the energy associated with bending and stretching vibrations can be seen in the infrared (IR) region of the spectrum, and this provides information about what bonds are connected to what nuclei. This is, for example, how we know that the molecule C_2H_5OH that corresponds to ethanol is different than the molecule CH_3OCH_3 that corresponds to dimethyl ether even though they both possess two carbons, six hydrogens, and one oxygen (C_2H_6O). Similarly, diethyl ether (ether, $CH_3CH_2OCH_2CH_3$) and 1-butanol ($CH_3CH_2CH_2CH_2OH$) both possess four carbons, ten hydrogens, and one oxygen ($C_4H_{10}O$), but they are clearly different molecules. How do we know (not guess) which is which in these pairs of isomeric molecules?

As shown in Figure A1.50 for the butanol/diethyl ether pair, the alcohol has an intense absorption between 3000 and 4000 cm^{-1}, while the ether does not. These bands (one hydrogen bonded and one not) correspond to the stretching of the bond between oxygen and hydrogen. Since there is no such bond in the ether, it is clearly distinguished from the alcohol.

Similarly, ethanol (CH_3CH_2OH) has such an intense band in its IR spectrum. Thus, it too is an alcohol.

Oxidation of ethanol (CH_3CH_2OH) to acetaldehyde (CH_3CHO) results in the disappearance of the O–H band centered at about 3600 cm^{-1} and the introduction of a C=O band as shown in the IR spectrum of acetaldehyde (CH_3CHO); see Figure A1.51.

There is another form of spectroscopy currently in use that is very valuable in determining exactly what molecules are involved in any process or present in any mixture (once separated). This form of spectroscopy (NMR) is used for nuclei (such as ^1H and ^{13}C) that have either (or both) odd mass and/or odd atomic number and depends upon such nuclei behaving as if they had a property called "spin."

When nuclei (which it will be recalled are charged particles consisting of protons and neutrons) possessing "spin"[16] are placed in an external magnetic field, two possibilities states are developed because "spinning" charges generate magnetic fields (and vice versa as in a dynamo). The states develop from nuclei with spin in the same direction as the external field (lower energy) or those with spin opposite to the external field (higher energy). The difference in energy is very small, and some of the nuclei spinning with the field can be induced to spin against it by application of a pulse of energy at the appropriate frequency. Subsequent relaxation of the excited nuclei back to the lower energy spin state results in a signal that can be detected.

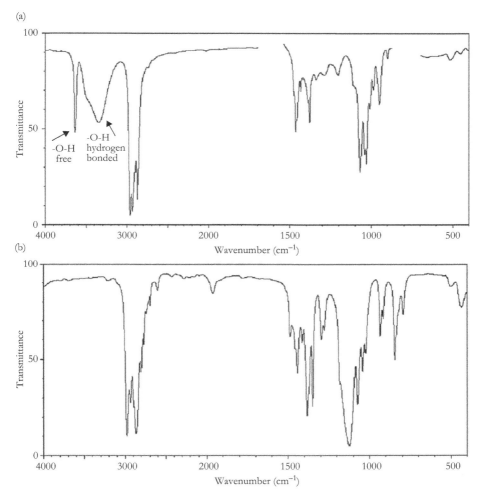

FIGURE A1.50 Representations of (a) the infrared spectrum of neat (no solvent) 1-butanol ($CH_3CH_2CH_2CH_2O-H$) (i.e., $C_4H_{10}O$ with the –OH group lying between 3000 and 4000 cm^{-1}) labeled. Both the free and hydrogen-bonded –OH functionalities are seen; and (b) the infrared spectrum of neat ethyl ether ($CH_3CH_2OCH_2CH_3$) (i.e., $C_4H_{10}O$ where there is no band between 3000 and 4000 cm^{-1} indicating the absence of any –OH group). The spectra are plotted with % Transmittance on the vertical (ordinate) axis, while the frequency of light used for the spectrum (cm^{-1}) is plotted on the horizontal (abscissa) axis. In place of cm^{-1} (frequency units), wavelength can be used. The values are reciprocal, so 4000 cm^{-1} = 2.5 microns (μ) = 2500 nm at the left and 400 cm^{-1} = 25 microns (μ) = 25000 nm on the right. Neat liquid spectra obtained on a Digilab FTS-40 spectrometer by the author.

Water, with two protons (^1H), can be detected in this fashion, and water in different cells (the human body is about 60% water) is held differently. Therefore, examining the protons in water in humans can help determine the state of the health of the tissue in which the water is found (once a database with enough information about how water is held in different tissues under different

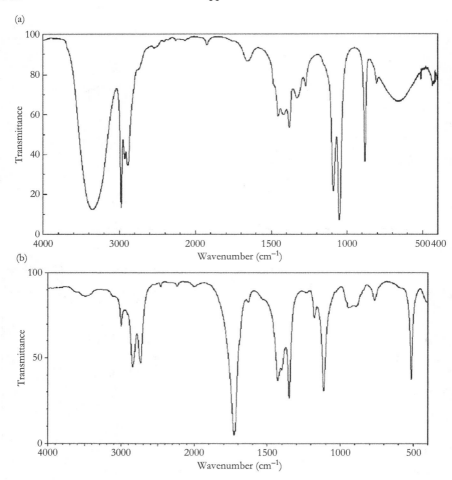

FIGURE A1.51 Representations of (a) the infrared spectrum of ethanol (ethyl alcohol CH_3CH_2OH) (i.e., C_2H_6O) with the broad hydrogen bonded –OH group lying about 3400 cm^{-1}. Only hydrogen-bonded –OH functionality is seen; and (b) the infrared spectrum of neat acetaldehyde (ethanal CH_3CHO) (i.e., C_2H_4O) with the –OH group missing (at about 3400 cm^{-1}) but the carbonyl group (C=O) seen at about 1730 cm^{-1}. The spectra are plotted with % Transmittance on the vertical (ordinate) axis, while the frequency of light used for the spectrum (cm^{-1}) is plotted on the horizontal (abscissa) axis. In place of cm^{-1} (frequency units), wavelength can be used. The values are reciprocal, so 4000 cm^{-1} = 2.5 microns (μ) = 2500 nm at the left and 400 cm^{-1} = 25 microns (μ) = 25000 nm on the right. Neat liquid spectra obtained on a Digilab FTS-40 spectrometer by the author.

health conditions is built). The process is called MRI (Magnetic Resonance Imaging) and typical MRI images of a human brain are shown below as Figure A1.52.

Just as MRI is used for protons (^{1}H) for water in different cells, so more finely tuned imaging can be used for smaller organic molecules in order to determine their structures. In these circumstances, for proton Nuclear Magnetic Resonance (^{1}H NMR), it is observed that protons attached to carbon and other nuclei are found at well-defined frequency and magnetic field strengths.

A Chemistry Primer   345

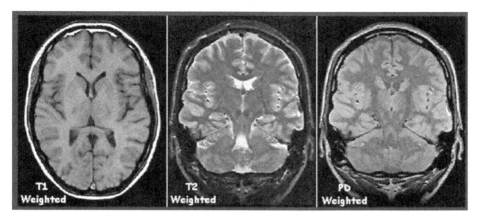

FIGURE A1.52 MRI images of the human brain. Wikipedia public domain image subsequently manipulated. https://commons.wikimedia.org/wiki/File:T1t2PD.jpg.

At a given magnetic field strength, different protons absorb and then emit the absorbed energy at different frequencies.

The same is true of carbon (^{13}C NMR). While the frequency at which carbon NMR works is different and does not overlap with proton NMR in the same magnetic field, different carbon atoms also absorb (and then emit the absorbed) energy at different frequencies. So NMR, both carbon and proton, has become a very useful tool in the determination of the structure of organic compounds.

For example, considering ^{13}C NMR, the decoupled[17] **carbon** spectrum of ethanol (CH_3CH_2OH) and diethyl ether ($CH_3CH_2OCH_2CH_3$) one might correctly, ignoring intensities of the peaks, conclude that there are only two kinds of carbons in both compounds. There are two methyl groups ($-CH_3$) in diethyl ether and only one methyl group ($-CH_3$) in ethanol, but in both, the methyl groups are attached to a methylene ($-CH_2-$) and that methylene ($-CH_2-$) is attached to an oxygen. Thus, as expected, although the two compounds are very different with one having an $-OH$ group in its IR spectrum and the other not, and one having almost twice the mass (so that the m/z of the parent ion is about twice the other) the ^{13}C spectra should be very similar. These spectra are shown in Figure A1.53.

On the other hand, if one compares the ^{13}C decoupled NMR spectrum of the $C_4H_{10}O$ isomers diethyl ether (ethyl ether, $CH_3CH_2OCH_2CH_3$) to 1-butanol ($CH_3CH_2CH_2CH_2OH$), it should be clear that although their masses are identical, the former still has only two (2) different carbon atoms, while the latter has four (4) different carbon atoms. Therefore these two compounds, should have very different ^{13}C decoupled NMR spectra (Figure A1.54). It also remains true that the two compounds are very different with one having an $-OH$ group that will appear in its IR spectrum and the other lacking that functionality.

Finally, the more complicated issue of proton NMR (^{1}H NMR) is briefly considered. There are two (2) compounds C_2H_6O, one of which is potable and has been referred to frequently, *viz.* ethanol (ethyl alcohol, CH_3CH_2OH) and another, dimethyl ether (CH_3OCH_3). As shown in Figure A1.55, in (a) all six of the protons in dimethyl ether are equivalent.

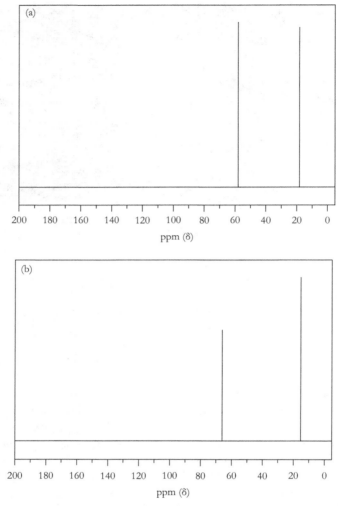

FIGURE A1.53 In (a) is presented the decoupled ^{13}C spectrum of ethanol. The signal at about 19 ppm corresponds to the $-CH_3$ group, and the one at about 59 ppm to the $-CH_2-$ group adjacent to oxygen. In (b) is presented the decoupled ^{13}C spectrum of diethyl ether. The signal at about 18 ppm corresponds to the $-CH_3$ groups, and the one at about 66 ppm to the $-CH_2-$ groups adjacent to oxygen. Spectra taken by the author on an Anasazi modified Varian FT-60 MHz spectrometer.

However, in (b) while the three hydrogen atoms on the methyl group ($-CH_3$) of ethanol are equivalent to each other, they are different than the two hydrogen atoms on the methylene ($-CH_2-$) group. The two hydrogen atoms of the methylene group are equivalent to each other. And the one hydrogen attached to oxygen is unique. So, in undecoupled (coupled) proton spectra the number of lines corresponds to n+1 where n = the number of identical protons on the adjacent carbon(s). Since there are three identical protons on the methyl group ($-CH_3$) there should be four lines (n + 1 = 3 + 1 = 4) for the methylene ($-CH_2-$). Similarly, the methyl group

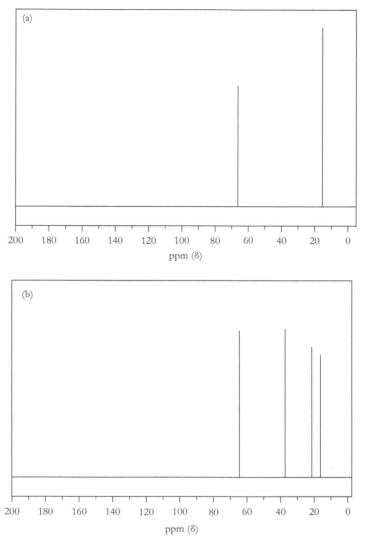

FIGURE A1.54 In (a) is presented the decoupled ^{13}C spectrum of diethyl ether. The signal at about 18 ppm corresponds to the –CH$_3$ groups and the one at about 66 ppm to the –CH$_2$– groups adjacent to oxygen. In (b) is presented the decoupled ^{13}C spectrum of 1-butanol. The signal at about 16 ppm corresponds to the –CH$_3$ group, and the one at about 62 ppm to the –CH$_2$– group adjacent to oxygen. The two other –CH$_2$ groups are also seen as separate signals. Spectra taken by the author on an Anasazi modified Varian FT-60 MHz spectrometer.

itself appears as a triplet (three lines) with lines in the ratio 1:2:1, because there are two methylene protons (–CH$_2$–) on the adjacent carbon and (n + 1 = 2 + 1 = 3). Thus, in addition to the obvious hydroxyl group on the alcohol, and the fact that there is broad –OH absorption around 3400 cm^{-1} in the IR for alcohols while there is no such absorption for ethers, the proton ^{1}H NMR spectrum also allows one to distinguish between these isomers.

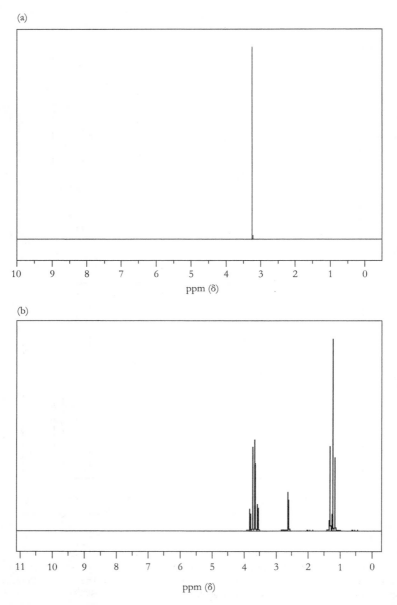

FIGURE A1.55 In (a) dimethyl ether, CH_3OCH_3, all six protons are identical, and thus there is only one line (a singlet). The signal appears at a chemical shift of 3.3 ppm, because the carbons bearing the protons are attached to oxygen. The chemical shift of such protons is usually between 3.3 and 4.0 ppm (δ). In (b), ethanol, CH_3CH_2OH, the single proton on the –OH appears at about 2.6 ppm, the protons on the methylene (–CH_2–) carbon bearing the oxygen again appear at a higher value (3.6 ppm), and they appear as a quartet (four lines) with lines in the ratio 1:3:3:1. The methyl group (–CH_3) itself appears as a triplet (three lines) with lines in the ratio 1:2:1. Spectra taken by the author on an Anasazi modified Varian FT-60 MHz spectrometer.

A Chemistry Primer 349

Utilizing the tools that allow gathering information beyond that seen in the visible region of the spectrum allows definitive information about structure and plays a critical role in the identification of substances present in wine. It is not necessary to guess; it is necessary to know.

CARBOCYCLES AND HETEROCYCLES

In the foregoing material there have been both acyclic (no ring) and cyclic (one or more rings) examples of compounds. If a ring is made up only of carbon atoms attached to each other (and appended hydrogens or other atoms which are not part of the cycle) then that cyclic structure is a *carbocycle*. Such carbocycles (cyclopropane (C_3H_6), cyclobutane (C_4H_8), cyclohexane (C_6H_{12}) (Figure A1.56), etc.) are quite common petroleum derivatives and are found in the environment. But they are uncommon in living systems (plants, fungi, and animals).

Life produces compounds where other atoms, commonly oxygen (O), nitrogen (N), and sulfur (S), are also part of the ring (Figures A1.57, A1.58, and A1.59, respectively). These kinds of compounds are called *heterocycles*. Typical examples include the oxygen heterocycles, the

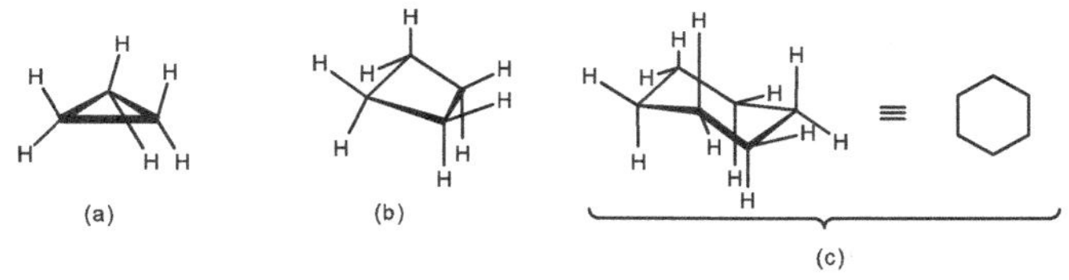

FIGURE A1.56 Representations of common carbocyclic compounds: (a) cyclopropane; (b) cyclobutane; and (c) cyclohexane.

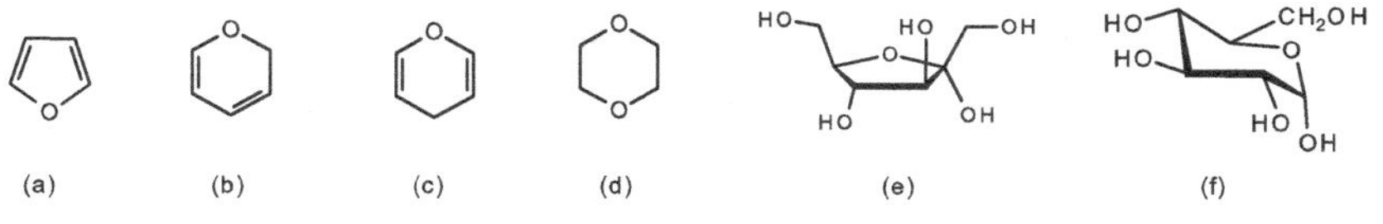

FIGURE A1.57 Representations of a few common oxygen heterocyclic compounds: (a) furan; (b) 2*H*-pyran; (c) 4*H*-pyran; (d) 1,4-dioxane; (e) e) α-D-fructofuranose; and (f) α-D-glucopyranose.

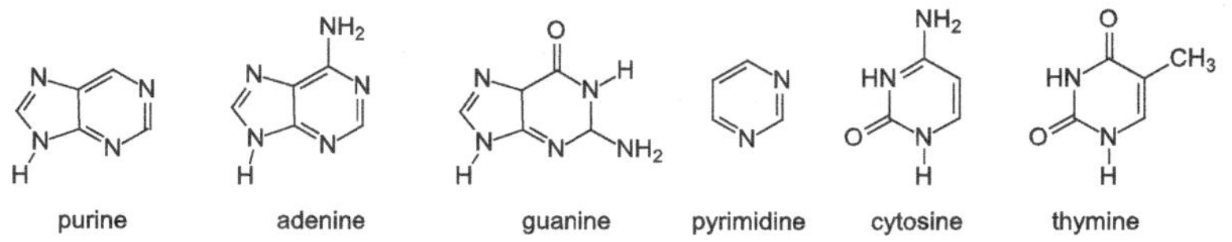

FIGURE A1.58 Representations of a few common nitrogen heterocyclic compounds: (a) pyrrole; (b) pyridine; (c) piperidine; (d) pyridazine; (e) pyrimidine; (f) pyrazine; and (g) purine.

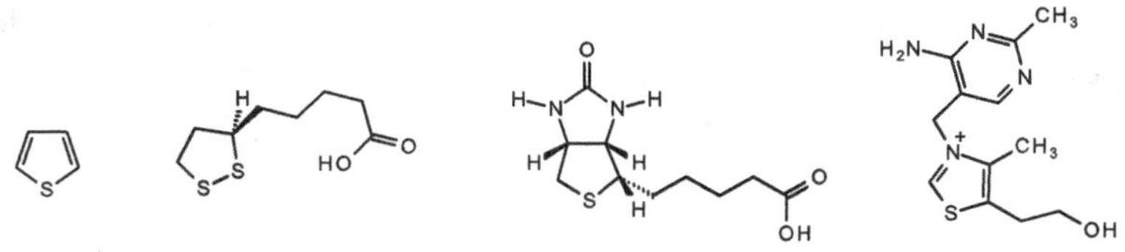

FIGURE A1.59 Representations of a few common sulfur-containing heterocyclic compounds: (a) thiophene; (b) lipoic acid; (c) biotin; and (d) thiamine (vitamin B_1).

five-membered ring in furan, the six-membered pyrans and dioxane, as well as the sugars fructose and glucose. The nitrogen heterocycles pyrrole, pyridine, piperidine, pyridazine, pyrimidine, pyrazine, and purine, where there are two co-joined nitrogen-containing rings, each containing two nitrogen nuclei are also shown. And then there are the sulfur heterocycles (e.g., thiophene and lipoic acid) and heterocycles such as biotin, with a dinitrogen ring and a ring containing sulfur, and thiamine (*aka* thiamin, vitamin, B_1), where both nitrogen and sulfur can be found in one ring (a 1,3-thiazole). In thiamine, there is also a pyrimidine ring.

Now it should be clear both the grape-bearing vine and the yeast are living systems. As living systems they need to process the food in the environment in which they find themselves. In that processing they must remain true to their form as dictated by their respective genetic codes so their progeny can carry on afterward.

While the subject of maintaining that code and building the forms of life it dictates can be considered the realms of biology and biochemistry, both of those disciplines rely on organic chemistry at their respective foundations. So, it is important a few things be said about the genetic code in general.

The millennia needed for establishment of pathways in living systems for the synthesis of peptides (proteins), required for life, have resulted in dynamic processes that are under active investigation as we hope to be able to modify them to our own ends. In plants and yeast, which are, for obvious reasons, much more commonly studied than animals, the amino acids discussed earlier in this Appendix are synthesized by (a) utilization of nitrogen in the soil combining with small organic fragments followed by reactions among those; or (2) degradation of proteins already present. So, for example, proteins in the grape will be used by yeast. The amino acids from these proteins are used for the synthesis of new proteins. Some of that process is discussed in the main body of the work (Chapters 2, 3, and 7). The place to start is with the genetic code and the carrier of the code, deoxyribonucleic acid (DNA).[18] For the most general case, DNA is composed of four deoxyribonucleotides (i.e., monophosphate ester derivatives of deoxynucleosides, shown in Figure A1.60) held in chains (Figure A1.61) by the formation of phosphate diesters, linearly from one deoxyribose (5′ end) to the next (3′ end). The chains are paired, in the form of a three-dimensional double helix (Figure A1.61), the now classical structure that James Watson and Francis Crick, standing on the backs of giants of previous decades, concluded in 1953 best represented DNA The two strands are held together by hydrogen bonding between the purine and pyrimidine bases (A to T and C to G) on opposite chains (Figure A1.62).

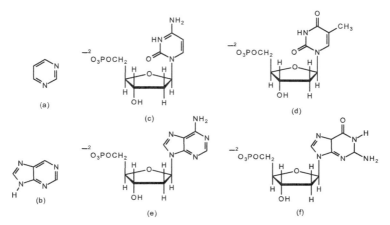

FIGURE A1.60 The two heterocyclic systems that serve as the fundamental ring systems upon which the bases in DNA and RNA are built are: (a) pyrimidine and (b) purine. The four bases built on those ring systems by suitable substitution and attached to the anomeric carbon of the sugar, 2-deoxyribose, and esterified with phosphoric acid at C5 of the deoxyribose unit are provided. They are: (c) cytosine, C, which yields deoxycytidine 5′-monophosphate; (d) thymine, T, which yields deoxythymidine 5′-monophosphate; (e) adenine, A, to produce deoxyadenosine 5′-monophosphate; and (f) guanine, G, producing deoxyguanosine 5′-monophosphate.

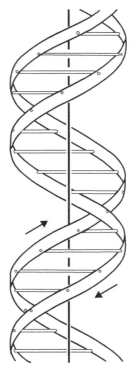

FIGURE A1.61 A cartoon representation of the DNA double strand woven into a helix. Although each strand continues from 5′ to 3′ phosphate bonding they may be thought of as "moving" in opposite directions. The diagram was proposed by James Watson and Francis Crick. © 1953 Nature Publishing Group, Watson, J. D. et. al., Molecular structure of nucleic acids, Nature **1953**, *171*(4356), 737–738. Used with permission.

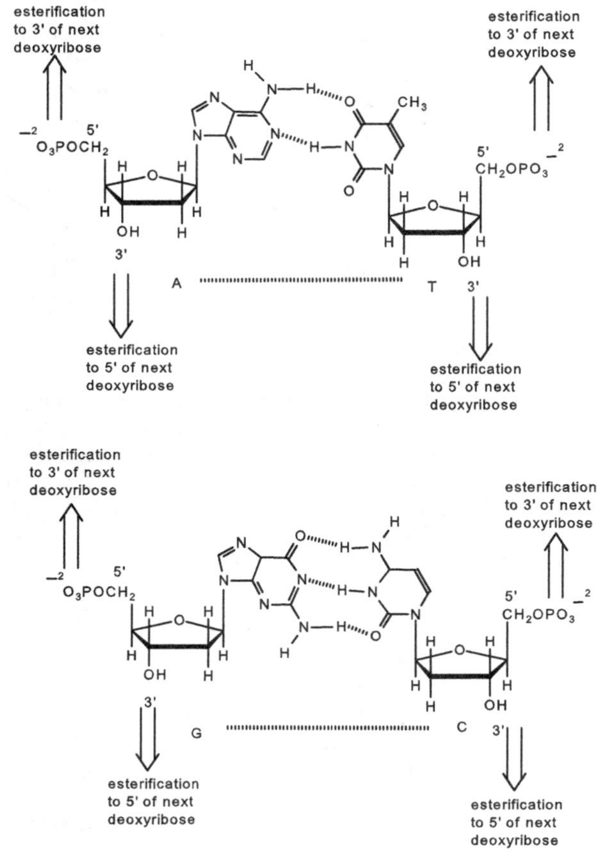

FIGURE A1.62 A representation of base pairs in adjacent strands of double helical DNA. The individual strands are composed of deoxyribosyl units bearing the pyrimidine and purine bases and linked through phosphate diesters (5′ to 3′ and 3′ to 5′) from one deoxyribose to the next. The chains are paired, in the classical helical structure (Figure A1.61), by hydrogen bonding between the purine and pyrimidine bases (A to T and C to G) on facing chains.

NOTES AND REFERENCES

1. Since this sentence was written in 2015, the unnamed elements discussed have been made. See http://www.rsc.org/chemistryworld/2016/01/new-elements-periodic-table-seventh-row-iupac.

2. The use of models has been commented upon by George E. P. Box (1919–2013). "Essentially, all models are wrong, but some are useful." Much of chemistry and almost all of the chemistry encountered here makes use of idealized models because atoms and compounds are smaller than the wavelength of light with which we might hope to see them. Thus, since we cannot see what they are doing, we rely upon their behavior to construct visual representations which we manipulate in controlled ways that allow us to (generally) make predictions about them and their interactions. Now, more often than not, the predictions lead to the expected outcomes!

3. In 1913, Neils Bohr (1885–1962) was able to mathematically account for light emitted by atomic hydrogen by proposing that, when excited, electrons would move from a specific, well-defined orbit around the nucleus (as the earth rotates about the sun) to another higher-energy orbit (energy absorbed) and then drop back again (energy emitted as light). The idea of such orbits was accepted, and cartoons showing the electrons moving in circular orbits are still used. However, the Bohr model was subsequently discarded when it was found that, except for hydrogen (and the helium cation He^+), no other element could be made to fit the formulas that were developed. Now, the newer model used involves regions of three-dimensional space around the nucleus where there is a probability of finding electrons. These regions are called *"orbitals."*

4. The "newer model" (*vide supra*) uses concepts within "quantum mechanics" and was developed over a period of years by Erwin Schrödinger (1887–1961), Werner Heisenberg (1901–1976), Paul Dirac (1902–1984), and Richard Feynman (1918–1968) among others.

5. An aside about units. Generally, it is agreed (the International System of Units [SI]) that length is measured in meters (m), weight in kilograms (kg), time in seconds (s) and temperature in kelvins (K). These fundamental units are frequently used with *scientific notation* where powers of 10 are used to express values. Simply, the power defines the position of a decimal point; so $10^3 = 1000$, $10^2 = 100$, $10^1 = 10$, $10^0 = 1$, $10^{-1} = 0.1$, $10^{-2} = 0.01$, etc. SI derivatives are used similarly, a centimeter (cm) = 10^{-2} m (one-hundredth of a meter); a nanometer (nm) = 10^{-9} m (one-billionth of a meter); a pico meter (pm) = 10^{-12} (one-trillionth of a meter; a gram (g) = 10^{-3} kg. A *mole* (mol) is currently defined as the amount of substance in a system which contains as many elementary entities as there are atoms in 0.012 kg (12 g) of carbon-12 (^{12}C)—which has 6 electrons, 6 protons, and 6 neutrons. So, 12 g (exactly) of ^{12}C will have Avogadro's number (*NA*) of carbon atoms or 6.02×10^{23} atoms or 1 mole. Indeed, one mole of anything has the same number (6.02×10^{23}) of entities. An *electron volt* (eV) is the amount of energy gained (or lost) by the charge of a single electron moved across an electric potential difference of one volt (1 eV ≈ 3.9×10^{-23} kcal/electron). One *calorie* (cal) = 10^{-3} kcal, and 1 cal = the amount of heat required to raise the temperature of 1 gram of water 1 K, where K is the temperature in *kelvins*. The kelvin (K) is defined as the temperature at which all thermal motion ceases and 0 K is taken as –273.15 degrees Celsius (°C). Some conversion factors (which you do not need to know) are 1 eV for a mol of electrons ≈ 23 kcal/mol ≈ 96 kJ/mol, and a Joule (J) is equal to the energy expended (or work done) in applying a force of one newton (N) through a distance of one meter. A newton (N) is equal to the force that produces an acceleration of one meter per second per second on a mass of one kilogram.

6. Laidler, K. J.; Meiser, J. H.; Sanctuary, B. C. *Physical Chemistry,* 4th Ed.; Houghton Mifflin Co.: Boston, 2003; p 921 ff.

7. Isaacs, E. D.; Shukla, A.; Platzman, P. M.; Hamann, D. R.; Barbiellini, B.; Tulk, C. A. *Phys. Rev. Lett.* **1999,** *82,* 600.

8. There are a number of amusing stories concerning the chlorination of hydrocarbons in general and wax candles in particular. These involve some detective work by Alexander Dumas and ruined dinner parties of French royalty. See, for example, Friedman, H. B. *J. Chem. Edu.* **1930,** *7,* 633.

9. Just as hydrogen exists as isotopes [(^1H) a proton with one electron and no neutrons; (^2H) deuterium, one electron, one proton, and one neutron; (^3H) tritium, one electron, one proton, and two neutrons], so does oxygen. The most common naturally occurring isotope of oxygen, ^{16}O,

has eight electrons, eight protons, and eight neutrons. Two other isotopes are also naturally occurring. Thus, ^{17}O has eight electrons, eight protons, and nine neutrons while ^{18}O has eight electrons, eight protons, and ten neutrons. Additional very short-lived isotopes have been isolated.

10. Henry Le Châtelier (1850–1936, Paris-Sorbonne). He is best known for formulating the principle bearing his name, which allows one to predict the effect a changing condition has on a system in equilibrium.

11. Sorensen, S. P. L. *Biochem. Zeitschr.* **1909**, *21*, 131. It has subsequently developed that a function of the hydrogen ion concentration called its "activity" is more appropriately used. The concentration term will be used here.

12. A catalyst is generally defined as a substance in whose presence the barrier between reactants and products is changed but which, itself, is not consumed in the process. The participation of a catalyst in a reaction does not generally change the course of the reaction, but rather changes the rate at which the reaction occurs. In organic chemistry where more than one outcome of a reaction is possible (e.g., salt formation *versus* amide formation), a suitable catalyst can change the rate of one of a plethora of possible outcomes to make a normally less favored process, favored.

13. The names of the individual amino acids are all "trivial" names, since they were named before a pattern of systematic nomenclature was devised. It is clearly not necessary to learn them. The three-letter abbreviation was devised to save the time and space required by writing the full name. The one-letter abbreviation became common when long chains (peptides and then proteins) of amino acids were finally being analyzed, and writing such long chains with anything but single letters was too cumbersome.

14. Wine that has been oxidized may smell of acetaldehyde (ethanal) or acetic acid (ethanoic acid), compounds which have been described earlier. The odor of acetaldehyde (ethanal) is commonly reported to be moldy and unpleasant, and that of acetic acid (ethanoic acid) is that of vinegar—for that is how vinegar is prepared. "Corked" wine, about which more will be written later, has a characteristic odor, variously described as resembling a moldy newspaper, wet dog, or damp basement, but regardless, the wine's native aromas are masked. These and other changes lead to wines no longer considered potable. The compound considered to be responsible for the flavor of "corked" wine, 2,4,6-trichloroanisole (TCA), may arise in different ways. They include metabolism of chlorine in the soil from earlier ocean deposition or as a consequence of pesticides used to protect cork trees or, perhaps, from the use of chlorinating agents in cork processing.

15. The parent ion should be at m/z = 58 amu, and hopefully it will be noticed that there is a very small peak at m/z = 59 amu. The latter can arise from the presence of ^{13}C in the environment, where it is found to the extent of 1.1% of all carbon atoms; it therefore makes a small contribution to the mass spectrum.

16. The word "spin" is in quotation marks because the actual spinning of the nuclei, if they are spinning, may not mean what the word spin normally could be interpreted as meaning. Further, measuring the property of spinning may actually result in the process occurring so it is not known whether spinning occurs until it is measured. However, the process is conveniently thought of in the traditional sense of a spinning top or other object.

17. A more substantive discussion (which will not be held here) is required to describe the interaction of protons attached to carbon and their role in the observed spectrum. It is not uncommon to simplify carbon spectra by uncoupling the interaction. That uncoupling results in a "decoupled" spectrum where each different carbon produces a unique signal. It is also common in NMR spectroscopy, regardless of the nucleus, to divide the frequency at which the signal is

A Chemistry Primer 355

observed by the frequency at which the spectrometer operates. Thus, frequency divided by frequency results in "unitless" numbers. The values are expressed as arbitrary delta (δ) units or in parts-per-million (ppm) from some standard defined as zero. The usual zero is, by general agreement, tetramethylsilane, $(CH_3)_4Si$.

18. There are three levels of complexity. At the lowest level, there are four bases, two purine (adenine, A, and guanine, G) and two pyrimidine (thymine, T, and cytosine, C). At the next level, each of these is attached to the anomeric carbon of 2-deoxyribose and they are now called deoxyribonucleosides; their names are deoxyadenosine, deoxyguanosine, deoxythymidine, and deoxycytidine, respectively. The addition of phosphate esters at C5 of the deoxyribose units changes the name to deoxyribonucleotide (e.g., deoxyadenosine 5′-monophosphate).

Appendix 2

BIOSYNTHETIC PATHWAYS OF ODOR, COLOR, AND FLAVOR COMPOUNDS

The function of this Appendix is to outline, in condensed form, paths to compounds that account for odors, colors, and flavors found in grapes and the flowers from which they come. These compounds include the terpenes, sesquiterpenes, diterpenes, and polyenes as well as phenols and some nitrogenous derivatives (anthocyanins and their sugar-free counterparts, anthocyanidins, etc.). In order to accomplish that end the carbon dioxide (CO_2) absorbed by the plant needs to be converted to acetate, the anion ($CH_3CO_2^-$) of acetic acid (CH_3CO_2H). Along the way we will pass through some simple sugars (as shown in the Calvin cycle, Figure A2.1) and other interesting compounds (such as pyruvic acid [CH_3COCO_2H], malonic acid [$H_2C(CO_2H)_2$]) and, later, from the Krebs cycle, fumaric acid [*trans*- or (E)-$HO_2CCH=CHCO_2H$].

To begin, the Calvin cycle is shown below (Figure A2.1). The Enzyme Data Base is a good general reference for reactions in the Cycle (http://www.enzyme-database.org). The cycle itself can be examined in more detail from a variety of sources. A good version is seen at http://www.chem.qmul.ac.uk/iubmb/enzyme/reaction/polysacc/Calvin2.html.

Details of the reaction between CO_2 and D-ribulose-1,5-bisphosphate on the enzyme Rubisco have been studied.[1] Schemes for a pathway are provided in Figures A2.2 and A2.3 (following that of the Calvin cycle, Figure A2.1). At this writing (2016) there are eighty-four (84) entries in the Protein Data Bank (PDB) for Rubisco as isolated from a variety of plant sources. The PDBs can be accessed via www.ebi.ac.uk/pdbsum/.

The details of the reaction between CO_2 and D-ribulose-1,5-bisphosphate on the enzyme Rubisco (EC 4.1.1.39) are divided here into two parts, The first (Figure A2.2) shows a representation in concert with what is known for the addition of carbon dioxide to the sugar diphosphate derivative. The second (Figure A2.3) provides a representation in concert with the cleavage of the adduct to generate 3-phosphoglycerate.[1]

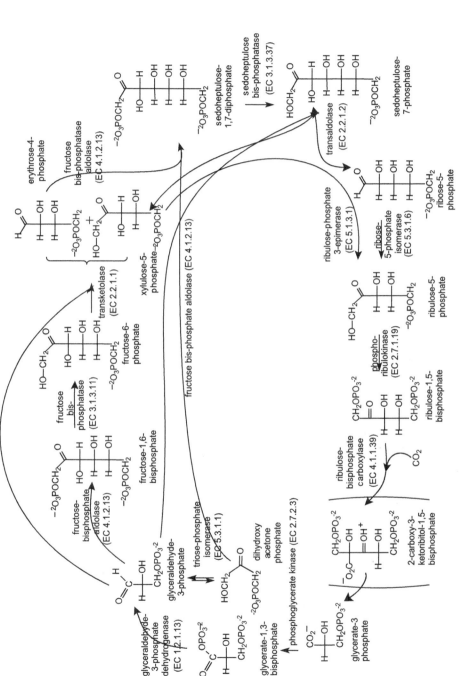

FIGURE A2.1 A representation of the Calvin cycle.

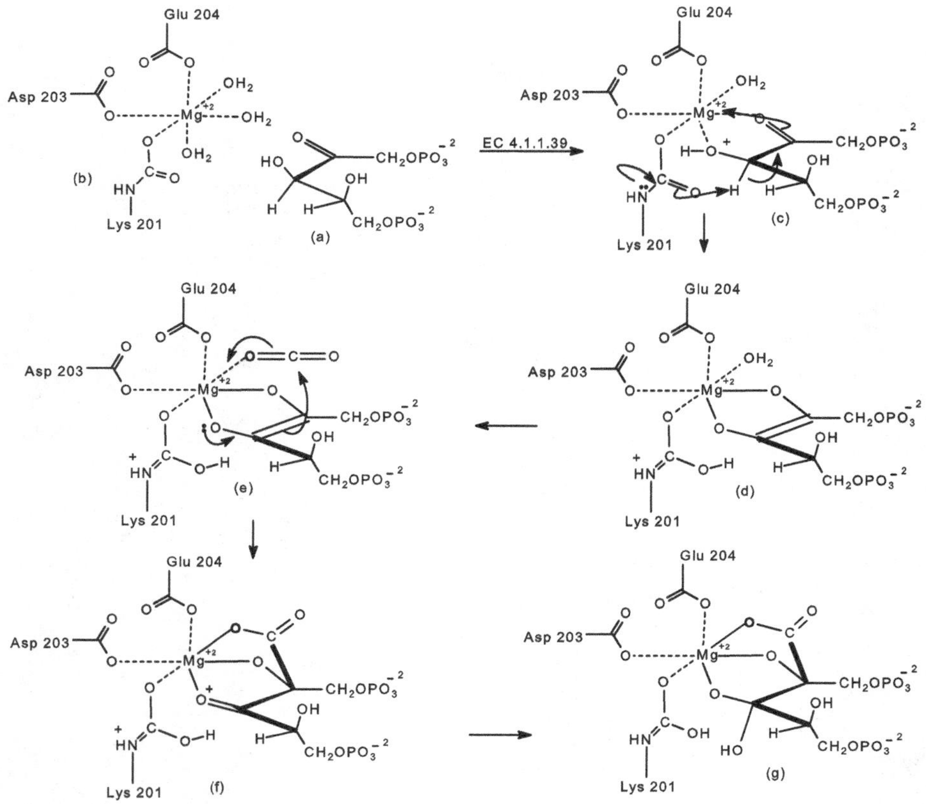

FIGURE A2.2 A cartoon representation of the addition of carbon dioxide (CO_2) to (a) ribulose 1,5-bisphosphate in the active site of the enzyme Rubisco (EC 4.1.1.39). The active site (b) is said to bear a magnesium cation (Mg^{2+}) with three waters of hydration which is held in place by a glutamic acid residue (Glu, E), an aspartic acid (Asp, D) residue, and a carbon dioxide carbamate residue from a lysine (Lys, K) on the enzyme. With loss of two waters, the ribulose 1,5-bishosphate is attached (c) to the active site and undergoes enolization to a fully bound enol (d). Carbon dioxide is captured to yield (e) which rearranges to (f) with the formation of the new carbon bond. Water adds in the correct orientation to (f) to yield (g).

Next, the 3-phosphoglycerate is isomerized to 2-phosphoglycerate.[2] The cartoon shown (Figure A2.4) represents the isomerization using a phosphoglycerate mutase.

Loss of water from 2-phosphoglycerate to generate phosphoenolpyruvate is effected through the participation of phosphopyruvate hydratase, EC 4.2.1.11. A cartoon depiction of one of the enzymes (carbon-oxygen lyase), utilizing two Mg^{2+} ions capable of effecting the dehydration, is seen in PDB 1ebh (Figure A2.5).[3,4]

In the next step, phosphoenolpyruvate is converted to pyruvate (Figure A2.6). The current (2016) Protein Date Bank (PDB) has seventy-one (71) pyruvate kinase enzyme (EC 2.7.1.40) entries, and it is clear that, while there is a requirement of magnesium (Mg^{2+}) ion, significant

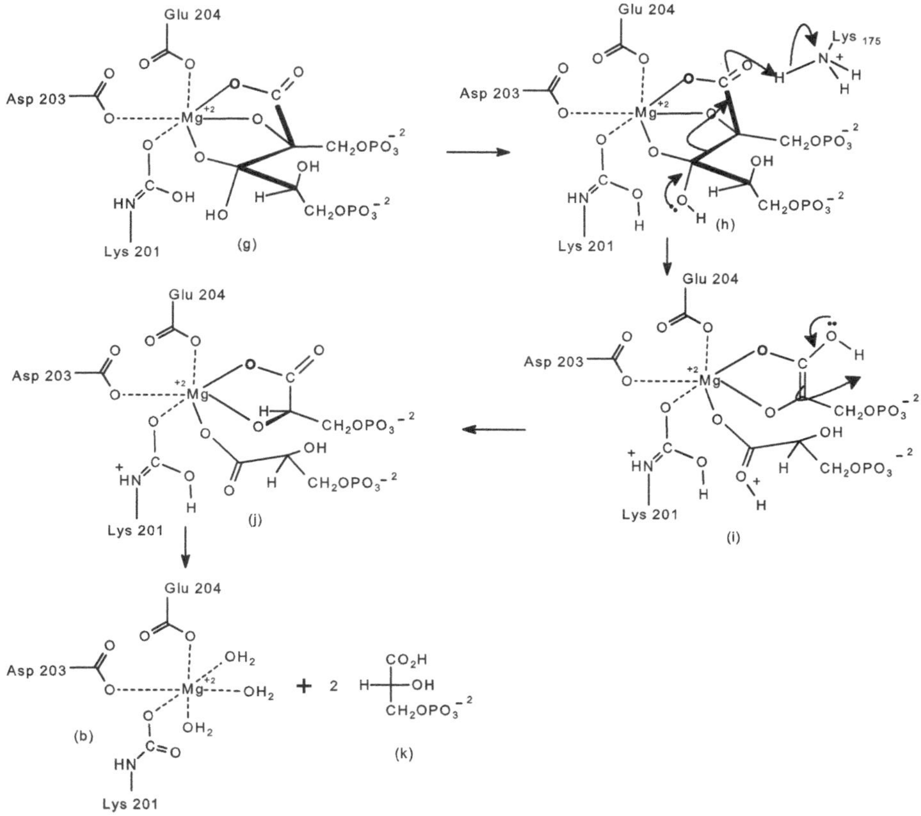

FIGURE A2.3 Continuation of the cartoon representation of the addition of carbon dioxide (CO_2) to ribulose 1,5-bisphosphate in the active site of the enzyme Rubisco (EC 4.1.1.39). The adduct of carbon dioxide and water (g) (vide supra, Figure A2.2) apparently undergoes decomposition as in (h) with carbon–carbon bond breaking while picking up a proton from an adjacent lysine (Lys, K), forming two three-carbon fragments that remain attached to the magnesium (Mg^{2+}) as in (i). Addition of a proton to the appropriate face of the alkene in (i) produces the bis-ester (j), which on hydrolysis leads to (b) the original active site magnesium cation (Mg^{2+}) with three waters of hydration and its attendant amino acids of the enzyme rubisco (EC 4.1.1.39) and two equivalents of the three-carbon D-3-phosphoglyceric acid (k).

differences obtain depending upon the source. The cartoon of Figure A2.6 does not show the role of the magnesium (Mg^{2+}) ion, but presumably, it is coordinated with the phosphate anions. The proton adding to the double bond is shown adding to the "si," face because in cases where it has been investigated, it appears preferred.

And now in order to build isopentenyl diphosphate and its equilibrium partner, dimethylallyl diphosphate (precursors to the needed terpenes, sesquiterpenes, diterpenes, etc.), two different sequences utilizing pyruvate must be considered.

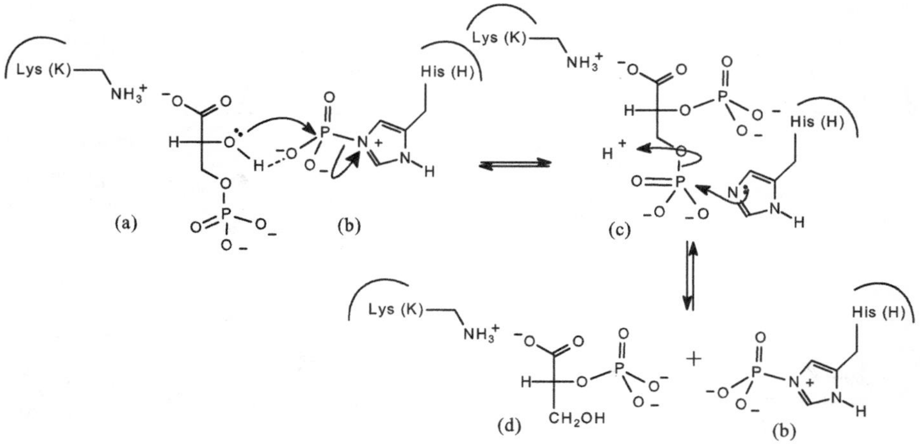

FIGURE A2.4 A cartoon representation of conversion of (a) 3-phosphoglycerate by phosphorylation using an enzyme-held N-phosphohistidine (b) to a 2,3-bisphosphate (c) and subsequent loss of the phosphate from C-3 of the bisphosphate (c) to yield 2-phosphoglycerate (d) and regenerated the N-phosphohistidine (b).

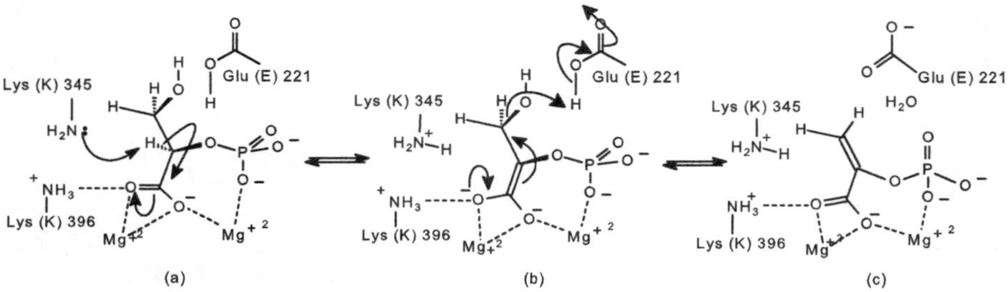

FIGURE A2.5 A cartoon representation of the dehydration of 2-phosphoglycerate (a) in the active site of carbon-oxygen lyase, EC 4.2.1.11, where initial proton loss generates an enolate (b) which rearranges, in the step where water is lost, to the magnesium-coordinated phosphoenol pyruvate (c).

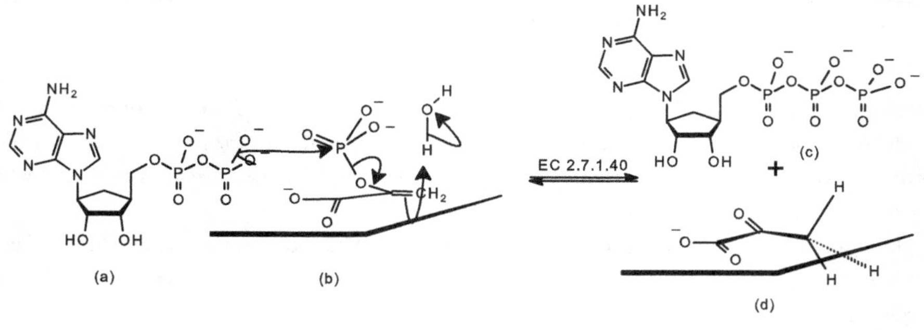

FIGURE A2.6 A cartoon representation of the transfer of a phosphate to adenosine diphosphate (a) from phosphoenol pyruvate (b), generating adenosine triphosphate (c) and pyruvate (d). A kinase enzyme such as EC 2.7.1.40 (pyruvate kinase) is utilized.

THE CYTOPLASM AND THE PLASTID

General Comment Concerning the Different Pathways for the Biosynthesis of Dimethylallyl Diphosphate and Isopentenyl Diphosphate

In the *cytoplasm* of the cell, pyruvate undergoes decarboxylation to yield, ultimately, acetyl coenzyme A ($CH_3CO-S-CoA$). Three equivalents of acetyl coenzyme A, *in the cytoplasm,* then go on. The path proceeds to an equilibrium mixture of isopentenyl diphosphate and dimethylallyl diphosphate (*via* hydroxymethylglutaryl-CoA synthase, EC 2.3.3.10, and hydroxymethylglutaryl-CoA reductase, EC 1.1.1.34) to the six-carbon intermediate mevalonate. The latter, *in the cytoplasm,* is a precursor to farnesol diphosphate, which subsequently leads to sesquiterpenes (C-15), plant sterols (triterpenes, C-30), and ubiquinones.

On the other hand, in the *plastid* compartments in the same cell,[5] pyruvate undergoes decarboxylation and reaction with glyceraldehyde 3-phosphate to produce 1-deoxyxylulose 5-phosphate (1-deoxy-D-xylulose-5-phosphate synthase, EC 2.2.1.7). On rearrangement, this five-carbon fragment leads to dimethylallyl diphosphate and isopentenyl diphosphate, and *in the plastid*, these same precursors yield monoterpenes (C-10), diterpenes (C-20), carotenoids (C-40), and their C_{13} degradation products. These compounds result from geranyl diphosphate (C-10) and geranylgeranyl diphosphate (C-20).[6]

As shown in Figure A2.7, in the cytoplasm, the conversion of pyruvate to acetate (in the form of acetyl coenzyme A or $CoASCOCH_3$) requires a multienzyme oxo-acid dehydrogenase complex (consisting of the enzymes EC 1.2.4.1, EC 2.3.1.12, and EC 1.8.1.4). In the enzyme complex there are five distinct reactions, the last two of which are required to return participants to their starting state for reuse. Acetyl coenzyme A ($CoASCOCH_3$) results from the first three. In the first reaction, the enzyme pyruvate dehydrogenase catalyzes the addition of thiamine diphosphate to the carbonyl of pyruvate and the subsequent decarboxylation (loss of CO_2) of the resulting product to form hydroxyethylthiamine diphosphate. Magnesium (Mg^{2+}) is required and is seen coordinated to the phosphate anions.

In the second reaction, the acyl group is transferred from the hydroxyethylthiamine diphosphate to a lipoic acid amide. The ring of the cyclic disulfide functional group of lipoic acid, attached as an amide (through a lysine [Lys (K)]) of dihydrolipoyltransacetylase is opened. The thiamine diphosphate is liberated.

The final reaction involves the transfer of the acetyl group from the lipoyl sulfur to the sulfur of coenzyme A.[7,8]

With acetyl coenzyme A available, we now have a fundamental building block that can be used to build all of the organic compounds that are needed.

In the cytoplasm (as noted above), acetyl coenzyme A is used to build mevalonate, the precursor to dimethylallyl diphosphate and its equilibrium isomer isopentenyl diphosphate. A typical process uses a thiolase such as acetyl-CoA *C*-acetyltransferase, EC 2.3.1.9.[9]

In this, as in some other thiolases, at least three (3) catalytic residues appear to be directly involved, *viz.* cysteine (Cys, C) 89, histidine (His, H) 384, and cysteine (Cys, C) 378, along with strategically placed waters (not shown in Figure A2.8) that allow acyl transfer in a way that resembles an acyl carrier protein (ACP).[10]

Continuing in the cytoplasm with acetoacetyl-CoA, the next (and first!) step in the synthesis of sesquiterpenes and plant steroids (occasionally referred to as *sterols*) is catalyzed by a 3-hydroxy-3- methylglutaryl -CoA (HMG-CoA) synthase enzyme.[11,12]

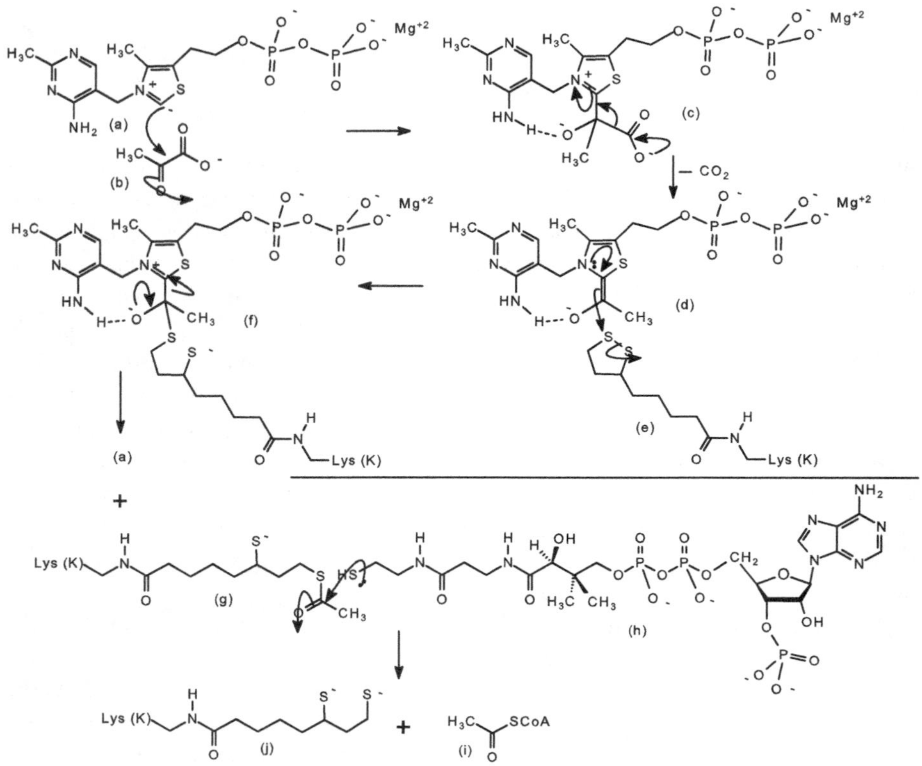

FIGURE A2.7 A cartoon representation of a pathway to acetyl CoA as described in the text. As shown, thiamine diphosphate (a) adds to the ketone carbonyl of pyruvate (b) to produce an intermediate (c) which undergoes decarboxylation to produce an enolate anion (d). Addition of the enolate anion (d) to enzyme-bound lipoic acid (e) leads to a presumed intermediate (f) which undergoes loss of thiamine (a) and formation of acetyl-bound lipoic acid anion (g). The two-carbon acetyl fragment is now in the right oxidation state to react via a simple addition–elimination sequence at the carbonyl with coenzyme A (h) to produce acetyl coenzyme A ($CH_3COSCoA$) (i) and the lipoate dianion (j), which needs to be reoxidized back to lipoic acid (usually accomplished with flavin adenine dinucleotide, FAD, which is reduced to $FADH_2$ in the process. $FADH_2$ is reoxidized with NAD^+).

Figure A2.9 outlines, in cartoon fashion, a pathway for the conversion of acetyl coenzyme A to the (S)-3-hydroxy-3-methylglutaryl-CoA.

The reduction of (S)-3-hydroxy-3-methylglutaryl-CoA to (R)-mevalonate by nicotinamide adenine dinucleotide (NADPH), with the oxidation of the latter to $NADP^+$, has been studied in conjunction with human steroid formation. There are currently (2016) 22 entries in the Protein Data Bank (PDB) for various 3-hydroxy-3-methylglutaryl-coenzyme A reductases.

As clearly pointed out[13] the coenzyme A is reductively removed with the formation of an aldehyde by hydride donation from NADPH, and then the second reduction to the alcohol occurs. Thus two equivalents of NADPH are required for each equivalent of (R)-mevalonate produced (Figure A2.10).

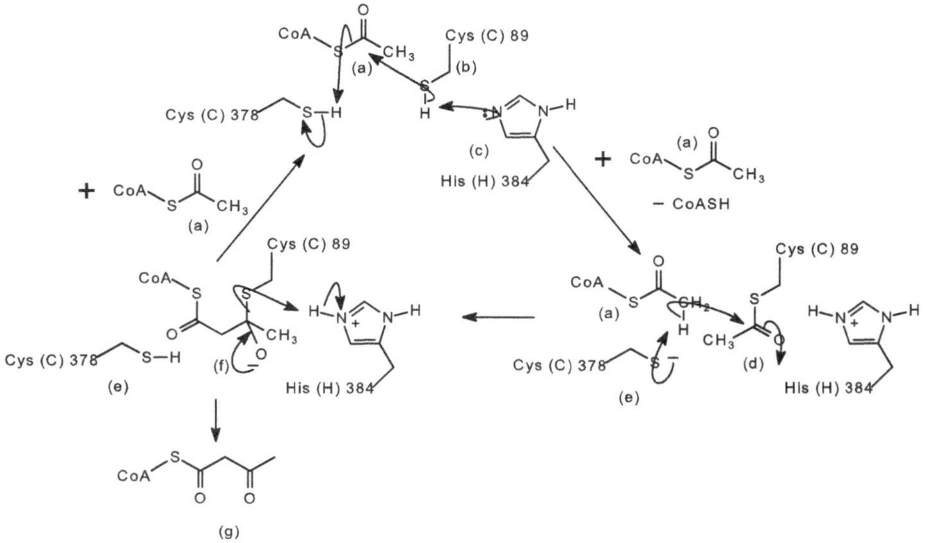

FIGURE A2.8 After Modis, Y.; Wierenga, R. K. *J. Mol. Biol.* **2000**, *297*, 1171. The cartoon shows the coordination of acetyl-CoA (a) with Cys-89 (b) of acetyl-CoA transferase and the participation of His-384 (c) abstracting the proton. The product (d), with the acetyl group now bonded to the sulfur of Cys-89, is set to be attacked by the anion from a second acetyl CoA (a) generated through the participation of the tholate anion from Cys-378 (e). The product of that process (f) is retained attached to the sulfur of Cys-89, and liberating Cys-89 (b) to be used again, is itself liberated as the coenzyme A derivative of acetoacetate (2-oxobutanoate), acetoacetyl-CoA (g).

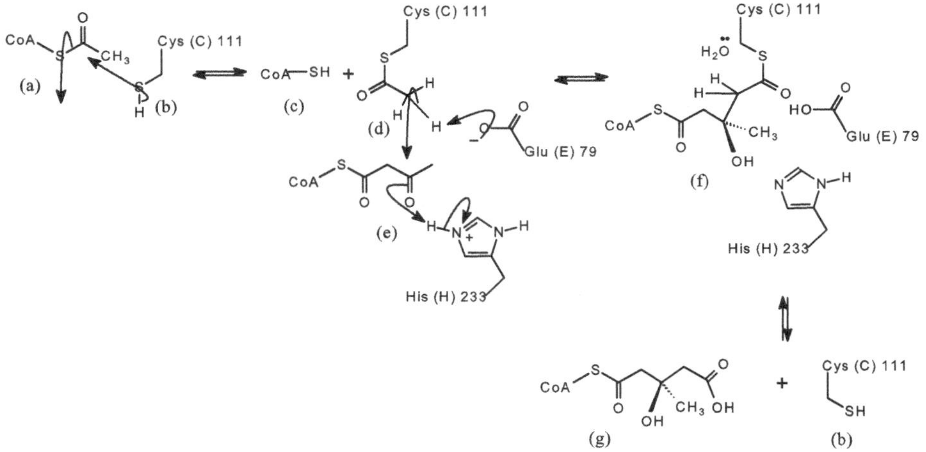

FIGURE A2.9 As pointed out by Theisen, M.H.; Misra, I.; Saadat, E.; Campobasso, N.; Miziorko, H. M.; Harrison, D. H. T. *Proc. Natl. Acad. Sci. U. S. A.* **2004**, *101*, 16442, it appears that acetyl-CoA (a) is attacked by the sulfur of Cys-111 of the enzyme (b) at the carbon of the carbonyl, resulting in the displacement of CoA-SH (c) and leaving the acetyl group bound to the cysteine as (d). Then, removal of a proton from the methyl group of the bound acetyl by glutamate anion (from Glu-79) and addition of the carbanion so generated to the carbon of the keto-carbonyl of acetoacetyl-CoA (e) (from Figure Appendix_2_10) produces a tetrahedral adduct (f). That adduct (f) is thought to pick up a proton on oxygen from a protonated histidine (His, H-233), and the geometry of the resulting alcohol after hydrolysis to remove the (*S*)-3-hydroxy-3-methylglutaryl-CoA (g) from the cysteine (Cys, C-111) is fixed while Cys-111 (b) is regenerated.

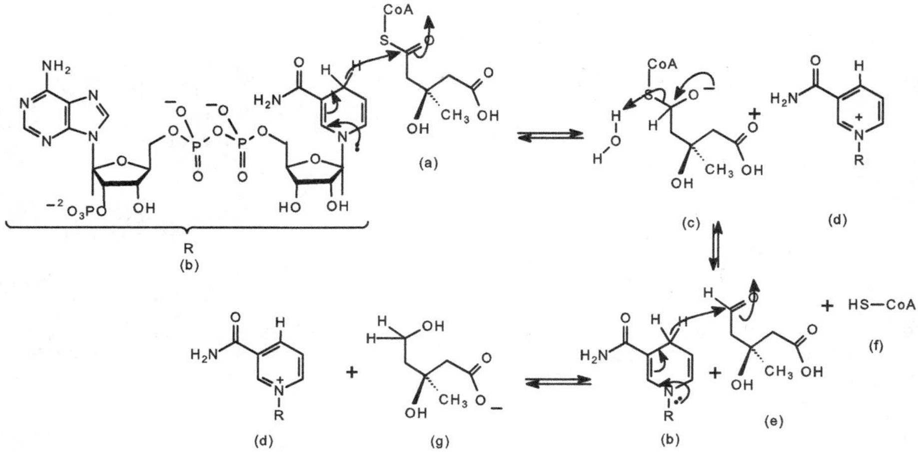

FIGURE A2.10 A vastly oversimplified cartoon attempting to depict (in the least cluttered way) the conversion of (S)-hydroxy-3-methylglutaryl-CoA (a) reacting with reduced nicotinamide adenine dinucleotide (NADPH) (b) to effect the reduction of the thioate ester to the corresponding thiohemiacetal (c) and NADP$^+$ (d). Then, after hydrolysis of the thiohemiacetal (c) to the corresponding aldehyde, (R)-mevaldate (e) and coenzyme A (f), further reduction, with a second equivalent of NADPH (b) the aldehyde (e) is converted to (R)-mevalonate (g) and NADP$^+$ (d).

In order to go from (R)-mevalonate to isopentenyl diphosphate (in catalyzed equilibrium with dimethylallyl diphosphate) it has become clear that three sequential phosphorylations, each utilizing the conversion of adenosine triphosphate (ATP) to adenosine diphosphate (ADP)[14] must occur, and in the last step a decarboxylation (loss of CO_2) must also obtain.

The first phosphorylation of the alcohol of (R)-mevalonate is catalyzed by the enzyme mevalonate kinase (EC 2.7.1.36), and (R)-5-phosphomevalonate results. The second phosphorylation is effected by phosphomevalonate kinase (EC 2.7.4.2), and (R)-5-diphosphomevalonate results. The final step, producing adenosine diphosphate, phosphate, carbon dioxide, and isopentenyl diphosphate, utilizes the enzyme diphosphomevalonate decarboxylase (EC 4.1.1.33).[14]

While the first two phosphorylations are considered straightforward transfers of phosphate from an "anhydride" to an alcohol to produce an ester (first transfer) or new anhydride (second transfer), the final transfer also involves the decarboxylation. A cartoon representation of the process to form isopentenyl diphosphate is shown in Figure A2.11.

Finally, isopentenyl diphosphate is in enzyme-catalyzed (EC 5.3.3.2) equilibrium with dimethylallyl diphosphate; it is this equilibrium mixture that leads to isoprene and isoprenoids. The process has been studied,[15] and the most recent data[16] suggests that it may be possible that a carbocation intermediate forms. Thus, for the isomerase isopentenyl diphosphate Δ-isomerase (EC 5.3.3.2, PDB 1hx3), it appears that a proton is abstracted from a cysteine (Cys, C61) to produce the carbocation. Subsequently, glutamate anion (Glu, E116) removes a proton from the substrate to generate the product alkene. The process, as shown in Figure A2.12, is written with loss of the pro-R hydrogen as is known and as occurring without passing through the carbocation. The isomerase itself apparently requires a metal for structural

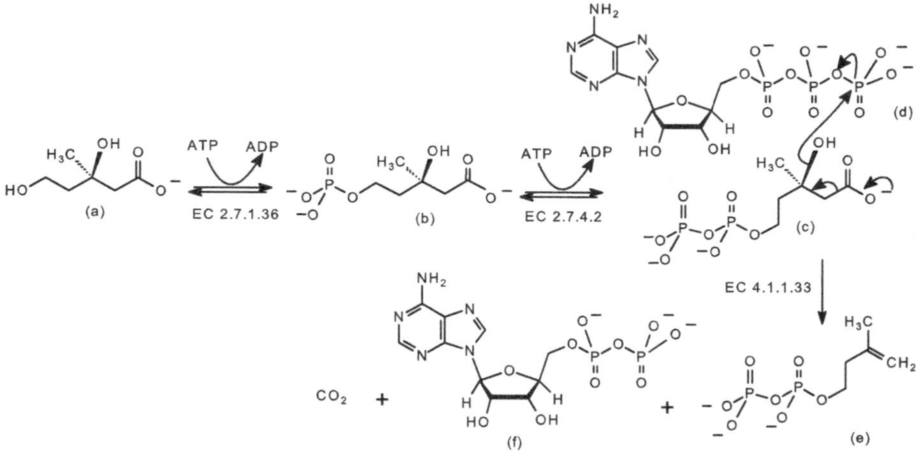

FIGURE A2.11 A simplified cartoon representation of the sequential conversion of (R)-mevalonate (a) first to a monophosphate ester (R)-5-phosphomevalonate (b), followed by a second phosphorylation to (R)-5-diphosphomevalonate (c). A final phosphorylation by adenosine triphosphate (ATP) (d) at the tertiary hydroxyl and elimination accompanied by loss of carbon dioxide (drawn here as one step), produces isopentenyl diphosphate (e) and adenosine diphosphate (ADP) (f) as well as inorganic phosphate (not shown).

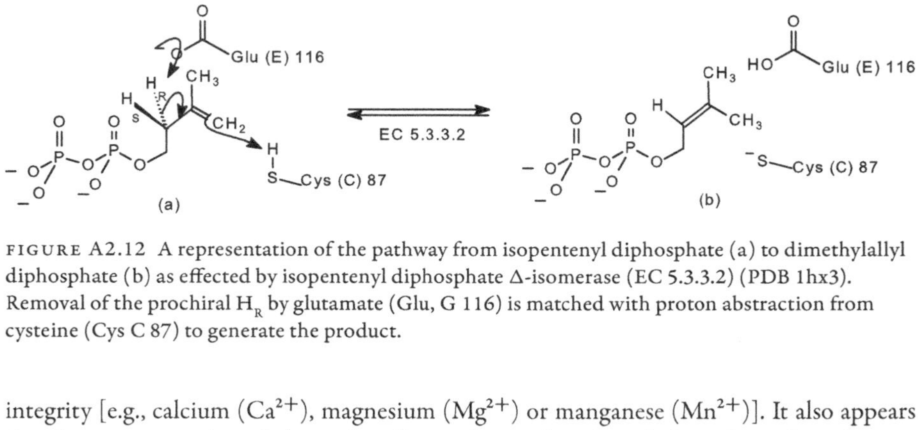

FIGURE A2.12 A representation of the pathway from isopentenyl diphosphate (a) to dimethylallyl diphosphate (b) as effected by isopentenyl diphosphate Δ-isomerase (EC 5.3.3.2) (PDB 1hx3). Removal of the prochiral H_R by glutamate (Glu, G 116) is matched with proton abstraction from cysteine (Cys C 87) to generate the product.

integrity [e.g., calcium (Ca^{2+}), magnesium (Mg^{2+}) or manganese (Mn^{2+})]. It also appears that flavin mononucleotide (FMN) or flavin adenine dinucleotide (FAD) may be cofactors, although their specific function(s) is not clear.

As noted earlier, in the cytoplasm, isopentenyl diphosphate and dimethylallyl diphosphate are precursors to farnesol diphosphate (C-15) which subsequently leads to sesquiterpenes (C-15) and plant sterols (triterpenes, C-30).

Dramatically different chemistry is found in the plastid compartments of the same cell.[17–20] In the plastid, pyruvate, already seen on the path to acetate, undergoes decarboxylation to produce a two-carbon fragment attached to thiamine as it did before (Chapter 17). Now, however, rather than generating acetate, the two-carbon fragment undergoes reaction with glyceraldehyde 3-phosphate (from the Calvin cycle) to generate the five-carbon derivative, 1-deoxyxylulose 5-phosphate. Then, on rearrangement, reduction, and loss of phosphate,

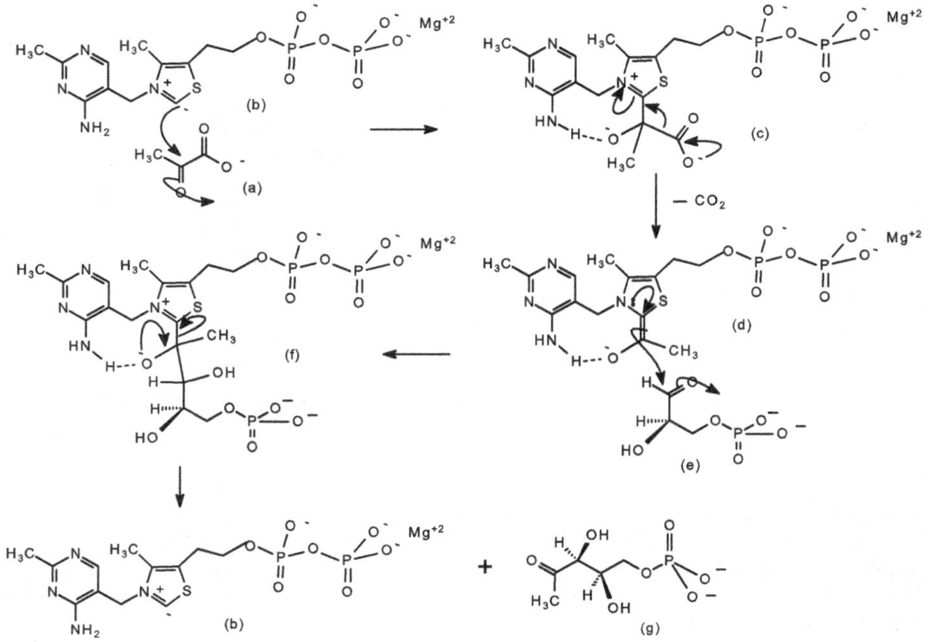

FIGURE A2.13 A cartoon representation of the conversion of pyruvate (a) on reaction with magnesium-coordinated thiamine diphosphate (b) to undergo decarboxylation (c) and produce a two-carbon fragment (d), still attached to thiamine diphosphate which is capable of being added to D-glyceraldehyde 3-phosphate (e). The resulting five-carbon fragment (as in f) is then cleaved to yield the thiamine diphosphate anion (b) again and 1-deoxy-D-xylulose 5-phosphate (g).

dimethylallyl diphosphate and isopentenyl diphosphate are formed. In the plastid, these same precursors yield monoterpenes (C-10), diterpenes (C-20), and carotenoids (C-40). These compounds result from geranyl diphosphate (C-10) and geranylgeranyl diphosphate (C-20).

In detail, for the 1-deoxy-D-xylulose 5-phosphate synthase (DXS) EC 2.2.1.7 (PDB 2o1s), Figure A2.13, it appears that the thiamine diphosphate lies bound with the phosphate in one of the two domains of which the synthase is composed while the aminopyrimidine sidechain is in the other. In this way the carbon lying between nitrogen and sulfur, prepared to attack the carbonyl of the glyceraldehyde 3-phosphate, is exposed in the cleft.

A cartoon representation of the process generating 1-deoxy-D-xylulose-5-phosphate is provided as Figure A2.13.

In the next step (Figure A2.14) rearrangement and reduction are effected by the same enzyme, *viz.*, 1-deoxy-D-xylulose-5-phosphate reductoisomerase [EC 1.1.1.267 (PDB 1jvs)], which catalyzes a classical retro-aldol-like process followed by reduction of the resulting carbonyl by NADPH. Interestingly, it appears that NADH fails to produce the 2-methyl-D-erythritol in good yield.[21,22]

To prepare for eventual elimination reactions to produce dimethylallyl diphosphate and isopentenyl diphosphate, the phosphate at C-4 of the 2-methyl-D-erythritol 4-phosphate is modified by elaboration with the addition of cytidine diphosphate. This is accomplished

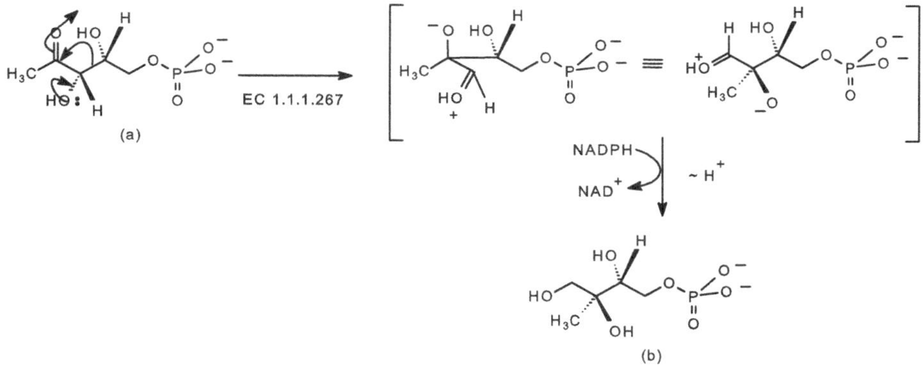

FIGURE A2.14 A cartoon representation of the utilization of 1-deoxy-D-xylulose 5-phosphate reductoisomerase [EC 1.1.1.267 (PDB 1jvs)] in the conversion of 1-deoxy-D-xylulose 5-phosphate (a) via a retro-aldol-type rearrangement followed by reduction to 2-methyl-D-erythritol 4-phosphate (b).

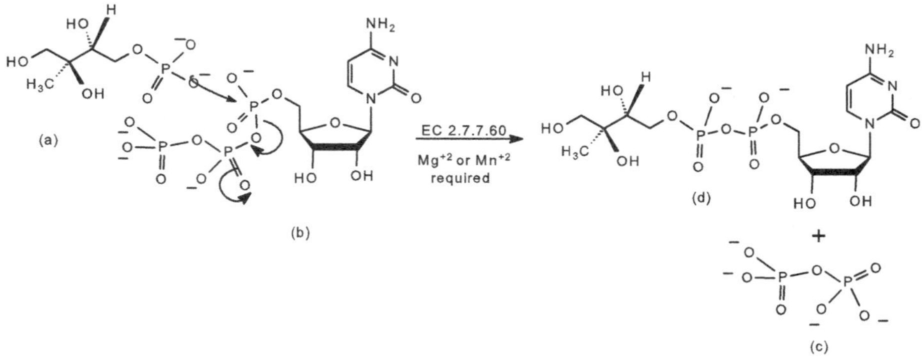

FIGURE A2.15 A cartoon representation of the possible process occurring with participation of cytidylyltransferase. As shown, 2-methyl-D-erythritol 4-phosphate (a) reacts with cytidine triphosphate (b) and removes two equivalents of inorganic phosphate (c), producing a diphosphocytidine derivative of 2-methyl-D-erythritol (d) from the cytidine triphosphate (b).

with the aid of the enzyme 2-C-methyl-D-erythritol 4-phosphate cytidylyltransferase (EC 2.7.7.60, pdb1w77).[23] A cartoon representation of the potential pathway is shown as Figure A2.15.

With the primary hydroxyl distinguished from the secondary and tertiary hydroxyl groups, further transformation of the backbone is accomplished by phosphorylation of the tertiary alcohol. In the presence of the phosphorylating enzyme, 4-(cytidine 5′-diphospho)-2-C-methyl-D-erythritol kinase (EC 2.7.1.148, PDB 2v8p) and magnesium (Mg^{2+}) or manganese (Mn^{2+}), adenosine triphosphate (ATP) transfers a phosphate to the tertiary hydroxyl at C-2. As shown in Figure A2.16, adenosine diphosphate (ADP) and 2-phospho-4-(cytidine 5′-diphospho)-2-C-methyl-D-erythritol result.

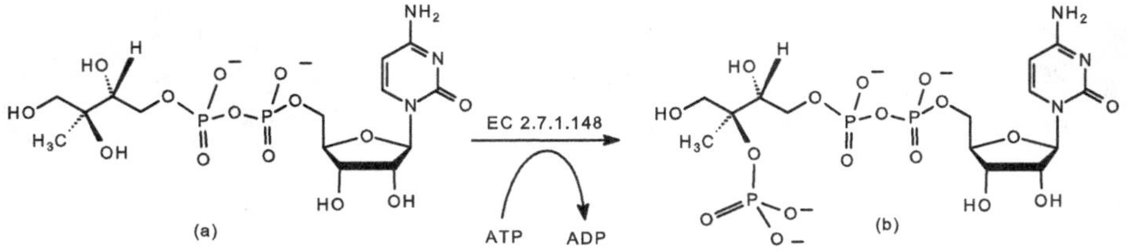

FIGURE A2.16 A cartoon representation of the conversion of 4-(cytidine 5′-diphospho)-2-*C*-methyl-D-erythritol (a) to the diphosphate 2-phospho-4-(cytidine 5′-diphospho)-2-*C*-methyl-D-erythritol (b) on the enzyme 4-(cytidine 5′-diphospho)-2-*C*-methyl-D-erythritol kinase (EC 2.7.1.148).

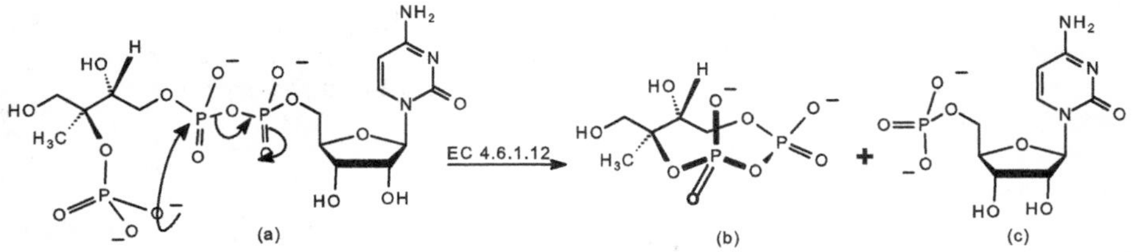

FIGURE A2.17 A representation of the conversion of 2-phospho-4-(cytidine 5′-diphospho)-2-*C*-methyl-D-erythritol (a) to the cyclic 2-*C*-methyl-D-erythritol 2,4-cyclodiphosphate (b) under the influence of 2-*C*-methyl-D-erythritol 2,4-cyclodiphosphate synthase (EC 4.6.1.12, PDB 4c8i). Cytidine monophosphate (c) is released having served as a leaving group.

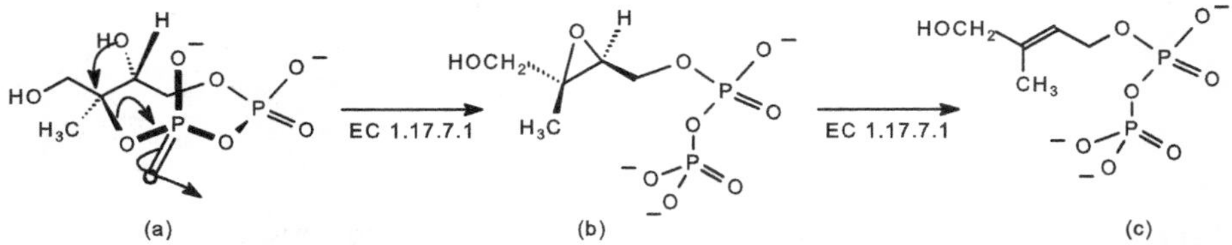

FIGURE A2.18 A representation of the opening of 2-*C*-methyl-D-erythritol 2,4-cyclodiphosphate (a) to a presumed epoxide intermediate (b) which subsequently loses oxygen to produce the product *E*-1-hydroxy-2-methylbut-2-enyl-4-diphosphate (c). The PDB entry for the enzyme EC 1.17.7.1 at this writing (February 2016) lacks a structure although an iron-sulfur (Fe_4S_4) cluster is known to be present.

And now, with the cytidine posed to leave, a cyclic bisphosphate, as shown in Figure A2.17, can form. The formation of the product, 2-*C*-methyl-D-erythritol 2,4-cyclodiphosphate, is catalyzed by the enzyme 2-*C*-methyl-D-erythritol 2,4-cyclodiphosphate synthase (EC 4.6.1.12, PDB 4c8i).

Despite the availability of the X-ray crystal structure of *E*-1-hydroxy-2-methylbut-2-enyl-4-diphosphate synthase (EC 1.17.7.1, PDB 4mwa), the mechanism of action (and/or inhibition) of conversion of 2-*C*-methyl-D-erythritol 2,4-cyclodiphosphate to *E*-1-hydroxy-2-methylbut-2-enyl-4-diphosphate, as shown in Figure A2.18, is "subject to controversy."[24,25]

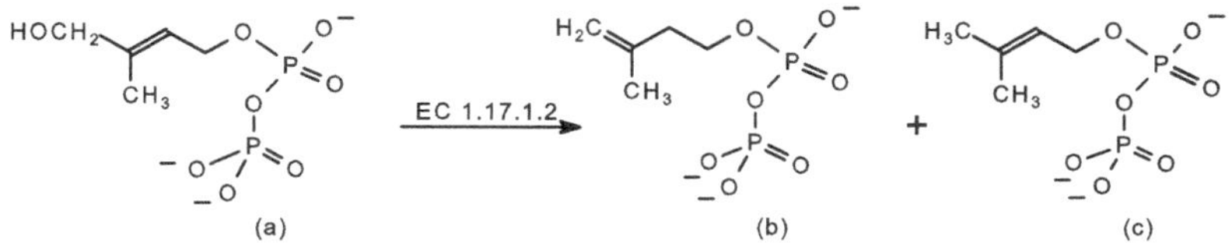

FIGURE A2.19 A representation for which details remain obscure of the conversion of (*E*)-1-hydroxy-2-methylbut-2-enyl-4-diphosphate (a) under the influence of (*E*)-1-hydroxy-2-methylbut-2-enyl-4-diphosphate reductase (EC 117.1.2) to produce isopentenyl diphosphate (b) and (c) dimethylallyl diphosphate.

In the final step, Figure A2.19, *E*-1-hydroxy-2-methylbut-2-enyl-4-diphosphate reductase (EC 1.17.1.2, PDB 3dnf), which also contains an iron-sulfur cluster, brings about the reductive conversion to isopentenyl diphosphate and dimethylallyl diphosphate. It appears that after the reduction (by the iron-sulfur cluster), re-oxidation is effected by a $NADP^+$ cofactor. Interestingly, it has been reported that a 5:1 mixture of isopentenyl diphosphate and dimethylallyl diphosphate results, but further information about the details of the reductive loss of the hydroxyl group have not been made available.[26,27]

TERPENES, SESQUITERPENES AND MORE—WHY PLANT FLOWERS HAVE ODOR

Isopentenyl diphosphate and dimethylallyl diphosphate have been singled out because it has become clear that they serve as the starting materials for the more than (at this writing) fifty-five thousand (55,000) known terpenes (C_{10}), sesquiterpenes (C_{15}), diterpenes (C_{20}), sesterterpenes (C_{25}), and triterpenes (C_{30}), as well as mixed metabolites (e.g., with amino acids), degraded metabolites (i.e., norisoprenoids formed by carotenoid breakdown) and other metabolites. These metabolites are built up in living systems using *polyprenyl transferases*.

There are many already identified (and certainly many not yet found) polyprenyl transferases which together form a superfamily that exists throughout the plant, fungal, and animal kingdoms and that produce the wealth of isolated products. A recent large-scale study was undertaken, only to find more work was necessary to assign function to members of the major trans-polyprenyl transferase subgroup in the isoprenoid synthase superfamily.[28]

A number of the compounds identified[29,30] in the headspace above the flowers of varieties of red grapes are sesquiterpenes. Their presence in varying concentrations defines the "perfume" of the grape flowers.[31]

As shown in Figure A2.20 it is argued that, initially, dimethylallyl diphosphate reacts with isopentenyl diphosphate in the active site of the isoprenoid synthase, dimethylallyl*trans*-transferase (EC 2.5.1.1, for which there are 81 PDB entries! [January 2016]). Although it is possible that the phosphate ester is hydrolyzed in the first step to a carbocation with enzyme-restricted rotation, it is also possible that a concerted process (such as that shown) obtains. The *pro-R* hydrogen loss is in concert with steroid biosynthetic processes.[32]

With the same or similar enzyme, (e.g., EC 2.5.1.10) the (2*E*,6*E*)-farnesyl diphosphate is generated. This is also shown in Figure A2.20. Interestingly, as far as is known, it appears that all of

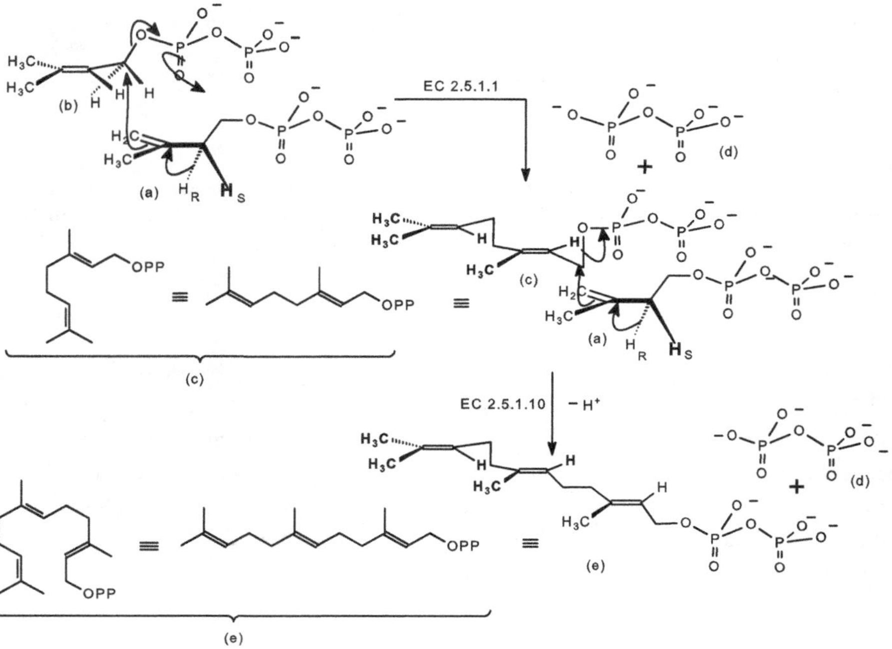

FIGURE A2.20 A simplified version of the result of the attack of isopentenyl diphosphate (a) onto dimethylallyl diphosphate (b) with proton loss (H_R) in the presence of dimethylallyl*trans*-transferase (EC 2.5.1.1) to generate geranyl diphosphate (c) and inorganic phosphate (d). Then in a subsequent step, the reaction is repeated with isopentenyl diphosphate (a) attacking the phosphate-bearing carbon of geranyl diphosphate (c), again in the presence of a transferase, *viz.* (2*E*,6*E*)-farnesyl diphosphate synthase, EC 2.5.1.10, to produce farnesyl diphosphate (e) and, again, inorganic phosphate.(d).

FIGURE A2.21 Representations of elimination of phosphate with alkene formation for farnesyl diphosphate (a) in the presence of the enzyme α-farnesene synthase (EC 4.2.3.46) to yield (b) α-farnesene and inorganic phosphate (c) and the same starting material, farnesyl diphosphate (a), in the presence of the enzyme β-farnesene synthase (EC 4.2.3.47) to yield (d) β-farnesene and, again, inorganic phosphate (c).

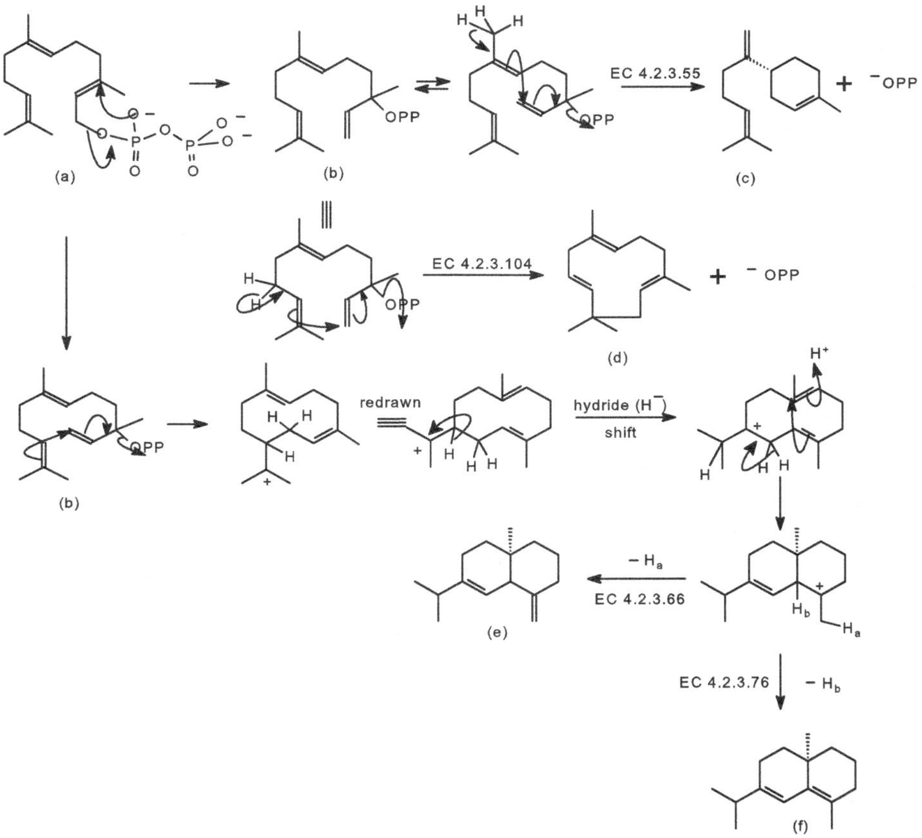

FIGURE A2.22 Representations of some potential rearrangements from farnesyl diphosphate (a) and its phosphate-rearranged isomer nerolidyl diphosphate (b) in which the phosphate has been moved to a tertiary carbon and the double bond to the terminus. Under the influence of the enzyme (S)-β-bisabolene synthase (EC 4.2.3.55) the sesquiterpene β-bisabolene (c) is formed. Alternatively, in the presence of α-humulene synthase (EC 4.2.3.104), α-humulene (d) is isolated. Further, a series of hydride shifts and proton loss in the presence of β-selinene cyclase (EC 4.2.3.66) leads to β-selinene (e), whereas in the presence of (+)-δ-selinene synthase (EC 4.2.3.76), δ-selinene (f) results.

the sesquiterpenes yielding, together, the aroma of the flowers are derived from (2E,6E)-farnesyl diphosphate. With (2E,6E)-farnesyl diphosphate in hand it is now clear how the sesquiterpenes listed as being found in the headspace of the flowers of some Austrian red wines are generated.

Current, proposed (http://www.enzyme-database.org/) biosynthetic pathways are provided in Figures A2.21, A2.22, A2.23, and A2.24, (with some emendation). A more complete list of what was found by the investigators[25,26] appears in the text.

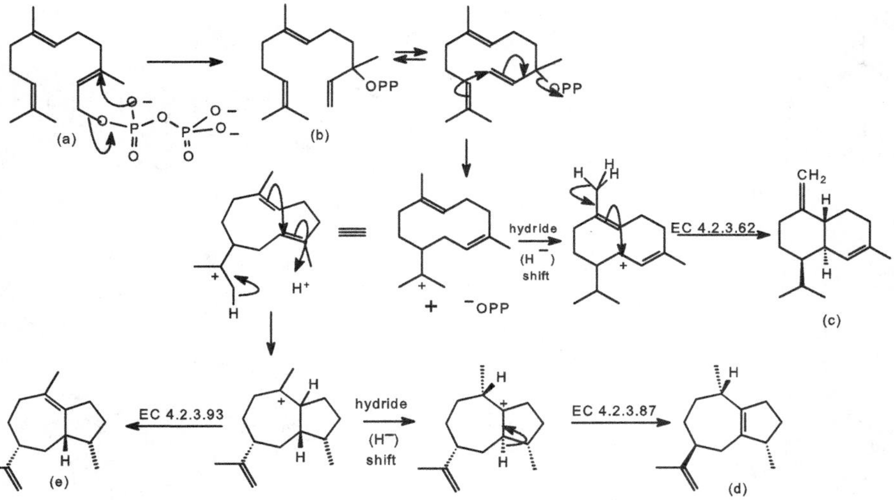

FIGURE A2.23 Representations of some potential rearrangements from farnesyl diphosphate (a) and its phosphate-rearranged isomer nerolidyl diphosphate (b) in which the phosphate has been moved to a tertiary carbon and the double bond to the terminus. Cyclization and rearrangement in the presence of the enzyme (−)-γ-cadinene synthase [(2Z,6E)-farnesyl diphosphate cyclizing] (EC 4.2.3.62) yields (−)-γ-cadinene (c), while different guaiene isomers are formed, α-guaiene (d) with α-guaiene synthase (EC 4.2.3.87) and δ-guaiene (e) with δ-guaiene synthase (EC 4.2.3.93).

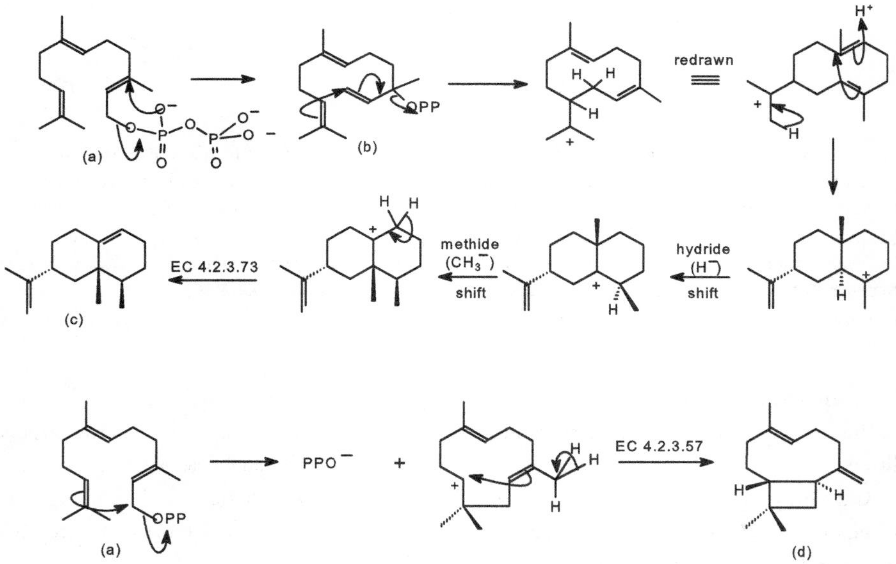

FIGURE A2.24 Representations of some potential rearrangements from farnesyl diphosphate (a) and its phosphate-rearranged isomer nerolidyl diphosphate (b) in which the phosphate has been moved to a tertiary carbon and the double bond to the terminus. Cyclization and rearrangement in the presence of the enzyme valencene synthase (EC 4.2.3.73) generates valencene (c). Reactions directly from farnesyl diphosphate (a) in the presence of (−)-β-caryophyllene synthase (EC 4.2.3.57) generate β-caryophyllene (d).

NOTES AND REFERENCES

1. Cleland, W. W.; Andrews, T. J.; Gutteridge, S.; Hartman, F. C.; Lorimer, G. H. *Chem. Rev.* **1998**, *98*, 549. See also the Uppsla Rubisco Pages; http://xray.bmc.uu.se/~tom/rubisco.html (accessed April 10, 2017); and Taylor, T. C.; Andersson, I. *Biochemistry* **1997**, *36*, 4041.

2. A variety of isomerases (e.g., EC 5.4.2.11 and 5.2.4.12) and the "transferase" PDB 3pgm, are also available. Many are involved primarily in glycolysis (use of glucose). See also Winn, S. I.; Watson, H. C.; Harkins, R. N.; Fothergill, L. A. *Phil. Trans. R. Soc. London* B **1981**, *293*, 121.

3. Reed, G. H.; Poyner, R. R.; Larsen, T. M.; Wedekind, J. E.; Rayment, I. *Curr. Opin. Struct. Biol.* **1996**, *6*, 736.

4. Wedekind, J. E.; Reed, G. H.; Rayment, I. *Biochemistry* **1995**, *34*, 4325.

5. "Two billion years ago, an early cell swallowed an energy-producing microbe, giving birth to the mitochondria that are the hallmarks of all eukaryotes, from protists to people." "Mitochondria . . . have their own set of genes which replicate and mutate faster than the cell's known complement in the nucleus . . . " (Pennisi, E. *Science* **2016**, *353*, 336). Following that general line of thought, it is not too much of a surprise to learn that from glycolysis, for example, there are two different pathways to isopentenyl diphosphate and dimethylallyl diphosphate and their *different* eventual isoprenoid products commonly found in the same cell!

6. Lichtenthaler, H. K. *Ann. Rev. Plant Physiol. Plant Mol. Biol.* **1999**, *50*, 47.

7. Arjunan, P.; Nemeria, N.; Brunskill, A.; Chandrasekhar, K.; Sax, M.; Yan, Y.; Jordan, F.; Guest, J. R.; Furey, W. *Biochemistry* **2002**, *41*, 5213.

8. Ciszak, E. M.; Korotchkina, L. G.; Dominiak, P. M.; Sidju, S.; Patel, M. S. *J. Biol. Chem.* **2003**, *278*, 21240. See also Proteopedia. http://www.proteopedia.org/wiki/index.php/Pyruvate_dehydrogenase (accessed April 10, 2017).

9. Modis, Y.; Wierenga, R. K. *J. Mol. Biol.* **2000**, *297*, 1171.

10. Meriläinen, G.; Poikela, V.; Kursula, P.; Wierenga, R. K. *Biochem.* **2009**, *48*, 11011.

11. Theisen, M. H.; Misra, I.; Saadat, E.; Campobasso, N.; Miziorko, H. M.; Harrison, D. H. T. *Proc. Natl. Acad. Sci. U. S. A.* **2004**, *101*, 16442.

12. Bahnson, B. J. *Proc. Natl. Acad. Sci. U. S. A.* **2004**, *101*, 16399.

13. Istvan, E. S.; Palnitkar, M.; Buchanan, S. K.; Deisenhofer, J. *EMBO J.* **2000**, *19*, 819.

14. Bonanno, J. B.; Edo, C.; Eswar, N.; Pieper, U.; Romanoski, M. J.; Ilyn, V.; Gerchman, S. E.; Kycia, H.; Studier, F. W.; Sali, A.; Burley, S. K. *Proc. Natl. Acad. Sci. U. S. A.* **2001**, *98*, 12896.

15. Cornforth, R. H.; Popják, G. *Methods Enzymol.* **1969**, *15*, 359.

16. Durbecq, V.; Sainz, G.; Oudjama, Y.; Clantin, B.; Bompard-Gilles, C.; Tricot, C.; Caillet, J.; Stalon, V.; Droogmans, L.; Villeret, V. *EMBO J.* **2001**, *20*, 1530.

17. Eisenreich, W.; Bacher, A.; Arigoni, D.; Rohdich, R. CMLS, *Cell. Mol. Life Sci.* **2004**, *61*, 1401. DOI:10.1007/s00018-004-3381-z.

18. Eubanks, L. M.; Poulter, C. D. *Biochemistry* **2003**, *42*, 1140.

19. Xiang, S.; Usnow, G.; Lang, B.; Busch, M. Tong, L. *J. Biol. Chem.* **2007**, *282*, 2676.

20. Patel, H.; Nemeria, N. S.; Brammer, L. A.; Meyers, C. L. F.; Jordan, F. *J. Am. Chem. Soc.* **2012**, *134*, 18374. DOI:10.1021/ja307315u.

21. Takahashi, S., Kuzuyama, T., Watanabe, H.; Seto, H. *Proc. Natl. Acad. Sci. U. S. A.* **1998**, *95*, 9879.

22. Munos, J. W.; Pu, X.; Mansoorabadi, S. O.; Kim, H. J.; Liu, H. W. *J. Am. Chem. Soc.* **2009**, *131*, 2048.

23. Björkelid, C.; Bergfors, T.; Henriksson, L. M.; Stern, A. L.; Unge, T.; Mowbray, S. L.; Jones, T. A. *Acta Crystallogr. D Biol. Crystallogr.* **2011**, *67,* 403.

24. Lee, M.; Gräwert, T.; Quitterer, F.; Rohdich, F.; Eppinger, J.; Eisenreich, W.; Bacher, A.; Groll, M. *J. Mol. Biol.* **2010**, *404,* 600.

25. Wang, W.; Li, J.; Wang, K.; Huang, C.; Zhang, Y.; Oldfield, E. *Proc. Natl. Acad. Sci. U. S. A.* **2010**, *107,* 11189.

26. Rekittke, I.; Wiesner, J.; Röhrich, R.; Demmer, U.; Warkentin, E.; Xu, W.; Troschke, K.; Hintz, M.; No, J. H.; Duin, E. C.; Oldfield, E.; Jomaa, H.; Ermler, U. *J. Am. Chem. Soc.*, **2008**, *130,* 17206.

27. Gräwert, T.; Kaiser, J.; Zepeck, F.; Laupitz, R.; Hecht, S.; Amslinger, S.; Shramek, N.; Schleicher, E.; Weber, S.; Haslbeck, M.; Buchner, J.; Rieder, C.; Arigoni, D.; Bacher, A.; Eisenreich, W.; Rohdich, F. *J. Am. Chem. Soc.* **2004**, *126,* 12847.

28. Wallrapp, F. H.; Pan, J.-J.; Ramamoorthy, G.; Almonacid, D. E.; Hillerich, B. S.; Seidel, R.; Patskovsky, Y.; Babbitt, P. C.; Almo, S. C.; Jacobson, M. P.; Poulter, C. D. *Proc. Nat. Acad. Sci. USA* **2013**, *110*(13), *E1196*. Published ahead of print March 14, 2013, DOI:10.1073/pnas.1300632110.

29. Buchbauer, G.; Jirovetz, L.; Wasicky, M.; Nikiforov, A. *J. Essent. Oil Res.* **1994**, *6,* 311.

30. Buchbauer, G.; Jirovetz, L.; Wasicky, M.; Nikiforov, A. *Z. Lebensmittel-Untersuchung und-Forschung*, **1995**, *200,* 443.

31. In addition to the classical lines of the Rubaiyat of Omar Khayyam (https://en.wikipedia.org/wiki/Rubaiyat_of_Omar_Khayyam) extolling the mystical and lyrical poetry of wine, others have praised the perfume of the flowers. See, e.g., Martin, D. M.; Toub, O.; Chiang, A.; Lo, B. C.; Ohse, S.; Lund, S. T.; Bohlmann, J. *Proc. Nat. Acad. Sci. USA* **2009**, *106*(17), 7245. DOI:10.1073/pnas.0901387106.

32. Popják, G.; Cornforth, J. W. *Biochem. J.* **1966**, *101,* 553.

Appendix 3

LIST OF ESTER ODORANTS

Many carboxylic acid esters have distinctive fruit-like odors, and many occur naturally in the essential oils of plants. This has also led to their commonplace use in artificial flavorings and fragrances when those odors aim to be mimicked. The reaction involved requires the loss of water from the combination of carboxylic acid with alcohol. The −OH is lost from the carboxylic acid.

$$R-COOH + HO-R' \rightleftharpoons R-COO-R' + H_2O$$

Ester Name	Formula	Odor and/or occurrence
allyl hexanoate (allyl caproate)	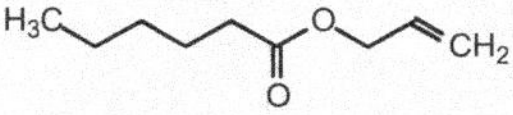	pineapple
benzyl acetate	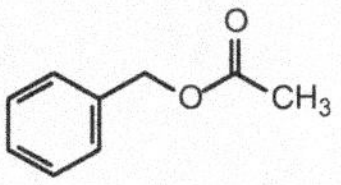	pear and strawberry
bornyl acetate (1S,2R,4S)-1,7,7-tri-methylbicyclo[2.2.1]-hept-2-yl acetate	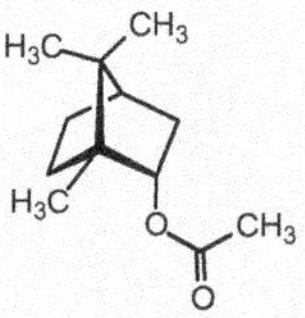	pine
butyl acetate	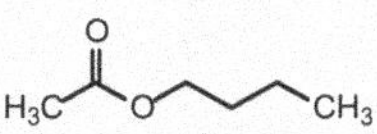	apple, honey

butyl butyrate	structure	pineapple
butyl propanoate	structure	pear
ethyl acetate	structure	nail polish remover, model paint, model airplane glue
ethyl butyrate	structure	banana, pineapple, strawberry
ethyl hexanoate	structure	pineapple
ethyl cinnamate	structure	cinnamon
ethyl formate	structure	lemon, rum, strawberry
ethyl heptanoate	structure	apricot, cherry, grape, raspberry

List of Ester Odorants

ethyl 3-methyl-butanoate (ethyl isovalerate)		apple
ethyl lactate		butter, cream
ethyl nonanoate		grape
ethyl pentanoate		apple
geranyl acetate		geranium
geranyl butyrate		cherry

Name	Structure	Odor
geranyl pentanoate		apple
isobutyl acetate		cherry, raspberry, strawberry
isobutyl formate		raspberry
isoamyl acetate		pear, banana
isopropyl acetate		fruity
linalyl acetate		lavender, sage

List of Ester Odorants

Name	Structure	Odor
linalyl butanoate (linalyl butyrate)		peach
linalyl formate		apple, peach
methyl acetate		glue
methyl anthranilate		grape, jasmine
methyl benzoate		fruity, ylang ylang
methyl butanoate (methyl butyrate)		pineapple, apple, strawberry

Name	Structure	Scent
methyl cinnamate	C6H5-CH=CH-C(=O)-O-CH3	strawberry
methyl pentanoate (methyl valerate)	H3C-CH2CH2CH2-C(=O)-O-CH3	flowery
methyl phenyl acetate	C6H5-CH2-C(=O)-O-CH3	honey
methyl salicylate (oil of wintergreen)	2-(HO)C6H4-C(=O)-O-CH3	wintergreen
nonyl octanoate (nonyl caprylate)	CH3(CH2)6-C(=O)-O(CH2)8CH3	orange
octyl acetate	H3C-C(=O)-O(CH2)7CH3	fruity-orange
octyl butanoate (octyl butyrate)	CH3CH2CH2-C(=O)-O(CH2)7CH3	parsnip
pentyl acetate (amyl acetate)	H3C-C(=O)-O(CH2)4CH3	apple, banana

List of Ester Odorants

Name	Structure	Odor
pentyl butyrate (amyl butyrate)	$CH_3CH_2CH_2C(=O)O(CH_2)_4CH_3$	apricot, pear, pineapple
pentyl hexanoate (amyl caproate)	$CH_3(CH_2)_4C(=O)O(CH_2)_4CH_3$	apple, pineapple
pentyl pentanoate (amyl valerate)	$CH_3(CH_2)_3C(=O)O(CH_2)_4CH_3$	apple
propyl acetate	$H_3C-C(=O)-O-CH_2CH_2CH_3$	pear
propyl hexanoate	$H_3C(CH_2)_4C(=O)O-CH_2CH_2CH_3$	blackberry, pineapple, cheese, wine
propyl 2-methylpropionate (propyl isobutyrate)	$(CH_3)_2CHC(=O)O-CH_2CH_2CH_3$	rum
2-(4-methyl-3-cyclohexen-1-yl)-2-propanyl butyrate (terpenyl butyrate)	(terpenyl butyrate structure)	cherry

Appendix 4

COMPOUNDS AND COLORS

Figure A4.1 provides representations of two compounds, typical of distinct classes, whose structures allow them to absorb radiation in the visible region of the spectrum. They are distinct from the long-chain unsaturates such as carotene (Chapter 10) and generally have more intense colors. It will be remembered that absorption in the visible determines which colors can be seen (i.e., those not absorbed) by the human eye. Both sets of representative pigments, betalains and anthocyanidins, are derived from the amino acid phenylalanine (Phe, F) and other species. For malvidin 3-glucoside (oenin) and other anthocyanin pigments, a polyketide synthase and carbohydrates are also required. For betalain pigments, phenylalanine (Phe, F), carbohydrates and other amino acids are used.

Many carbohydrates can be obtained from the Calvin cycle (Chapter 11 and Appendix 2), and the genesis of the others, through transformations and rearrangements of those and other carbohydrates, is beyond this discussion.

The story of the betalain and anthocyanidin pigments begins with the formation of phenylalanine. Phenylalanine (Phe, F) biosynthesis itself starts with carbohydrate chemistry. The reaction between phosphoenol pyruvate and erythrose 4-phosphate in the presence of 3-deoxy-7-phosphoheptulonate synthase (EC 2.5.1.54) yields a sedoheptulose (a seven-carbon carbohydrate) which undergoes cyclization to a tri-hydroxy substituted ketocyclohexane-carboxylic acid, *viz*. 3-dehydroquinate (Figure A4.2).

Continuing in the biosynthetic pathway of phenylalanine (Phe, F), in the presence of 3-dehydroquinate dehydratase (EC 4.2.1.10), loss of water from the carboxylate-bearing carbon of 3-dehydroquinate (Figure A4.3) results in the introduction of a carbon–carbon double bond and formation of 3-dehydroshikimate.

Reduction of the carbonyl of 3-dehydroshikimate to shikimate, followed by phosphorylation at C-3, and transfer of phosphoenolpyruvate to the hydroxyl at C-5, yields a carboxyvinyl derivative

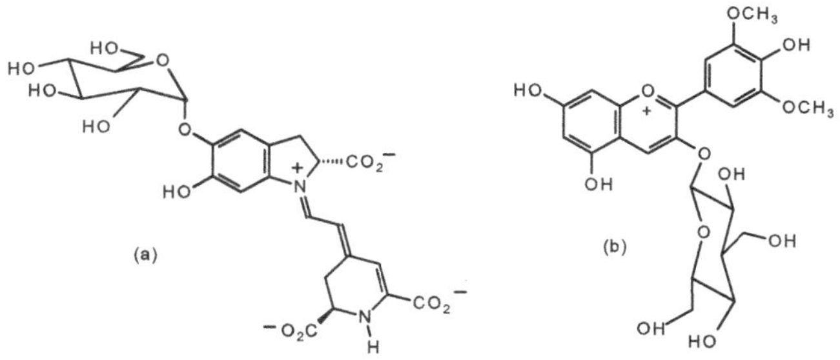

FIGURE A4.1 Representations of some pigments found in flowers and fruit skins. While not known to occur in the same plants, both (a) betanin, a betalain pigment, and (b) malvidin 3-glucoside (oenin), an anthocyanin pigment, have phenylalanine (Phe, F) as their parent amino acid progenitor.

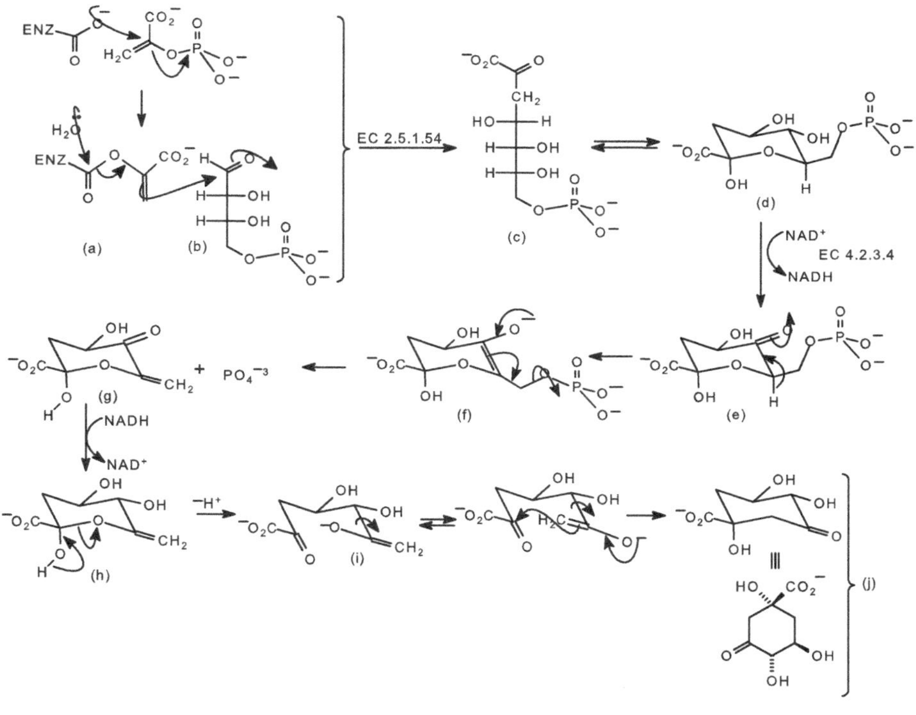

FIGURE A4.2 A cartoon representation of the enzyme [3-deoxy-7-phosphoheptulonate synthase (EC 2.5.1.54)] catalyzed reaction between phosphoenolpyruvate (a) and D-erythrose 4-phosphate (b) to yield 3-deoxy-D-arabinohept-2-ulosonate 7-phosphate (c). Continuing, cyclization of (c) to the corresponding pyranose (d) and oxidation by NAD+ results in the formation of the ketone (e) which, with the enzyme 3-dehydroquinate synthase (EC 4.2.3.4) participating, presumably forms the enolate anion (f), which then undergoes phosphate elimination to the exocyclic alkene (g). Reduction of the carbonyl of (g) by NADH leads to the hemiketal (h) which opens to the α-keto-acid (i) and recloses with the formation of a new carbon–carbon bond to 3-dehydroquinate (j).

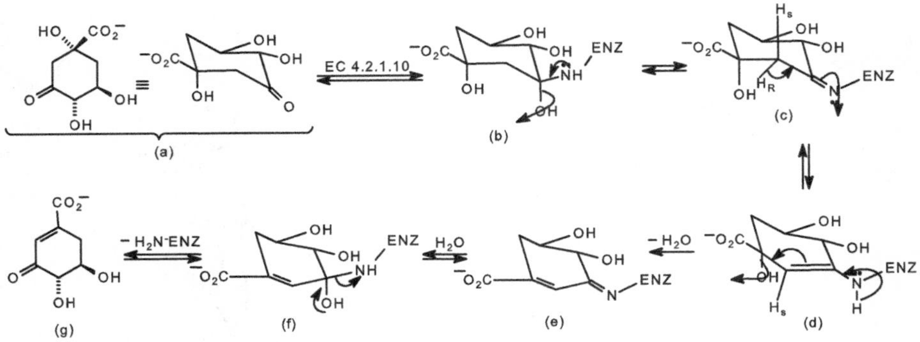

FIGURE A4.3 A continuation of the representation of the biosynthesis of phenylalanine (Phe, F). It is presumed that a lysine (Lys, K) on the enzyme 3-dehydroquinate dehydratase (EC 4.2.1.10) reacts with the carbonyl of 3-dehydroquinate (a) to yield a carbonolamine (b) which undergoes loss of water to an imine (c) that specifically loses the H_R proton on C-2 to produce the enamine (d). Dehydration with reformation of an imine produces (e) which then re-adds water to (f) and loses the enzyme bound nitrogen to produce 3-dehydroshikimate (g).

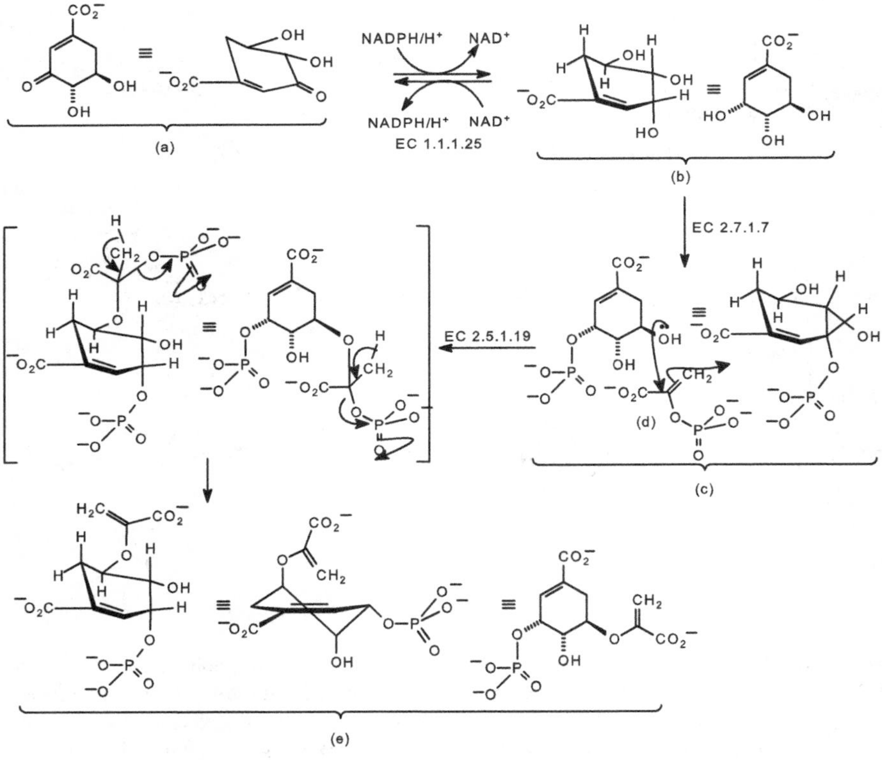

FIGURE A4.4 A continuation of the representation of the biosynthesis of phenylalanine (Phe, F). Having 3-dehydroshikimate (a) in the presence of shikimate dehydrogenase (EC 1.1.1.25) results in the formation shikimate (b) and the conversion of NADPH to NADP+. Then, in the presence of shikimate kinase (EC 2.7.1.7) the hydroxyl at C-3 of (b) undergoes phosphorylation to the corresponding ester (c), and in a subsequent step, the C-5 hydroxyl adds a carboxyvinyl group (from phosphoenolpyruvate) (d) to yield 5-(1-carboxyvinyl)-3-phosphoshikimate (e) in the presence of 3-phosphoshikimate-1-carboxyvinyl transferase (EC 2.5.1.19).

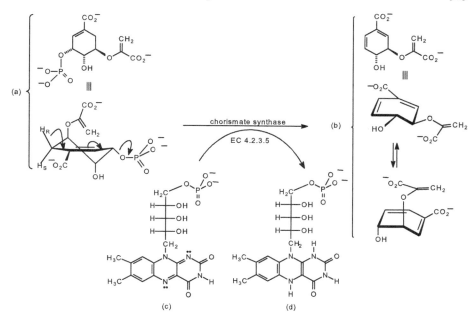

FIGURE A4.5 A continuation of the representation of the biosynthesis of phenylalanine (Phe, F). This representation depicts a cartoon showing the chorismate synthase (EC 4.2.3.5) catalyzed conversion of 5-(1-carboxyvinyl)-3-phosphoshikimate (a) to chorismate (b). Although this depiction appears to show a simple conjugate elimination, the requirement of flavin mononucleotide, FMN (c), and its conversion to the reduced form, $FMNH_2$ (d), as well as other data, suggests that the reaction proceeds via a radical mechanism. See Osborne, A.; Thorneley, R. N.; Abell, C.; Bornemann, S. *J. Biol. Chem.* **2000**, *275*, 35825.

ready for elimination to chorismate (Figures A4.4 and A4.5). Figure A4.6 recapitulates in summary fashion Figures A4.2, A4.3, and A4.4.

With chorismate in hand, rearrangement to prephenate follows, and then decarboxylation to tyrosine (Tyr, Y) or dehydration and decarboxylation to phenylalanine (Phe, F) result. The products are shown in Figure A4.7. Although conversion of phenylalanine (Phe, F) to tyrosine (Tyr, Y) is known, the reverse does not generally occur.

A pathway from phenylalanine (Phe, F) to tyrosine (Tyr, Y) catalyzed by phenylalanine 4-monooxygenase (EC 1.14.16.1), and often referred to as using the "NIH shift" as the reaction was initially reported from the National Institutes of Health, is also known. In that process, it appears that at the active site containing an iron II (Fe^{2+}) there is formation of an arene oxide, with the oxygen attached between the carbon atoms at C-3 and C-4. Subsequent bond breakage between the oxygen–carbon C-3 bond and migration of the hydrogen (partially retained) at C-4 to C-3 followed by tautomerization of the resulting ketone lead to tyrosine (Tyr, Y).[1] The process is depicted in Figure A4.8.

With phenylalanine (Phe, F) having been made available, discrimination between the betalains and anthocyanidins can be undertaken.

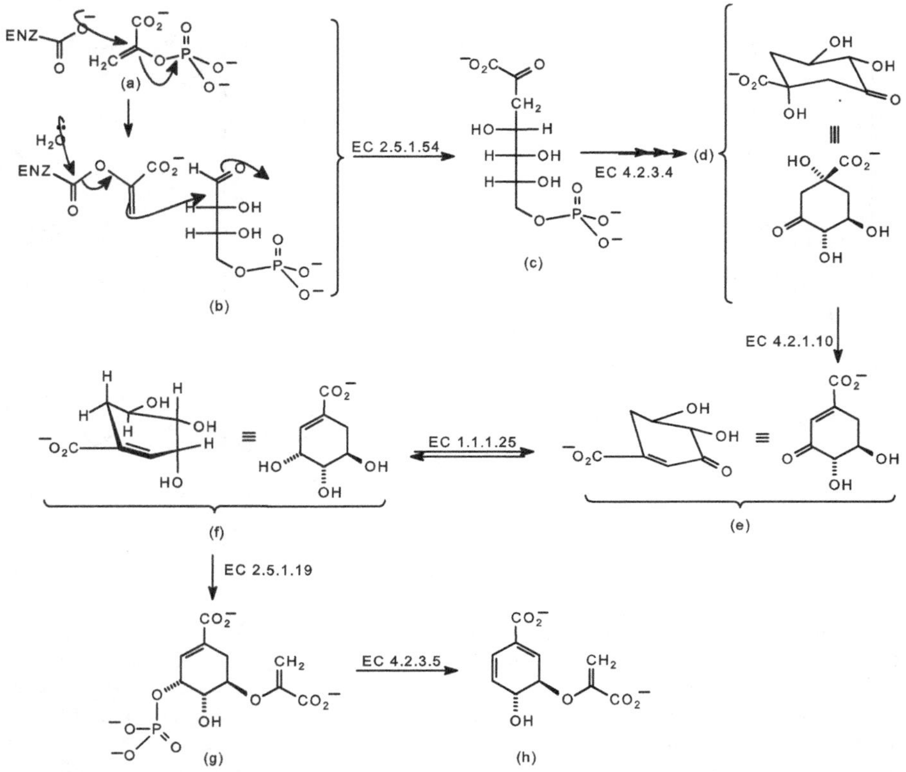

FIGURE A4.6 A recapitulation without details of Figures A4.2 through A4.4 for the conversion of phosphoenolpyruvate (a) and erythrose 4-phosphate (b) to 2-dehydro-3-deoxy-7-phospho-D-arabinoheptanoate (c), 3-dehydroquinate (d), 3-dehydroshikimate (e), shikimate (f), 5-(1-carboxyvinyl)-3-phosphoshikimate (g), and finally chorismate (h).

For the betalains, tyrosine (Tyr, Y) formed from chorismate, or via oxidation of phenylalanine (Phe, F) with phenylalanine 4-monooxygenase (EC 1.14.16.1), is known to undergo oxidation to form dihydroxyphenylalanine (using tyrosine-3-monooxygnease, EC 1.14.16.2).

Then, the dihydroxyphenylalanine undergoes further oxidation. The oxidation apparently leads to aldehydes (see, e.g., EC 1.13.11.29, *stizolobate synthase*) which, depending upon the system, can be processed in distinct pathways. One process, which requires oxidation [NADP+], leads to a pair of dicarboxylic acid lactones, stizolobic acid and stizolobinic acid.[2,3] The other avoids oxidation and produces a pair of (nonisolable) cyclized aldehydes that subsequently react at the aldehyde carbonyl to produce imines.[4] With regard to the latter, one of the pair is the aldehydic betalamic acid, parent of the brightly colored betalains (both betacyanins—the red to violet pigments—and betaxanthins—the yellow to orange pigments). The processes are outlined in Figures A4.9 and A4.10.

Interestingly, as also shown, and depending upon the specific way the oxidation occurs on the enzyme(s) surface, the oxidation that yields betalamic acid can also be seen (Figure A4.9) to yield muscaflavine and both stizolobic and stizolobinic acids.

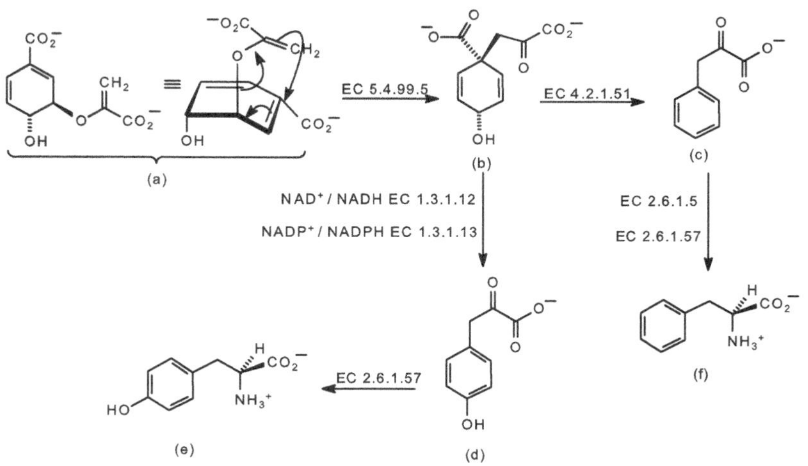

FIGURE A4.7 A representation of the conversion of chorismate (a), in the presence of chorismate mutase (EC 5.4.99.5) to prephenate (b) and thence, with prephenate dehydratase (EC 4.2.1.51) to (c) phenylpyruvate. Also shown is the conversion of prephenate (b) to 4-hydroxyphenylpyruvate (d) as catalyzed by prephenate dehydrogenase (EC 1.3.1.12 and EC 1.3.1.13) and then to tyrosine (Tyr, Y) (e), with catalysis by the transaminase (EC 2.6.1.57) using pyridoxal and glutamate (Glu, E) as the source nitrogen. Interestingly, the same transaminase, also using pyridoxal and glutamate (Glu, E) as the source nitrogen, as well as tyrosine transaminase (EC 2.6.1.5), can be used to catalyze the conversion of phenylpyruvate (c) to phenylalanine (Phe, F) (f).

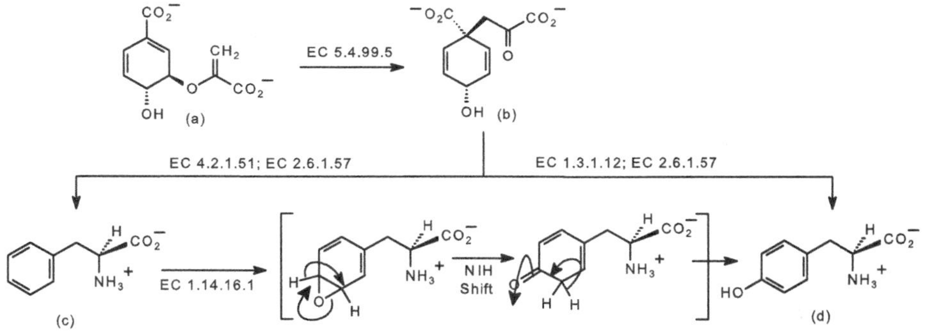

FIGURE A4.8 A overall representation of the conversion of chorismate (a) to prephenate (b) as catalyzed by chorismate mutase (EC 5.4.99.5) and thence to phenylpyruvate and phenylalanine (Phe, F) (c) with prephenate dehydratase (EC 4.2.1.51) and transaminase (EC 2.6.1.57) as well as from prephenate (b) to 4-hydroxyphenylpyruvate with prephenate dehydrogenase (EC 1.3.1.12) and thence to tyrosine (Tyr, Y) (d) with transaminase (EC 2.6.1.57). Also provided is the representation of the NIH shift under the influence of phenylalanine 4-monooxygenase (EC 1.14.16.1) for the conversion of phenylalanine (Phe, F) (c) to tyrosine (Tyr, Y) (d). (Udenfriend, S.; Cooper, J.R. *J. Biol. Chem.* **1952**, *194*, 503).

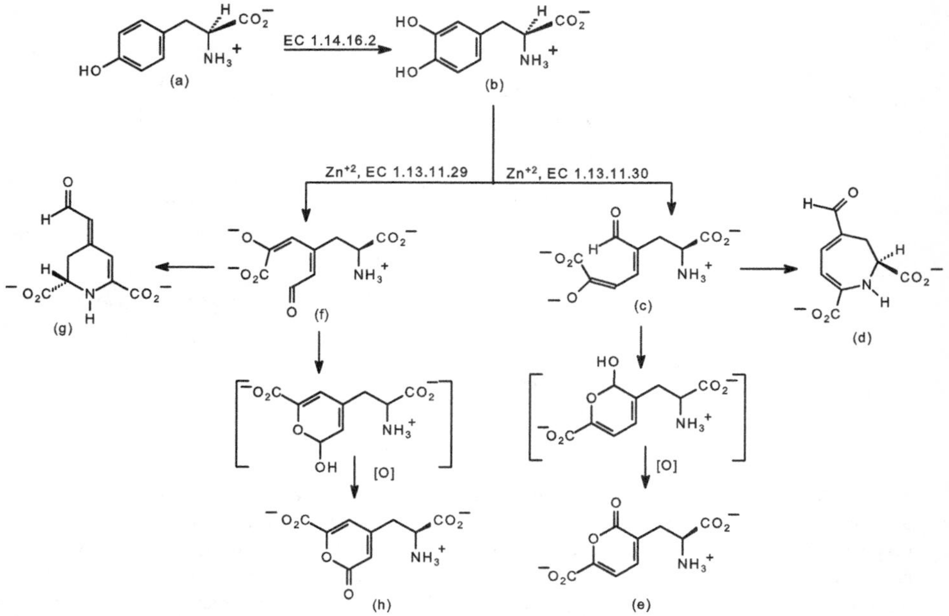

FIGURE A4.9 A representation of the conversion of tyrosine (Tyr, Y) (a) to 3,4-dihydroxyphenylalanine (DOPA) (b) with oxygen and the iron-containing enzyme tyrosine-3-monooxygnease, EC 1.14.16.2. Further oxidation of the aromatic ring results in cleavage. With the zinc-containing stizolobinate synthase (EC 1.13.11.30) the presumed intermediate 5-(L-alanin-3-yl)-2-hydroxy-*cis,cis*-muconate 6-semialdehyde (c) can go on to yield muscaflavine (d) as well as stizolobinic acid (e). The alternative, with the zinc-containing stizolobate synthase (EC 1.13.11.29) leads to the presumed intermediate 4-(L-alanin-4-yl)-2-hydroxy-*cis,cis*-muconate 6-semialdehyde (f) which can go on to betalamic acid (g) or stizolobic acid (h).

Reaction of betalamic acid with an unsaturated system such as 5-glcyosolated 5,6-dihydroxy-2,3-dihydroindole-2-carboxylate, itself formed by oxidation of 3,4-dihydroxy-phenylalanine (DOPA) produces the brightly colored red betacyanin, betanin (Figure A4.10). Alternatively reaction of betalamic acid with a simple amino acid such as proline produces the yellow betaxanthine, indicaxanthine (Figure A4.11).

The biosynthesis of the anthocyanidin pigments and related structures are quite different. They also begin with phenylalanine which, after loss of the nitrogen (with introduction of a double bond) adds the additional ring by a polyketide pathway (Figure A4.12).

Figure A4.13 shows a cartoon representation of a typical pathway to the appropriate cinnamic acid which could, as shown, be oxidized to coumaric acid (with one phenolic hydroxyl) or caffeic acid (with two –OH groups). Further elaborated by methylation (using *S*-adenosylmethionine) to ferulic acid can then occur. The oxidation and methylation patterns are typical.

Then, it is likely that phosphorylation at the carboxyl group occurs followed by replacement of the phosphate with coenzyme A (EC 6.2.1.12) which activates this portion of the final construct

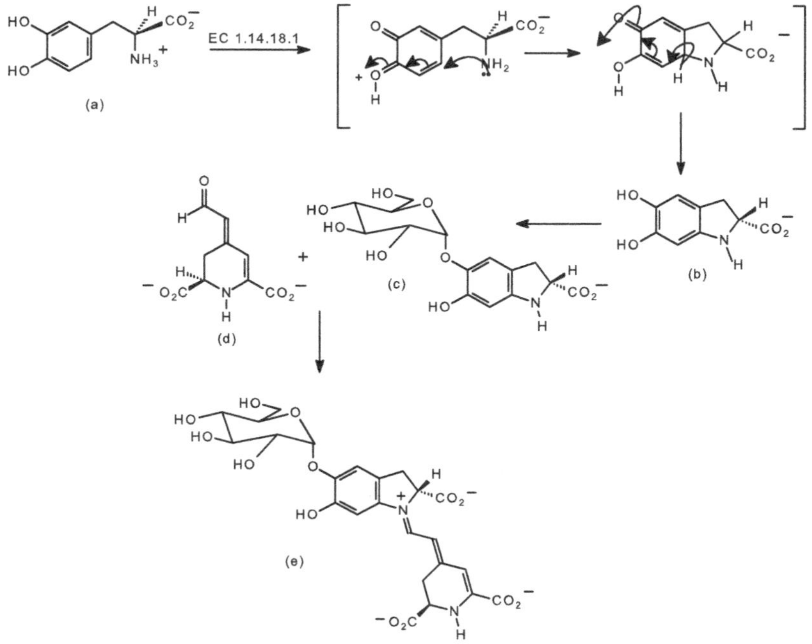

FIGURE A4.10 A representation of the further oxidation of 3,4-dihydroxyphenylalanine (DOPA) (a) by the copper-containing enzyme tyrosinase (EC 1.14.18.1) leads to (2R)-2,3-dihydro-5,6-dihydroxy-1H-indole-2-carboxylic acid (cyclo-DOPA) (b). It is presumed that phosphorylation of glucose followed by reaction with cyclo-DOPA (b) leads to the glycosylated derivative (c) which on reaction with betalamic acid (d) generates the brightly colored betacyanin pigment, betanin (e).

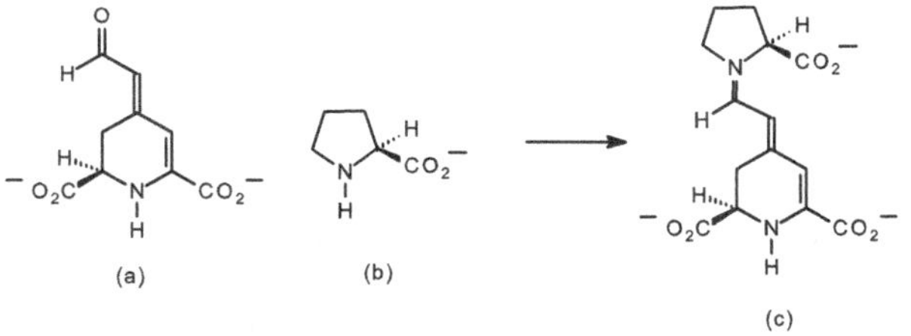

FIGURE A4.11 A cartoon depiction of the result of the reaction between the betalamic acid (a) and the amino acid proline (Pro, P) (b). Imine formation (with loss of water) leads to the betaxanthin pigment, indicaxanthine (c).

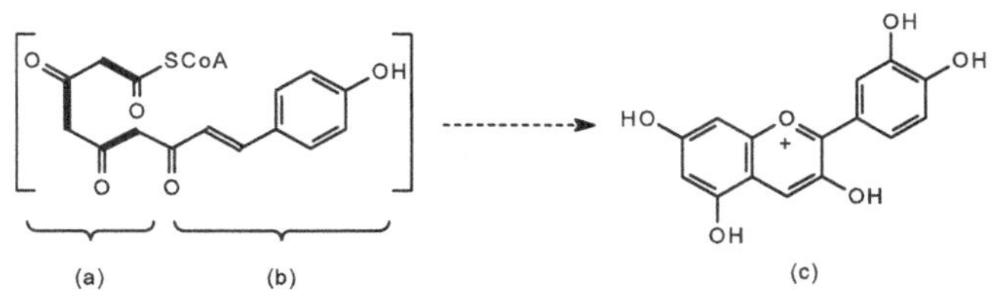

FIGURE A4.12 A cartoon picture ignoring the formation of the starting material and the details of the conversion to product but giving the overall idea that three acetate units (a) coupled to one phenylalanine (or tyrosine) derived cinnamate unit can give rise to an anthocyanidin type pigment. Cyanidin (c) is the trivial name of the anthocyanidin shown.

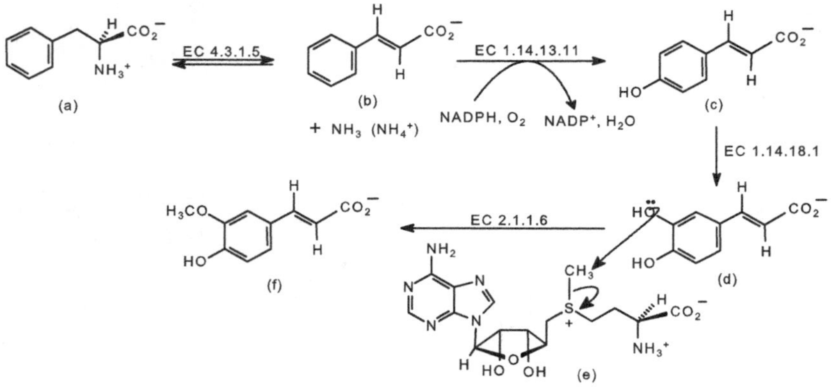

FIGURE A4.13 A cartoon representation of the deamination of phenylalanine (Phe, F) (a) with phenylalanine ammonia lyase (EC 4.3.1.5) to (E)-cinnamic acid (b). The *trans*-cinnamic acid (b) is shown to undergo oxidation (O_2) as catalyzed by *trans*-cinnamate 4-monooxygenase (EC 1.14.13.11) to *para*-coumaric acid (c) and, in the presence of monophenol monooxygenase (EC 1.14.18.1), to caffeic acid (d). Methylation of caffeic acid (d) through transfer of the methyl group of S-adenosylmethionine (e) in the presence of the enzyme catechol O-methyltransferase (EC 2.1.1.6) yields ferulic acid (f).

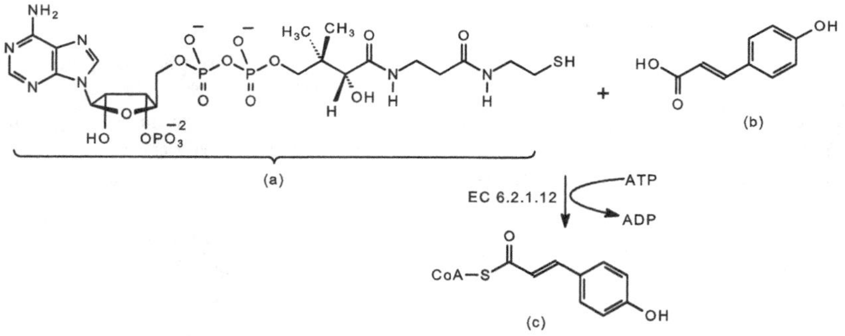

FIGURE A4.14 A representation of coenzyme A (a) and *para*-coumaric acid (b) forming the coenzyme A thioester, 4-coumaryl-CoA (c). Initial phosphorylation with ATP is followed by loss of phosphate in the 4-coumarate-CoA ligase (EC 6.2.1.12) catalyzed process.

(Figure A4.14). In this way both resveratrol and other stilbene derivatives can be produced as well as flavinols which go on to anthocyanidins (Figure A4.15).

Interestingly, but not surprisingly, different results, as a function of different enzymes obtain. Thus, for example, the structure (g) on the right in Figure A4.15, in the presence of trihydroxystilbene synthase (EC 2.3.1.95), leads to resveratrol and its O-methyl derivatives (Figure A4.16).

Referring again to Figure A4.15, the representation of the triketothioester of Figure A4.15(g) on the left, in naringenin-chalcone synthase (EC 2.3.1.74), undergoes a different cyclization to a 1,3,5-triketone. This trione is a tautomer of the corresponding phenol which is the naturally occurring naringenin chalcone. Then, as shown in Figure A4.17, addition of a phenol to the α,β-unsaturated ketone leads to the flavanone naringenin as well as a host of other compounds that include: apigenin (a flavone); aromadendrin and taxifolin (dihydroflavonols); and

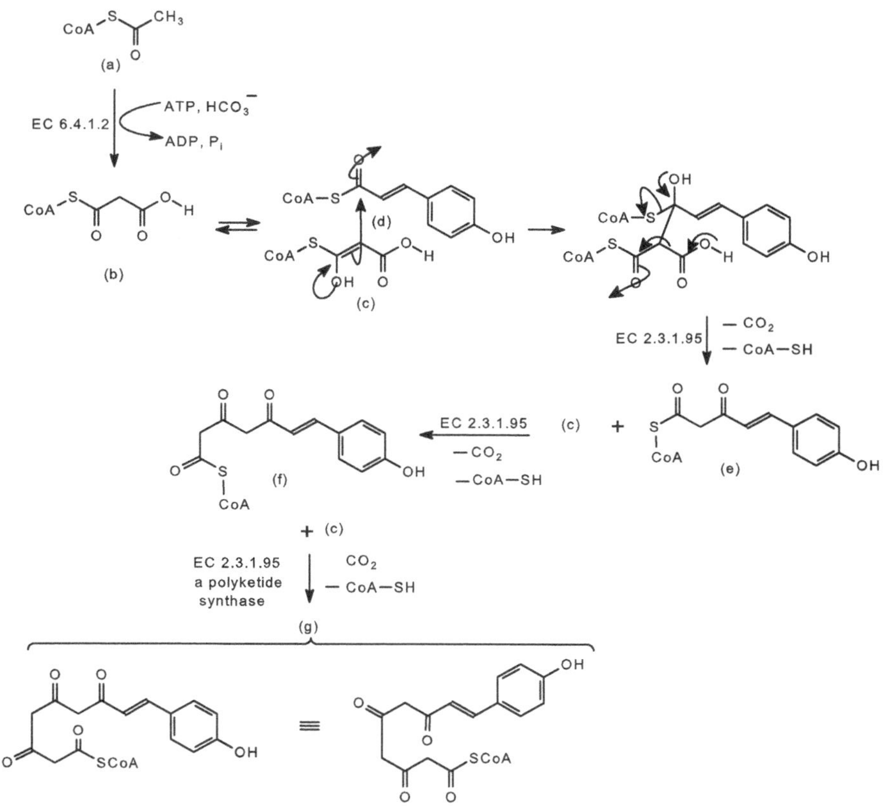

FIGURE A4.15 A representation of the beginning of the formation of flavanols and anthocyanidins. The reaction of acetyl-CoA with bicarbonate as a source of carbon dioxide (CO_2) in the presence of acetyl-CoA carboxylase (EC 6.4.1.2), a biotin-containing enzyme which utilizes adenosine triphosphate (ATP) and produces malonyl-CoA (b). Enolization of malonyl-CoA (c) and reaction with 4-coumaryl-CoA (d) in the presence of EC 2.3.1.95, a polyketide synthase, results in loss of carbon dioxide (CO_2) and acetyl CoA and produces the keto-thioester (e). Repeating the reaction with malonyl-CoA on (e) yields (f), and a third repetition produces the triketo-CoA rotational isomers (g) It is important to note that the two different representations of (g) are identical except for the way they are drawn.

leucopelagonidin (a flavan-3,4-diol). These are, of course, only a few possible examples of what can obtain through the utilization of the reactions shown. Additional possibilities result from methylation, glycosylation, and further oxidation.

And, finally, oxidative pathways lead to the anthocyanidins (Figures A4.18 and A4.19) which are further elaborated by glycosylation to anthocyanins. Both anthocyanidins and anthocyanins undergo further alkylation on oxygen to a wealth of isomers.

Interestingly, as noted in the text, cyanidin (as an example) changes color as a function of the acidity of the solution (pH) into which it is put (Figure A4.20).

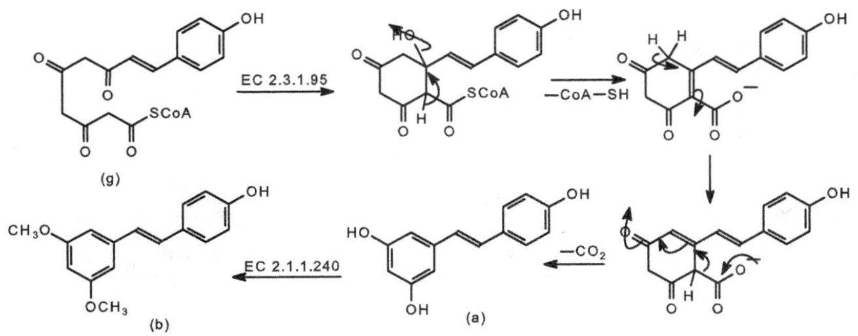

FIGURE A4.16 A representation of the result of cyclization of the triketothioester of Figure A4.15(g) in the presence of trihydroxystilbene synthase (EC 2.3.1.95), followed by decarboxylation to yield 3,4′,5-trihydroxy-*trans*-stilbene (resveratrol) (a). On methylation in the presence of *trans*-resveratrol di-*O*-methyl-transferase (EC 2.1.1.240), (a) yields pterostilbene (b).

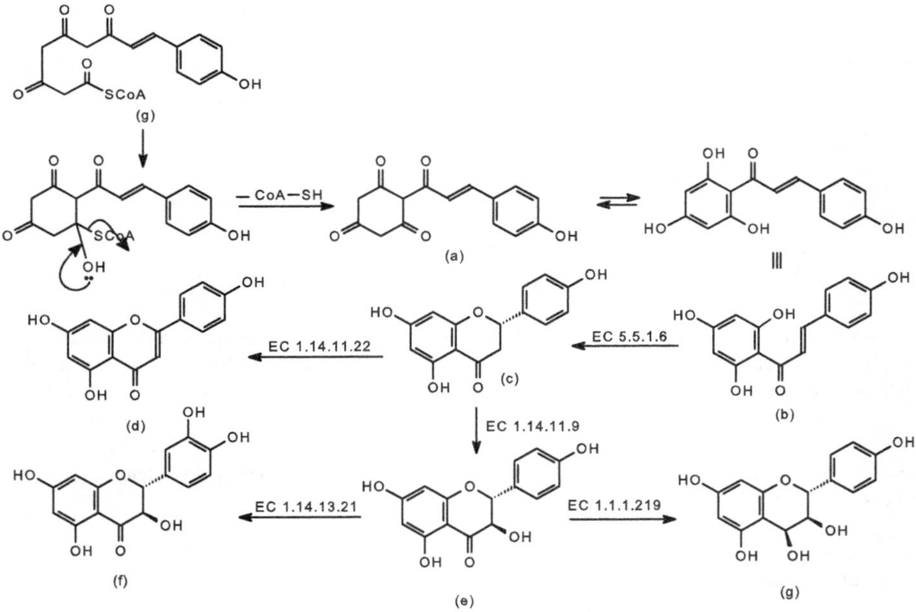

FIGURE A4.17 A representation of the formation of flavanols and related compounds as derived from the triketothioester (g) product (on the left) of the reactions described in Figure A4.15. Cyclization of (g) with loss of acetyl-CoA yields the 1,3,5-cyclohexanetrione (a) which is a tautomer of the 1,3,5-triphenol naringenin chalcone (b). Attack of a phenolic hydroxyl on the α,β-unsaturated ketone in the presence of the enzyme chalcone isomerase (EC 5.5.1.6) leads to the flavanone naringenin (c). In the presence of flavone synthase (EC 1.14.11.22) the flavone apigenin (d) is formed from the flavanone (c). Oxidation of the flavanone naringenin (c) with oxygen in the presence of flavone 3-dioxygenase (EC 1.14.11.9) generates aromadendrin (e), and from there both the dihydroflavonol taxifolin (f), formed by oxidation of the aromatic ring with the flavonoid 3′-monooxygenase (EC 1.14.13.21) as well as to the flavan-3,4-diol leucopelagonidin (g) with dihydrokaempferol 4-reductase (EC 1.1.1.219) result.

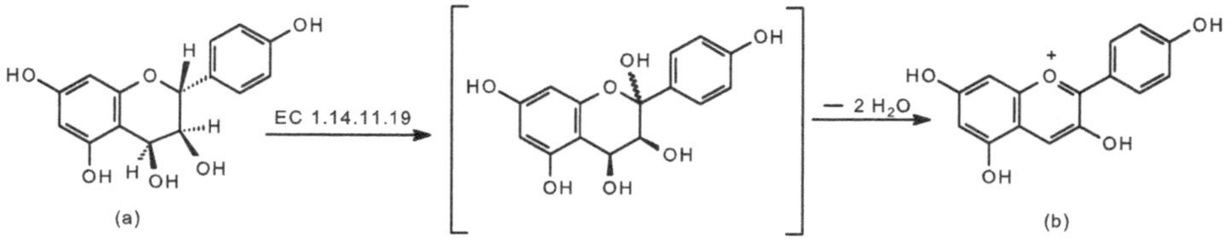

FIGURE A4.18 A cartoon representation of the oxidative conversion of leucopelagonidin (a flavan-3,4-diol) (a) from Figure A4.17 to pelargonidin (b), an anthocyanidin, in the presence of leucocyanidin oxygenase (EC 1.14.11.19). Interestingly, the enzyme requires iron [Fe(II)] and ascorbic acid (vitamin C) to effect the conversion.

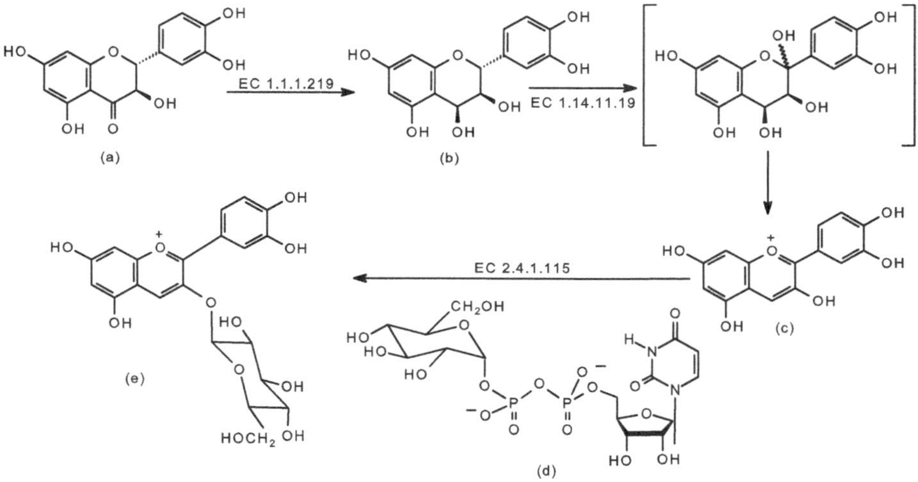

FIGURE A4.19 A cartoon representation of the oxidative conversion of taxifolin (dihydroflavonol) (a) from Figure A4.17 to leucocyanidin (b) in the presence of dihydrokaempferol 4-reductase (EC 1.1.1.219) which, presumably using the same leucocyanidin oxygenase (EC 1.14.11.19), is then converted to the anthocyanidin cyanidin (c). Transfer of glucose to (c) from uridine diphosphate glucose (UDP-glucose) (d) with anthocyanidin 3-O-glucosyltransferase (EC 2.4.1.115), a typical glycosylating agent, then generates the anthocyanin, cyanidin 3-O-β-D-glucoside (e).

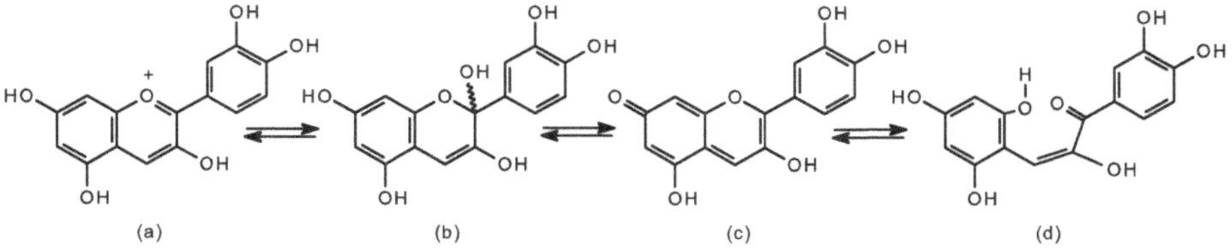

FIGURE A4.20 A representation of the color changes occurring when acidity (pH) changes in the solution in which the anthocyanidin, cyanidin, is dissolved. Reading from left to right, the flavylium cation (a) (gegenion unspecified but probably chloride, Cl^-) at a pH of 1–2 is red. The carbinol (b) which results as the pH is raised to between 4 and 5 is colorless. Carefully added base deprotonates the phenol, water is lost, and the quinoidal base (c) results at a pH between 6 and 6.5. The quinoidal base is blue. Finally, on the far right, as the pH becomes greater than pH = 7 and the solution becomes basic, the pale yellow chalcone (d) is found.

NOTES AND REFERENCES

1. Udenfriend, S.; Cooper, J. R. *J. Biol. Chem.* **1952,** *194,* 503.
2. Saito, K.; Komamine, A. *Eur. J. Biochem.* **1976,** *68,* 237.
3. Senoh, S.; Immoto, S.; Maeno, Y.; Yamashita, K.; Matsui, M.; Tokuyama, T.; Sakan, T.; Komamine, A.; Hattori, S. *Tet. Lett.* **1964,** *46,* 3437.
4. Terradas. F.; Wyler, H. *Helv. Chim. Acta* **1991,** *74,* 124.

Appendix 5

IMPACT ODORANTS

As reported by Polášková, P.; Herszage, J.; Ebeler, S. E. *Chem. Soc. Rev.* **2008,** *37,* 2478 there is a group of "impact odorants" that contribute to a given grape varietal. Table A5.1 provides information from that work. These odorants are detected as the *bouquet* (a distinctive, characteristic, and often subtle aroma) of the wine after it has been decanted into the glass.

It is also clear that aside from "impact odorants" there are other components, isolated individually and which, when mixed, produce a recognizable *bouquet*. A method called GC-O (which separates the compounds by gas chromatography [Appendix 1] and then uses experts to detect and classify the olfactory components) can be used for many vintages. These odorants, isolated from the thousand or so components in rich wines, are different from those "impact odorants" of Table A5.1 and are listed as an "olfactory component" in Table A5.2.

TABLE A5.1

Impact odorants (after Polášková, P.; Herszage, J.; Ebeler, S. E. Chem. Soc. Rev. **2008**, *37*, 2478).

Variety	Characteristic Odorant	Structure	Descriptor
Muscat	S-(+)-linalool, R-(-)-linalool, nerol, geraniol		Floral and Citrus
Riesling	1,1,6-trimethyl-1,2-dihydronaphthalene		kerosene
Cabernet Sauvignon, Sauvignon blanc, Cabernet franc, Merlot, Carmenere	3-isobutyl-2-methoxypyrazine and related pyrazines.		Bell pepper
Gewürztraminer	*cis*-tetrahydro-4-nethyl-2-(2-methylpropenyl)2H-pyran(*cis*-Rose oxide) *and* 3,6-dimethyl-3a,4,5,6a-tetrahydro-2H-1-benzofuran-2-one (wine lactone)		Geranium oil, carrot leaves and coconut, sweet
Sauvignon blanc, Scheurebe	4-methyl-4-mercaptopentan-2-one		Blackcurrant
Grenache rosé, Sauvignon blanc, Semillon	3-mercapto-1-hexanol *and* R-2-(4-methylcyclohex-3-enylpro-ane-2-thiol (grapefruit mercaptan) *and* (4R,4aS, 6R)-4,4a,5,6,7,8-hexahydronaphthl-ene-2(3H)one [(+)-nootkatone] *and* R-(+)-1-methyl-4-(1-methyl-ethenyl)-cyclohexane [(+)-limonene]		Grapefruit,/citrus peel (R-enantiomer of 3-mercapto-1-hexanol) *and* Passion fruit (S-enantiomer)
Shiraz	(3S,5R,8S)-5-isopropenyl-3,4-dimethyl-3,4,5,6,7,8-hexahydro-1(2H) azulene (rotundone)		Black pepper

TABLE A5.2

Some odorants detected using the GC-O technique (after Polášková, P.; Herszage, J.; Ebeler, S. E. Chem. Soc. Rev. **2008**, *37*, 2478).

Wine	Odorants	Structures
Gewürztraminer	cis-tetrahydro-4-nethyl-2-(2-methyl-propenyl)-2H-pyran (cis-Rose oxide) and 3,6-dimethyl-3a,4,5,6a-tetrahydro-2H-1-benzofuran-2-one (wine lactone) and ethyl-2-methylbutanoate and 3-methylbutanol and 2-phenylethanol and 3-ethylphenol and 3-hydroxy-4,5-dimethyl-2(5H)-furanone(sotolon)	cis - Rose oxide; wine lactone; ethyl 2-methylbutanoate; 3-methylbutanol; 2-phenylethanol; 3-ethylphenol; sotolon
Grenache rose	3-mercapto-1-hexanol and 4-hydroxy-2,5-dimethyl-3(2H)-furanone (furaneol) and 2-ethyl-4-hydroxy-5-methyl-3(2H)furanone (homofuraneol)	3-mercapto-1-hexanol; furaneol; homofuraneol
Chardonnay	ethyl butanoate and octanoic acid and 2-phenylacetaldehyde and 4-vinylphenol and δ-decalactone and 2-methyltetrahydrothiophen-3-one and 3-methylbutyl acetate and decanoic acid and 4-vinyl-2-methoxyphenol and 3,7-dimethylocta-1,6-dien-3-ol (linalool)	octanoic acid; ethyl butanoate; 4-vinylphenol; phenylacetaldehyde; δ-decalactone; 2-methyltetrahydro-thiophene-3-one; 3-methylbutyl acetate; 4-vinyl-2-methoxy-phenol; decanoic acid; S-(+)-linalool
Pinot noir	2-phenylethanol and 3-methyl-1-butanol and 2-methyl propanoate and ethyl butanoate and 3-methylbutyl acetate and ethyl hexanoate and benzaldehyde	2-phenylethanol; 3-methy-1-butanol; methyl 2-methylpropanoate; ethyl butanoate; 3-methylbutyl acetate; ethyl hexanoate; benzaldehyde
Cabernet Sauvignon and Merlot from Bordeaux	methylbutanols and 2-phenylethanol and 2-methyl-3-sulfanylfuran and acetic acid and 3-(methylsulfanyl)-propanal and methylbutanoic acids and (E)-1-(2,6,6-trimethyl-1-cyclo-hexa-1,3-dienyl)but-2-en-1-one (β-damascenone) and 3-sulfanylhexan-1-ol and 4-hydroxy-2,5-dimethyl-3(2H)-furanone (furaneol) and 2-ethyl-4-hydroxy-5-methyl-3(2H)-furanone (homofuraneol)	3-methylbutanoic acid; 3-methyl-1-butanol; 2-phenylethanol; 2-methyl-3-sulfanylfuran (2-methyl-3-furanthiol); acetic acid; 3-methylsulfanylpropanal [3-(methythio)propionaldehyde]; β-damascenone; 3-mercapto-1-hexanol; furaneol; homofuraneol
Madera	3-hydroxy-4,5-dimethyl-2(5H)-furanone(sotolon) and 2-phenylacetaldehyde and (4S,5S)-5-butyl-4-methyldihydrofuran-2(3H)-one (cis-whisky lactone)	sotolon; phenylacetaldehyde; cis-whisky lactone

(continued)

TABLE A5.2

Continued

Cabernet Sauvignon and Merlot from USA and Australia	3-methyl-1-butanol *and* 3-hydroxy-2-butanone *and* octanal *and* ethyl hexanoate *and* ethyl 2-methylbutanoate *and* (E)-1-(2,6,6-trimethyl-1-cyclo-hexa-1,3-dienyl)but-2-en-1-one (β-damascenone) *and* 2-methoxyphenol *and* 4-vinyl-2-methoxyphenol *and* ethyl 3-methylbutanoate *and* acetic acid *and* 2-phenylethanol	3-methyl-1-butanol, 3-hydroxy-2-butanone, octanal, ethyl hexanoate, β-damascenone, 2-methoxyphenol, 4-vinyl-2-methoxyphenol, acetic acid, ethyl 2-methylbutanoate, ethyl 3-methylbutanoate, 2-phenylethanol
Riesling (from USA)	(E)-1-(2,6,6-trimethyl-1-cyclo-hexa-1,3-dienyl)but-2-en-1-one (β-damascenone) *and* 2-phenylethanol *and* 3,7-dimethylocta-1,6-dien-3-ol (linalool) *and* ethyl 2-methylbutanoate *and* E-2-hexen-1-ol *and* Z-3-hexen-1-ol *and* E-3,7-dimethyl-2,6-octadien-1-ol (geraniol) *and* ethyl butanoate *and* S-(+)-2-methyl-5-(1-methyl-ethenyl)-2-cyclohexenone (carvone) *and* ethyl hexanoate *and* isoamyl acetate *and* "fatty acids" *e.g.*, C_{14}, C_{16}, C_{18} acids.	β-damascenone, 2-phenylethanol, S-(+)-linalool, ethyl 2-methylbutanoate, E-2-hexene-1-ol, Z-3-hexene-1-ol, geraniol, ethyl butanoate, carvone, ethyl hexanoate, isoamyl acetate

GLOSSARY

abscisic acid	(2Z,4E)-5-[(1S)-1-hydroxy-2,6,6-trimethyl-4-oxo-2-cyclohexen-1-yl]-3-methyl-2,4-pentadienoic acid	11,12,20,73,83,108,137,141,154, 209,210
acetaldehyde	ethanal	138,167,188,194,198,208,217,234, 252,265,278,282,300,343,355
acetaldehyde diethyl acetal	1,1-diethoxyethane	208,290
acetamide, N-(3-ethylphenyl)	N-(3-ethylphenyl)acetamide	287
acetamide, N-(3-methylbutyl)	N-(3-methylbutyl)acetamide	287
acetamide, N-(3-methylthiopropyl)	N-(3-methylthiopropyl)acetamide	262
acetic acid	ethanoic acid	58,61,196,205,217,234,252,261, 300,327,355
acetoin	3-hydroxy-2-butanone	214,217
acetolactic acid	2-hydroxy-2-methyl-3-oxobutanoic acid	214
acetone	propanone	217,343
acetosyringone	1-(4-hydroxy-3,5-dimethoxyphenyl)ethanone	167
acetovanillone	1-(4-hydroxy-3-methoxyphenyl)ethanone	159,211,271

Glossary

acetyl-1H-pyrrole, 2-	1-(1H-pyrrol-2-yl)ethanone	247
acetyl coenzyme A	S-[2-[3-[[[(2R)-4-[[[(2R,3S,4R,5R)-5-(6-aminopurin-9-yl)-4-hydroxy-3-phosphono-oxyoxolan-2-yl]methoxyhydroxy-phosphoryl]-oxy-hydroxyphosphoryl]-oxy-2-hydroxy-3,3-dimethylbutanoyl]amino]-propanoylamino]ethyl]ethanethioate	61,137,194,206,300,361,362
acetylfuran, 2-	1-(furan-2-yl)ethanone	218,247
acetyltetrahydropyridine, 2-	1-(1,2,3,4-tetrahydropyridin-6-yl)ethanone	217
aconitic acid (cis)	(1Z)-1-propene-1,2,3-tricarboxylic acid	96
actinidols	2,4,7,7a-tetrahydro-α,4,4,7a-tetramethyl-2-benzofuranmethanol	164,168,220,221
adenine (A)	6-amino-1H-purine	12,355
adenosine diphosphate	[(2R,3S,4R,5R)-5-(6-amino-9H-purin-9-yl)-3,4-dihydroxytetrahydro-2-furanyl]methyl trihydrogen diphosphate	23,24,40,41,45,52,55,56,62,94,197,360,364
adenosine monophosphate, cyclic	(4aR,6R,7R,7aS)-6-(6-amino-9H-purin-9-yl)tetrahydro-4H-furo[3,2-d][1,3,2]dioxaphosphinine-2,7-diol 2-oxide	301
adenosine triphosphate	[(2R,3S,4R,5R)-5-(6-amino-9H-purin-9-yl)-3,4-dihydroxytetrahydro-2-furanyl]methyl trihydrogen triphosphate	39,23,40,44,45,46,52,5,56,62,94,360,367
aesculetin	6,7-dihydroxycoumarin	125
alanine (Ala, A)	2-aminopropanoic acid	28,48,49,50,247,330,337
alloaromadendrene	(1aR,4aS,7R,7aR,7bS)-1,1,7-trimethyl-4-methylidene-2,3,4a,5,6,7,7a,7b-octahydro-1aH-cyclopropa[e]azulene	275,276
aloin	(10S)-10-glucopyranosyl-1,8-dihydroxy-3-(hydroxymethyl)-9(10H)-anthracenone	37
amorphene, α-	4,7-dimethyl-1-propan-2-yl-1,2,4a,5,6,8a-hexahydronaphthalene	275,277
ampelopsin-D	(1E,2R,3R)-3-(3,5-dihydroxyphenyl)-2-(4-hydroxyphenyl)-1-[(4-hydroxyphenyl)-methylidene]-2,3-dihydroindene-4,6-diol	153

ampelopsin-H	(1*R*,2*R*,6*R*,6a*R*,7*R*,8*R*,12*R*,12a*R*)-*rel*-(+)-1,7-bis(3,5-dihydroxyphenyl)-1,2,6,6a,7,8,12,12a-octahydro-2,6,8,12-tetrakis(4-hydroxyphenyl)-pentaleno[1,2-*e*:4,5-*e'*]bisbenzofuran-5,11-diol	153
anethole	1-methoxy-4-[(*E*)-prop-1-enyl]benzene	206
anhydrolinalool oxide, *cis*-	(2*S*,5*R*)-2-ethenyltetrahydro-2-methyl-5-(1-methylethenyl)furan	219
anhydrolinalool oxide, *trans*-	(2*S*,5*S*)-2-ethenyltetrahydro-2-methyl-5-(1-methylethenyl)furan	219
anorthite	$CaAl_2Si_2O_8$	158
apigenin	5,7-dihydroxy-2-(4-hydroxyphenyl)-4*H*-1-benzopyran-4-one	225,391,393
apricolin	5-pentyloxolan-2-one	210,211
apocynin	1-(4-hydroxy-3-methoxyphenyl)ethanone	211
arabonic acid lactone	arabonic acid lactone	238
arabinofuranosyl-β-D-glucoside, α-L-	α-L-arabinofuranosyl-β-D-glucoside	209
arabinose	(2*S*,3*R*,4*R*)-2,3,4,5-tetrahydroxypentanal	253,277
arachidonic acid	(5*Z*,8*Z*,11*Z*,14*Z*)-icosa-5,8,11,14-tetraenoic acid	32
arginine (Arg, R)	(2*S*)-2-amino-5-(diaminomethylideneamino)-pentanoic acid	206,276
aromadendrene	(1*aR*,4a*R*,7*R*,7a*R*,7b*S*)-1,1,7-trimethyl-4-methylidene-2,3,4a,5,6,7,7a,7b-octahydro-1*aH*-cyclopropa[e]azulene	275,276
ascorbic acid	(5*R*)-5-[(1*S*)-1,2-dihydroxyethyl]-3,4-dihydroxyfuran-2(5*H*)-one	91,92,97,99,100,101,104,180,234,394
asparagine (Asn, N)	(2*S*)-2,4-diamino-4-oxobutanoic acid	276,278,333
aspartic acid (Asp, D)	2-aminobutanedioic acid	48,49,358
apiofuranosyl-β-D-glucoside, β-D-	β-D-apiofuranosyl-β-D-glucoside	209
astaxanthin	(6*S*)-6-hydroxy-3-[(1*E*,3*E*,5*E*,7*E*,9*E*,11*E*,13*E*,15*E*,17*E*)-18-[(4*S*)-4-hydroxy-2,6,6-trimethyl-3-oxocyclohexen-1-yl]-3,7,12,16-tetramethyloctadeca-1,3,5,7,9,11,13,15,17-nonaenyl]-2,4,4-trimethylcyclohex-2-en-1-one	203

astilbin	(2*R*,3*R*)-3-[(6-deoxy-α-L-mannopyranosyl)-oxy]-2-(3,4-dihydroxyphenyl)-2,3-dihydro-5,7-dihydroxy-4*H*-1-benzopyran-4-one	130,132
aurantinidin	3,5,6,7-tetrahydroxy-2-(4-hydroxyphenyl)-1-benzopyrylium ion	78,145
bentonite clay	cation interspersed silicate	231,232,233
benzaldehyde	benzencarboxyaldehyde	127,140,155,208,217
benzoic acid, 2-methyl	2-methylbenzoic acid	238
benzothiazol	benzothioazole	261,262
benzothiazol-2-oxyacetic acid	2-(2-benzothiazolyloxy)-acetic acid	83
benzyl acetate	benzyl acetate	134,376
benzyl alcohol	benzyl alcohol	127,144,155,156,268
benzyl benzoate	benzyl benzoate	70
betanin	(1*E*,2*S*)-2-carboxy-1-[(2*E*)-2-[(2*S*)-2,6-dicarboxy-2,3-dihydro-4(1*H*)-pyridinylidene]ethylidene]-2,3-dihydro-6-hydroxy-1*H*-indolium-5-yl-β-D-glucopyranoside inner salt	77,384,387,390
biochanin A	5,7-dihydroxy-3-(4-methoxyphenyl)chromen-4-one	225
bicyclogermacrene	(4*E*,8*E*)-4,8,11,11-tetramethylbicyclo-[8.1.0]undeca-4,8-diene	275,277
biotin	5-[(3*aS*,4*S*,6*aR*)-2-oxo-1,3,3a,4,6,6a-hexahydrothieno[3,4-d]imidazol-4-yl]pentanoic acid	93,216,217,225,226,350,392
bisabolene, α	1-methyl-4-[(2*Z*)-6-methylhepta-2,5-dien-2-yl]cyclohexene	275,277
bisabolene, β	(4*S*)-1-methyl-4-(6-methyl-1,5-heptadien-2-yl)cyclohexene	70,371
bisphosphoglycerate	2-hydroxy-1,3-bis(phosphonooxy)-1-propanone	59,62
bisulfite anion	bisulfite anion	113,162,183,185,186,187,188,189
blumenol A	(4*S*)-4-hydroxy-4-[(*E*,3*R*)-3-hydroxybut-1-enyl]-3,5,5-trimethylcyclohex-2-en-1-one	129,209
blumenol B	(4*S*)-4-hydroxy-4-[(3*R*)-3-hydroxybutyl]-3,5,5-trimethylcyclohex-2-en-1-one	129
blumenol C	4-(3-hydroxybutyl)-3,5,5-trimethylcyclohex-2-en-1-one	129

botcinic acid	(2R,3S)-3-hydroxy-3-[(2S,3S,4S,5R,6S)-3-hydroxy-5-[(E,4S)-4-hydroxyoct-2-enoyl]oxy-2,4,6-trimethyloxan-2-yl]-2-methylpropanoic acid	280
boctinolide	[(2S,3R,4S,5S,6R,7R,8R)-5,6,7-trihydroxy-2,4,6,8-tetramethyl-9-oxooxonan-3-yl] (E)-4-hydroxyoct-2-enoate	280
botrydial	[(1S,3aR,4S,6R,7S,7aS)-1,7-diformyl-7a-hydroxy-1,3,3,6-tetramethyl-2,3a,4,5,6,7-hexahydroinden-4-yl] acetate	280
Bourbonene, α-	1-(propan-2-yl)-1,2,3,3a,3b,4,6a,6b-octahydrocyclobuta[1,2-a:3,4-a']dicyclopentene	166
bourbonene, β-	[1S-(1a,3aa,3bb,6ab,6ba)]-decahydro-3a-methyl-6-methylene-1-(1-methylethyl)cyclobuta-[1,4]dicyclopentene	275,276
brassinolide	(3aS,5S,6R,7aR,7bS,9aS,10R,12aS,12bS)-10-[(2S,3R,4R,5S)-3,4-dihydroxy-5,6-dimethyl-2-heptanyl]-5,6-dihydroxy-7a,9a-dimethylhexadecahydro-3H-benzo[c]indeno[5,4-e]oxepin-3-one	210
brassinolide, epi-	(3aS,5S,6R,7aR,7bS,9aS,10R,12aS,12bS10-[(1S,2R,3R,4R)-2,3-dihydroxy-1,4,5-trimethylhexyl]hexadecahydro-5,6-dihydroxy-7a,9a-dimethyl)-3H-benz[c]indeno[5,4-e]oxepin-3-one	109,110
but-1,3-diene, (E)-1-(2,3,6-trimethylpheny)-	(E)-1-(2,3,6-trimethylphenyl)-but-1,3-diene	325,326,338
butanal	butanal	217,325,326
butanediol, 1,3-	1,3-butanediol	208
butanediol, 2,3	2,3-butanediol	208,214
butanediol, 2,3-(2R,3R)-	(2R,3R)-2,3-butanediol	138
butanediol, 2,3-(2R,3S)-	(2R,3S)-2,3-butanediol	138
butanediol, 2,3-(2S,3S)-	(2S,3S)-2,3-butanediol	138
butanedione, 2,3-	2,3-butanedione	271
butanoic acid	butanoic acid	11,139,205
butanol, 1-	1-butanol	273,282,323,324,342,348

butanol, 2-	2-butanol	323,325
butanol, 4-(methylthio)-1-	4-(methylthio)-1-butanol	261,262
butanone, 1-(2,3,6-trimethylphenyl)-2-	1-(2,3,6-trimethylphenyl)-2-butanone	159,279
butanethioic acid	S-butyl butanethioate	217
buten-2-one, 1-(2,3,6-trimethylphenyl)-3-	1-(2,3,6-tri-methylphenyl)-3-buten-2-one	159,325
butyl acetate	butyl acetate	217
butyrolactone, γ-	dihydrofuran-2(3H)-one	219,261
cadalene	1,6-dimethyl-4-propan-2-ylnaphthalene	277
cadaverine	pentane-1,5-diamine	278
cadinene, δ-	(1S,8aR)-4,7-dimethyl-1-propan-2-yl-1,2,3,5,6,8a-hexahydronaphthalene	166,275,277
cadinene, γ-	(1S,4aR,8aR)-1-Isopropyl-7-methyl-4-methylene-1,2,3,4,4a,5,6,8a-octahydro-naphthalene	70,166,275
caffeic acid	(2E)-3-(3,4-dihydroxyphenyl)-2-propenoic acid	104,107,114,135,169,223,240,260, 265,268,269,270,389,391
caftaric acid, cis-	(2R,3R)-2-[(Z)-3-(3,4-dihydroxyphenyl)prop-2-enoyl] oxy-3-hydroxybutanedioic acid	126,130
caftaric acid, trans-	(2R,3R)-2-[(E)-3-(3,4-dihydroxyphenyl)prop-2-enoyl] oxy-3-hydroxybutanedioic acid	126,130,132,169,180,269,270
calacorene, α-	(1S)-4,7-dimethyl-1-propan-2-yl-1,2-dihydronaphthalene	277
calamenene	1,2,3,4-tetrahydro-1,6-dimethyl-4-(1-methylethyl)-naphthalene	128,156,275,277
calarene	(1aR,7R,7aR,7bS)-1,1,7,7a-tetramethyl-2,3,5,6,7,7b-hexahydro-1aH-cyclopropa[a]naphthalene	166
carbon dioxide	carbon dioxide	26,36,55,58,60,61,63,93,180,182, 183,185,186,189,190,194,198,206, 214,228,288,289,326,356,392
carotenal, 10′-apo-β	(2E,4E,6E,8E,10E,12E,14E)-4,9,13-trimethyl-15-(2,6,6-trimethylcyclohexen-1-yl)pentadeca-2,4,6,8,10,12,14-heptaenal	150

Glossary

carotene, -β	1,1'-[(1E,3E,5E,7E,9E,11E,13E,15E,17E)-3,7,12,16-tetramethyl-1,3,5,7,9,11,13,15,17-octadecanonaene-1,18-diyl]bis(2,6,6-trimethylcyclohexene)	47,76,77,106,116,128,146,148,149, 172,203,263,339,383
carvomenthenol, 4-	4-methyl-1-propan-2-ylcyclohex-3-en-1-ol	275
carvone	2-methyl-5-prop-1-en-2-ylcyclohex-2-en-1-one	275
caryophyllene, -β	(1R,4E,9S)-4,11,11-trimethyl-8-methylenebicyclo(7.2.0)undec-4-ene	70,128,156,158,160,166,275,276, 277,372
caryophyllene oxide, -β	(1R,4R,6R,10S)-4,12,12-trimethyl-9-methylene-5-oxatricyclo[8.2.0.0^{4,6}]dodecane	70
castalagin	(1R,2R,20R,42S,46R)-7,8,9,12,13,14,25,26,27,30,31,32,35,36,37,46-hexadecahydroxy-3,18,21,41,43-pentaoxanonacyclo[27.13.3.1^{38,42}.0^{2,20}.0^{5,10}.0^{11,16}.0^{23,28}.0^{33,45}.0^{34,39}]hexatetraconta-5,7,9,11,13,15,23,25,27,29(45),30,32,34,36,38-pentadecaene-4,17,22,40,44-pentone	235,239,242
catechin, D-(+)-	(2R,3S)-2-(3,4-dihydroxyphenyl)-3,4-dihydro-2H-1-benzopyran-3,5,7-triol	125,131,169,223,241,264,266
cellulose	glucose polymer	7,23,27,63,66,116,179,235,236, 237,245
cerin	(2R,4R,4aS,6aS,6bR,8aR,12aR,12bS,14aS,14bS)-2-hydroxy-4,4a,6b,8a,11,11,12b,14a-octamethylicosahydro-3(2H)-picenone	254,255
chlorogenic acid	(1S,3R,4R,5R)-3-[(E)-3-(3,4-dihydroxyphenyl)prop-2-enoyl]oxy-1,4,5-trihydroxycyclohexane-1-carboxylic acid	107,269,270
chlorophyll-a	magnesium (3S,4S,21R)-14-ethyl-13-formyl-21-(methoxycarbonyl)-4,8,18-trimethyl-20-oxo-3-(3-oxo-3-{[(2E,7R,11R)-3,7,11,15-tetramethyl-2-hexadecen-1-yl]oxy}propyl)-9-vinyl-23,25-didehydrophorbine-23,25-diide	42,47,51,91,128,160,340

cholesterol	2,15-dimethyl-14-(1,5-dimethylhexyl)-tetracyclo[8.7.0.0^{2,7}.0^{11,15}]heptadec-7-en-5-ol	29,31,110
choline	N,N,N-trimethylethylethanolamine	29,42,330
chorismic acid	(3R,4R)-3-[(1-carboxyethenyl)oxy]-4-hydroxy-1,5-cyclohexadiene-1-carboxylic acid	102,103,386,387,388
chroman	3,4-dihydro-2H-chromene	221,222
chromanone, 4-	2,3-dihydrochromen-4-one	221,222
chromene, 2H-	2H-chromene	221,222
chromene, 4H-	4H-chromene	221,222
chromone	chromen-4-one	221,222
chrysanthemin	2-(3,4-dihydroxyphenyl)-5,7-dihydroxy-3-{[(2S,3R,4S,5S,6R)-3,4,5-trihydroxy-6-(hydroxymethyl)tetrahydro-2H-pyran-2-yl]oxy}chromenium ion	78,79
cineol, 2-hydroxy-1,8-	2,2,4-trimethyl-3-oxabicyclo[2.2.2]octan-6-ol	163
cinnamic acid	(E)-3-phenylpropenoic acid	28,107,113,123,237,240,265,268,389,391
cinnamyl isovalerate	[(E)-3-phenylprop-2-enyl] 3-methylbutanoate	154
citral	(2E)-3,7-dimethylocta-2,6-dienal	130,131
citronellol, (−)	(−)-3,7-dimethyloct-6-en-2-ol	125,125,134,274
citronellol, (+)	(+)-3,7-dimethyloct-6-en-2-ol	134,137,274,290
citric acid	2-hydroxy-1,2,3-propanetricarboxylic acid	93,113,183,206,207,216,307,321
coniferyl alcohol	4-[(E)-3-hydroxyprop-1-enyl]-2-methoxyphenol	237,238
copaene, (−)-α-	(1R,2S,6S,7S,8S)-8-isopropyl-1,3-dimethyl-tricyclo[4.4.0.0^{2,7}]dec-3-ene	128,165,166
coumaric acid, *cis*-	(Z)-3-(4-hydroxyphenyl)-2-propenoic acid	103,269
coumaric acid, *trans*-	(E)-3-(4-hydroxyphenyl)-2-propenoic acid	91,101,102,103,107,114,159,237,240,264,265,268,270,389,391
coumaryl alcohol	4-[(E)-3-hydroxyprop-1-enyl]phenol	237,238
coumesterol	3,9-dihydroxy-[1]benzofuro[3,2-c]chromen-6-one	225

Glossary

coutaric acid, *cis*	(2*R*,3*R*)-2-hydroxy-3-[[(2*E*)-3-(4-hydroxy-phenyl)-1-oxo-2-propen-1-yl]oxy]-butane-dioic acid	126,269,270
coutaric acid, *trans*	(2*R*,3*S*)-2-hydroxy-3-[[(2*E*)-3-(4-hydroxy-phenyl)-1-oxo-2-propen-1-yl]oxy]butanedioic acid	126,269,270
cresol, *ortho*	2-methylphenol	285
cresol, *para*	4-methylphenol	271,285
creosol, *para*	2-methoxy-4-methylphenol	247
cubebene, (−)-α-	(3*aS*,3*bR*,4*S*,7*R*,7*aR*)-3a,3b,4,5,6,7-hexahydro-3,7-dimethyl-4-(1-methylethyl)-1*H*-cyclopenta[1,3]cyclopropa[1,2]benzene	128,275,276
cubebene, -(−)-β-	(3*aS*-(3aa,3bb,4*S*,7*R*,7*aS*))-octahydro-7-methyl-3-methylene-4-(1-methylethyl)-1*H*-cyclopenta(1,3)cyclopropa(1,2)benzene	275,276
cumin aldehyde	4-propan-2-ylbenzaldehyde	275
curcumene	1-methyl-4-(6-methylhept-5-en-2-yl)benzene	206
cyanidin	2-(3,4-dihydroxyphenyl)-3,5,7-trihydroxy-1-benzopyrylium ion	77,78,79,106,145,150,151,153,211,212,221,232,241,392
cyanidin-3-*O*-glucoside	(3*R*,4*S*,5*S*,6*R*)-2-[2-(3,4-dihydroxyphenyl)-5,7-dihydroxychromenylium-3-yl]oxy-6-(hydroxymethyl)oxane-3,4,5-triol	394
cyclocitral, β-	trimethylcyclohexene-1-carbaldehyde	275,278
cyclohexane, 1,2,3,5-tetrahydroxy	1,2,3,5-tetrahydroxycyclohexane	238
cyclotene	2-hydroxy-3-methyl-2-cyclopenten-1-one	245
cymene, *para*	1-methyl-4-propan-2-ylbenzene	140,157,275
cymenene, *para*	1-methyl-4-prop-1-en-2-ylbenzene	275
cysteine (Cys, C)	2-amino-3-sulfanylpropanoic acid	51,52,196,204,227,283,361, 363,364
cysteine, (*S*)-3-(1-heptanol)-L-	*S*-[(1*S*)-1-(2-hydroxyethyl)pentyl]-L-cysteine	135
cysteine, (*S*)-3-(1-pentanol)-L-	*S*-[(1*S*)-1-ethyl-3-hydroxypropyl]-L-cysteine	135
cysteine, (*S*)-3-(2-methyl-1-butanol)-L-	*S*-(3-hydroxy-1,2-dimethylpropyl)-L-cysteine	135
cytosine C	6-amino-2(1*H*)-pyrimidinone	12,189,190,226,351,355
cytosine sulfonate	6-amino-2(1*H*)-pyrimidinone-4-sulfonate	190

damsacenone, β-	(2*E*)-1-(2,6,6-trimethyl-1,3-cyclohexadien-1-yl)-2-buten-1-one	135,154,159,164,168,206,211,220, 262,274,279,283
damsacenone, 3-hydroxy-β-	(*E*)-1-(3-hydroxy-2,6,6-trimethylcyclohexen-1-yl)but-2-en-1-one	373
decadienal, (2*E*,4*E*)-2,4-	(2*E*,4*E*)-2,4-decadienal	130
decalactone, γ-	5-hexyloxolan-2-one	279
decalactone, δ-	6-pentyltetrahydro-2*H*-pyran-2-one	279,398
decanoic acid	decanoic acid	139,238,271,282
dehydroascorbic acid	(5*R*)-5-[(1*S*)-1,2-dihydroxyethyl]oxolane-2,3,4-trione	180,181,234
dehydrodiconiferyl alcohol	4-[3-(hydroxymethyl)-5-[(*E*)-3-hydroxyprop-1-enyl]-7-methoxy-2,3-dihydro-1-benzofuran-2-yl]-2-methoxyphenol	241
dehydroshikimic acid	3,4-dihydroxy-5-oxo-1-cyclohexene-1-carboxylic acid	101
delphinidin	3,5,7-trihydroxy-2-(3,4,5-trihydroxyphenyl)-1-benzopyrylium ion	115,239,241,243
delphinidin-3-*O*-glucoside	(2*S*,3*R*,4*S*,5*S*,6*R*)-2-[5,7-dihydroxy-2-(3,4,5-trihydroxyphenyl)chromenylium-3-yl]oxy-6-(hydroxymethyl)oxane-3,4,5-triol	263
dendranthemoside A	(1*S*,4*S*,5*R*)-4-hydroxy-4-[(1*E*,3*R*)-3-hydroxy-1-buten-1-yl]-3,3,5-trimethyl-cyclohexyl-β-D-glucopyranoside	164
diacetyl	2,3-butanedione	205,213,214,215,216,217
diadzein	7-hydroxy-3-(4-hydroxyphenyl)chromen-4-one	224
diadzein-7-*O*-glucoside	3-(4-hydroxyphenyl)-7-[(2*S*,3*R*,4*S*,5*S*,6*R*)-3,4,5-trihydroxy-6-(hydroxymethyl)oxan-2-yl]oxychromen-4-one	223
dichlofluanid*	*N*-{[dichloro(fluoro)methyl]sulfanyl}-*N'*,*N'*-dimethyl-*N*-phenylsulfuric diamide	166,169
dichlorophenoxyacetic acid, 2,4-	2-(2,4-dichlorophenoxy)acetic acid	106
didehydroshikimic acid	(4*S*)-4-hydroxy-3,5-dioxocyclohexene-1-carboxylate	103

diendiol-1	(3*E*)-2,6-dimethyl-3,7-octadiene-2,6-diol	137
diethoxyethane, 1,1-	acetaldehyde diethyl acetal	271,290
diethyl-3-methylene-1-oxetan-2-one, 4,4-	4,4-diethyl-3-methylene-1-oxetan-2-one	138
diethyl adipate	diethyl hexanedioate	140
diethyl disulfide	diethyl disulfide	218
diethyl glutarate	diethyl pentanedioate	140
diethyl hydroxyglutarate, 2-	diethyl 2-hydroxypentanedioate	284
diethyl malate	diethyl 2-hydroxybutanedioate	140,284
diethyl pentanedioate	diethyl pentanedioate	284
diethyl pimelate	diethyl heptanedioate	140
diethyl subarate	diethyl octanedioate	140
diethyl succinate	diethyl butanedioate	140,144,217,284
diethyl sulfide	diethyl sulfide	218
dihydroactinidiolide.	(7*aS*)-4,4,7a-trimethyl-6,7-dihydro-5*H*-1-benzofuran-2-one	129
dihydromyrcenol	2,6-dimethyloct-7-en-2-ol	275
dihydronaphthalene, 1,1,6-trimethyl-1,2-	1,1,6-trimethyl-1,2-dihydronaphthalene	159,164,219,220,261,262,275,397
dihydronaphthalene, 1,5,8-trimethyl-1,2-	1,5,8-trimethyl-1,2-dihydronaphthalene	159
dihydrophaseic acid	(2*Z*,4*E*)-5-[(1*R*,3*S*,5*R*,8*S*)-3,8-dihydroxy-1,5-dimethyl-6-oxabicyclo[3.2.1]oct-8-yl]-3-methyl-2,4-pentadienoic acid	137,138,141
dihydroxyacetone	1,3-dihydroxy-2-propanone	61,62,63,94,195
dihydroxyacetone monophosphate	1,3-dihydroxy-2-propanone-1-phosphate	62,63,66,93,94
dihydroxybutane, 2,3-	2,3-dihydroxybutane	217
dihydroxycoumarin, 6,7-	6,7-dihydroxycoumarin	125
dimethoxyphenol, 4-allyl-2,6-	4-allyl-2,6-dimethoxyphenol	167
dimethylallyl diphosphate	(3-methyl-2-buten-1-yl) diphosphoric acid monoester	87,109,110,148,201,202,203,219,359,361,366,369
dimethylbenzaldehyde 1,3-	1,3-dimethylbenzene-carboxyaldehyde	272
dimethyl-3-ethylpyrazine, 2,5-	2,5-dimethyl-3-ethylpyrazine	248,249
dimethyl-1,4-octadiene-3,7-diol, (*Z*)-3,7-	(*Z*)-3,7-dimethyl-1,4-octadiene-3,7-diol	151
dimethyl-1,5-octadiene-3,7-diol, (*E*)-3,7-	(*E*)-3,7-dimethyl-1,5-octadiene-3,7-diol	165

dimethyl-2,7-octadiene-1,6-diol, (E)-2,6-	(E)-2,6-dimethyl-2,7-octadien-1,6-diol	165
dimethyl-2,7-octadiene-1,6-diol, (Z)-2,6-	(Z)-2,6-dimethyl-2,7-octadien-1,6-diol	151
dimethyl-3,7-octadiene-2,6-diol, (E)-2,6-	(E)-2,6-dimethyl-3,7-octadiene-2,6-diol	151
dimethyl-1,7-octanediol, 3,7-	3,7-dimethyl-1,7-octanediol	272
dimethyl disulfide	dimethyl disulfide	205,218
dimethylpyrazine, 2,5-	2,5-dimethylpyrazine	248
dimethylpyrazine, 2,6-	2,6-dimethylpyrazine	205
dimethyl sulfide	dimethyl sulfide	218,261
dimethyl sulfone	dimethyl sulfone	261
diphosphoglycerate	(2R)-1,3-bis(phosphonooxy)-propanoic acid	94
disulfide, bis-(2-hydroxydiethyl)	bis-(2-hydroxydiethyl)disulfide	261
dodecanol, 1-	1-dodecanol	144
edulan, (5S,9R)-3,4-dihydro-3-oxo-	(2S-trans)-2,3,5,6,8,8a-hexahydro-2,5,5,8a-tetramethyl-7H-1-benzopyran-7-one	129
egosterol	(3S,9S,10R,13R,14R,17R)-17-[(E,2R,5R)-5,6-dimethylhept-3-en-2-yl]-10,13-dimethyl-2,3,4,9,11,12,14,15,16,17-decahydro-1H-cyclopenta[a]phenanthren-3-ol	203
ellagic acid	6,7,13,14-tetrahydroxy-2,9-dioxatetracyclo[6.6.2.0.^{4,16}0^{11,15}] hexadeca-1(15),4,6,8(16),11,13-hexaene-3,10-dione	169,244,245
enterodiol	(2S,3S)-2,3-bis[(3-hydroxyphenyl)methyl] butane-1,4-diol	224
enterolactone	(3S,4S)-3,4-bis[(3-hydroxyphenyl)-methyl]oxolan-2-one	224
epi-bicyclosesqiphellandrene	(1R,4S)-4-methyl-7-methylidene-1-propan-2-yl-2,3,4,4a,5,6-hexahydro-1H-naphthalene	275
epicatechin, (−)	(2R,3R)-2-(3,4-dihydroxyphenyl)-3,4-dihydro-2H-chromene-3,5,7-triol	125,126,131,132,161,169,223,261, 264,266
epicatechin gallate	[(2R,3R)-2-(3,4-dihydroxyphenyl)-5,7-dihydroxy-3,4-dihydro-2H-chromen-3-yl] -3,4,5-trihydroxybenzoate	124,131,132
epizonarene	(1S,8aR)-1,6-dimethyl-4-propan-2-yl-1,2,3,7,8,8a-hexahydronaphthalene	275,277

equol	(3S)-3-(4-hydroxyphenyl)-3,4-dihydro-2H-chromen-7-ol	225
ergosterol	(3S,9S,10R,13R,14R,17R)-17-[(E,2R,5R)-5,6-dimethylhept-3-en-2-yl]-10,13-dimethyl-2,3,4,9,11,12,14,15,16,17-decahydro-1H-cyclopenta[a]phenanthren-3-ol	201,202,286,287
eriodictyol	(2S)-2-(3,4-dihydroxyphenyl)-5,7-dihydroxy-4-chromanone	224
erythrose	2,3,4-trihydroxybutanal	63,103,383,384,387
estragol	1-methoxy-4-prop-2-enylbenzene	206
ethanol	ethanol	21,29,32,55,93,112,135,142,177,182,183,192,194,199,207,216,217,237,260
ethanol, 1-(4-methyl-1,3-thiazol-5-yl)	1-(4-methyl-1,3-thiazol-5-yl)ethanol	205
ethanol, 2-(methylthio)	2-methylthioethanol	261,262
ethanolamine	2-aminoethanol	32
ethyl acetate	ethyl ethanoate	144,208,217,282,290,327
ethyl benzoate	ethyl benzoate	140,144
ethyl butanoate	ethyl butanoate	139,144,208,217,271,282,290,298
ethyl butenoate, (E)-2-	ethyl (E)-2-butenoate	139
ethyl carbamate	ethyl carbamate	207
ethyl cinnamate	ethyl (E)-3-phenylprop-2-enoate	210,211,282, 285,377
ethyl coumarate	(2E)-3-(4-hydroxyphenyl)-2-propenoic acid ethyl ester	285
ethyl decanoate	ethyl decanoate	144,208,217,282
ethyl decenoate, 9-	ethyl 9-decenoate	284
ethyl 3,5-dimethyl-3-pyrazinylacetate	ethyl 3,5-dimethyl-3-pyrazinylacetate	248
ethyl dodecanoate	ethyl dodecanoate	144,217
ethyl furan-2-carboxylate	ethyl furan-2-carboxylate	218
ethyl geranate	ethyl (2E)-3,7-dimethyl-2,6-octadienoate	137
ethyl glycine	ethyl 2-aminoacetate	248
ethyl heptanoate	ethyl heptanoate	144,377
ethyl hexadeconate	ethyl hexadecanoate	144
ethyl hexanoate	ethyl hexanoate	132,134,144,208,217,271,279,282,290,377
ethylhexanoic acid, 2-	2-ethylhexanoic acid	139
ethyl hexenoate, (E)-2-	ethyl (E)-2-hexenoate	139
ethyl-1-hexanol, 2-	2-ethyl-1-hexanol	138,273

ethyl hydroxybutanoate, 3-	ethyl 3-hydroxybutanoate	284
ethyl hydroxybutanoate, 4-	ethyl 4-hydroxybutanoate	284
ethyl hydroxy-3-methylbutanoate, 2-	ethyl 2-hydroxy-3-methylbutanoate	284
ethyl hydroxy-3-methylpentanoate, (2S,3S)-2-	ethyl 2-hydroxy-4-methylpentanoate	271,284
ethyl hydroxy-4-methylpentanoate, 2-	ethyl 2-hydroxy-4-methylpentanoate	271
ethyl isoamyl succinate	ethyl 3-methylbutyl ester butanedioic acid	144
ethyl isobutyl succinate	ethyl 2-methylpropyl ester butanedioic acid	144
ethyl isopropyl succinate	ethyl 1-methylethyl ester butanedioic acid	144
ethyl lactate	ethyl (2S)-2-hydroxypropanoate	208,217,284,378
ethyl laurate	ethyl dodecanoate	208
ethyl mercaptan	ethanethiol	218
ethyl methylbutanoate, 2-	ethyl 2-methylbutanoate	134,217,271,398
ethyl methylbutanoate, 3-	ethyl 3-methylbutanoate	139,144,217,271
ethyl methylpropionate, 2-	ethyl 2-methylpropionate	271
ethyl methyl succinate	ethyl methyl succinate	140
ethyl-3-(2-methyl-2-propanyl)benzene, 1-	1-ethyl-3-(2-methyl-2-propanyl)benzene	140
ethyl-4-methyl-1-pentanol, 3-	3-ethyl-4-methyl-1-pentanol	273
ethyl 3-(methylthio)propanoate	ethyl 3-(methylthio)propanoate	140
ethyl (methylthio)acetate	ethyl (methylthio)acetate	262
ethyl nonanoate	ethyl nonanoate	370
ethyl octanoate	ethyl octanoate	208,217,271,282,290,291
ethyl palmitate	ethyl hexadecanoate	248
ethyl pentanoate	ethyl pentanoate	378
ethyl phenylacetate	ethyl 2-phenylacetate	284
ethyl propionate	ethyl propionate	139,271
ethyl propyl succinate	ethyl propyl ester butanedioic acid	284
ethyl salicylate	ethyl 2-hydroxybenzoate	284
ethyl sorbate	ethyl (2E,4E)hexa-2,4-dienoate	139,140
ethyl tetradeconate	ethyl tetradeconate	144
ethyl tetrahydro-5-oxo-2-furancarboxylate	tetrahydro-5-oxo-2-furancarboxylic acid ethyl ester	287
ethyl vanillate	ethyl 4-hydroxy-3-methoxybenzoate	154,284
ethyl vanillyl ether	4-(ethoxymethyl)-2-methoxyphenol	245
ethylene	ethene	161,325
ethylguaiacol, 4-	4-ethyl-2-methoxyphenol	154,206,244,247,271,285
ethylphenol, 4-	4-ethylphenol	217,244,271,285

eucalyptol	1,3,3-trimethyl-2-oxabicyclo[2.2.2]octane	138,157
eugenol	2-methoxy-4-prop-2-enylphenol	167,211,244,246,247,271,279,282
europinidin	2-(3,4-dihydroxy-5-methoxyphenyl)-3,7-dihydroxy-5-methoxy-1-benzopyrylium ion	78,145
farnescene (alpha)	(3E,6E)-3,7,11-trimethyl-1,3,6,10-dodecatetraene	158,160
farnescene (beta)	(6Z)-7,11-dimethyl-3-methylene-1,6,10-dodecatriene	158
farnesol, (2E,6E)	(2E,6E)-3,7,11-trimethyldodeca-2,6,10-trien-1-ol	202,203,274,275,361
farnesyl diphosphate	(2Z,6Z)-3,7,11-trimethyl-2,6,10-dodecatrien-1-yl trihydrogen diphosphate	361,365
ferulic acid, cis-	(Z)-3-(4-hydroxy-3-methoxyphenyl)-2-propenoic acid	125
ferulic acid, $trans$-	(E)-3-(4-hydroxy-3-methoxyphenyl)-2-propenoic acid	107,114,125,169,223,236,240,260,265,391
flavanone	2-phenyl-2,3-dihydrochromen-4-one	221,222
flavin adenine dinucleotide (FAD)	[[(2R,3S,4R,5R)-5-(6-aminopurin-9-yl)-3,4-dihydroxyoxolan-2-yl]methoxy-hydroxy-phosphoryl] [(2R,3S,4S)-5-(7,8-dimethyl-2,4-dioxobenzo[g]pteridin-10-yl)-2,3,4-trihydroxypentyl] hydrogen phosphate	33,44,52,53,362,365
flavin mono nucleotide (FMN)	(2R,3S,4S)-5-(7,8-dimethyl-2,4-dioxo-3,4-dihydrobenzo[g]pteridin-10(2H)-yl)-2,3,4-trihydroxypentyl dihydrogen phosphate	226
flavone	2-phenylchromen-4-one	263,391
folic acid	pteroyl-L-glutamate	33
formic acid	methanoic acid	328
formonetin	6,7-dihydroxy-3-(4-methoxyphenyl)chromen-4-one	225
friedelin	(4R,4aS,6aS,6bR,8aR,12aR,12bS,14aS,14bS)-4,4a,6b,8a,11,11,12b,14a-octamethylicosahydro-3(2H)picenone	254,255
fructose	(3S,4R,5R)-1,3,4,5,6-pentahydroxy-2-hexanone	55,59,61,62,63,64,93,104,108,179,188,290

fructose 1,6-bisphosphate	(3*S*,4*R*,5*R*)-1,3,4,5,6-pentahydroxy-2-hexanone-1,6-bisphosphate	61,62,63,64,65,94,95
D-fructosonic acid	D-arabino-2-hexulosonic acid	238
fumaric acid	(2*E*)-2-butenedioic acid	206,356
furaneol	4-hydroxy-2,5-dimethyl-3(2*H*)furanone	134,154,219,278,285
furanone, dihydro-5-methyl-3(2*H*)-	dihydro-5-methyl-3(2*H*)-furanone	205
furanone, 4-hydroxy-2,5-dimethyl-3(2*H*)-	4-hydroxy-2,5-dimethyl-3(2*H*)-furanone	247
furfural	furan-2-carbaldehyde	205,218,244,245,273,282,285
furfural, 5-methyl	5-methylfuran-2-carbaldehyde	247,273,284
furfuranol	2-furanmethanol	205,284
furfuryl alcohol	furan-2-ylmethanol	245,284
furfuryl ethyl ether	2-(ethoxymethyl)furan	245
furylmethanol, 5-methyl-2-	5-methyl-2-furylmethanol	205
furylmethanethiol, 2-	furan-2-ylmethanethiol	205
galactose	(2*R*,3*S*,4*S*,5*R*)-2,3,4,5,6-pentahydroxyhexanal	42,99,100,235,236,253
galestro	CaAl$_2$Si$_2$O$_8$	157
gallic acid	3,4,5-trihydrocybenzoic acid	37,92,101,103,104,107,116,125, 132,146,169,222,239,242,269
gallocatechin	(2*R*,3*S*)-2-(3,4,5-trihydroxyphenyl)-3,4-dihydro-2*H*-chromene-3,5,7-triol	222,241,243
gallocatechin gallate	(2*R*,3*S*)-2-(3,4-dihydroxyphenyl)-3,4-dihydro-3,7-dihydroxy-2*H*-1-benzopyran-5-yl 3,4,5-trihydroxybenzoate	222,243
genistein	5,7-dihydroxy-3-(4-hydroxyphenyl) chromen-4-one	223,225
genistein-7-*O*-glucoside	5-hydroxy-3-(4-hydroxyphenyl)-7-[(2*S*,4*S*,5*S*)-3,4,5-trihydroxy-6-(hydroxymethyl)oxan-2-yl] oxychromen-4-one	223
gentiobioside	β-D-glucopyranosyl-β-D-glycoside	209
geranial	(2*E*)-3,7-dimethylocta-2,6-dienal	131,136
geranic acid	3,7-dimethyl-2,6-ocadienoic acid	124,125,137,165,272,275
geraniol	(2*E*)-3,7-dimethyl-2,6-octadien-1-ol	71,128,130,131,134,137,143,151, 154,156,157,158,163,165,168,202, 210,272,272,282,290,387

Glossary

geranyl acetate	(2*E*)-3,7-dimethyl-2,6-octadien-1-ol acetate	137,278
geranylacetone	(5*Z*)-3,6-diemthyl-5,9-undecadien-2-one	71,137
geranyl diphosphate	mono[(2*E*)-3,7-dimethyl-2,6-octadien-1-yl]diphosphoric acid ester	84,87,110,148,202,361,366,370
geranylgeranyl diphosphate	(2*E*,4*E*,6*E*)-3,5,7,11-tetramethyldodeca-2,4,6,10-tetraene-1-diphosphate	84,87,148,361,366
germacrene D	(1*E*,6*E*,8*S*)-1-methyl-5-methylidene-8-propan-2-ylcyclodeca-1,6-diene	275,277
gibberellic acid	(2*S*,4a*R*,4b*R*,7*S*,9a*S*,10*S*,10a*R*)-1,2,4b,5,6,7,8,9,10,10a-decahydro-2,7-dihydroxy-1-methyl-8-methylene-13-oxo-4a,1-(epoxymethano)-7,9a-methanobenz[a]azulene-10-carboxylic acid	11,84,85,109
gibberellin A$_1$	(1*R*,2*R*,5*S*,8*S*,9*S*,10*R*,12*S*)-5,12-dihydroxy-11-methyl-6-methylene-16-oxo-15-oxapentacyclo[9.3.2.1^{5,8}.0^{1,10}.0^{2,8}]hepta-decane-9-carboxylic acid	84
gibberellin A$_2$	(1*R*,2*R*,5*R*,6*R*,8*R*,9*S*,10*R*,11*S*,12*S*)-6,12-dihydroxy-6,11-dimethyl-16-oxo-15-oxapentacyclo[9.3.2.1^{5,8}.0^{1,10}.0^{2,8}]hepta-decane-9-carboxylic acid	84
gibberellin A$_3$ (same as gibberellic acid)	(2*S*,4a*R*,4b*R*,7*S*,9a*S*,10*S*,10a*R*)-1,2,4b,5,6,7,8,9,10,10a-decahydro-2,7-dihydroxy-1-methyl-8-methylene-13-oxo-4a,1-(epoxymethano)-7,9a-methanobenz[a]azulene-10-carboxylic acid	11,84,85,109
gibberellin A$_4$	(1*R*,2*R*,5*R*,8*R*,9*S*,10*R*,12*S*)-12-hydroxy-11-methyl-6-methylene-16-oxo-15-oxapentacyclo[9.3.2.1^{5,8}.0^{1,10}.0^{2,8}]heptadecane-9-carboxylic acid	84
gibberellin A$_{12}$	(1*R*,2*S*,3*S*,4*R*,8*S*,9*S*,12*R*)-4,8-dimethyl-13-methylidenetetracyclo[10.2.1.0.^{1,9} 0^{3,8}] pentadecane-2,4-dicarboxylic acid	84

gingerone	4-(4-hydroxy-3-methoxyphenyl)butan-2-one	159
gluconic acid	(2R,3S,4R,5R)-2,3,4,5,6-pentahydroxyhexanoic acid	101,102,130,132,238,281,282,290
gluconic acid lactone	(3R,4S,5S,6R)-3,4,5-trihydroxy-6-(hydroxymethyl)tetrahydro-2H-pyran-2-one	238
glucopyranuronic acid, β-D-	(2S,3S,4S,5R,6R)-3,4,5,6-tetrahydroxyoxane-2-carboxylic acid	304
glucose	(3R,4S,5S,6R)-6-(hydroxymethyl)oxane-2,3,4,5-tetrol	27,58,60,63,64,65,93,97,183,192,194,204,218,235,236,237,253,394
glutamic acid (Glu, E)	2-aminopentanedioic acid	48,49,197,204,358
glutathione	(2S)-2-amino-5-[[(2R)-1-(carboxymethyl-amino)-1-oxo-3-sulfanylpropan-2-yl]amino]-5-oxopentanoic acid	129,134,152,156,160,180
glyceraldehyde	2,3-dihydroxypropanal	41,63
glyceraldehyde 3-phosphate	2,3-dihydroxyacetaldehyde-3-phosphate	41,59,61,66,94,95,196,361,365
glyceric acid	2,3-dihydroxypropanoic acid	238
glycerol	1,2,3-trihydroxypropane	31,32,42,206,214,217,228,254,281,290,329
glycine (Gly, G)	2-aminoacetic acid	28,106,249,332,333
glycitein	7-hydroxy-3-(4-hydroxyphenyl)-6-methoxychromen-4-one	225
glycoaldehyde	2-hydroxyacetaldehyde	101,102
glyphos	N-phosphonomethyl)glycine	106
grasshopper ketone	4-[(2R,4S)-2,4-dihydroxy-2,6,6-trimethyl-cyclohexylidene]but-3-en-2-one	129,164
guaiacol	2-methoxyphenol	159,167,245,247,261,271,273,279,285
guaiazulene	1,4-dimethyl-7-propan-2-ylazulene	375
guaiene (alpha)	(1S,4S,7R)-1,2,3,4,5,6,7,8-octahydro-1,4-dimethyl-7-(1-methylethenyl)azulene	93,372
guaiene (delta)	(1S,7R,8aS)-1,2,3,5,6,7,8,8a-octahydro-1,4-dimethyl-7-(1-methylethenyl)azulene	372
guanine (G)	2-amino-3,7-dihydro-6H-purin-6-one	12,226,351,355
guanosine diphosphate	2-amino-9-{5-O-[hydroxy(phosphonooxy)phosphoryl]-β-D-ribofuranosyl}-3,9-dihydro-6H-purin-6-one	75,76,97

guanosine triphosphate	2-amino-1,9-dihydro-9-[5-O-[hydroxy[-[hydroxy(phosphonooxy)phosphinyl]oxy]phosphinyl]-β-D-ribofuranosyl]-6H-purin-6-one	75,76,98
gurjunene, α-	(1aR,4R,4aR,7bS)-1a,2,3,4,4a,5,6,7b-octahydro-1,1,4,7-tetramethyl-1H-cyclo-prop[e]azulene	128
gurjunene, β-	(1aR,4R,4aR,7aR,7bR)-1,1,4-trimethyl-7-methylidene-2,3,4,4a,5,6,7a,7b-octahydro-1aH-cyclopropa[e]azulene	276
heme-b	iron(2+) salt (1:1) 7,12-diethenyl-3,8,13,17-tetramethyl-21H,23H-porphine-2,18-dipropanoic acid	47,49
heptanol, 1-	1-heptanol	144,273
heptanol, 2-	2-heptanol	138
heptanone, 2-	2-heptanone	138
hepten-2-one (6-methyl-5-)	6-methyl-5-hepten-2-one	72
hesperetin	(2S)-5,7-dihydroxy-2-(3-hydroxy-4-methoxyphenyl)-2,3-dihydrochromen-4-one	225
hexadecanoic acid	hexadecanoic acid	31,32,42
hexadienal, 2,4-	2,4-hexadienal	128
hexanal	hexanal	128,138,156
hexanoic acid	hexanoic acid	139,271,273,282
hexanol, 1-	1-hexanol	128,130,131,138,144,156,162,208,261,282
hexanol, 3-(methylthio)-1-	3-(methylthio)-1-hexanol	261,262,
hexenal, cis-2-	(Z)-2-hexanal	130,131
hexenal, trans-2-	(E)-2-hexanal	124,125,128,131,156
hexene	1-hexene	71
hexen-1-ol, cis-3-	(Z)-3-hexen-1-ol	128,130,131,138,273,
hexen-1-ol, trans-2-	(E)-2-hexen-1-ol	130,273
hexen-1-ol, trans-3-	(E)-3-hexen-1-ol	124,138
hexen-3-ol	1-hexene-3-ol	71
hexenoic acid, (E)-2-	(E)-2-hexenoic acid	273
hexenyl acetate, cis-3	(Z)-3-hexenyl acetate	139
hexenyl butanoate, cis-3-	(Z)-3-hexenyl butanoate	128,156
hexyl acetate	1-hexanol acetic acid ester	139,144,156,208,217,284
hexyl isobutyrate	hexyl 2-methylpropanoate	144
histidine (His, H)	2-amino-3-(1H-imidazol-4-yl)propanoic acid	48,49,50,51,100,226,361,363
homocysteine	2-amino-4-sulfanylbutanoic acid	227

homofuraneol	2-ethyl-4-hydroxy-5-methylfuran-3-one	278,285
homovanillic acid	2-(4-hydroxy-3-methoxyphenyl) acetic acid	159
homovanillic alcohol	4-(2-hydroxyethyl)-2-methoxyphenol	284
hotrienol	3,7-dimethyl-1,5,7-octatrien-3-ol	136,143,158,165,275,285
humulene, α-	(1*E*,4*E*,8*E*)-2,6,6,9-tetramethyl-1,4,8-cycloundecatriene	71,166,275,277,371
hydrocaffeic acid	3-(3,4-dihydroxyphenyl)propanoic acid	269,270
hydrogen sulfide	hydrogen sulfide	216,218
hydroxybenzaldehyde, *para*	*para*-hydroxybenzaldehyde	125,269
hydroxybenzoic acid, *para*	*para*-hydroxybenzoic acid	125,269
hydroxydihydroedulans	3,4,6,8a-tetrahydro-2,5,5,8a-tetramethyl-2*H*-1-benzopyran-4a(5*H*)-ols	220
hydroxydihydrolinalool, *cis*-8-	(*Z*)-2,6-dimethyl-2-octene-1,6-diol	285
hydroxydihydrolinalool, *trans*-8-	(*E*)-2,6-dimethyl-2-octene-1,6-diol	285
hydroxyethyl butyrate, 2-	2-hydroxyethyl butyrate	144
hydroxyferulic acid, 5-	3-(3,4-dihydroxy-5-methoxyphenyl) prop-2-enoic acid	240
hydroxylinalool, *cis*-8-	(2*Z*)-2,6-dimethyl-2,7-octadien-1,6-diol	163,165
hydroxymegastigma-7-en-9-one, 5,6-epoxy-3-	(3*E*)-4-[(1*S*,4*S*,6*R*)-4-hydroxy-2,2,6-trimethyl-7-oxabicyclo[4.1.0]hept-1-yl]-3-buten-2-one	164
hydroxymethyl-2-furaldehyde, 3-	3-(hydroxymethyl)-2-furancarboxaldehyde	269
hydroxymethylfurfural, 5-	5-(hydroxymethyl) furan-2-carbaldehyde	218,245
hydroxypyruvic acid	3-hydroxy-2-oxopropanoic acid	214
hydroxytheaspirane, 3-	2,6,6,10-tetramethyl-1-oxaspiro[4.5] dec-9-en-8-ol	164
hydroxytheaspirane, 8-	8-hydroxy-2,6,10,10-tetramethyl-1-oxaspiro-[4.5]dec-6-ene	129
hypocrolide A	(1*S*,3*S*,4*R*,5a*S*,12a*S*,14a*S*)-11-hydroxy-1,3-dimethoxy-4,13,13,14a-tetramethyl-4,5,5a,13,14,14a-hexahydro-1*H*,3*H*,6*H*-cyclopenta[4′,5′] isochromeno[6′,5′:4,5]furo[3,2-c] chromen-6-one	280,281

Glossary

idonic acid	(2*R*,3*S*,4*R*,5*S*)-2,3,4,5,6-pentahydroxyhexanoic acid	101,102
indole-3-acetic acid	1*H*-indole-3-acetic acid	11,23,24,33,34,83,109,210
indole-3-butanoic acid	1*H*-indole-3-butanoic acid	11
indole-3-ethanol	2-(1*H*-indol-3-yl)ethanol	217
inositol	(1*R*,2*R*,3*S*,4*S*,5*R*,6*S*)-cyclohexane-1,2,3,4,5,6-hexol	32,237
ionol, 3-hydroxy7,8-dihydro-β-	4-(3-hydroxy-1-butyn-1-yl)-3,5,5-trimethyl-2-cyclohexen-1-ol	287
ionol, 3-oxo-α-	4-(3-hydroxy-1-buten-1-yl)-3,5,5-trimethyl-2-cylohexen-1-one	129,209,287
ionene, α-	4,4,7-trimethyl-2,3-dihydro-1*H*-naphthalene	275
ionone, β-	(3*E*)-4-(2,6,6-trimethyl-1-cyclohexen-1-yl)-3-buten-2-one	128,131,150,155,210,211,219,220, 262,274
ionone, *trans*-3-dehydro-, β-	4-(2,6,6-trimethyl-1,3-cyclohexadien-1-yl)-3-buten-2-one	129,220
ionone, 3-oxo-β-	2,4,4-trimethyl-3-[(*E*)-3-oxobut-1-enyl]cyclohex-2-en-1-one	168
isoamyl acetate	3-methyl-1-butyl ester acetic acid	230
isoamyl alcohol	3-methyl-1-butanol	138
isoamyl hexanoate	3-methylbutyl hexanoae	144
isoamyl isovalerate	3-methylbutyl 3-methylbutanoate	144
isoamyl lactate	3-methylbutyl 2-hydroxypropanoate	144
isoamyl octanoate	3-methylbutyl octanoate	144
isoamyl pyruvate	3-methylbutyl 2-oxopropanoate	144
isobutanol	2-methyl-1-propanol	138,144
isobutyl acetate	2-methylpropyl ester acetic acid	144,208
isobutyl hexanoate	2-methylpropyl ester hexanoic acid	144
isobutyl lactate	2-methylpropyl ester 2-hydroxypropanoic acid	144
isobutyric acid	2-methylpropanoic acid	217
isocitric acid	3-carboxy-2,3-dideoxypentaric acid	93
isoeugenol, (*E*)-	2-methoxy-4-[(*E*)-prop-1-enyl]phenol	167,246,247,272
isohopeaphenol	(1*R*,1′*R*,6*R*,6′*S*,7*S*,7′*S*,11*bR*,11*b′R*)-1,1′,6,6′,7,7′,11b,11′b-octahydro-1,1′,7,7′-tetrakis(4-hydroxyphenyl)[6,6′-bibenzo[6,7]cyclohepta[1,2,3-cd]benzofuran]-4,4′,8,8′,10,10′-hexol	153
isoledene	(1*aR*,4*R*,7*R*,7*bS*)-1,1,4,7-tetramethyl-1a,2,3,4,5,6,7,7b-octahydrocyclopropa[e]azulene	275,276
isopentanol, 1-	3-methylbutan-1-ol	144

isopentenyl diphosphate	(3-methyl-3-buten-1-yl)-monodiphosphoric acid ester ion	84,109,110,148,201,203,359,361, 364,369
isopropyl disulfide	bis(1-methylethyl)disulfide	218
isopulegone, *cis*-	(2*R*,5*R*)-5-methyl-2-prop-1-en-2-ylcyclo-hexan-1-one	130,131
isorhamnetin	3,5,7-trihydroxy-2-(4-hydroxy-3-methoxyphenyl)-4*H*-chromen-4-one	124,126
jasmonic acid	{(1*R*,2*R*)-3-oxo-2-[(2*Z*)-2-penten-1-yl]cyclopentyl}acetic acid	20
kaempferol	3,5,7-trihydroxy-2-(4-hydroxyphenyl)-4*H*-1-benzopyran-4-one	124,126,130,153,393,394
kaurene (*ent*)	5,5,9-trimethyl-14-methylidenetetracyclo-[11.2.1.0^{1,10}.0^{4,9}]hexadecane	84
ketoglutaric acid, alpha	2-oxopentanedioic acid	140
ketone, methyl vanillyl	1-(4-hydroxy-3-methoxyphenyl)propan-2-one	159
kinetin	*N*-(2-furanylmethyl)-3*H*-purin-6-amine	83
lactic acid	2-hydroxypropanoic acid	58,61,143,146,184,206,214,217
lanosterol	(3*S*,5*R*,10*S*,13*R*,14*R*,17*R*)-4,4,10,13,14-pentamethyl-17-[(2*R*)-6-methyl-5-hepten-2-yl]-2,3,4,5,6,7,10,11,12,13,14,15,16,17-tetradecahydro-1*H*-cyclopenta[a]phenanthren-3-ol	109,110
lavandulol	(2*S*)-5-methyl-2-(1-methylethenyl)-4-hexen-1-ol	137
leucine (Leu, L)	(2*S*)-2-amino-4-methylpentanoic acid	108
lilial	3-(4-tert-butylphenyl)-2-methylpropanal	274,275
limonene	1-methyl-4-(prop-1-en-2-yl)cyclohexene	136,143,157,158,206,275,397
limetol	2-ethenyltetrahydro-2,6,6-trimethyl-2*H*-pyran	157
linalool	3,7-dimethyl-1,6-octadien-3-ol	130,136,143,151,157158,163,210, 219,274,285,397
linalool, 6,7-dihydro-7-hydroxy	6,7-dihydo-7-hydroxylinalool	158
linalool acetate	3,7-dimethyl-1,6-octadien-3-ol acetate	157

linalool oxide, cis	(2*R*,5*S*)-5-ethenyltetrahydro-α,α,5-trimethyl-2-furanmethanol	143,151,165,168,210,219, 272,285
linalool oxide, *trans*	(2*S*,5*S*)-5-ethenyltetrahydro-α,α,5-trimethyl-2-furanmethanol	136,137,143,151,163,165,168,210, 219,272,285
linalool oxide, (*Z*)-pyranoid	(3*S*,6*S*)-6-ethenyl-2,2,6-trimethyloxan-3-ol	158,285
linalool oxide, (*E*)-pyranoid	(3*R*,6*S*)-6-ethenyl-2,2,6-trimethyloxan-3-ol	158,285
linoleic acid	(9*Z*,12*Z*)-9,12-octadecadienoic acid	238
lipoic acid	5-(dithiolan-3-yl)pentanoic acid	190,194,216,350,361
loliolide	(6*S*,7a*R*)-6-hydroxy-4,4,7a-trimethyl-6,7-di-hydro-5*H*-1-benzofuran-2-one	129
lutein	(1*R*,4*R*)-4-{(1*E*,3*E*,5*E*,7*E*,9*E*, 11*E*,13*E*,15*E*,17*E*)-18-[(4*R*)-4-hydroxy-2,6,6-trimethyl-1-cyclohexen-1-yl]-3,7,12,16-tetramethyl-1,3,5,7,9,11,13,15,17-octadecanonaen-1-yl}-3,5,5-trimethyl-2-cyclohexen-1-ol	76,77,261,263
luteolin	2-(3,4-dihydroxyphenyl)-5,7-dihydroxychromen-4-one	224
lycopene, *trans*-	(6*E*,8*E*,10*E*,12*E*,14*E*,16*E*,18*E*,20*E*,22*E*,24*E*,26*E*)-2,6,10,14,19,23,27,31-octamethyldotri-aconta-2,6,8,10,12,14,16,18,20,22,24,26,30-tridecaene	149,203,338,339
lysine (Lys, K)	2,6-diaminohexanoic acid	62,64,80,81,195,198,358,361,385
malic acid	(*S*)-hydroxybutanedioic acid	91,92,93
malonyl-β-D-glucoside	malonyl-β-D-glucoside	209
maltol	3-hydroxy-2-methyl-4*H*-pyran-4-one	245
malvidin	3,5,7-trihydroxy-2-(4-hydroxy-3,5-dimethoxyphenyl)-1-benzopyrylium ion	114,153,159,161,263,340
malvidin-3-*O*-glucoside	(2*S*,3*R*,4*S*,5*S*,6*R*)-2-[5,7-dihydroxy-2-(4-hydroxy-3,5-dimethoxyphenyl)chromenylium-3-yl]oxy-6-(hydroxymethyl)oxane-3,4,5-triol	113,114,161,262,263,266,383
malvidin-3-*O*-(6′-acetyl) glucoside	malvidin-3-*O*-(6′-acetyl)glucoside	153
malvidin-3-*O*-(6′-coumaroyl) glucoside	malvidin-3-*O*-(6′-coumaroyl) glucoside	153
mannose	(2*S*,3*S*,4*R*,5*R*)-2,3,4,5,6-pentahydroxyhexanal	98,99,206,235,253

matairesinol	(3*R*,4*R*)-3,4-bis[(4-hydroxy-3-methoxyphenyl) methyl] oxolan-2-one	224
megastigma-4,7-diene-3,6,9-triols	1-(3-hydroxy-1-buten-1-yl)-2,6,6-trimethyl-2-cyclohexene-1,4-diol	220,221
megastigma-4-ene-3,6,9-triols	1-(3-hydroxybutyl)-2,6,6-trimethyl-2-cyclohexene-1,4-diols	220,221
megastigma-7-ene-3,6,9-triol, (3*S*,5*R*, 6*S*, 9)-	(1*S*,4*S*,6*R*)-1-[(1*E*,3*R*)-3-hydroxy-1-buten-1-yl]-2,2,6-trimethyl-1,4-cyclohexanediol	164
megastigma-4,6,8-trien-3-one, (6*Z*,8*E*)	4-[(2*E*,4*E*)-buta-2,4-dienyl]-3,5,5-trimethyl-cyclohex-2-en-1-one	129,164
megastigmatrienone D	(4*E*)-4-[(2*E*)-2-buten-1-ylidene]-3,5,5-trimethyl-2-cyclohexen-1-one	220
menth-1-ene-4,8-diol	1-(2-hydroxy-2-propanyl)-4-methyl-3-cyclohexen-1-ol	165
menthol	(1*R*,2*S*,5*R*)-5-methyl-2-(1-methylethyl)-cyclohexanol	136,275
mercaptoethanol, 2-	2-mercaptoethanol	261
mercapto-1-heptanol, 3-	3-mercapto-1-heptanol	283
mercapto-1-hexanol, 3-	3-mercapto-1-hexanol	162,279,283
mercapto-1-hexanol acetate, 3-	3-mercapto-1-hexanol acetate	283
mercapto-1-pentanol, 3-	3-mercapto-1-pentanol	283
mercaptohexan-1-ol-L-cysteine, 3-	3-mercaptohexan-1-ol-L-cysteine	127,129,134,138,142,152,162
mercaptohexan-1-ol-L-glutathione, 3-	3-mercaptohexan-1-ol-L-glutathione	130,135,160,162
mercapto-2-methyl-1-butanol, 3-	3-mercapto-2-methyl-1-butanol	283
mercapto-2-methyl-1-pentanol, 3-	3-mercapto-2-methyl-1-pentanol	283
mercapto-4-methylpentan-2-ol, 4-	4-methyl-4-sulfanylpentan-2-ol	218
mercapto-4-methyl-2-pentanone, 4-	4-methyl-4-sulfanylpentan-2-one	134,156,160,162,278,283
mercapto-4-methyl-1-pentanone-L-cysteine, 4-	4-mercapto-4-methyl-1-pentanone-L-cysteine	135,160,162
mercapto-4-methyl-2-pentanone-L-glutathione, 4-	4-mercapto-4-methyl-2-pentanone-L-glutathione	135,160,162
metabisulfite, potassium	potassium metabisulfite	113,182,183,185,187
metabisulfite, sodium	sodium metabisulfite	185

Glossary

metalaxyl*	2-[(2,6-dimethylphenyl)-(2-methoxy-1-oxoethyl)-amino] propanoic acid methyl ester	166,169
metam sodium	N-methylcarbamodithoic acid sodium salt	105
methanol	methanol	259,270,314,328
methional	3-(methylthio)propanal	154,205,278
methionine	(2S)-2-amino-4-methylsulfanylbutanoic acid	201,204,216,217,227
methionol	3-(methylthio)-1-propanol	154,282
methyl-4-(1-methylethenyl)-2-cyclohexane-1-hydroperoxide, (1S,4R)-1-	(1S,4R)-1-methyl-4-(1-methylethenyl)-2-cyclohexane-1-hydroperoxide	137
methyl acetate	methyl acetate	280,380
methylbutanamine, 3-	3-methylbutanamine	276,278
methyl dihydrojasmonate	methyl 2-[(1S,2R)-3-oxo-2-pentylcyclopentyl]acetate	274,275
methyl-3-furanthiol, 2-	2-methylfuranthiol	278,279,398
methyl hexadecanoate	methyl hexadecanoate	144
methyl hexanoate	methyl hexanoate	140
methyl mercaptan	methyl hydrogen sulfide	218
methyl 2-methyl-2-hydroxybutyrate	methyl 2-hydroxy-2-methylbutanoate	140
methyl octanoate	methyl octanoate	140,144
methyl salicylate	methyl 2-hydroxybenzoate	158,160
methyl thioacetate	S-methyl ethanethioate	218
methyl-1-butanol, 2-	2-methyl-1-butanol	208,217,271
methyl-1-butanol, 3-	3-methyl-1-butanol	138,208,217,271,383
methyl-1-propanol, 2-	2-methyl-1-propanol	138,144,217,282
methyl-5-(methylsulfanyl)furan, 2-	2-methyl-5-(methylsulfanyl)furan	205
methylanisol, 4-	4-methylanisol	140
methylbutanal, 2-	2-methylbutanal	205,217,271
methylbutanal, 3-	3-methylbutanal	205,217,271
methylbutanoic acid, 2-	2-methylbutanoic acid	271
methylbutanoic acid, 3-	3-methylbutanoic acid	205,271,285
methyl-1-butanol, 2-	2-methyl-1-butanol	208,217,271
methyl-1-butanol, 3-	3-methyl-1-butanol	138,208,217,271,303
methyl-2-butanol, 2-	2-methyl-2-butanol	273
methyl-2-butanol, 3-	3-methyl-2-butanol	273
methylbutyl acetate, 3-	3-methylbutyl acetate	139,217
methylbutyl butanoate, 3-	3-methylbutyl butanoate	240
methylbutyl decanoate, 3-	3-methylbutyl decanoate	240
methylbutyl hexanoate, 3-	3-methylbutyl hexanoate	240
methylbutyl lactate, 3-	3-methylbutyl lactate	284

methylbutyl octanoate, 3-	3-methylbutyl octanoate	140
methylfurfural, 5-	5-methyl-2-furancarboxaldehdye	205,244,247,284
methylfurfuryl alcohol, 5-	(5-methyl-2-furyl)methanol	245
methylfurfuryl ethyl ether, 5-	2-(ethoxymethyl)-5-methylfuran	245
methylguaiacol, 4-	2-methoxy-4-methylphenol	245
methyl-5-(1-methylethenyl) cyclohexanol, 2-	2-methyl-5-(1-methylethenyl) cyclohexanol	275
methylpropanal, 2-	2-methylpropanal	217
methylpropanoic acid, 2-	2-methylpropanoic acid	217
methylpropyl acetate, 2	2-methylpropyl acetate	217
methyltetrahydrothiophene-3-ol, *cis*-2-	*cis*_2,5-anhydro-1,4-dideoxy-2-thiopetitol	398
methyltetrahydrothiophene-3-ol, *trans*-2-	*trans*_2,5-anhydro-1,4-dideoxy-2-thiopetitol	398
methyltetrahydrothiophene-3-one, 2-	2-methyldihydro-3(2*H*)-thiophenone	398
(methylthio)propanal, 2-	2-(methylthio)propanal	271
(methylthio)peopanol, 3-	3-(methylthio)-1-propanol	271
methyl vanillate	methyl 4-hydroxy-3-methoxybenzoate	284
mevalonic acid	(3*R*)-3,5-dihydroxy-3-methylpentanoic acid	104
MSG (sodium glutamate)	sodium;(2*S*)-2-amino-5-hydroxy-5-oxopentanoate	204
muscone	(*R*)-3-methylcyclopentadecanone	142,172
muurolene, α-	(1*S*,4a*S*,8a*R*)-1,2,4a,5,6,8a-hexahydro-4,7-dimethyl-1-(1-methylethyl)-naphthalene	128,156,166,275,276
muurolene, γ-	(1*S*,4a*S*,8a*R*)-1-isopropyl-7-methyl-4-methylene-1,2,3,4,4a,5,6,8a-octahydro-naphthalene	128,275,276
myrcene, α-	2-methyl-6-methyene-1,7-octadiene	157
myrcene, β-	7-methyl-3-methylideneocta-1,6-diene	136
myrcenol	2-methyl-6-methyleneoct-7-en-2-ol	158
myricetin	3,5,7-trihydroxy-2-(3,4,5-trihydroxyphenyl)-4*H*-1-benzopyran-4-one	124,126,169,223
myricetin-3-rhamnoside	3-[(6-deoxy-α-L-mannopyranosyl)oxy]-5,7-dihydroxy-2-(3,4,5-trihydroxyphenyl)-4*H*-1-benzopyran-4-one	124
myristic acid	tetradecanoic acid	31

naphthylacetic acid (α-)	1-naphthylacetic acid, naphthalene-1-acetic acid	11
naringenin	5,7-dihydroxy-2-(4-hydroxyphenyl)-2,3-dihydrochromen-4-one	137,141,153,224,391,393
naringin	(2S)-7-[(2S,3R,4S,5S,6R)-4,5-dihydroxy-6-(hydroxymethyl)-3-[(2S,3R,4R,5R,6S)-3,4,5-trihydroxy-6-methyloxan-2-yl]oxyoxan-2-yl]oxy-5-hydroxy-2-(4-hydroxyphenyl)-2,3-dihydrochromen-4-one	224
neoxanthin	(1R,3S)-6-[(3E,5E,7E,9E,11E,13E,15E,17E)-18-[(1R,3S,6S)-3-hydroxy-1,5,5-trimethyl-7-oxabicyclo[4.1.0]heptan-6-yl]-3,7,12,16-tetramethyloctadeca-1,3,5,7,9,11,13,15,17-nonaenylidene]-1,5,5-trimethylcyclohexane-1,3-diol	261,263
neral	(2Z)-3,7-dimethylocta-2,6-dienal	397
neric acid	(2E)-3,7-dimethylocta-2,6-dienoic acid	168
nerol	(2Z)-3,7-dimethyl-2,6-octadien-1-ol	137,143,151,154,155,158,163,210,272,274,397
nerolidol, cis-	(6Z)-3,7,11-trimethyl-1,6,10-dodecatrien-3-ol	144,275,277
nerolidol, trans-	(6E)-3,7,11-trimethyl-1,6,10-dodecatrien-3-ol	167
nerol oxide	3,6-dihydro-4-methyl-2-(2-methyl-1-propen-1-yl)-2H-pyran	165,219,275
neryl acetate	(2Z)-3,7-dimethyl-2,6-octadien-1-ol acetate	143
nicotinamide adenine diphosphate (NAD+)	nicotinamide adenine diphosphate (NAD+)	42,44,45,52,53,59,84,94,100,181,196,215,299,362,364,387
nicotinamide adenine diphosphate (NADH)	nicotinamide adenine diphosphate (NADH)	42,44,45,52,53,59,84,94,100,181,196,215,299,362,364,387
nonadecane	nonadecane	138
nonadecene, 1-	1-nonadecene	138
nonalactone, γ-	5-pentyloxolan-2-one	210
nonanal	nonanal	138
nonanoic acid	nonanoic acid	273
nonanol, 1-	1-nonanol	138,144
nonanone, 2-	2-nonanone	138,273

nonatriene, (*E*)-4,8-dimethyl-1,3,7-	(*E*)-4,8-dimethyl-1,3,7-nonatriene	158
norfuraneol	4-hydroxy-5-methylfuran-3-one	278,279,285
oak lactone, *cis*	(4*S*,5*S*)-5-butyl-4-methyldihydrofuran-2(3*H*)-one	244,246,247
oak lactone, *trans*	(4*S*,5*R*)-5-butyl-4-methyldihydrofuran-2(3*H*)-one	244,247
ocimene, α-	(3*E*)-3,7-dimethyl-1,3,7-octatriene	143
ocimene, *cis*-β-	(3*Z*)-3,7-dimethyl-1,3,6-octatriene	157
ocimene, *trans*-β-	(3*E*)-3,7-dimethyl-1,3,6-octatriene	136,257
ocimene quintoxide	tetrahydro-2,2-dimethyl-5-(1-methyl-1-propenyl)furan	157
ocimenol, cis	(5*Z*)-2,6-dimethyl-2,5,7-octatrien-4-ol	158
ocimenol, *trans*-	(5*E*)-2,6-dimethyl-2,5,7-octatrien-4-ol	143,158
octadecanoic acid	octadecanoic acid	32,238
octanal	octanal	128,138
octanoic acid	octanoic acid	138,139,217,271,273,282
octanol, 1-	1-octanol	138,144,273
octanol, 3-	3-octanol	138
octanone, 3-	3-octanone	138
octene-3-ol, 1-	1-octene-3-ol	282,285
octyl acetate	octyl acetate	303
oenin (mlavidin-3-*O*-glucoside)	(2*S*,3*R*,4*S*,5*S*,6*R*)-2-[5,7-dihydroxy-2-(4-hydroxy-3,5-dimethoxyphenyl)chromenylium-3-yl]oxy-6-(hydroxymethyl)oxane-3,4,5-triol	262,264,265,266,267,383,384
oestragole	1-methoxy-4-prop-2-enylbenzene	206
oleic acid	(9*Z*)-octadec-9-enoic acid	31,32,35,36,42,238,254
ornithine	(2*S*)-2,5-diaminopentanoic acid	207
oxaloacetic acid	2-oxobutanedioic acid	215,216
oxaloglycolic acid	2-hydroxy-3-oxosuccinic acid	214
oxalosuccinic acid	1-oxo-1,2,3-propanetricarboxylic acid	96
oxoedulan	(5*S*,9*R*)-2,3,5,6,8,8a-hexahydro-2,5,5,8a-tetramethyl-7*H*-1-benzopyran-7-one	129
pallidol	(4*bR*,5*R*,9*bR*,10*R*)-5,10-bis(4-hydroxyphenyl)-4b,5,9b,10-tetrahydroindeno[2,1-a]indene-1,3,6,8-tetrol	153
palmitic acid	hexadecanoic acid	31,32,35,36,42,254

pantolactone	3-hydroxy-4,4-dimethyloxolan-2-one	154,167
paraquat	1,1′-dimethyl-4,4′bipyridinium dichloride	106
pelargonaldehyde	nonanal	74
pelargonidin	3,5,7-trihydroxy-2-(4-hydroxyphenyl)-1-benzopyrylium ion	225,394
penconazole*	1-[2-(2,4-dichlorophenyl)-pentyl]-1H-1,2,4-triazole	167,169
pentadecene-1	1-pentadecene	74
pentanal	pentanal	217
pentanoic acid	pentanoic acid	217
pentanol, 1-	1-pentanol	217,273
pentanol, 2-	2-pentalol	273
pentanol, 3-	3-pentanol	273
pentanone, 2-	2-pentanone	273
pentyldihydro-2(3H)-butryolactone, (5S)-5-	(5S)-5-pentyldihydro-2(3H)-furanone	279
peonidin	3,5,7-trihydroxy-2-(4-hydroxy-3-methoxyphenyl)-1-benzopyrylium ion	78
peonidin-3-O-glucoside	(2S,3R,4S,5S,6R)-2-[5,7-dihydroxy-2-(4-hydroxy-3-methoxyphenyl)chromenylium-3-yl]oxy-6-(hydroxymethyl)oxane-3,4,5-triol	224
petrostilbene, *trans*-	4-[(E)-2-(3,5-dimethoxyphenyl)ethenyl]phenol	223
petunidin	2-(3,4-dihydroxy-5-methoxyphenyl)-3,5,7-trihydroxy-1-benzopyrylium ion	78,153
petunidin-3-O-glucoside	(2S,3R,4S,5S,6R)-2-[2-(3,4-dihydroxy-5-methoxyphenyl)-5,7-dihydroxychromenylium-3-yl]oxy-6-(hydroxymethyl)oxane-3,4,5-triol	224
phellandrene, α-	2-methyl-5-(1-methylethyl)-1,3-cyclohexadiene	136
phenethyl acetate, 2-	2-phenylethyl acetate	139
phenethylamine, 2-	2-phenylethanamine	278
phenethyl butyrate	2-phenylethyl butanoate	217
phenethyl formate	2-phenylethyl formate	217
phenol	phenol	391,394

Glossary

phenol, 2,6-dimethyl-4-(2-propenyl)-	2,6-dimethyl-4-(2-propenyl)phenol	159
phenolphthalein	3,3-bis(4-hydroxyphenyl)-1(3H)-isobenzo-furanone	79
phenoxyethanol, 2	2-phenoxyethanol	167
phenylacetaldehyde, 2-	2-phenylacetaldehyde	167,279
phenylacetic acid	2-phenylacetic acid	217
phenylalanine (Phe, F)	(2S)-2-amino-3-phenylpropanoic acid	28,102,103,141,240,383,386,388
phenylethanol, 2-	2-phenylethanol	134,138,144,217,272,282
pheophytin	(3S,4S,21R)-14-ethyl-13-formyl-21-(methoxycarbonyl)-4,8,18-trimethyl-20-oxo-3-(3-oxo-3-{[(2E,7R,11R)-3,7,11,15-tetramethyl-2-hexadecen-1-yl]oxy}propyl)-9-vinyl-23,25-didehydrophorbine-23,25-diide	47
phloroglucinol	1,3,5-benzenetriol	238
phosphate anion	Phosphate anion (P_i)	23,23,331,32,41
phosphatidylcholine	2-[[hydroxy[(2R)-3-[(1-oxohexadecyl)oxy]-2-[[(9Z)-1-oxo-9-octadecen-1-yl]oxy]propoxy]phosphinyl]oxy]-N,N,N-trimethyl ethanaminium inner salt	29,31,40,42
phosphoenolpyruvate	2-(phosphonooxyl)-2-propenoic acid	61,96,97,107,197,358,383,384
phosphoglycerate, 2-	3-hydroxy-2-(phosphonooxy) propanoic acid	197,198,358,360
phosphoglycerate, 3-	2-hydroxy-3-(phosphonooxy) propanoic acid	58,60,63,93,94,356,358,360
phosphatidylinositol	(2R)-2,3-bis(formyloxy)propyl (2R,3R,5S,6R)-2,3,4,5,6-pentahydroxycyclohexyl phosphoric acid ester	29,32
phytoene, cis, 15-	(6E,10E,14E,16Z,18E,22E,26E)-2,6,10,14,19,23,27,31-octamethyldotriaconta-2,6,10,14,16,18,22,26,30-nonaene	148,149
piceatannol	4-[(E)-2-(3,5-dihydroxyphenyl) ethenyl]benzene-1,2-diol	223
pinanol, cis-2-	2,6,6-trimethylbicyclo[3.1.1] heptan-1-ol	143

pinene, β-	(1S,5S)-6,6-dimethyl-2-methylenebicyclo[3.1.1]heptane	143,206,275
pinene oxide, α-	(1S,6S)-2,7,7-trimethyl-3-oxatricyclo-[4.1.1.0^{2,4}]octane	136
pinoresinol	4-[(3S,3aR,6S,6aR)-6-(4-hydroxy-3-methoxyphenyl)-1,3,3a,4,6,6a-hexahydro-furo[3,4-c]furan-3-yl]-2-methoxyphenol	241
plastoquinol	2,3-dimethyl-5-[(2E,6E,10E,14E,18E,22E,26E,30E)-3,7,11,15,19,23,27,31,35-nonamethyl-2,6,10,14,18,22,26,30,34-hexatriacontanonaen-1-yl]-1,4-benzenediol	48,50
plastoquinone	2,3-dimethyl-5-[(2E,6E,10E,14E,18E,22Z,26E,30E)-3,7,11,15,19,23,27,31,35-nonamethyl-2,6,10,14,18,22,26,30,34-hexatriacontanonaen-1-yl]-1,4-benzoquinone	43,46,47,48,49,51
prephenic acid	cis-1-carboxy-4-hydroxy-α-oxo-2,5-cyclohexadiene-1-propanoic acid	104
presilphiperfolanol	(2aS,4aS,5R,7aS,7bR)-1,1,2a,5-tetramethyl-decahydro-7bH-cyclopenta[cd]inden-7b-ol	280
primeveroside	(2R,3R,4S,5S,6R)-2-[(1R)-1-phenylethoxy]-6-[[(2S,3R,4S,5R)-3,4,5-trihydroxyoxan-2-yl]oxymethyl]oxane-3,4,5-triol	209
procyanidin A1	2,8-bis(3,4-dihydroxyphenyl)-3,4-dihydro-8,14-methano-2H,14H-1-benzopyrano[7,8-d][1,3]benzodioxocin-3,5,11,13,15-pentol	222
procyanidin B1	(2R,2′R,3R,3′S,4R)-2,2′-bis(3,4-dihydroxy-phenyl)-3,3′,4,4′-tetrahydro-[4,8′-Bi-2H-1-benzopyran]-3,3′,5,5′,7,7′-hexol	126,222

procyanidin B2	(2R,2'R,3R,3'R,4R)-2,2'-bis(3,4-dihydroxy-phenyl)-3,3',4,4'-tetrahydro-[4,8'-Bi-2H-1-benzopyran]-3,3',5,5',7,7'-hexol	126
procyanidin B3	(2R,2'R,3S,3'S,4S)-2,2'-bis(3,4-dihydroxy-phenyl)-3,3',4,4'-tetrahydro-[4,8'-Bi-2H-1-benzopyran]-3,3',5,5',7,7'-hexol	126
propanal	propanal	265,267
propanal, 3-(methylsulfanyl)	3-(methylsulfanyl)propanal	261
propanoic acid	propanoic acid	166
propanoic acid, 3-(methylthio)	3-(methylthio)propanoic acid	261,262
propanol, 1-	1-propanol	138,217,282,329
propanol, 3-(ethylthio)-1-	3-(ethylthio)-1-propanol	427
propanol, 3-(methylsulfanyl)	3-(methylsulfanyl)propanol	205
propanol, 3-(methylthio)-1-	3-(methylthio)-1-propanol	261,262
propanol, 3-(methylthio)-1, acetate	3-(methylthio)-1-propanol acetate	262
propyl acetate	propyl acetate	139
propyl sorbate	propyl sorbate	140
protocatachuic acid	3,4-dihydroxybenzoic acid	125
purine	imidazo[4,5-d]pyrimidine	192,201,225,350,355
pyrazine, 2-methoxy-3-(1-methylethyl)-	2-methoxy-3-(1-methylethyl)pyrazine	163
pyrazine, 2-methoxy-3-(1-methylpropyl)-	2-methoxy-3-(1-methylpropyl)pyrazine	163
pyrazine, 2-methoxy-3-(2-methylpropyl)-	2-methoxy-3-(2-methylpropyl)pyrazine	163
pyrazine, 2,6-dimethyl	1,6-dimethylpyrazine	205
pyrazine, 2,3,5-trimethyl	2,3,5-trimethylpyrazine	205
pyrimethanil°	4,6-dimethyl-N-phenylpyrimidin-2-amine	166,169
pyrimidine	pyrimidine	188,192,225
pyrrole, 2-acetyl	1-(1H-pyrrol-2-yl)ethanone	247
pyruvic acid	2-oxopropanic acid	58,61,188,206,213,262,263,268,356
quercetin	2-(3,4-dihydroxyphenyl)-3,5,7-trihydroxy-4H-1-benzopyran-4-one	124,136,130,132,169,170,224
quercetin-3-galactoside	2-(3,4-dihydroxyphenyl)-3-(β-D-galacto-pyranosyloxy)-5,7-dihydroxy-4H-1-benzo-pyran-4-one	124
quercetin-3-glucoside	2-(3,4-dihydroxyphenyl)-3-[(4ξ)-β-D-xylo-hexofuranosyloxy]-5,7-dihydroxy-4H-1-benzopyran-4-one	124

Glossary

quercetin-3-rhamnoside	3-[(6-deoxy-α-L-mannopyranosyl) oxy]-2-(3,4-dihydroxyphenyl)-5,7-dihydroxy-4H-1-benzopyran-4-one	124
quinic acid	(1α,3α,4α,5β)-1,3,4,5-tetrahydroxycyclohexanecarboxylic acid	104,107
raspberry ketone	4-(4-hydroxyphenyl)butan-2-one	279
resveratrol, *trans*	(E)-5-(2-(4-hydroxyphenyl) ethenyl)-1,3-benzenediol	130,132,169,170,223,226,234, 301,393
resveratrol, *cis*	(Z)-5-(2-(4-hydroxyphenyl) ethenyl)-1,3-benzenediol	153,235
retinal-11 *cis*	(2E,4Z,6E,8E)-3,7-dimethyl-9-(2,6,6-trimethyl-1-cyclohexen-1-yl)-2,4,6,8-nonatetraenal	81
rhamnose	(2R,3R,4R,5R,6S)-6-methyloxane-2,3,4,5-tetrol	153,253
ribose	(2R,3R,4R)-2,3,4-trihydroxy-5-oxopentane	63,192,218,300
ribulose	(3R,4R)-1,3,4,5-tetrahydroxypentan-2-one	58,356,359
ribulose 1,5-bisphosphate	1,5-bis(dihydrogen phosphate)-2-pentulose	58,59,61,62,63,93
riesling acetal	(2R,4aR,8aR)-3,4,6,8a-tetrahydro-2,5,5,8a-tetramethyl-5H-2,4a-epoxy-2H-1-benzopyran	129,164,168
roburin A	32H-29,5,9-(epoxyethanylylidyne)-1,30-methano-11H,16H-dibenzo[h,o]dibenzo[7,8:9,10][1,5]dioxacycloundecino[3,2-b][1,6]dioxacycloheptadecine	240
rose furan	3-methyl-2-(3-methyl-2-buten-1-yl) furan	219
rose oxide	4-methyl-2-(2-methyl-1-propen-1-yl)-tetra-hydro-2H-pyran	134,136,168
rosinidin	3,5-dihydroxy-2-(4-hydroxy-3-methoxy-phenyl)-7-methoxy-1-benzopyrylium ion	78
rotundone	(3S,5R,8S)-5-isopropenyl-3,8-dimethyl-3,4,5,6,7,8-hexahydro-1(2H)-azulenone	165,166

Roundup®	*N*-(phosphonomethyl)glycine	106
rutin	2-(3,4-dihydroxyphenyl)-5,7-dihydroxy-3-[(2*S*,3*R*,4*S*,5*S*,6*R*)-3,4,5-trihydroxy-6-[[[(2*R*,3*R*,4*R*,5*R*,6*S*)-3,4,5-trihydroxy-6-methyloxan-2-yl]oxymethyl]oxan-2-yl]oxychromen-4-one	124,153,224
rutinoside	α-L-rhamnopyranosyl-β-D-glucoside	209
S-adenosylmethionine (SAM)	(2*S*)-2-amino-4-[{[(2*S*,3*S*,4*R*,5*R*)-5-(6-amino-9*H*-purin-9-yl)-3,4-dihydroxytetrahydro-2-furanyl]methyl}(methyl)sulfonio]-butanoate	389,391
sabinene	4-methylidene-1-propan-2-ylbicyclo[3.1.0]hexane	206
salicylic acid	2-hydroxybenzoic acid	20
sedoheptulose	(3*S*,4*R*,5*R*,6*R*)-1,3,4,5,6,7-hexahydroxy-2-heptanone	59,63,383
selinene, β-	decahydro-4a-methyl-1-methylene-7-(1-methylethenyl)-naphthalene	72
selinene, δ-	(4*aR*)-2,3,4,4a,5,6-hexahydro-1,4a-dimethyl-7-(1-methylethyl)-naphthalene	72
selinene, γ-	(4*aS*,8*aR*)-8a-methyl-4-methylidene-6-propan-2-ylidene-2,3,4a,5,7,8-hexahydro-1*H*-naphthalene	275
serine (Ser, S)	2-amino-3-hydroxypropanoic acid	97,196,214,333
sherry lactone	dihydro-5-(1-hydroxyethyl)-2(3*H*)-furanone	287
shikimic acid	(3*R*,4*S*,5*R*)-3,4,5-trihydroxy-1-cyclohexene-1-carboxylic acid	104
sinapinic acid	(*E*)-3-(4-hydroxy-3,5-dimethoxyphenyl)prop-2-enoic acid	240
sodium glutamate (MSG)	sodium;(2*S*)-2-amino-5-hydroxy-5-oxopentanoate	204
sorbic acid	(2*E*, 4*E*)-hexa-2,4-dienoic acid	139
sorbitol	(2*R*,3*R*,4*R*,5*S*)-hexan-1,2,3,4,5,6-hexol	238,307
sotolon	4-hydroxy-2,3-di(methyl)-2*H*-furan-5-one	154,219,271,278,398

squalene	(6*E*,10*E*,14*E*,18*E*)-2,6,10,15,19,23-hexamethyl-2,6,10,14,18,22-tetracosahexaene	109,110,201,202
starch	1→4-glucose polymer	8,18,23,31,41,63
stearic acid	octadecanoic acid	32,36,254
strigol	(3*E*,3a*R*,5*S*,8b*S*)-5-hydroxy-8,8-dimethyl-3-({[(2*R*)-4-methyl-5-oxo-2,5-dihydro-2-furanyl]oxy}methylene)-3,3a,4,5,6,7,8,8b-octahydro-2*H*-indeno[1,2-b]furan-2-one	154,155
succinic acid	1,4-butanedioic acid	93
sucrose	(2*R*,3*R*,4*S*,5*S*,6*R*)-2-{[(2*S*,3*S*,4*S*,5*R*)-3,4-dihydroxy-2,5-bis(hydroxymethyl)tetrahydro-2-furanyl]oxy}-6-(hydroxymethyl)tetrahydro-2*H*-pyran-3,4,5-triol	38,55,63,65,91,93,96,104,107, 112,184
sulfite anion	sulfite anion	184,186,188
sulfoglucopyanose	6-deoxy-6-sulfo-D-glucopyranose	42
sulfur dioxide	sulfur dioxide	113,182,185
sulfurous acid	sulfurous acid	185
synapyl alcohol	4-[(*E*)-3-hydroxyprop-1-enyl]-2,6-dimethoxyphenol	269
syringaldehyde	4-hydroxy-3,5-dimethoxybenzaldehyde	125,159,269,270,283
syringic acid	4-hydroxy-3,5-dimethoxybenzoic acid	104,153,169,238,270
syringol	2,6-dimethoxyphenol	159,246,247
tannic acid	glucose esterified with gallic acid	37,92,101,143,146,235
tartaric acid L-(+)-	(2*R*,3*R*)-2,3-dihydroxybutanedioic acid	91,92,93,94,97,101,107,146, 213, 335
taxifolin	(2*R*,3*R*)-2-(3,4-dihydroxyphenyl)-3,5,7-trihydroxy-2,3-dihydrochromen-4-one	153,391,393
TDN	1,1,6-trimethyl-1,2-dihydronaphthalene	146,159,164,220
terpin	4-(2-hydroxypropan-2-yl)-1-methylcyclohexan-1-ol	158,165
terpin hydrate	4-(2-hydroxypropan-2-yl)-1-methylcyclohexan-1-ol + water	158
terpineol, 1-	1-methyl-4-(1-methylethenyl)-cyclohexanol	143

terpineol, α-	2-(4-methylcyclohex-3-en-1-yl)propan-2-ol	130,163,165,168,202,211,272,274
terpineol, β-	1-methyl-4-prop-1-en-2-ylcyclohexan-1-ol	143
terpineol, 4-	4-methyl-1-(1-methylethyl)-3-cyclohexen-1-ol	274,282,293
terpinene, α-	1-methyl-4-(1-methylethyl)-1,3-cyclohexadiene	143
terpinene, γ-	1-methyl-4-(1-methylethyl)-1,4-cyclohexadiene	275,136
terpinolene	1-methyl-4-(1-methylethylidene)-cyclohexene	136
terpinyl acetate, α-	2-(4-methyl-3-cyclohexen-1-yl)-2-propanyl acetate	158
tetrahydroxycyclohexane,1,2,3,5-	1,2,3,5-tetrahydroxycyclohexane	238
tetrathiocane, 1,3,5,7-	1,3,5,7-tetrathiacyclooctane	130,131,226
teturonic acid	(2R,3R)-2,3-dihydroxy-4-oxobutanoic acid	102
theaspiranes	2,6,10,10-tetramethyl-1-oxaspiro[4.5]dec-6-en-8-ols	220
thiamine	3-[(1,6-dihydro-6-imino-2-methyl-5-pyrimidinyl)-methyl]-5-(2-hydroxyethyl)-4-methylthiazolium ion	63,188,199,215,352,362,366
thiazoline, 2-acetyl	1-(4,5-dihydro-1,3-thiazol-2-yl)ethanone	205
threuronic acid	(2R,3R)-2,3-dihydroxy-4-oxobutanoic acid	102
thymine (T)	5-methyl-2,4(1H,3H)-pyrimidinedione	12,226,351
trichloroanisol, 2,4,6-	1,3,5-trichloro-2-methoxybenzene	253
trichlorophenol, 2,4,6-	2,4,6-trichlorophenol	255
tridecanone, 2-	2-tridecanone	73
tridecene, 1-	1-tridecene	72
trihydroxystearic acid	9,10,18-trihydroxyoctadecanoic acid	35,254
trimethyl-3-cyclohexen-1,2-diols, 4-(3-hydroxy-1-buten-1-yl)-3,5,5	4-(3-hydroxy-1-buten-1-yl)-3,5,5-trimethyl-3-cyclohexen-1,2-diols	220
trimethyl-3-cyclohexen-1-ones, 2-(1,3-butadien-1-yl)-2,6,6-	2-(1,3-butadien-1-yl)-2,6,6-trimethyl-3-cyclohexen-1-ones	220
trimethylcyclohexanone, 2,2,6-	2,2,6-trimethylcyclohexanone	128,261,262
trimethylcyclohex-2-ene-1,4-dione, 2.6,6-	2,6,6-trimethylcyclohex-2-ene-1,4-dione	129
trimethylcyclohex-3-en-1-one, 2-(3-hydroxybut-1-enyl)-2,6,6-	2-(3-hydroxybut-1-enyl)-2,6,6-trimethylcyclo-hex-3-en-1-one	164

Glossary

trimethylpyrazine, 2,3,5-	2,3,5-trimethylpyrazine	249
trithiolane, trans-3,5-dimethyl-1,2,4-	trans-3,5-dimethyl-1,2,4-trithiacyclopentane	130,131,225
tyrosine (Tyr, Y)	2-amino-3-(4-hydroxyphenyl)propanoic acid	48,50,102,237,240,386,388,389
tyrosol	2-(4-hydroxyphenyl)ethanol	169,269,270
uracil (U)	2,4(1H,3H)-primidinedione	12,189,190
uracil sulfonate	2,4(1H,3H)-primidinedione-6-sulfonate	190
urea	urea	206,207
uridine	1-(3,4-dihydroxy-5-hydroxymethyl-tetrahydrofuran-2-yl)-1H-pyrimidine-2,4-dione	226,394
valencene	(1R,7R,8aS)-1,2,3,5,6,7,8,8a-octahydro-1,8a-dimethyl-7-(1-methylethenyl)-naphthalene	372
vanillic acid	4-hydroxy-3-methoxybenzoic acid	125,159,169,223,238,259
vanillin	4-hydroxy-3-methoxybenzaldehyde	38,125,134,269,270,271,279
vanillyl alcohol	4-(hydroxymethyl)-2-methoxyphenol	245
vescalagin	(46S)-7,8,9,12,13,14,25,26,27,30,31,32,35,36,37,46-hexadecahydro-3,18,21,41,43-pentaoxanonacyclo[27.13.3.1^{38,42}.0^{2,20}.0^{5,10}.0^{11,16}.0^{23,28}.0^{33,45}.0^{34,39}]hexatetraconta-5,7,9,11,13,15,23,25,27,29(45),30,32,34,36,38-pentadecaene-4,17,22,40,44-pentone	239
vicianoside	α-L-arabinopyranosyl-β-D-glucoside	209
viniferin, α-	(2R,2aR,7R,7aR,12S,12aS)-2,7,12-tris(4-hydroxyphenyl)-2,2a,7,7a,12,12a-hexahydro-bis[1]benzofuro[3′,4′:4,5,6;3″,4″:7,8,9]cyclonona[1,2,3-cd][1]benzofuran-4,9,14-triol	225
viniferin, ε-	5-[(2R,3R)-6-hydroxy-2-(4-hydroxyphenyl)-4-[(E)-2-(4-hydroxyphenyl)ethenyl]-2,3-dihydro-1-benzofuran-3-yl]benzene-1,3-diol	153,226
vinylguaiacol, 4-	4-ethenyl-2-methoxyphenol	154,206,211,245,279
vinylphenol, 4-	4-ethenylphenol	206,207,217,245

violaxanthin	(1*R*,3*S*,6*S*)-6-[(1*E*,3*E*,5*E*, 7*E*,9*E*,11*E*,13*E*,15*E*,17*E*)-18- [(1*R*,3*S*,6*S*)-3-hydroxy-1,5,5- trimethyl-7-oxabicyclo[4.1.0] heptan-6-yl]-3,7,12,16- tetramethyloctadeca- 1,3,5,7,9,11,13,15,17-nonaenyl]- 1,5,5-trimethyl-7-oxabicyclo[4.1.0] heptan-3-ol	261,263
vitisin A	structure in question	159,161,264,265,268
vitisin B	structure in question	159,161,264,265,268
vitispirane	2,6,6-trimethyl-10-methylidene- 1-oxaspiro[4.5]dec-8-ene	129,155,181,164,262,275,278
vomifoliol	(4*S*)-4-hydroxy-4-[(*E*,3*R*)- 3-hydroxybut-1-enyl]-3,5,5- trimethylcyclohex-2-en-1-one	129,164,209
whisky lactone	(4*S*,5*R*)-5-butyl-4-methyloxolan- 2-one	261,271,279,398
wine lactone	(3*S*,3a*S*,7a*R*)-3,6-dimethyl-3a,4,5,7a- tetrahydro-1-benzofuran-2(3*H*)-one	219,271,397
xylulose	(3*S*,4*R*)-1,3,4,5-tetrahydroxypentan- 2-one	63,154,361,366
ylangene, α-	(1*S*,2*R*,6*R*,7*R*,8*S*)-8-isopropyl-1,3- dimethyltricyclo[4.4.0.0^{2,7}]dec-3-ene	165,166,275,276
zeatin (*cis*)	(2*Z*)-2-methyl-4-(3*H*-purin-6- ylamino)-2-buten-1-ol	83
zeatin (*trans*)	(2*E*)-2-methyl-4-(3*H*-purin-6- ylamino)-2-buten-1-ol	12,83,210
zeatin (trans) riboside	(2*R*,3*S*,4*R*,5*R*)-2-(hydroxymethyl)- 5-[6-[[(*E*)-4-hydroxy-3-methylbut- 2-enyl]amino]-9-purinyl] tetrahydrofuran-3,4-diol	83
zingerone	4-(4-hydroxy-3-methoxyphenyl) butan-2-one	167,211

INDEX

Note: Page references followed by a "*t*" indicate table; "*f*" indicate figure.

abscisic acid, 11, 12*f*, 19, 20*f*, 108, 109*f*, 137, 141*f*, 155*f*, 210*f*
absorption, 47
acetal, 243*f*
2-acetal-1*H*-pyrrole, 247*f*
acetaldehyde, 138*f*, 188, 188*f*, 194, 196, 198, 199*f*, 208*f*, 217*f*, 228n8, 234, 242, 243, 243*f*, 299, 299*f*, 300, 307, 343, 354n14
acetaldehyde dehydrogenase, 300
acetaldehyde diethyl acetal, 208*f*
acetate, 74*f*, 300, 361, 365
acetic acid, 58, 61*f*, 196, 205*f*, 207*f*, 208*f*, 217*f*, 234, 300, 307, 327, 328*f*, 354n14, 397*t*, 398*t*
acetoacetate, 363*f*
acetoacetyl-CoA, 363*f*
acetoin, 215*f*, 216, 217*f*
acetolactate, 215*f*
acetolactate synthase, 214, 215*f*
acetone, 341, 343*f*
acetosyringone, 167*f*
acetovanillone, 159*f*, 211*f*
acetyl choline, 330, 330*f*
acetyl-CoA. *See* acetyl-coenzyme A
acetyl-CoA carboxylase, 392*f*

acetyl-coenzyme A (acetyl-CoA), 141*f*, 189, 189*f*, 190*f*, 194, 197–98, 199*f*, 216, 221*f*, 228n8, 300, 361, 362*f*, 363*f*, 392*f*
2-acetylfuran, 218*f*, 247*f*
2-acetylpyrrole, 205*f*
2-acetylthiazoline, 205*f*
acidic material, 320
acidity, 113
ACS. *See* American Chemical Society
actinidols, 129*f*, 159*f*, 164*f*, 168*f*, 220*f*
acyclic hydrocarbons, 321, 322
adenine, 12*f*, 13n9, 226*f*, 300*f*, 351*f*, 355n18
adenosine diphosphate (ADP), 24*f*, 41, 41*f*, 44–45, 45*f*, 52–55, 56*f*, 59, 62*f*, 94, 198*f*, 300*f*, 360*f*, 364, 365*f*, 367
adenosine triphosphate (ATP), 24*f*, 41, 41*f*, 44, 45*f*, 52–54, 56*f*, 59, 62*f*, 94, 198*f*, 360*f*, 364, 365*f*, 367, 392*f*
ADP. *See* adenosine diphosphate
ADP-glucose, 65*f*
aged Pinotage wine, 263
aging process, 232
  oak in, 235
  vessels for, 235
aglycones, 156, 209, 209*f*, 210*f*

agricultural pest management, 105
Airén grape, 123–24, 124f, 125f
air locks, 182f
air space, 37
alanine, 50f, 247f, 330, 331f, 332f
albumin, 232
alcohol acetates, 206
alcohol concentration, 288
alcohol dehydrogenase, 199f, 299
alcohol levels, 112
alcohols, 307
   benzyl, 144f
   coniferyl, 238f
   para-coumaryl, 238f
   (+)-dehydrodiconiferyl, 241f
   ethyl, 302n2, 327
   furfuryl, 245f
   in Gewürztraminer wines, 138f
   measuring quantities of, 302n1
   5-methylfurfuryl, 245f
   in Muscat wine, 144f
   n-butyl, 150
   norisoprenoid, 155, 209f, 210
   in Riesling grapes, 156f
   sinapyl, 238f
   sugars and levels of, 112
   vanillyl, 245f
   wood, 314–15
   yeast and levels of, 112
alcohol weight, 307
aldehyde, 138f, 156f, 195, 232, 307, 326, 328f, 364f
   bisulfite and, 188
aldehyde dehydrogenases, 300
aldehydes, 326
aldehydic betalamic acid, 386
aldol-type condensation, 94, 95f
alkane, 324f
alkenes, 248f, 321, 327, 359f
alleles, 5n10
allotropes, 328
all-*trans*-lycopene, 149f, 203f
4-allyl-2,6-dimethoxyphenol, 167f
4-allyl-2-methoxyphenol, 210
allyl hexanoate (allyl caproate), 375
aloin, 37f
alpha-amino acids, 330
Alzheimer disease, 301, 302
American Chemical Society (ACS), xviin6

amide bonds, 330
amines, 328
aminoacetone, 248f
amino acids, 48, 50f, 102, 112, 179, 240f, 307, 330, 331, 332f, 354n13, 359f
   carbohydrate reactions to, 246
   common, 331f
   2-methoxy-3-isobutylpyrazine pathways from, 108, 108f
   oak-treated wine and, 246, 248f
   sulfur containing, 227f
   sulfur metabolism of, 204, 225
ammonia, 16, 103f, 207f
ammonium nitrate, 16
ammonium nitrogen, 201
ammonium phosphate, 196
Amontillado, 265, 269f, 270f, 271f
Amontillado sherry, 267
ampelography, 4
ampelopsin-D, 153f
ampelopsin-H, 153f
AMUs. *See* atomic mass units
amyl acetate, 380
amyl butyrate, 381
amyl caproate, 381
amyloplasts, 8, 18, 23, 24f, 39
amyl valerate, 381
analytical chemistry, xv–xvi, 4n1, xvin1
Andalusia, Spain, 265
anethole, 206f
cis-anhydrolinalool oxide, 219f
trans-anhydrolinalool oxide, 219f
anion, 8n5
anlagen, 68
antenna chlorophylls, 47
anther, 69, 69f
anthesis, 81, 82f
anthocyanidin-3-ol, 209f
anthocyanidin pigments, 340f, 388, 389f, 390
anthocyanidins, 77, 78, 78t, 79f, 106, 107, 116, 145t, 149, 150, 211, 221, 221f, 222f, 243, 243f, 382, 392, 392f
   ring systems, 141f
anthocyanins, 107, 112, 113, 113f, 116, 123, 142, 150, 159, 160, 166, 168t, 221, 392
   biosynthesis regulation, 149
antioxidants, 180, 244
apical meristem, 7, 33f, 39, 67, 67f
apigenin, 225f, 390, 392f

Index

β-D-apiofuranosyl-β-D-glucoside, 209*f*
10′-apo-β-carotenal, 150*f*
appellation d'origine contrôlée, 287
apricolin, 210, 211*f*
aquaporin, 25, 25*f*
Arabidopsis thaliana, 27, 31, 73, 74*f*, 211, 281
β-D-(−)-arabinofuranose, 238*f*
α-L-arabinofuranosyl-β-D-glucoside, 209*f*
β-D-(−)-arabinopyranose, 238*f*
α-L-arabinopyranosyl-β-D-glucoside, 209*f*
arabonic acid lactone, 238*f*
arginine, 206, 207*f*, 331*f*
aroma compounds, 216
aromadendrin, 390
aromatic bonds, 325
aromatic compounds, in Riesling grapes, 159*f*
aromatic hydrocarbon benzene, 325
aryl rings, 221
ascorbic acid (vitamin C), 97, 99–101, 101*f*, 102*f*, 104*f*, 180, 181*f*, 234*f*
asparagine, 331*f*, 333*f*
aspartic acid, 50*f*, 331*f*, 358*f*
astaxanthin, 203*f*
astilbin, 132*f*
atmospheric nitrogen, 16
atomic mass units (AMUs), 340
ATP. *See* adenosine triphosphate
ATP synthase (ATPase), 52, 53, 54*f*, 55*f*, 56*f*
aurantinidin, 145*t*
Aureusidin synthase, 183
autogamy, 83
auxins, 11, 11*f*, 24*f*, 31, 33*f*, 83*f*

banana odor, 303n4
Bandol AOC, 142
barriers, 44
basal leaf plucking, 38
base peak, 342*f*
bases, 320
bentonite, 231, 232, 233*f*
benzaldehyde, 140*f*, 208*f*, 397*t*
benzene, 325, 327*f*, 338
benzenoids, 155
benzodihydropyran, 222*f*
benzoic acids, 123
benzopyran systems, 222*f*
benzothiazol-2-oxyacetic acid, 83*f*
benzyl acetate, 134*f*, 375
benzyl alcohol, 144*f*

berries. *See* grape berries
betacyanin pigment, 389*f*
betacyanins, 386
betalains, 77, 78, 382
betalamic acid, 388, 389*f*
betanin, 77*f*, 383*f*, 389*f*
betaxanthins, 386, 389*f*
bicarbonate, 293
bicarbonate anion, 186
biochanin A, 225*f*
Biot, Jean-Baptiste, 334
biotin, 217*f*, 226*f*, 350, 350*f*
biotinylated proteins, 93
bis-ester, 359*f*
1,6-bisphosphate, 195, 195*f*
2,3-bisphosphate, 360*f*
1,3-bisphospho-D-glycerate, 59
1,3-bisphosphoglycerate, 62*f*
bisulfite, 187*f*, 188, 189
bisulfite adducts, 188
bisulfite anion, 185, 186, 186*f*, 188*f*, 190*f*
bisulfite salts, 113
bitter principles, 37
blumenol A, 129*f*
blumenol B, 129*f*
blumenol C, 129*f*
Boal grape, 269, 273, 274*f*
Bohr, Neils, 353n3
bonds
    aromatic, 325
    breaking, 318–21
    conjugated, 325
    covalent, 310, 317, 318
    double, 325
    hydrogen, 312, 318*f*
    ionic, 310
    peptide, 330
bornyl acetate, 375
Botrytis cinerea ("Noble Rot"), 130, 132*f*, 162–63, 213, 259, 276–86, 279*f*, 284*f*, 285*f*, 287*f*, 288
botrytized wine volatiles, 278, 279*f*
bouquet, 396
α-bourbonene, 166*f*
Box, George E. P., 353n2
branches, 29
    buds on, 68*f*
brassinolide, 210*f*
brassinosteroids, 108, 109
brightly colored compounds, 340*f*

brix, 111
bubbles, in Champagne, 289
buds, 67–69, 68f, 73, 75–76
Busby, James, 164–65
1,2-butadiene, 326f
1,3-butadiene, 325, 326f
butanal, 328f
2,3-butandione, 205f
butane, 322f, 324f
(2R,3S)-2,3-butanediol, 138f
1,3-butanediol, 208f
2,3-butanediol, 208f, 214, 216
butanoic acid, 139f, 205f, 328f
butanol, 208f, 323, 324
1-butanol, 138f, 144f, 323, 323f, 324f, 325f, 342, 344f, 346, 348f
2-butanol, 323f, 324, 324f, 325f
butanol/diethyl ether pair, 342, 344f
2-butanone, 325, 325f
1-butene, 326f
cis-2-butene, 327f
trans-2-butene, 325
trans-2-butene, 327f
(4S,5S)-5-butyl-4-methyldihydrofuran-2(3H)-one, 397t
butyl acetate, 139f, 375
butyl butyrate, 376
butyl propanoate, 376
2-butyne, 325
butyrolactone, 219f

C-5 hydroxyl, 384f
C-13 norisoprenoids, 285f
C-13 norterpenoids, 275, 278f
Cabernet franc, 6, 396t
Cabernet Sauvignon, 6, 124–27, 126f, 128f, 396t, 397t, 398t
γ-cadinene, 166f, 372f
δ-cadinene, 166f
(−)-γ-cadinene synthase, 372f
caffeic acid, 104f, 107f, 169f, 223f, 240f, 388, 390f
trans-caffeic acid, 125f
caftaric acid, 132f, 169f, 180, 181f
cis-caftaric acid, 126f
trans-caftaric acid, 126f
Calabrese di Montenuovo, 157
calamenene, 128f, 156f
calarene, 166f
calorie, 354n5

Calvin-Benson-Bassham cycle (CBB cycle), 58–63
Calvin cycle, 58, 59f, 93, 101, 356, 357f, 382
calyptra, 68–69, 69f
cAMP. See cyclic adenosine monophosphate
canopy work, 105, 107, 112
canteiros, 272
cap, in red wine making, 183
caraway, 333f
carbamic acid, 207
carbocycles, 348–53, 349f
carbocyclic compounds, 350f
carbocyclic hydrocarbons, 323f
carbohydrates, 39, 41, 112, 201, 209f
  amino acid reactions with, 246
  in oak, 236, 237
  structural units, 236f
carbon, 321, 327
carbonate anion, 186
carbon dioxide, 55, 58, 59, 60f, 61f, 63, 182, 190f, 206, 216, 216f, 228n8, 288–89, 293, 328f, 358f, 359f
  absorption of, 356
carbonic acid, 185
carbon nuclear magnetic resonance, 345
carbonolamine, 384f
carbon spectra, 355n17
carbonyl group, 328f
carbonyl of 3-dehydroshikimate to shikimate, 382
carbonyl reduction, 383f
2-carboxy-3-ketoribitol-1,5-bisphosphate, 58, 60f, 61f
carboxylic acid esters, 375
carboxylic acids, 73, 104, 125f, 179, 196f, 206, 231, 236, 326, 328, 330
  levels of, 113–14
5-(1-carboxyvinyl)-3-phosphoshikimate, 384f, 385f, 386f
Carmenere, 396t
β-carotene, 76, 77f, 106, 149f, 150f, 203f, 338, 339f
ζ-carotene isomerase, 149f
carotenes, 209f, 382
carotenoid-9′,10′-cleaving dioxygenase, 150f
carotenoids, 76, 106, 112, 128, 154, 220
carvone, 398t
caryophyllene, 128f, 156f, 160f
(E)-caryophyllene, 158
β-caryophyllene, 166f
(−)-β-caryophyllene synthase, 372f

Index

CAS. *See* Chemical Abstracts Service
castalagin, 235, 242*f*
catalyst, 354n12
catalytic oxidation, 44
catalyzed hydrolysis, 243*f*
catechin, 132*f*, 169*f*
(+)-catechin, 125*f*, 126*f*, 223*f*, 241*f*
catechol *O*-methyltransferase, 390*f*
cation, 8n5
CBB cycle. *See* Calvin-Benson-Bassham cycle
cell membrane, 24, 24*f*, 25*f*, 32*f*
cells
    epidermal, 18
    eukaryotic, 17n1, 192, 193*f*
    guard, 36
    harvesting light, 44–57
    leaves, 35–39, 36*f*
    light on leaves, 39–43
    mesophyll, 35, 37
    outside membrane, 31*f*
    parenchyma, 8, 36–37
    plant, 24*f*, 40*f*, 192, 193*f*
    primordial, 33–34
    roots, shoots, leaves, and grapes, 23–35
    Saccharomyces, 200*f*
    surface, 18
cellulose, 27, 236, 236*f*
    formation of, 63, 64*f*, 66
    glucose units in, 27*f*
    hemicellulose, 235, 245
cell wall, 23
Celsius, Anders, 4n1
Celsius scale, 4n1
Centigrade scale, 4n1
chalcone isomerase, 392*f*
Champagne, 286–89
Chardonnay, 38, 127–29, 129*f*, 130*f*, 283*f*, 286, 397*t*
Chemical Abstracts Service (CAS), xvin6
chemistry, 353n2
chemistry primer, 307–17, 354–55
    breaking bonds, 318–21
    carbocycles and heterocycles, 348–53
    oxidation and reduction, 321–48
Chenin blanc, 129–31, 225
chirality, 332, 335
chlorination, 316*f*
    of methane, 322
chloroalkanes, 316

chlorobutane, 322–23
1-chlorobutane, 323, 323*f*
2-chlorobutane, 323*f*
chlorogenic acid, 107*f*
chlorophyll, 43, 160, 340*f*
chlorophyll-a, 47–48, 51
chlorophyll-bearing-proteins, 47, 47*f*
chloroplast envelope, 42*f*
chloroplasts, 23, 24*f*, 31, 36, 39, 41*f*, 44, 192
    components of, 40–43
chloroplast stroma, 46*f*
chloroplast wall, 44, 46*f*
cholesterol, 29, 110, 110*f*
choline, 29, 31*f*, 330, 330*f*
chorismate, 103*f*, 385, 385*f*, 386*f*, 387*f*
chroman, 221*f*, 222*f*
4-chromanone, 222*f*
2*H*-chromene, 221*f*
4*H*-chromene, 221*f*
chromone, 222*f*
chromophore, 33, 34n15, 78–79
chrysanthemin, 78, 79*f*
trans-cinnamate, 141*f*
trans-cinnamate 4-monooxygenase, 141*f*, 390*f*
cinnamates, 210
cinnamic acid, 107, 107*f*, 123, 237, 240*f*
(*E*)-cinnamic acid, 390*f*
trans-cinnamic acid, 390*f*
trans-cinnamyl isovalerate, 154*f*
Cinsaut, 152
citral, 131
citrate, 216
citric acid, 93, 206, 207*f*, 307, 321, 321*f*
(−)-citronellol, 125*f*, 134*f*
(+)-citronellol, 134*f*
(*R*)-(+)-citronellol, 137*f*, 290*f*
citrus fruit, 321
Clarke, Oz, 140, 152, 155
clay fining agents, 231–32, 233*f*
climate, 3
coenzyme A (CoA), 189*f*, 361, 364
colors, 336–38, 337*f*, 339*f*
    appearing on grapes, 106
    compounds and, 382–95
    of flowers, 76–81
    of light, 76, 77*f*
    phenol pH and, 78
    of product, 116

colors (Cont.)
   skin and, 116
   wavelength regions, 80f
column chromatography, 338
common enol, 94, 94f
compounds, 312. *See also specific compounds*
   aroma, 216
   brightly colored, 340f
   colors and, 382–95
   heterocyclic, 350f
   odor-active, 268
   truly odiferous, 278–79
   yeast and presence of, 217–18
concentration gradients, 18
coniferyl alcohol, 238f
conjoined acetate units, 221f
conjugated bonds, 325
Conservation of Energy law, 319
(−)-α-copaene, 128f
α-copaene, 165, 166f
ent-copalyldiphosphate, 84f
ent-copalyl diphosphate synthase, 84f
corked wine, 253, 255, 288, 332, 354n14, 355
corolla, 69, 69f
cotyledons, 7, 31
4-coumarate, 141f
para-coumarate, 141f
4-coumarate-CoA ligase, 390f
coumaric acid, 101, 102, 103f, 107f, 237, 240f
para-coumaric acid, 104f, 245f, 390f
para-coumaryl alcohol, 238f
4-coumaryl-CoA, 390f, 392f
cis-para-coumeric acid, 125f
trans-para-coumeric acid, 125f
coumesterol, 225f
coutaric acid, 132f
cis-coutaric acid, 126f
trans-coutaric acid, 126f
covalent bonding, 310, 317–18
creosol, 247f
Crick, Francis, 351f, 353
cross-fertilization, 83
cross-linked polysaccharides, 179
cross-linking, 35
crushing, 180f
   for red wine, 183
   for white wine, 179–80
cryptochromes, 31
crystallographic methods, 342

(−)-α-cubebene, 128f
cultivars, 5n7, 15n1
cumulative diene,1,2-butadiene, 327f
cumulative double bonds, 325
curcumene, 206f
cuticle, 35
cutin, 35
cuttings, hardwood, 14–15, 14f, 15n1
cyanidin, 77f, 79f, 145t, 151f, 153f, 211, 224f, 239, 241f, 389f
cyanidin 3-*O*-β-D-glucoside, 393f
cyclic adenosine monophosphate (cAMP), 301, 301f
cyclic hydrocarbons, 321, 322
cyclized aldehydes, 386
cyclobutane, 324, 348, 350f
cyclohexane, 237, 323f, 348, 350f
1,3,5-cyclohexanetrione, 392f
cyclopentane, 323f
cyclopropane, 323f, 348, 350f
cyclotene, 245f
para-cymene, 140f, 157f
Cys-89, 363f
Cys-111, 363f
cysteine, 51f, 52f, 227f, 331f, 363f, 364
4-(cytidine 5′-diphospho)-2-*C*-methyl-D-erythritol, 368f
4-(cytidine 5′-diphospho)-2-*C*-methyl-D-erythritol kinase, 368f
cytidine diphosphate, 366, 368
cytidine monophosphate, 368f
cytidine triphosphate, 367f
cytidylyltransferase, 366, 367f
cytochrome b6f complex, 49
cytochrome P_{450} hemoproteins, 299
cytochrome P_{450} mono-oxygenases, 85
cytochrome P_{450} oxidase, 84f
cytokinins, 11, 83f, 109f
cytoplasm, 24f, 26, 361–71
cytosine, 12f, 13n9, 189, 190f, 226f, 351f, 355n18

2,4-D. *See* 2,4-dichlorophenoxyacetic acid
damascenone, 164f, 397t
(*E*)-β-damascenone, 138f
β-damascenone, 397t, 398t
trans-β-damsacenone, 206f
dark reactions, 39
Darwin, Charles, 7, 23, 31
Darwin, Francis, 7, 23, 31

(2E,4E)-2,4-decadienals, 130f
δ-decalactone, 397t
decanoic acid, 139f, 397t
decarboxylation, 213f, 214, 215f, 216, 361, 362f, 364, 366f
   to phenylalanine, 385
   to tyrosine, 385
decoupled carbon spectrum of ethanol, 346, 347f, 348f
decoupled spectra, 355n17
degraded metabolites, 369
dehydration, 213, 213f, 219, 360f, 384f, 385
2-dehydro-3-deoxy-7-phospho-D-arabinoheptanoate, 386f
dehydroascorbic acid, 180, 181f, 234f
(+)-dehydrodiconiferyl alcohol, 241f
3-dehydroquinate, 383f, 386f
3-dehydroquinate dehydratase, 384f
3-dehydroshikimate, 103f, 384f, 386f
3-dehydro-β-ionone, 220f
trans-3-dehydro-β-ionone, 129f, 164f
delphinidin, 145t, 153f, 225f, 239, 241f, 243, 243f
dendranthemoside A, 164f
3-deoxy-7-phosphoheptulonate synthase, 383f
deoxyadenosine 5′-monophosphate, 351f
deoxycytidine 5′-monophosphate, 351f
3-deoxy-D-arabinohept-2-ulosonate 7-phosphate, 383f
1-deoxy-D-xylulose 5-phosphate, 366f
1-deoxy-D-xylulose 5-phosphate reductoisomerase, 366, 367f
1-deoxy-D-xylulose 5-phosphate synthase, 154, 366
deoxyguanosine 5′-monophosphate, 351f
deoxyribonucleic acid (DNA), 225, 351f, 352f, 353
deoxythymidine 5′-monophosphate, 351f
1-deoxyxylulose 5-phosphate, 361, 365
dessert wines, 277
destemming, 115, 116, 179
detectors, 339
diacetyl, 213f, 214, 215f, 216, 217f
diadzein, 224f
diadzein-7-O-glucoside, 223f
diatomaceous earth, 182
dicarboxylic acid lactones, 386
Dichlofluanid, 166, 169f
N-{[Dichloro(fluoro)methyl]sulfanyl}-N',N'-dimethyl-N-phenylsulfuric diamide, 166

dichloromethane, 316, 316f
   tetrahedral configuration, 317f
2,4-dichlorophenoxyacetic acid (2,4-D), 106, 106f
1-[2-(2,4-dichlorophenyl)-pentyl]-1H-1,2,4-triazole, 167
9,9′-$dicis$-ζ-carotene desaturase, 149f
dicotyledons, 30
3,5-didehydroshikimate, 103f
diendiol-1, 137f
1,1-diethoxyethane, 290f
4,4-diethyl-3-methylene-1-oxetan-2-one, 138f
diethyl esters, 140f
diethyl ether, 342, 346
(S)-diethylmalate, 140f
diethyl pentanedioate, 144f
diethyl succinate, 140f, 144f
diffusional reorientation, 97
diglycosides, 208
dihydro-2(3H)-furanone, 261f
(5S,9R)-3,4-dihydro-3-oxoedulan, 129f
(2R)-2,3-dihydro-5,6-dihydroxy-1H-indole-2-carboxylic acid, 389, 390f
dihydro-5-methyl-3(2H)-furanone, 205f
6,7-dihydro-7-hydroxylinalool, 158f
dihydroactinidiolide, 129f
dihydroflavonols, 390
dihydroflavonol taxifolin, 392f
dihydrokaempferol 4-reductase, 392f
dihydrolipoyl dehydrogenase, 189, 190f
dihydrolipoyllysine-residue acetyltransferase, 189, 190f
dihydrophaseic acid, 138, 141f
dihydroxyacetone monophosphate, 62f, 63, 64f, 195, 195f
2,3-dihydroxybutane, 217f
6,7-dihydroxycoumarin, 125f
3,4-dihydroxyphenylalanine, 388, 388f, 389f, 390f
diketopiperazine, 108f
2,6-dimethoxy-4-(2-propenyl)phenol, 159f
trans-3,5-dimethyl-1,2,4-trithiolane, 131, 226f
(E)-4,8-dimethyl-1,3,7-nonatriene, 158, 160f
3,7-dimethyl-1,3-octadien-7-ol, 158f
(Z)-3,7-dimethyl-1,4-octadiene-3,7-diol, 151f
(E)-2,6-dimethyl-3,7-octadiene-2,6-diol, 165f
3,6-dimethyl-3a,4,5,6a-tetrahydro-2H-1-benzofuran-2-one, 396t, 397t
2,5-dimethyl-3-ethylpyrazine, 248f, 249

1,1′-dimethyl-4,4′-bipyridinium dichloride (paraquat), 106, 106f
dimethylacetylene, 325
dimethylallyl diphosphate, 109, 110f, 148f, 201, 202f, 359f, 364, 365f, 369, 369f, 370f, 373n5
dimethylallyl transferase, 202f
dimethylallyl *trans*-transferase, 369, 370f
dimethyl diphosphate, biosynthesis pathways, 361–69
dimethyl disulfide, 205f
dimethyl ether, 342, 349f
4,6-dimethyl-*N*-phenylpyrimidin-2-amine, 166
3,7-dimethylocta-1,6-dien-3-ol, 397t, 398t
2-[(2,6-dimethylphenyl)-(2-methoxy-1-oxoethyl)-amino]propanoic acid methyl ester, 166
2,5-dimethylpyrazine, 248f, 249
2,6-dimethylpyrazine, 205f
dinitrogen ring, 350
dioxane, 350, 350f
3-dioxygenase, 392f
diphosphate, 110f
diphosphate 2-phospho-4-(cytidine 5′-diphospho)-2-C-methyl-D-erythritol, 368f
diphosphocytidine, 367f
(*R*)-5-diphosphomevalonate, 364, 365f
Dirac, Paul, 353n4
distilled beverage, 259–60
diterpenes, 369
dithiocarbamate, 105f
DNA. *See* deoxyribonucleic acid
1-dodecanol, 144f
double bonds, 325
double-headed arrows, 325
double helical DNA, 352f
double ion, 330
Douro Valley, Portugal, 260
drainage, 16
drinking wine, 297–304
Dumas, Alexander, 354n8

E-1-hydroxy-2-methylbut-2-enyl-4-diphosphate reductase, 369
E-2-butene, 325
E-2-hexen-1-ol, 398t
E-3,7-dimethyl-2,6-octadien-1-ol, 398t
egg white, 231, 232

electromagnetic radiation, 336
electromagnetic spectrum, 337, 337f
electrons, 308
  two-spin pairs, 311, 311f
electron volt, 354n5
ellagic acid, 169f, 245f
Embden-Meyerhof-Parnas pathway, 194
embryogenesis, 31
embryonic leaves, 32f
emendation, 183
emission lines, 334
enamine, 195f
enantiomers, 332
endoplasmic reticulum, 24f, 26
energy, 310
enol, 248f
enolase, 197f
enolate, 360f, 362f
enolpyruvate, 213f
(+)-enterodiol, 224f
enterolactone, 224f
entgegen (opposed), 324
enzymatic processes, 27, 34n5
enzyme-catalyzed reduction, 94
Enzyme Database, 100, 356
enzymes, 6, 8n6
  kinase, 360f
epi-brassinolide, 109, 110, 110f
epicatechin, 132f, 169f
(−)-epicatechin, 125f, 126f, 223f
epicatechingallate, 124f, 132f
epidermal cells, 18
epidermis, 35
epigenetic, 39
5,6-epoxy-3-hydroxymegastigma-7-en-9-one, 164f
equilibrium, 4n1, 319, 319f
equol, 225f
ergosterol, 202, 202f, 287f
eriodictyol, 224f
erythrose 4-phosphate, 103f, 386f
esterification, 197f, 222
esters, 112, 139f, 140f, 142, 156f, 206, 307, 326. *See also specific esters*
  hydrolysis of, 232
estufagem process, 272, 276, 278f
ethanal, 188, 299, 354n14
ethane, 322f
ethanoic acid, 327, 354n14

Index

ethanol, 93, 182, 192, 206–7, 207f, 217f, 227, 234, 299f, 307, 323, 327, 328f, 329, 341–43, 342f, 345f, 349f
  boiling point of, 312
  decoupled carbon spectrum of, 346, 347f, 348f
  glucose conversion to, 194–98
  metabolism of, 299
  oxidation of, 242
  toxicity, 300
  in wine, 299
  yeast tolerance of, 177–78
ether, 243f
2-ethyl-1-hexanol, 138f
ethyl 2,5-dimethyl-3-pyrazinylacetate, 248f, 249
ethyl (E)-2-butenoate, 139f
ethyl (E)-2-hexenoate, 139f
ethyl 2-methylbutanoate, 134f, 139f, 397t, 398t
1-ethyl-3-(2-methyl-2-propanyl)benzene, 140f
ethyl 3-methylbutanoate, 139f, 144f, 397t
ethyl 3-(methylthio)propanoate, 140f
2-ethyl-4-hydroxy-5-methyl-3(2H)-furanone, 397t
(R)-ethyl 5-oxopyrrolidene-2-carboxylate, 261f
ethyl acetate, 139f, 144f, 208f, 290f, 327, 328f, 376
ethyl alcohol, 302n2, 327
ethyl benzoate, 140f, 144f
ethyl butanoate, 139f, 144f, 208f, 291f, 397t, 398t
ethyl butyrate, 376
ethyl carbamate, 207, 207f
ethyl cinnamate, 211f, 376
ethyl decanoate, 144f, 208f
ethyl dodecanoate, 144f
ethylene, 20, 20f, 85, 108, 161
ethyl ester of acetic acid, 328f
ethyl esters, 139f, 206, 207, 248f, 249
ethyl ethanoate, 327
ethyl ether, 123
ethyl formate, 376
ethyl furan-2-carboxylate, 218f
ethyl geranate, 137f
ethyl glycine, 248f
4-ethylguaiacol, 154f, 206, 207f, 244f, 247f
ethyl heptanoate, 144f, 376
ethyl hexadecanoate, 144f
ethyl hexanoate, 134f, 144f, 208f, 291f, 377, 396t, 397t
2-ethylhexanoic acid, 139f
ethyl isovalerate, 377
ethyl lactate, 208f, 377
ethyl (R)-(+)-lactate, 144f
ethyl (S)-(-)-lactate, 140f
ethyl laurate, 208f
ethyl methyl succinate, 140f
ethyl nonanoate, 377
ethyl octanoate, 208f, 291f
ethyl palmitate, 208f
ethyl pentanoate, 377
3-ethylphenol, 397t
4-ethylphenol, 206, 207f, 244f
para-ethylphenol, 154f
ethyl propionate, 139f
ethyl sorbate, 139f
ethyl tetradecanoate, 144f
ethyl vanillate, 154f
ethyl vanillyl ether, 245f
eucalyptol, 128f, 157f
eudicots, 154
eugenol, 167f, 211f, 244f, 247f
eukaryotes, 192
eukaryotic cells, 17n1, 192, 193f
European Bioinformatics Institute, 333f
European grapevine moth, 158
europinidin, 145t
exocarp, 91
exocyclic alkene, 383f
EXTOXNET, 105

FAD. *See* flavin adenine dinucleotide
FADH. *See* reduced flavin adenine dinucleotide
Fahrenheit, Daniel Gabriel, 4n1
Fahrenheit scale, 4n1
(E,E)-α-farnescene, 158, 160f
farnesol, 109
(2E,6E)-farnesol, 203f
farnesyl diphosphate, 84f, 85, 110f, 148f, 202, 202f, 369, 370f, 371, 371f, 372f
(2E,6E)-farnesyl diphosphate, 369, 370f
fatty acid ethyl esters, 206
fatty acids, 398t
fermentation
  end of, 231
  malolactic, 184, 214
  measuring process of, 182
  of red wine, 183–84
  second, 288
  secondary, 214
  vessels for, 181–82
  of white wine, 181–82
  yeasts for, 192

fermentation response genes (FRS), 227
ferredoxin, 51, 52
fertilization, 81, 82f, 83
ferulic acid, 107f, 223f, 240f, 390f
(E)-ferulic acid, 169f
cis-ferulic acid, 125f
trans-ferulic acid, 125f
Feynman, Richard, 353n4
field work, 105
filtering, 182
fining, 212, 231–32, 233f, 235, 242, 249
Fino, 265, 269f, 270f
Fino sherry, 267
flavan-3,4-diol leucopelagonidin, 392f
flavan-3-ols, 123, 209f, 222, 222f
flavanols, 236, 243, 392f
flavanone, 221, 222f
flavin adenine dinucleotide (FAD), 33, 33f, 44, 45f, 52, 53f, 362f, 365
flavin mononucleotide (FMN), 226f, 365, 385f
flavon-3-ol, 222f
flavones, 221, 222f
  ring systems, 141f
flavor precursor fraction, 209, 210
flavylium cation, 393f
flesh, of grape berries, 91
flor, 266, 267
floral notes, 210
flowers, 6
  bloom sequence, 68–69, 69f
  colors of, 76–81
  formation of, 67–69, 73, 75–76
  grapes from, 81–83, 85
  headspace above, 69, 70t–73t, 73
  opening, 81
  smells of, 369–72
FMN. *See* flavin mononucleotide
folic acid, 33f
formaldehyde, 328f
formic acid, 328f
formonetin, 225f
fortified beverages, 259
Fourier transform infrared (FT-IR), 260
free phenols, 223
Freon 11, 244
FRS. *See* fermentation response genes
α-D-fructofuranose, 109f, 194f, 290f, 350f
β-D-fructofuranose, 109f, 194f
β-D-fructofuranose 6-phosphate, 65f, 195, 195f

D-fructopyranose, 238f
fructose, 63, 188, 194, 350, 350f
fructose 1,6-bisphosphate, 61, 62f, 63, 65f
fructose-6-phosphate, 194
fructose bisphosphate aldolase, 61, 62f, 64f, 94
fructose-bisphosphate aldolase, 195, 195f
fructosonic acid, 238f
fruit, defining, 82
FT-IR. *See* Fourier transform infrared
fumaric acid, 206, 207f
fumigants, 105
functional groups, 321
fungicides, 166–67, 169f, 177
fungus, 277
furan, 350, 350f
furan-2-carboxaldehyde, 218f
furan derivatives, 218f, 219, 249
furaneol, 134f, 154f, 219f, 397t
furfural, 205, 218f, 244f, 247f, 273
furfuranol, 205f
furfuryl alcohol, 245f
furfuryl ethyl ether, 245f
2-furylmethanethiol, 205f
fused-ring aryl-pyran systems, 221f, 222f

L-galactono-1,4-lactone, 100, 101f
L-galactono-1,4-lactone dehydrogenase, 100, 101f
galactonolactone, 100f
GDP-1-α-galactopyranose, 100f
L-α-galactopyranose, 100
L-α-galactopyranose 1-phosphate, 101f
galactose, 101f, 236f
L-galactose, 99, 99f
L-galactose 1-dehydrogenase, 101f
L-galactose-1-phosphate, 99
galestro soil, 157
gallic acid, 103f, 104f, 107, 125f, 146f, 169f, 223f, 238f, 242f, 243f, 245f
gallic acid ester of epicatechin, 132f
gallic acid esters, 239
gallocatechin, 243f
(+)-gallocatechin gallate, 222f, 243f
gallocatechol, 241, 243
Garganega, 283f
gas, 288
gas chromatographic separation, 278
gas chromatography, 340
gas chromatography with mass spectrometry (GC-MS), 130, 134, 140, 147f, 167

Index

gas-liquid partition chromatography (GLPC), 339
GC-MS. *See* gas chromatography with mass spectrometry
GC-O technique odorants, 396, 397*t*
GDP-1-L-galactopyranose, 99, 99*f*, 100
GDP-mannopyranose, 98*f*
GDP-mannose-3,5-epimerase, 99*f*
gegenion, 8n5
gel electrophoresis, 339
gel permeation chromatography (GPC), 339
generic plant cell, 40*f*
genetically dictated timer, 6
genetic code, 352–53
genistein, 225*f*
genistein-7-O-glucoside, 223*f*
genome, 4n1, 26
genomes, xx
geranial, 131, 136*f*
geranic acid, 125*f*, 137*f*, 165*f*
geraniol, 128*f*, 131, 134*f*, 137*f*, 143*f*, 151*f*, 155*f*, 156*f*, 158*f*, 163*f*, 165*f*, 168*f*, 210*f*, 290*f*, 396*t*, 398*t*
geranyl acetate, 137*f*, 377
geranylacetone, 71*t*, 137*f*, 143*f*
geranyl butyrate, 377
geranyl diphosphate, 110*f*, 148*f*, 202, 202*f*, 361, 366
geranylgeranyl diphosphate, 84*f*, 148*f*, 361, 366
geranylgeranyl diphosphate synthase, 84*f*
geranyl pentanoate, 378
German Rieslings, 276–77
germination, 6, 7*f*, 8n2, 81
Gewürztraminer, 38, 131–32, 134–35, 134*f*, 135*f*, 136*f*, 138*f*, 139*f*, 140*f*, 396*t*, 397*t*
  iced wine, 289–90, 290*f*
  odorants in, 290*f*, 291*f*
gibberellic acid, 11, 11*f*, 84*f*, 109*f*
gibberellin A_{12}, 84*f*, 85, 85*f*
gibberellins, 11, 85, 85*f*
Gila monster, 332
gingerone, 159*f*
glass, light absorption by, 234
glasses, 297
globulin, 232
GLPC. *See* gas-liquid partition chromatography
β-D-glucofuranose, 218*f*
gluconeogenesis, 63, 64*f*, 65*f*, 66, 93
gluconic acid, 102*f*, 132*f*, 281–82, 282*f*
D-gluconic acid, 290*f*
D-gluconic acid lactone, 238*f*
D-glucopyranose, 238*f*
α-D-glucopyranose, 109*f*, 194*f*, 242*f*, 350*f*
α-D-1→4-glucopyranose, 64*f*
β-D-glucopyranose, 98*f*, 109*f*, 194*f*, 218*f*, 290*f*
β-D-glucopyranose-1-phosphate, 96*f*, 98*f*
glucopyranoside, 146*f*
β-D-glucopyranoside, 195*f*
β-D-glucopyranosyl-β-D-glycoside, 209*f*
glucose, 27*f*, 37*f*, 60*f*, 65*f*, 213, 236*f*, 246, 247*f*, 350, 350*f*
  ethanol conversion from, 194–98
  synthesis of, 63
D-glucose, 99
glucose-fructose enol, 195*f*
α-D-glucose phosphate, 65*f*
glucosidases, 268
β-D-glucoside, 209*f*
glutamate, 198*f*, 363*f*, 364, 365*f*, 387*f*
glutamic acid, 50*f*, 197, 331*f*, 358*f*
glutamine, 331*f*
glutathione (GSH), 180, 181*f*
glycans, 27
glyceraldehyde, 248*f*, 249, 282
D-Glyceraldehyde, 63
glyceraldehyde-3-phosphate, 59–61, 62*f*, 64*f*, 195*f*, 196, 196*f*, 214, 214*f*, 228n8, 361, 365, 366
D-glyceraldehyde 3-phosphate, 366*f*
glyceraldehyde-3-phosphate dehydrogenase, 196, 196*f*
glycerate, 214, 214*f*
glyceric acid, 58, 61*f*, 238*f*
glycerol, 32*f*, 213, 214*f*, 217*f*, 238*f*, 281–82, 282*f*, 290*f*, 307, 330*f*
glycerol dehydrogenase, 214, 214*f*
glycerone phosphate, 228n8
glycine, 246, 249, 330, 331*f*, 332*f*
glycitein, 225*f*
glycoaldehyde, 102*f*
glycogen synthase, 65*f*
glycolipids, 29
glycolysis, 93
glycones, 206
glycosidases, 208, 209*f*
glycoside hydrolases, 236
glycosides, 222
glycosylated derivative, 390*f*
glycosylated malvidin, 213
glycosyltransferases, 236

glyoxylic acid, 248f
GMP. *See* guanosine monophosphate
Golgi bodies, 24f, 26
GPC. *See* gel permeation chromatography
GPCRs. *See* G-protein coupled receptors
G-protein, 75f, 79
G-protein coupled receptors (GPCRs), 73, 74f, 75, 80, 298
grafting, 9n13, 10–13
grana, 41, 42
grape berries, 92f, 105, 107–13, 115–17
    analysis of, 114
    color appearing on, 106
    composition of, 179
    fruit set, 91–104
    micronutrients in, 102
    plasticity of, xv
    seeds, 91
grape flowers, 6
    perfume of, 369
grapefruit mercaptan, 396t
grape growing, 3–5
    from grafting, 10–13
    grapevine from seed, 6–9, 7f
    from hardwood cuttings, 14–15, 14f, 15n1
    roots of the vine, 18–20
    soil, 16–17
grape-harvesting, science of, 3
grape juice
    components of, 179
    particulate matter in, 180–81
grape leaves, 35–39
grapes. *See also specific grapes*
    biology of raising, 4
    complexity of, 223
    crushing, 179–80, 180f, 183
    evenness of ripening, 112–13
    flowers to, 81–83, 85
    formation of, 69
    washing, 177
grapevine gene bank, xx
grape vines
    in North America, 10–11
    primordium parts of growing, 67f
    shoots, 67, 68
grasshopper ketone, 129f, 164f
Graves region, Bordeaux, 276–77
gravitropism, 7–8
gravity, 7–8

greenhouse gas, 17n2
Grenache rosé, 396t, 397t
GSH. *See* glutathione
GTP. *See* guanosine triphosphate
guaiacol, 154f, 159f, 167f, 211f, 244f, 247f, 273
guanine, 12f, 13n9, 226f, 351f, 355n18
guanine isomers, 372f
guanosine diphosphate, 76f
guanosine monophosphate (GMP), 100f
guanosine triphosphate (GTP), 76f, 98f
guard cells, 36
α-gurjunene, 128f
gynoecium, 69

haloalkane, 322
halogen, 323
Handbook of Enology, 281
handedness, 332, 333–34
hardwood cuttings, 14–15, 14f, 15n1
harvest
    beginning of, 110–14
    evaluation of timing, 112
    manual, 115, 115f
    mechanical, 114–15
    methods of, 114–16
    science of, 3
harvesting light, 44–57
harvest wagon, 115–16
HCl. *See* hydrochloric acid
headspace, 86n8, 170n2
    analysis of, 123
    of flowers, 69, 70t–73t, 73
    yeast, 205f, 207f
head trimming, 38
Heisenberg, Werner, 353n4
hemiacetal, 243f
hemicellulose, 235, 245
hemiketal, 383f
Henry's Gas Law Constant, 288–89
Henry's Law, 288–89
1-heptanol, 144f
2-heptanol, 138f
2-heptanone, 138f
herbicides, 105, 106
hermaphroditic reproduction, 6
hesperetin, 225f
heterocycles, 218–25, 218f, 219f, 226f, 227, 348–53
heterocyclic compounds, 350f
heterolysis, 320

heterozygous outcrossers, 3
(*E,E*)-hexa-2,4-dienal, 131
hexadecanoic acid, 32*f*
2,4-hexadienal, 128*f*
(4R,4aS, 6R)-4,4a,5,6,7,8-hexahydronaphthl-ene-2(3H)one, 396*t*
hexan-1-ol, 134, 138, 152, 156
hexanal, 128*f*, 138*f*, 156*f*
hexane, 322*f*
hexanoic acid, 139*f*
1-hexanol, 128*f*, 131, 138*f*, 144*f*, 156*f*, 208*f*
(*E*)-3-hexen-1-ol, 138*f*
(*Z*)-3-hexen-1-ol, 128*f*, 138*f*
3-hexen-1-ol, 144*f*
cis-3-hexen-1-ol, 131
trans-3-hexen-1-ol, 125*f*
(*E*)-2-hexenal, 128*f*, 156*f*
cis-2-hexenal, 131
trans-2-hexenal, 125*f*, 131
(*E*)-2-hexene-1-ol, 138*f*
(*Z*)-3-hexenenyl acetate, 156*f*
(*Z*)-3-hexenyl acetate, 139*f*
(*Z*)-3-hexenyl butanoate, 128*f*, 156*f*
hexyl acetate, 139*f*, 144*f*, 156*f*, 208*f*
high-performance liquid chromatography with tandem mass spectrometry (HPLC-MS/MS), 114
high pressure liquid chromatography (HPLC), 152, 159, 223, 339
His-383, 363*f*
histidine, 50*f*, 51*f*, 100, 226*f*, 331*f*, 363*f*
HMG-CoA. *See* 3-hydroxy-3-methylglutaryl-CoA
homocysteine, 227*f*
homofuraneol, 397*t*
homolosine projection of Earth, 4*f*
homovanillic acid, 159*f*
hormones, 11–12, 13n7, 20, 108, 137, 210*f*
hotrienol, 136*f*, 158*f*, 165*f*
(*R*)-(−)-hotrienol, 143*f*
HPLC. *See* high pressure liquid chromatography
HPLC-MS/MS. *See* high-performance liquid chromatography with tandem mass spectrometry
human brain, MRI images of, 346*f*
α-humulene, 166*f*, 371*f*
hydrocarbon chlorination, 354n8
hydrocarbons, 321, 322*f*
hydrochloric acid (HCl), 150

hydrogen, 321, 327
 bonds, 312, 318*f*
 ion concentration, 354n11
 molecule, 311, 311*f*
hydrogen bonded water, 316–17
hydrogen bonds, 312
 aggregate threads comparison, 318*f*
hydrogen phosphate, 198*f*
hydrogen sulfide, 216
hydrolysis, 196, 207*f*, 219, 359*f*
 catalyzed, 243*f*
 of esters, 232
 of sucrose, 109*f*, 194, 194*f*
hydronium ion, 185, 186, 319, 319*f*
hydrophobic, 29*f*
hydroxide anion, 319, 319*f*
2-hydroxy-1,8-cineole, 163*f*
2-(3-hydroxy-1-buten-1-yl)-2,6,6-trimethyl-3-cyclohexen-1-ones, 220*f*
4-(3-hydroxy-1-buten-1-yl)-3,5,5-trimethyl-3-cyclohexen-1,2-diols, 220*f*
4-hydroxy-2,5-dimethyl-3(2*H*)-furanone, 247*f*, 397*t*
3-hydroxy-2-butanone, 398*t*
(*E*)-1-hydroxy-2-methylbut-2-enyl-4-diphosphate, 129*f*, 134*f*, 154*f*, 155*f*, 159*f*, 168*f*, 210, 211*f*, 220*f*, 368, 369*f*
(*E*)-1-hydroxy-2-methylbut-2-enyl-4-diphosphate reductase, 369*f*
(*E*)-1-hydroxy-2-methylbut-2-enyl-4-diphosphate synthase, 368
(*S*)-hydroxy-3-methylglutaryl-CoA, 364*f*
3-hydroxy-3-methylglutaryl-CoA (HMG-CoA), 361, 362
3-hydroxy-4,5-dimethyl-2(5H)-furanone, 397*t*
para-hydroxybenzaldehyde, 125*f*
para-hydroxybenzoic acid, 125*f*
2-(3-hydroxybut-1-enyl)-2,6,6-trimethylcyclohex-3-en-1-one, 164*f*
hydroxycinnamic tartaric esters, 123
hydroxydihydroedulans, 220*f*
hydroxyethylthiamine diphosphate, 361
5-hydroxyferulic acid, 240*f*
hydroxyimine, 248*f*
hydroxyl groups, 330*f*, 367
cis-8-hydroxylinalool, 163*f*, 165*f*
cis-4-hydroxymethyl-2-methyl-1,3-dioxolane, 261*f*
hydroxymethylfurfural, 245*f*
5-hydroxymethylfurfural, 218*f*

hydroxymethylglutaryl-CoA reductase, 361
4-hydroxyphenylpyruvate, 387f
4-hydroxyphenylpyruvate with prephenate dehydrogenase, 387f
hydroxypyruvate, 214
hydroxypyruvic acid, 214f
3-hydroxytheaspirane, 164f
8-hydroxytheaspirane, 129f
hypocotyl, 7
hypodermis, 93

Ice wines, 259, 289–91, 290f
Icelandic spar, 334
L-idonic acid, 102f
imines, 64f, 195f, 248f, 389f, 396t
impact odorants, 396–99
indicaxanthine, 388, 389f
indole-3-acetic acid, 11f, 24f, 83f, 109f, 210f
indole-3-butanoic acid, 11f
inflorescence primordia, 68
infrared light, 234, 342
infrared spectroscopy, 114
infrared spectrum
   1-butanol, 344f
   ethanol, 345f
inorganic phosphate (Pi), 44, 45, 53–54, 56f, 76f, 84f, 110f, 367f, 370f
inositol, 238f
Institut national de l'origine et de la qualité, 287
International System of Units, 353n5
The International Union of Pure and Applied Chemistry (IUPAC), xviin6
internode, 37
intramolecular phosphoryl transfer, 97
ion channels, 298
ion exchange chromatography, 339
ionic bond, 310
ionization, 321, 321f
β-ionone, 128f, 131, 146, 150f, 155f, 156f, 210, 211f, 219, 220f
ions, 6, 8n5
   concentration of hydrogen, 354n11
   double, 330
   hydronium, 185, 186, 319, 319f
   tastes and, 298
iron-sulfur domain, 52f
irradiation, 31
isoamyl acetate, 208f, 378, 398t

isobaric tags for relative and absolute quantification (iTRAQ), 114
isobutanol, 323f, 324
3-isobutyl-2-methoxypyrazine, 396t
isobutyl acetate, 208f, 378
isobutyl formate, 378
isoeugenol, 247f
(E)-isoeugenol, 167f
isohopeaphenol, 153f
isoleucine, 331f
isomer, 316
isomeric molecules, 342
isomerization, 94, 196
isopentenyl diphosphate, 84f, 109, 110f, 148f, 201, 203f, 364, 365, 365f, 369, 369f, 370f, 373n5
   biosynthesis pathways, 361–69
isoprenoids, 165f, 203, 219, 285f
(3S,5R,8S)-5-isopropenyl-3,4-dimethyl-3,4,5,6,7,8-hexahydro-1(2H) azulene, 396t
isopropyl acetate, 378
cis-isopulegone, 131
isorhamnetin, 124f, 126f
isotopes, 354n9
iTRAQ. See isobaric tags for relative and absolute quantification
IUPAC. See The International Union of Pure and Applied Chemistry

jasmonic acid, 20, 20f
Joule, 354n5

kaempferol, 124f, 126f, 169f, 223f
kaempferol-3 glucoside, 132f
kaempferol 3-O-rutinoside, 153f
kaolin, 231
ent-kaurene, 84f
ent-kaurene oxidase, 84f
ent-kaurenoic acid, 84f
kelvins, 354n5
α-keto-acid, 383f
ketocarboxylic acid, 102f
2-keto-L-gulonic acid, 102f
ketone carbonyl, 362f, 363f
ketones, 99f, 113f, 138f, 156f, 188, 195, 326, 341, 383f
keto-sugar diphosphate, 60f
keto-thioester, 392, 392f
Khayyam, Omar, 374n31
killer yeasts, 212–13
kinase enzymes, 360f

Index

kinetin, 83*f*
Krebs cycle, 95–96, 96*f*, 97*f*, 216, 356

lactate, 216
(*S*)-lactate, 216*f*
L-lactate dehydrogenase, 216*f*
lactate-malate *trans*-hydrogenase, 215, 215*f*
lactic acid, 58, 61*f*, 143, 206, 215*f*
(*S*)-(+)-lactic acid, 146*f*, 207*f*
lactic acid bacteria, 214, 215
Lactobacillales bacteria, 214
lactones, 210
5-(l-alanin-3-yl)-2-hydroxy-*cis,cis*-muconate 6-semialdehyde, 388*f*
4-(l-alanin-4-yl)-2-hydroxy-*cis,cis*-muconate 6-semialdehyde, 388*f*
lamina, 36
lanosterol, 109, 110, 110*f*
lateral root formation, 28–29
(*S*)-lavandulol, 137*f*
L-cysteine, 283*f*, 284
leaf blade, 37
leaves, 33*f*
    cells, 35–39, 36*f*
    embryonic, 32*f*
    light on, 39–43
    protection for, 37–38
    true, 31
    vine health and growth of, 112
Le Chatelier, Henry, 354n10
Le Chatelier's Principle, 289
lees, 181, 182
left handed, 334
lemons, 321
leucine, 331*f*
leucocyanidin oxygenase, 393*f*
leucopelagonidin, 392
LHC. *See* light-harvesting complex
light
    colors of, 76, 77*f*
    effects of, 234
    harvesting, 44–57
    infrared, 234, 342
    on leaves, 39–43
    polarized, 334–36, 335*f*, 336*f*
    sunlight, 39, 44, 107
    ultraviolet, 234, 235*f*, 337–38, 339*f*
    white, 336, 337*f*
light absorption, 340*f*

light-harvesting complex (LHC), 52
light reactions, 39
lignan, 237, 241*f*, 245
lignanols, 237
lignin, 27, 28*f*, 235, 237, 238–39, 239*f*
limetol, 157*f*
limonene, 157*f*, 158, 160*f*
(*R*)-limonene, 143*f*
(*R*)-(+)-limonene, 136*f*
(*S*)-(−)-limonene, 206*f*
(+)-limonene, 396*t*
linalool, 131, 157*f*, 158, 160*f*, 163*f*, 168*f*, 203, 210, 397*t*, 398*t*
(*R*)-(−)-linalool, 143*f*, 151*f*, 155*f*, 210*f*, 211*f*
(*S*)-(+)-linalool, 134*f*, 136*f*
linalool acetate, 157*f*
linalool oxide, 158*f*, 165*f*, 210*f*
(2*R*,5*S*)-*cis*-linalool oxide, 151*f*, 157*f*
(2*S*,5*S*)-*trans*-linalool oxide, 151*f*, 157*f*
(3*R*,6*S*)-*trans*-linalool oxide, 137*f*
cis-linalool oxide, 143*f*, 165*f*, 168*f*, 210*f*, 219*f*
trans-linalool oxide, 136*f*, 143*f*, 163*f*, 165*f*, 168*f*, 210*f*, 219*f*
linalyl acetate, 378
linalyl butanoate (linalyl butyrate), 379
linalyl formate, 379
linoleic acid, 238*f*
lipid bilayer, 29, 30*f*
lipids, 29
lipoic acid, 217*f*, 350, 350*f*, 361, 362*f*
lipoic acid amide, 361
liposome, 29, 29*f*
lipoyl sulfur, 361
Lobesia botrana, 158, 160*f*
local soil microbiome, 17
locules, 91
loliolide, 129*f*
lutein, 76, 77*f*
luteolin, 224*f*
lychee, 132, 134*f*
lycopene, 338, 339*f*
lysine, 64*f*, 81*f*, 195*f*, 197, 198*f*, 331*f*, 358*f*, 359*f*, 384*f*

Madeira, 233, 259–60, 268–76, 273*f*, 275*f*, 277*f*, 278*f*, 397*t*
Madeira Islands, 268
magnesium cation, 358–59, 358*f*, 359*f*
magnetic resonance imaging (MRI), 342, 343
Maillard reaction, 236, 246–47, 247*f*

malate, 216
L-malate, 96, 97f, 104f
S-malate, 216f
malate dehydrogenase, 216f
L-(-)-malic acid, 238f
(S)-(−)-malic acid, 146f
malic acids, 93, 104, 107, 143, 206, 214, 215, 215f, 307
malolactic fermentation, 184, 214
malonyl-CoA, 141f, 221f, 392f
malonyl-β-D-glucoside, 209f
maltol, 245f
Malvaisa, 268, 269, 273, 274f
malvidin, 145t, 153f, 161f, 340f
  glycosylated, 213
malvidin 3-glucoside, 113f, 114, 263, 264f, 265f, 266f, 382, 383f
malvidin-3-glucoside-(4,8'-catechin), 161f
malvidin-3-glucoside-(8,8'-ethylepicatechin), 161f
mannitol, 307
D-(+)-mannopyranose, 238f
GDP-1-α-D-mannopyranose, 99f
mannose, 236f
D-mannose, 99, 99f
mannose 1-phosphate, 97
mannose-1-phosphate guanyltransferase, 98f
mannose 6-phosphate, 97
β-D-manopyranose 1-phosphate, 98f
β-D-manopyranose 6-phosphate, 98f
manual harvesting, 115, 115f
marigolds, 76
mass spectrometry (MS), 114, 152, 159, 223, 263, 264f, 291n15, 340, 341, 341f, 342f, 355n15
Mastigocladus laminosus, 49
(−)-matairesinol, 224f
maturation, xv
3-mecaptohexan-1-ol-L-cysteine, 135f
mechanical harvesting, 114–15
(6Z,8E)-megastigma-4,6,8-trien-3-one, 129f, 164f
megastigma-4,7-diene-3,6,9-triols, 220f
megastigma-4-ene-3,6,9-triols, 220f, 221
(3S,5R,6S,9)-megastigma-7-ene-3,6,9-triol, 164f
megastigmatrienone D, 220f
melanopsin, 80
membranes, of chloroplast, 40–41
menth-1-ene-4,8-diol, 165f
Mentha spicata, 333f
L-(−)-menthol, 136f
3-mercapto-1-hexanol, 162, 396t, 397t
4-mercapto-4-methyl-1-pentanone-L-cysteine, 135f, 160f, 162f
4-mercapto-4-methyl-2-pentanone, 134
S-4-mercapto-4-methyl-2-pentanone, 156, 162
4-mercapto-4-methyl-2-pentanone-L-glutathione, 135f, 160f, 162, 162f
3-mercaptohexan-1-ol-L-cysteine, 127f, 130f, 142f, 152, 162
3-mercaptohexan-1-ol-L-glutathione, 130f, 135f, 152, 160f, 162f
meristems, 9n7, 10, 12n1, 18, 19f, 31, 33f
Merlot, 38, 135–38, 396t, 398t
  from Bordeaux, 397t
mesocarp, 91
mesophyll cells, 35, 37
messenger molecules, 13n7
metabisulfite salts, 113
  potassium, 182, 183, 185, 187, 187f
  sodium, 185
metabolites, 369
metabolome, xvin3
Metalaxyl, 166, 169f
metals, 16
3-metcaptohexan-1-ol-L-cysteine, 160f, 162f
methanamine, 328, 329f
methane, 315, 315f, 318f, 322f, 328f
methanoic acid, 328f
methanol, 138f, 314–15, 328f
methional, 154f
methionine, 217f, 227f, 331f
methionol, 154f
2-methoxy-3-(1-methylethyl)-pyrazine, 163f
2-methoxy-3-(1-methylpropyl)pyrazine, 163f
1-methoxy-3-(2-methylpropyl)pyrazine, 164f
2-methoxy-3-(2-methylpropyl)pyrazine, 163f
2-methoxy-3-isobutylpyrazine, 108, 108f
2-methoxy-4-vinylphenol, 210
2-methoxyphenol, 210, 261f, 398t
1-(4-methyl-1,3-thiazol-5-yl)ethanol, 205f
2-methyl-1-butanol, 208f
3-methyl-1-butanol, 138f, 208f, 397t, 398t
3-methyl-1-butanol acetate, 303n4
3-methyl-1-butyl acetate, 144f
2-methyl-1-propanol, 138f, 144f
5-methyl-2-furfural, 273
methyl 2-methylpropanoate, 397t
5-methyl-2-furylmethanol, 205f
2-(4-methyl-3-cyclohexen-1-yl)-2-propanyl butyrate, 381

Index

2-methyl-3-sulfanylfuran, 397t
(1S,4R)-1-methyl-4-(1-methylethenyl)-2-cyclohexane-1-hydroperoxide, 137f
1-methyl-4-(1-methyl-ethenyl)-cyclohexane, 396t
4-methyl-4-mercaptopentan-2-one, 396t
2-methyl-5-(methyl-sulfanyl)furan, 205f
methyl acetate, 208f, 379
methylamine, 328, 329f
methylammonium butanoate, 328, 329f
4-methylanisol, 140f
methyl anthranilate, 379
methyl benzoate, 379
2-methylbenzoic acid, 238f
2-methylbutanal, 205f
3-methylbutanal, 205f
N-methylbutanamide, 329f
methyl butanoate (methyl butyrate), 379
3-methylbutanoic acid, 205f, 397t
methylbutanoic acids, 397t
3-methylbutanol, 397t
methylbutanols, 397t
3-methylbutyl acetate, 139f, 397t
3-methylbutyl butanoate, 140f
3-methylbutyl decanoate, 140f
3-methylbutyl hexanoate, 140f
3-methylbutyl octanoate, 140f
methyl cinnamate, 380
5-methylcytosine, 189
2-C-methyl-D-erythritol 2,4-cyclodiphosphate, 368, 368f
2-C-methyl-D-erythritol 2,4-cyclodiphosphate synthase, 368f
2-methyl-D-erythritol 4-phosphate, 366, 367f
2-C-methyl-D-erythritol 4-phosphate cytidylyltransferase, 367
methylene, 346
methylene carbon, 188
5-methylfurfural, 205f, 244f, 247f
5-methylfurfuryl alcohol, 245f
5-methylfurfuryl ethyl ether, 245f
methyl group of ethanol, 330f
4-methylguaiacol, 245f
methyl hexanoate, 140f
methyl octanoate, 140f, 144f
methyl pentanoate, 380
methyl phenyl acetate, 380
2-methyl propanoate, 397t
2-methylpropanol, 208f
2-methylpropyl acetate, 139f

methyl salicylate, 158, 160f, 380
3-(methylsulfanyl)-1-propanol, 205f
3-(methylsulfanyl)propanal, 205f
2-methyltetrahydrothiophene-3-one, 397t
methyl valerate, 380
methyl vanillyl ketone, 159f
β-methyl-γ-octalactone, 261f
(R)-mevaldate, 364f
(R)-mevalonate, 364, 365f
mevalonate kinase, 364
mevalonic acid, 104f
micelle, 29f
microbiome, 17, 17n4
microtubule, 24f, 26
middle leaf, 36
mineral deficiency, 16
minerals, 307
    in grape juice, 179
mirror symmetry, 332, 333, 334f, 335
mitochondria, 24f, 26, 373n5
monolignols, 239
3′-monooxygenase, 392f
monophenol monooxygenase, 390f
monopotassium salt, 335
monosodium salt of glutamic acid (MSG), 204, 204f
Moscato di Noto, 140
Moscato di Pantelleria, 140
Moscato di Siracusa, 140
Moscato rose del Trentino, 139
Mourvèdre, 142–43, 146f, 147f
mouth feel, 113, 298
    of Merlot, 135
MRI. See magnetic resonance imaging
MS. See mass spectrometry
MSG. See monosodium salt of glutamic acid
Muscat, 139–40, 142, 144f, 396t
Muscat Blanc à Petits Grains, 139
Muscat Giallo, 139
Muscat Hamburg, 139
Muscat Norway, 140
must, 170n2, 179, 183, 207
α-muurolene, 128f, 156f, 166f
γ-muurolene, 128f
myrcene, 131, 143f
α-myrcene, 157f
β-myrcene, 136f
myrcenol, 158f
myricetin, 124f, 126f, 169f, 223f

NAD+. *See* nicotinamide adenine dinucleotide
NADH, 199*f*, 299*f*, 366
NADP+. *See* nicotinamide adenine dinucleotide 2'-phosphate
NADP+, 384*f*
NADPH. *See* nicotinamide adenine dinucleotide; reduced nicotine adenine dinucleotide 2'-phosphate
naphthalene, 339, 339*f*
naphthalene-1-acetic acid, 83*f*
α-naphthylacetic acid, 11*f*
naringenin, 153*f*, 224*f*, 392*f*
naringenin chalcone, 390
naringenin-chalcone synthase, 141*f*, 390
naringin, 224*f*
nasal receptors, 298
National Pesticide Information Center, 106
Naturalis Historia (Pliny the Elder), 143
n-butanol, 150
n-butyl alcohol, 150
neat ethyl ether, 344*f*
neat liquid spectra, 344*f*
Nebbiolo, 143, 145–46, 219, 220*f*
necrotrophs, 280
neolignans, 239
nephelometer, 181
neric acid, 168*f*
nerol, 137*f*, 143*f*, 151*f*, 155*f*, 158*f*, 163*f*, 210*f*, 396*t*
(S)-(+)-(Z)-nerolidol, 144*f*
trans-nerolidol, 137*f*
nerolidyl diphosphate, 371*f*, 372*f*
nerol oxide, 157*f*, 165*f*
neryl acetate, 143*f*
neutral, 320
neutrons, 308
newton, 354n5
Newton, Isaac, 76
Nicol, William, 334
nicotinamide adenine dinucleotide (NADPH/NAD+), 199*f*, 300*f*, 362*f*, 364, 364*f*, 366, 384*f*
nicotinamide adenine dinucleotide 2'-phosphate (NADP+), 44, 45*f*, 52, 53*f*, 54, 196, 369
night blindness, 80, 80*f*
nitrogen, 16, 327–28, 350
nitrogen heterocyclic compounds, 350*f*
nitrous oxide, 17n2
NMR. *See* nuclear magnetic resonance
Noble Gases, 309
"Noble Rot." *See Botrytis cinerea*
node, 37
1-nonanol, 138*f*, 144*f*
2-nonanone, 138*f*
nonyl caprylate, 380
nonyl octanoate, 380
(+)-nootkatone, 396*t*
norisoprenoids, 219
norisoprenoid β-ionone, 150*f*
norterpenes, 205
   in Riesling grapes, 159*f*
norterpenoids, in Port, 262*f*
North America, grapevines in, 10–11
North American Oak barrels, 266
nose, 73, 74*f*
nouaison, 91
nuclear magnetic resonance (NMR), 260, 342, 343–47
nuclear magnetic resonance spectroscopy, 114
nucleic acids, 179
nucleoids, 39, 42
nucleolus, 24*f*, 26
nucleus, 24*f*, 26

oak
   aging in, 235
   amino acids and, 246, 248*f*
   carbohydrates in, 236, 237
   casks, 177, 272
   compounds in, 235–45, 238*f*
   tannins in, 235
   toasted, 244–47
   volatiles related to, 246, 247*f*
cis-oak lactone, 244*f*, 247*f*
trans-oak lactone, 244*f*, 247*f*
Ochagavia, Don Silvestre, 135
cis-ocimene, 157*f*
trans-ocimene, 157*f*
trans-β-ocimene, 136*f*
α-ocimene, 143*f*
cis-β-ocimenol, 158*f*
trans-ocimenol, 143*f*
trans-β-ocimenol, 158*f*
octadecanoic acid, 238*f*
octanal, 128*f*, 138*f*, 398*t*
octanoic acid, 139*f*, 397*t*
1-octanol, 138*f*, 144*f*
3-octanol, 138*f*
3-octanone, 138*f*

1-octen-3-ol, 138f, 144f
octyl acetate, 303n4, 380
octyl butanoate, 380
octyl butyrate, 380
odor-active compounds, 268
odorants, in Gewürztraminer, 290f, 291f
OEC. *See* oxygen-evolving complex
oenin, 382, 383f
oestragole, 206f
oil of wintergreen, 380
old tawny port, 261
oleic acid, 32f, 36f
olfaction, 75
olfactory receptors, 332
olfactory stimuli, 73, 271f
Oloroso, 265, 269f, 270f
  sherry, 267
O-methyl derivatives, 390
operational taxonomic units (OTUs), 117
opsins, 79, 80
orbitals, 310, 353n3
organelles, 23
organic compounds, functional groups and classes of, 321–48
ornithine, 207f
OTUs. *See* operational taxonomic units
outside cell membrane, 31f
ovary, 69, 69f
oxalic acid, 101, 281
oxaloacetate, 107, 213, 213f, 215, 216
oxaloacetate acetyl hydrolase, 281
oxaloacetate decarboxylase, 216f
oxaloacetic acid, 215, 215f, 216f
oxaloglycolate, 214
oxaloglycolic acid, 214f
oxidasic casse, 183
oxidation, 214, 214f, 215, 219, 234f, 321–48, 325f, 326f, 327f, 328f, 390f
  of ethanol, 242
  in finishing, 231
  of plastoquinol, 51
  in serving, 298
  of water, 48
oxidative phosphorylation system, 52
oxidoreductase, 52, 52f
oxo-acid dehydrogenase complex, 361
3-oxo-α-ionol, 129f, 209f
3-oxo-β-ionone, 168f
oxygen, 7, 59, 63, 302n3, 327, 346, 350

oxygen-evolving complex (OEC), 48–49, 49f, 50f
oxygen heterocycles, 350
oxygen transport, 7
ozone, 185, 186f

pallidol, 153f
palmitic acid, 31f, 32f, 35, 36f
Palomino Fino grape, 265, 267, 271f, 272f, 273f
pantolactone, 154f, 167f
paraquat, 106, 106f
parenchyma cells, 8, 36–37
particulate matter, 180–81
parts per million (ppm), 293n55
PAS. *See* photoacoustic spectroscopy
PDB. *See* Protein Data Bank
pectin decomposition, 281
pelargonidin, 145t, 225f
Penconazole, 167, 169f
pentane, 322f
1-pentanol, 144f
pentyl acetate, 380
pentyl butyrate, 381
(5*S*)-5-pentyldihydro-2(3*H*)-furanone, 219f
pentyl hexanoate, 381
pentyl pentanoate, 381
peonidin, 145t
PEP. *See* phosphoenolpyruvate
peptide bonds, 330
peptides, 307, 330
perfume, of grape flowers, 369
Perfumed Traminer, 131
pericarp, 91
Periodic Table, 308–10, 308f, 312f
pesticides, 177
pest management, 105
petroleum derivatives, 350
trans-petrosilbene, 225f
petunidin, 145t, 153f
pH
  adjustment of, 113
  measurement of, 114
  phenolic compound color and, 78
phaeophytins, 48
α-phellandrene, 136f
2-phenethyl acetate, 139f
phenolic acids, 244, 245f
phenolic aryl groups, 221
phenolic hydroxyl, 388

phenolic hydroxyl groups, 223
phenolphthalein, 78, 79*f*
phenols, 112, 113, 124, 206, 222, 223*f*, 224, 224*f*, 225*f*, 307. *See also specific phenols*
  free, 223
  in grape juice, 179
  pH and colors of, 78
  ripeness and, 114
  in Viognier grapes, 169*f*
  in white grapes, 169*f*
  wine color and, 116
phenotype, 5n6, 172n54
2-phenoxyethanol, 167*f*
phenylacetaldehyde, 397*t*
2-phenylacetaldehyde, 167*f*, 397*t*
phenylalanine, 103*f*, 141*f*, 237, 238*f*, 240*f*, 331*f*, 382, 383*f*, 384*f*, 387*f*
phenylalanine 4-monooxygenase, 385, 387*f*
phenylalanine ammonia lyase, 141*f*, 390*f*
2-phenylchromanone, 221
2-phenylchromone, 221
2-phenylethanol, 134*f*, 138*f*, 144*f*, 208*f*, 397*t*, 398*t*
2-phenylethyl acetate, 144*f*
phenyl groups, 221
phenylpropanoid biosynthesis pathway, 240*f*
phenylpyruvate, 387*f*
phloem, 38, 93, 104–5
phloem transport system, 29, 37
phloroglucinol, 238*f*
2-phophoglycerol phosphate, 198*f*
6-phosphate, 195*f*
phosphate anion (Pi), 24*f*, 45*f*, 197*f*, 330*f*
phosphate exchange, 59
phosphatidylcholine, 29, 31*f*, 32*f*, 42*f*
phosphatidylethanolamine, 29
phosphatidylinositol, 29, 32*f*
phosphatidylserine, 29
2-phospho-4-(cytidine 5'-diphospho)-2-*C*-methyl-D-erythritol, 367, 368*f*
phosphodiesterases, 301
phosphoenolpyruvate (PEP), 103*f*, 107, 197, 198*f*, 358, 360*f*, 384*f*, 386*f*
phosphofructokinase, 65*f*
3-phosphoglyceraldehyde, 94*f*, 194, 195
2-phosphoglycerate, 198*f*, 358, 360*f*
3-phosphoglycerate, 58–59, 60*f*, 62*f*, 94, 94*f*
phosphoglycerate kinase, 62*f*
phosphoglycerate mutase, 196, 197*f*

2-phosphoglyceric acid, 198*f*
3-phosphoglyceric acid, 61*f*, 196*f*
D-3-phosphoglyceric acid, 359*f*
2-phosphoglyceroyl phosphate, 197*f*
3-phosphoglyceroyl phosphate, 196*f*, 197*f*
2-phosphoglyceroyl phosphate anhydride, 196
phosphohexomutases, 97
N-phosphohistidine, 360*f*
phosphomannose mutase, 98*f*
(*R*)-5-phosphomevalonate, 364, 365*f*
N-(phosphonomethyl)glycine (Roundup), 106, 106*f*
phosphopyruvate hydratase, 196–97, 358
phosphoric acid, 103*f*, 196*f*
phosphorus, 327
phosphorylation, 59, 65*f*, 194, 195*f*, 360*f*, 364, 365*f*
3-phosphoshikimate-1-carboxyvinyl transferase, 384*f*
3-phosphyglycerate, 360*f*
photoacoustic spectroscopy (PAS), 114
photons, 40
photopsin I, 79
photopsin II, 79
photopsin III, 79
photoreception, 31
photosynthesis, 10, 23, 39, 40, 45, 68
photosynthetic mesophyll, 36
photosystem I (PSI), 51, 52
photosystem II (PSII), 45, 47, 47*f*, 49, 51
pH scale, 113, 320, 320*f*
phylloquinone, 52*f*
phylloxera plague, 9n13, 10–11, 135
Phylloxeridae, 10
15-*cis*-phytoene, 148*f*, 149*f*
5-*cis*-phytoene desaturase, 149*f*
phytohormones, 11–12, 20
phytotoxic metabolites, 280, 280*f*
P$_i$. *See* inorganic phosphate; phosphate anion
piceatannol, 223*f*
pigments, 382, 388
cis-2-pinanol, 143*f*
(+)-β-pinene, 206*f*
β-pinene, 143*f*
α-pinene oxide, 136*f*
(+)-pinoresinol, 241*f*
Pinotage, 152, 154, 155*f*, 263, 267*f*
Pinot blanc, 146, 148–50
Pinot gris, 148, 150–52, 151*f*
Pinot Muenier, 286

Pinot noir, 4n1, 148, 152, 153*f*, 154*f*, 210, 286, 397*t*
piperidine, 350, 350*f*
plane-polarized light, 334–36, 335*f*
plant cell, 24*f*, 40*f*, 192, 193*f*
plant embryo, 6, 7
plant flower odors, 369–72
plant-pollinator interactions, 29
plants
   loneliness of, 20n1
   primordium parts of growing, 67*f*
plant sterols, 36
plasma membrane, structure of, 25*f*, 29
plasticity, of grape berry, xv
plastids, 23, 39, 40, 361–71
plastocyanin, 51*f*
plastoglobuli, 42
plastoquinol, 49, 51
plastoquinones, 43, 48*f*, 49
Platonic solid, 314
Pliny the Elder, 143
polarimeter, 336*f*
polarized light, 334–36, 335*f*, 336*f*
pollen, 81
   production of, 69
pollination, 81, 82*f*
polycopene isomerase, 149*f*
polymeric tannin, 145*f*
polyphenolic polymers, 143
polyprenyl transferases, 369
polysaccharide chains, 27, 28*f*, 29
polysaccharides, cross-linked, 179
pomace, 170n2, 179, 184
pores, 36
Port, 259–65, 267*f*
portisins, 265, 268*f*
postgenetic, 39
potassium bisulfite, 188
potassium metabisulfite, 182, 183, 185, 187, 187*f*
pot residue, 260
pourri plein, 281
pourri rôti, 281
ppm. *See* parts per million
precipitated solids, 231
prephenate, 387*f*
prephenate dehydratase, 387*f*
prephenate dehydrogenase, 387*f*
prephenic acid, 104*f*
press wine, 184
primordia, 68

primordial cells, 33–34
proanthocyanidins, 150, 151*f*, 239, 241*f*
procyanidins, 123, 126*f*, 222, 222*f*, 239, 264*f*
procyanidin tannins, 239, 241*f*
proline, 331*f*, 388, 389*f*
3-(methylsulfanyl)-propanal, 397*t*
propane, 322*f*, 330*f*
1,2,3-propanetriol, 330*f*
propanol, 208*f*
1-propanol, 138*f*, 330*f*
2-propanol, 330*f*
2-propanone, 343*f*
proplastids, 39
propyl 2-methylpropionate, 381
propyl acetate, 139*f*, 381
propyl hexanoate, 381
propyl isobutyrate, 381
propyl sorbate, 140*f*
Protein Data Bank (PDB), 332, 333*f*, 356, 358, 362
proteins, 27, 28*f*, 112, 330, 331, 352–53
   analysis of, 114
   biotinylated, 93
   chlorophyll-bearing, 47, 47*f*
   cytochrome P_{450} hemoproteins, 299
   G-protein, 75*f*, 79
   in grape juice, 179
   1L2Y, 332, 333*f*
   TRP-Cage, 332, 333*f*
protocatechuic acid, 125*f*, 126*f*, 169*f*, 245*f*
proton nuclear magnetic resonance, 344, 346–47
PSI. *See* photosystem I
PSII. *See* photosystem II
pterion, 33
pterostilbene, 392*f*
purine, 350, 350*f*, 351*f*
pyranose, 383*f*
pyrans, 221, 350, 350*f*
pyrazines, 108, 108*f*, 112, 125, 162, 163*f*, 248*f*, 249, 350, 350*f*, 396*t*
pyridazine, 350, 350*f*
pyridine, 350, 350*f*
pyridinium ring, 299*f*
pyridoxal, 387*f*
Pyrimethanil, 166, 169*f*
pyrimidine, 350, 350*f*, 351*f*, 355n18
pyrimidine ring, 352
pyrrole, 350, 350*f*
pyrrole rings, 47

pyruvate, 190f, 194, 197, 199f, 213–16, 215f, 216f, 228n8, 360f, 361, 365, 366f
   ketone carbonyl of, 362f
pyruvate decarboxylase, 198, 199f
pyruvate dehydrogenase complex, 189, 190f
pyruvate kinase, 360f
pyruvic acid, 58, 61f, 188, 189f, 206, 207f, 213f, 215f, 217f

quantum mechanics, 310, 353n4
quercetin, 124f, 126f, 169f, 224f, 241f
quercetin-3 glucoside, 132f
quercoresinosides, 245, 246f
Quercus petraea wood, 245
quinic acid, 104f

R-(+)-1-methyl-4-(1-methyl-ethenyl)-cyclohexane, 396t
racking, 180, 182, 184, 231
radicle, 7
reactive oxygen species (ROS), 16, 281
redox systems, 52, 53f
reduced flavin adenine dinucleotide ($FADH_2$), 33, 53f
reduced nicotine adenine dinucleotide 2'-phosphate (NADPH), 44, 45f, 53f, 54, 84f
reduction, 48, 59, 99f, 321–48
red wine. *See also* Port; *specific grapes*
   cap in making, 183
   crushing process for, 183
   fermentation of, 183–84
refractometer, 111, 111f
(1S,2S)-*rel*-1-(4-hydroxy-3-methoxyphenyl)-2-[[4-[(1E)-3-hydroxy-1-propen-1-yl]-2-methoxyphenyl]methyl]-1,3-propanediol, 241f
resistant species, American native, 11
resonance, 326
resonance structures, 326
resveratrol, 113, 113f, 132f, 153f, 225, 300, 301, 302, 392f
cis-resveratrol, 153f, 235f
trans-resveratrol, 169f, 223f, 226f, 235f, 301f
trans-resveratrol di-*O*-methyl-transferase, 392f
11-*cis*-retinal, 81, 81f
11-*trans*-retinal, 81
retrotransposon, 172n54

α-L-rhamnopyranosyl-β-D-glucoside, 209f
rhizosphere, 18, 19, 20, 20n2
rhodopsin, 79, 80, 80f, 81f
β-D-ribofuranose, 218f
β-D-(–)-ribofuranose, 238f
ribonucleic acid (RNA), 225
β-D-ribopyranose, 218f
β-D-(–)-ribopyranose, 238f
ribose, 300f
ribosomes, 24f, 26–27, 42
ribulose-1,5-bisphosphate (RuBP), 58, 60f, 61f, 63, 358f, 359f
ribulose 6-phosphate, 63
riddling, 287–88
Riesling, 155–56, 156f, 157f, 158f, 159f, 276–77, 396t
Riesling acetal, 129f, 164f, 168f
right handed, 334
ring system, 141f, 221, 350, 351f
ripening. *See* véraison
(R)-(–)-linalool, 396t
RNA, 351f
RNS. *See* ribonucleic acid
Roburin A, 240, 242f
Roburins, 240
root apical meristem, 19f
root caps, 8
root feeders, 8
root hairs, 18, 37
root meristem, 7
root production, 6–7
roots
   biology of, 18–19
   defense process, 19–20, 20f
   lateral formation, 28–29
root-specific metabolism, 19
rootstock, 9n13, 14
root walls, 27
ROS. *See* reactive oxygen species
rose furan, 219f
rose oxide, 134f, 396t, 397t
cis-rose oxide, 136f, 168f, 290f, 396t, 397t
trans-rose oxide, 136f
rosé wine, 183, 184
rosinidin, 145t
rotundone, 166f, 396t
Roundup®, 106, 106f

Rubaiyat (Khayyam), 374n31
Rubisco, 356, 358f, 359f
RuBP. *See* ribulose-1,5-bisphosphate
Ruby Port, 260
rutin, 124f, 153f, 224f

S-(+)-2-methyl-5-(1-methylethenyl)-2-cyclohexenone, 398t
sabinene, 206f
saccharides, 27
Saccharomyces cerevisiae, xv, 177–78, 182, 185, 192, 198, 201–4, 207, 208f, 210, 211f, 213, 216, 227, 231, 260, 290, xivn4
   cell, 200f
   headspace, 205f
S-adenosylmethionine (SAM), 108, 108f, 390f
salicylic acid, 20, 20f
salt water, 204
SAM. *See* S-adenosylmethionine
Sangiovese, 156–61
Sauvignon blanc, 6, 131, 161–62, 162f, 163f, 284, 396t
Sauvignon blanc grapes, 276–77
S. cerevisiae, 288
Scheurebe, 396t
Schrödinger, Erwin, 353n4
scientific notation, 353n5
scion, 10
secondary fermentation, 214
second fermentation, 288
second messengers, 301
secretome, 205
sediment, 234
sedoheptulose 1,7-bisphosphate, 63
sedoheptulose 7-phosphate, 63
sedoheptulose-bisphosphatase, 63
seeds
   fruit size and, 82
   of grape berries, 91
   grapevine from, 6–9, 7f
self-fertilization, 83
self-pollination, 29
β-selinene, 371f
δ-selinene, 371f
β-selinene cyclase, 371f
δ-selinene synthase, 371f
seltzer, 288

Semillon, 162–64, 164f, 165f, 396t
Sercial grape, 269, 273, 274f
serine, 214, 214f, 331f
serine residue, 97
sesquiterpene phytohormones, 12, 12f, 165, 166f, 365
sesquiterpenes, 69, 112, 156f, 275, 276f, 277f, 280, 369–72
sesquiterpene β-bisabolene, 371f
sesquiterpenoids, 281
sesterterpenes, 369
Sherry, 259–60, 265–68, 269f, 270f
shikimate, 384f, 386f
shikimate kinase, 384f
shikimic acid, 104f
Shiraz, 164–65, 166f, 396t
shoot apical meristems, 31, 33, 33f
shoots, 67, 68
Sicilian Muscat, 140, 143f
signaling process, 37
sinapinic acid, 240f
sinister, 334
sites of unsaturation, 338
skin
   of grape berries, 91
   wine color and, 116
(S)-(+)-linalool, 396t, 397t
smell, 73
Smirnoff-Wheeler pathway, 97
smoother taste, 214
sodium metabisulfite, 185
sodium methyldithocarbamate, 105, 105f
soil, xix
   galestro, 157
   for growing, 16–17
   microbiome, 17
solids, removal of, 180–81
solvent, 319
somatic mutation, 149, 172n54
sorbic acid, 139f
sorbitol, 238f, 307
Sørensen, Søren, 320
sotolon, 154f, 219f, 397t
sparkling wines, 259
   Champagne, 286–89
specialized wines, 259–91
   Botrytis cinerea (Noble Rot), 276–86
   Ice wines, 289–91, 290f

specific rotation, 335, 335f
spectroscopy, 114
sphingomylein, 29
"spin," 343, 355n16
spinacia oleracea, 51f
squalene, 202f
squalene oxide, 109, 110f
stamen, 69, 69f
starter cultures, 200
starter yeasts, 206
statocytes, 23
statoliths, 23
stearic acid, 36f
steroids, 112, 201, 361
  formation of, 110f
stigma, 69, 69f
stipule, 37
stizolobate synthase, 386, 388f
stizolobic acid, 386
stizolobinate synthase, 388f
stizolobinic acid, 386, 388f
stomata, 36
stratoliths, 7
(+)-strigol, 155f
stroma, 41, 42, 44, 54
strong force, 309, 310
succinic acids, 93, 307
suckers, 67, 86n1
sucrose, 93, 104f
  formation of, 63, 65f, 66
  glucose from, 194
  hydrolysis of, 109f, 194, 194f
  measuring levels of, 111
sucrose 6F-phosphate, 65f
sugars, xix, 179, 288, 307
  alcohol levels and, 112
  keto-sugar diphosphate, 60f
3-sulfanylhexan-1-ol *and* 4-hydroxy-2,5-
  dimethyl-3(2H)-furanone, 397t
sulfate, 197f, 216
sulfite, 182, 183, 187f, 189, 216
sulfite anion, 186
sulfur, 350, 352
  amino acid metabolism, 204, 225
  in amino acids, 227f
sulfur-containing heterocyclic compounds, 350f
sulfur dioxide, 113, 182, 184, 186f, 187f
  adding, 185–89
sulfuric acid, 219

sulfurous acid, 185, 186f
sulfur trioxide, 187
sunlight, 39, 44
  ripening and, 107
surface cells, 18
surface tension, 302n2
sweetness rating, 288
synthesis, 342
Syrah, 164–65, 166f
syringaldehyde, 125f, 159f, 167f
syringic acid, 104f, 153f, 169f, 238f
syringol, 159f, 247f

table wine, 297
tannic acid, 37, 37f, 92f, 146f, 235
tannins, 37, 39, 112, 179, 222f, 223, 224
  gallic acid-associated, 242f
  hydrolyzable, 239–41
  in oak, 235
  polymeric, 145f
  procyanidin, 239, 241f
  ripeness and, 114
  stems and, 116
(2R,3R)-(+)-tartaric acid, 146f, 207f
tartaric acids, 93, 97, 100, 101, 102f, 104, 107, 113,
  125f, 143, 206, 213f, 214f, 307
  dehydration of, 213
D-(−)-tartrate, 213f
L-tartrate, 104f
L-(+)-tartrate, 99, 213f
Tawny Port, 260
taxifolin, 153f, 390
TC. *See* thermal conductivity detector
TDN. *See* 1,1,6-trimethyl-1,2-dihydronaphthalene
Tempranillo, 38, 165–67, 167f, 168f, 169f
tendril primordia, 68
tendrils, 37
terpenes, 136f, 137f, 142, 210, 268, 272f, 274,
  333f, 369–72
  in Riesling grapes, 157f, 159f
  in Shiraz/Syrah grape aroma, 166f
terpenoids, 129f, 130, 143f, 154, 156, 205, 206f,
  234, 332
  hydroxyl-bearing, 210f
  in Port, 263f
terpenols
  in Riesling grapes, 158f
  in Sauvignon blanc grapes, 163f
terpenyl butyrate, 381

Index

1,8-terpin, 158f, 165f
α-terpinene, 143f, 206f
γ-terpinene, 136f, 157f
4-terpinenol, 158f
α-terpineol, 131, 136f, 163f, 165f, 168f, 203f, 210, 211f
β-terpineol, 143f
γ-terpineol, 157f
1,8-terpin hydrate, 158f
4-terpinol, 136f
terpinolene, 136f, 143f, 157f
α-terpinolene, 168f
α-terpinyl acetate, 158f
territorial fitness, xix
territory, xix
terroir, xv, xix, 68, 102, 110–11, 124, 135–36, 259–60, 282, xviin2
  specificity of, 117
tetrahedra, 314, 315f
tetrahydro-2,2-dimethyl-5-(1-methyl-1-propenyl)furan, 157f
cis-tetrahydro-4-nethyl-2-(2-methylpropenyl)2H-pyran, 396t, 397t
1,2,3,5-tetrahydroxycyclohexane, 238f
tetrathiocane, 225
1,3,5,7-tetrathiocane, 131, 226f
theaspiranes, 220f
thermal conductivity (TC) detector, 340
Thermosynechococcus elongatus, 45, 45f, 48, 49f, 50f
Thermosynechococcus vulcanus, 50f
Thermus thermophilus, 53, 54f
thiamine, 188–89, 188f, 204, 226f, 350f, 352, 362f, 365
thiamine diphosphate, 63, 189f, 190f, 199f, 215f, 361, 366, 366f
thin-layer chromatography (TLC), 338–39
thiohemiacetal, 364f
thiol, 283f, 284
thiophene, 350, 350f
threonine, 331f
L-*threo*-tetruronate, 102f
thylakoid lumen, 46f
thylakoids, 41, 42, 44
thymine, 12f, 13n9, 226f, 351f, 355n18
timer, genetically dictated, 6
Tinta Negra, 269
TLC. *See* thin-layer chromatography
Tokaji Aszu, Hungary, 276–77

Touriga Nacional, 260
Traminer, 131, 135
transaminase, 387f
transcription, 30
transcription factor, 172n54
transcriptome, 5n8
transketolase, 63
transpiration, 36
tree of life, 192, 193f
trellis systems, 15
tricarboxylic acid cycle, 95–96, 96f
trichlorofluoromethane, 244
9,10,18-trihydroxystearic acid, 36f
trihydroxystilbene synthase, 390, 392f
3,4′,5-trihydroxy-*trans*-stilbene, 392f
triketo-CoA, 392f
triketothioester, 390, 392f
1,1,6-trimethyl-1,2-dihydronaphthalene (TDN), 146, 150f, 159f, 164f, 219, 220f, 396t
1,5,8-trimethyl-1,2-dihydronapthalene, 159f
(E)-1-(2,6,6-trimethyl-1-cyclohexa-1,3-dienyl)but-2-en-1-one, 397t, 398t
(2,10,10-trimethyl-6-methylene-1-oxaspiro[4.5]dec-7-ene, 261f
N,N,N-trimethylamine group, 330f
(1S,2R,4S)-1,7,7-trimethylbicyclo[2.2.1]-hept-2-yl acetate, 375
2,6,6-trimethylcyclohex-2-ene-1,4-dione, 129f
2,2,6-trimethylcyclohexanone, 128f
N,N,N trimethylethanolammonium cation, 330, 330f
1-(2,3,6-trimethylphenyl)-2-butanone, 159f
1-(2,3,6-tri-methylphenyl)-3-buten-2-one, 159f
(E)-1-(2,3,6-trimethylphenyl)-but-1,3-diene, 168f
2,3,5-trimethylpyrazine, 205f, 248f, 249
trimming, 38
  vines, before and after, 14f
triose phosphate, 41, 41f
triose phosphate isomerase, 195–96
1,3,5-triphenol naringenin chalcone, 392f
triphosphate isomerase, 61
trisphenol, 224
triterpenes, 369
trithiolane, 225
trivial names, 354n13, xviin6
TRP-Cage protein, 332, 333f
true leaves, 31
truly odoriferous compounds, 278–79
tryptophan, 331f

Tswett, M. S., 338, 339
tylose, 8
tyrosinase, 389f, 390f
tyrosine, 50f, 77, 103f, 237, 238f, 240f, 331f, 387f, 388f, 389f
tyrosine-3-monooxygnease, 388f
tyrosine ammonia lyase, 103f
tyrosine transaminase, 387f
tyrosol, 169f

ubiquinones, 202
UDP-glucose, 65f
ultraviolet (UV) light, 234, 235f, 337–38, 339f
ultraviolet spectroscopy, 114
umami, 204
uncoupling, 355n17
unpolarized light, 336f
uracil, 12f, 13n9, 189, 190f
urea, 206, 207f
uridine, 226f
uridine diphosphate, 65f
uridine diphosphate glucose, 393f
uridine triphosphate (UTP), 65f
UV. *See* ultraviolet light

vacuole, 24f, 27
vacuum UV, 338
valencene synthase, 372f
valine, 331f
vanillic acid, 125f, 169f, 223f, 238f
vanillin, 125f, 134f, 154f, 159f, 167f, 210, 211f, 245f, 247f
vanillyl alcohol, 245f
vascular cambium, 10
vascular joining, 10
vascular tissue, 33f, 93
Vegrodolce, 157
veins, 37
véraison, 69, 104–10
Verdelho grape, 269, 273, 274f
vernalization, 6, 20n4
vescalagin, 242f
vine-growing
 science of, 3
 training, 15
viniculture, 16
(+)-α-viniferin, 226f

ε-viniferin, 153f, 225, 226f
vintner practices, xx
4-vinyl-2-methoxyphenol, 397t, 398t
4-vinylguaiacol, 154f, 206, 207f, 211f, 245f, 247f, 291f
4-vinylphenol, 206, 207f, 245f, 247f, 261, 397t
vinylphenols, 210
Viognier, 167, 169f, 170
visceral flesh, 36
visible spectroscopy, 114
visual receptors, 79
vitamin A, 80
vitamin B1. *See* thiamine
vitamin C, 234. *See also* ascorbic acid
vitamin K, 51, 52f
viticulture, 3
vitifolia. See Phylloxeridae
vitisin A, 159, 161f
vitisin B, 159, 161f
vitispirane, 129f, 155f, 159f, 164f, 168f, 220, 220f
Vitis vinifera, xx, 3, 4n3, 11, 30, 281, 283
volume, 307
vomifoliol, 129f, 164f, 209f

water, 185, 186f, 187f, 313–14, 313f, 314f
 excess, 7
 hydrogen bonded, 316–17
 molecules, 317f
 oxidation, 48
 salt, 204
 in wine, 298
water balance, 105
water loss, 91–92
Watson, James, 351f, 353
wax candles, 354n8
weed control, 106
whiskey lactone, 167f, 219f
cis-whisky lactone, 397t
white light, 336, 337f
white wine, 116. *See also specific grapes*
 aging process, 232, 235
 crushing process for, 179–80
 fermentation of, 181–82
 final racking for, 231
wild yeasts, 177, 192, 200, 203, 206
wine. *See also* red wine; white wine
 corked, 253, 255, 288, 332, 354n14, 355

dessert, 277
drinking, 297–304
ethanol in, 299
iced, 259, 289–91, 290f
press, 184
for shipping, 259
table, 297
water in, 298
wineglasses, 297
wine lactone, 219f, 396t, 397t
winemaking, 177
wood alcohol, 314–15

xylans, 236, 237f, 238f
xylem, 10, 18, 37, 38, 104–5
β-D-(+)-xylopyranose, 238f
β-D-xylopyranosyl-β-D-glucoside, 209f
xylose, 236f

yeast, xix, 192–94, 205f, 288
    actions on other components, 205–11
    alcohol levels and, 112
    compounds present due to, 217–18
    ethanol tolerance, 177–78
    glucose conversion to ethanol, 194–98

head space, 205f, 207f
heritage strains, 203
killer, 212–13
living, 198–205
    in red wine making, 183
starter, 206
sulfur dioxide and, 185
    in white wine making, 182
wild, 177, 192, 200, 203, 206
    in wine production, 200
yeast extract, 204
yeast paste, 205, 206f, 207f
α-ylangene, 165, 166f
young leaf primordium, 33f
young ruby Port, 261

(Z)-3-hexen-1-ol, 398t
zeatin, 12, 12f, 109f, 210f
cis-zeatin, 83f
trans-zeatin, 83f
trans-zeatin riboside, 83f
zinc, 74f
zingerone, 167f, 211f
zusammen (together), 324
zwitterion, 330